ENCYCLOPEDIA OF
EARTHQUAKES AND VOLCANOES

THIRD EDITION

ENCYCLOPEDIA OF
EARTHQUAKES AND VOLCANOES

THIRD EDITION

ALEXANDER E. GATES, PH.D.
AND DAVID RITCHIE

Facts On File
An imprint of Infobase Publishing

Encyclopedia of Earthquakes and Volcanoes, Third Edition

Facts On File, Inc.
An imprint of Infobase Publishing
132 West 31st Street
New York NY 10001

Library of Congress Cataloging-in-Publication Data

Gates, Alexander E., 1957–
Encyclopedia of earthquakes and volcanoes.—3rd ed. / Alexander E. Gates and David Ritchie.
p. cm.
Ritchie's name appears first on the previous ed.
Includes bibliographical references and index.
ISBN 0-8160-6302-8 (acid-free paper)
1. Earthquakes—Encyclopedias. 2. Volcanoes—Encyclopedias. I. Ritchie, David, 1952 Sept. 18-
II. Title.
QE521.R58 2007
551.203—dc22   2005046619

Facts On File books are available at special discounts when purchased in bulk quantities for businesses, associations, institutions, or sales promotions. Please call our Special Sales Department in New York at (212) 967-8800 or (800) 322-8755.

You can find Facts On File on the World Wide Web at http://www.factsonfile.com

Text design by Joan McEvoy
Cover design by Anastasia Ple
Illustrations by Sholto Ainslie © Infobase Publishin
Photo research by Tobi Zausner, Ph.D.

Printed in the United States of America

VB Hermitage 10 9 8 7 6 5 4 3 2 1

This book is printed on acid-free paper.

This book is dedicated to my father, David L. Gates,
and to my mentor, Dr. Lynn Glover III

CONTENTS

PREFACE: An Essay on Plate Tectonics

Until the 1950s, the various branches of geology, including the study of volcanoes and earthquakes, appeared to have no real connection, and the progress of the science was toward continued divergence. What pulled everything back together was the radical concept of PLATE TECTONICS that came into fruition in the 1960s and 1970s. Plate tectonics is sometimes referred to as "the glue that holds geology together" to reflect this power. It explains virtually all volcanoes and the lion's share of earthquakes and even relates them to each other.

The following review of plate tectonics is a good place to start for anyone needing a refresher.

Earth Architecture

The Earth is a sphere that is flattened at the poles and bulging at the equator. Internally, it is like a hard-boiled egg. In the center, instead of a yolk, it has a CORE. The core is composed of iron-nickel and contains an INNER CORE that is solid and an OUTER CORE that is liquid. The spinning of the Earth causes the two to interact as a self-exciting dynamo that gives Earth its strong magnetic field. The egg white is equivalent to the Earth's MANTLE, which encases the core. The mantle is composed of dense minerals that are rich in iron and magnesium. It has several layers reflecting different minerals and mechanical properties. The shell of the egg is equivalent to the Earth's CRUST, a thin layer of light rock upon which humans live. Unlike the shell, which is uniform, there are two types of crust. Thin, dense, young crust is pulled toward the center of the Earth by gravity and sinks down, whereas thick, light, old crust floats higher. The deeper crust is covered by oceans and called OCEANIC CRUST, whereas the lighter crust forms the continents and is called CONTINENTAL CRUST. The concept of ISOSTASY is the balance between them.

If a person dropped the egg and the shell cracked into fragments that remained stuck to the egg, these would be the plates of the Earth. However, the plates are not only composed of crust. If the egg was not placed in cold water after it was boiled, a thin layer of egg white would be stuck to the shell. This sandwich of shell and white is equivalent to the sandwich of crust and rigid mantle and is called the LITHOSPHERE. Plates are considered lithospheric plates because they are not only composed of crust. The complication arises below the lithosphere because rather than the mantle staying rigid throughout, there is a gummy layer of mantle beneath the lithosphere called the ASTHENOSPHERE. (Imagine a layer within the egg in which the egg white remained runny.) It is the floating, moving, and interacting of the lithospheric plates that defines the science of plate tectonics.

Plate Margins

A plate margin, or plate boundary, occurs where lithospheric plates meet. Adjacent plates interact at plate margins in one of three ways:

1. They move away from each other in a divergent margin;
2. They move toward or into each other in a convergent margin; or
3. They slide past each other in a transform margin.

But why should they move at all? Why don't they just remain in one spot? The answer is convection. There are hotter and cooler areas in the mantle. The hotter mantle is less dense and tends to rise, just like hot air or hot water. The cooler mantle is more dense and tends to sink. When mantle is heated in the hotter areas, it rises to the upper mantle and spreads out, cooling as it moves away from the heat source and sinking back to the lower mantle where it can be reheated. In this way the mantle circulates in a convection cell similar to a boiling pot of soup. The lithospheric plates float on the flowing, circulating mantle.

Divergent Margins

Divergent margins typically begin on continental crust but quickly wind up on oceanic crust. There are several distinct stages in these margins. The process of pulling crust apart is called RIFTING. The initiation of rifting appears to involve the formation of TRIPLE JUNCTIONS. Mantle plumes strike the underside of the plates, leaving a hole with three cracks emanating from it at 120° angles. Two of the cracks on the Earth will become active rift zones and mature divergent margins that connect with other triple junctions, while the third crack will start to rift but eventually fail. The failed crack is called an aulocogen. The best example of a triple junction is at the southern tip of the ARABIAN CRUSTAL PLATE. The Gulf of Aden and Red Sea represent the two active cracks, and the East African rift system is the aulocogen.

The early stage of rifting is short-lived, if present at all, and involves bulging of the continental crust and uplift. The second stage involves thinning of the continental crust. The lower crust is ductile and stretches thinner, whereas the upper crust is brittle. Brittle deformation includes the development of active NORMAL FAULTS and consequently HORSTS and GRABENS (similar to the modern Basin and Range Province of the southwestern United States). Eventually, MAFIC magma from the upper mantle reaches the surface in FISSURE ERUPTIONS. FLOOD BASALTS cover the landscape with massive flows (such as the LAKI eruptions of ICELAND or the Columbia River Plateau of WASHINGTON). These lava plateaus are the largest accumulations of lava on the continents. In some cases RHYOLITE volcanoes may also be produced by the melting of continental crust from the elevated heat flow from the BASALT. These volcanoes can be violent, producing huge eruptions, such as at YELLOWSTONE NATIONAL PARK.

The next stage involves the development of a narrow ocean basin such as the Red Sea. Volcanic activity continues, but it is submarine and purely basalt. Earthquakes can continue on land, but they are less common and less intense. Most of the earthquake activity is also submarine. The final stage is the mature ocean basin such as the Atlantic Ocean. The coasts are PASSIVE MARGINS with no volcanic activity and rare seismic activity. All of the volcanic and seismic action is submarine and occurs at the MID-OCEAN RIDGE at the center of the basin. A mid-ocean ridge is where new ocean crust is continuously being formed. The Mid-Atlantic Ridge is a huge submarine mountain range that extends almost from pole to pole. It surfaces at Iceland, which provides a glimpse of the intense volcanic and seismic activity associated with these margins.

Convergent Margins

There are three types of convergent margins depending upon the type of crust on the colliding plates:

1. ocean-ocean convergent margins
2. ocean-continent convergent margins
3. continent-continent convergent margins

It is at convergent margins where ocean crust is consumed at the exact rate as it is being produced in the mid-ocean ridge. This balance is necessary or the Earth would be constantly changing size.

At an ocean-ocean convergent margin, ocean crust is driven beneath facing ocean crust in a feature called a SUBDUCTION ZONE. On the ocean floor, the top edge of the subduction zone is marked by a trench that contains the deepest ocean depths on Earth. The downgoing or subducting plate drives deeper into the asthenosphere, where it first partially melts and then is absorbed. Earthquakes occur all along the surface of the subducting plate and are thus of progressively deeper FOCUS away from the trench. The melted ocean crust forms MAGMA of INTERMEDIATE composition that rises through the overlying crust eventually to form submarine volcanoes. These volcanoes continue to grow until they breach the ocean surface to become volcanic islands. This chain of islands forms an arcuate map pattern called an ISLAND ARC. The ALEUTIAN ISLANDS are a good example of an island arc. These ANDESITE volcanoes are explosive and among the most dangerous on Earth. They are also very seismically active primarily as the result of movement on REVERSE and THRUST FAULTS. KRAKATOA is a subduction zone volcano. Subduction zones are the source of MEGATHRUSTS that produce such disasters as the 2004 BANDA ACEH tsunami.

Ocean-continent margins are geometrically similar to ocean-ocean margins. The difference is that the overriding plate is continental rather than oceanic crust. Instead of island arcs, there are volcanic mountain ranges called MAGMATIC ARCS. The Andes Mountain range is the best example of a magmatic arc. These margins also have subduction zones and BENIOFF ZONES with both seismic and volcanic activity. The volcanoes are at higher elevations than those in island arcs, and the magma chambers are much larger.

Connected to the subducting ocean crust, somewhere far behind, is another continent. Eventually, the ocean crust is completely consumed in the subduction zone, and the two continents face each other and collide. The subducting plate attempts to follow the ocean crust down the subduction zone, but it is too buoyant to enter the asthenosphere. Instead, the two plates collide. The subducting plate plows beneath the overriding plate and then large thrust faults develop on the subducting plate, moving large amounts of rock away from the collision zone accompanied by a series of massive earthquakes. The overriding plate crumples while building huge mountains that progressively grow away from the collision zone. This crumpling involves intense seismic activity. The only potential volcanic activity is leftover from the subduction zone activity on the old magmatic arc.

There is an odd effect that occurs in continent-continent collisions called EXTRUSION or escape TECTONICS. If one or both sides of the overriding plate (at a high angle to the collision zone) faces an open ocean basin, it is said to have a "free face." As the continental collision proceeds, chips of the overriding plate are shoved laterally and out of the way of the intense collision zone along large STRIKE-SLIP FAULTS. Basically, land is being squirted sideways, away from the zone of high force. The strike-slip faults accommodate up to 621 miles (1,000 kilometer) offset and produce regular, intense earthquakes. An example of extrusion tectonics is TURKEY, which is being squirted westward into the Mediterranean Sea along the strike-slip North Anatolian fault. Some of the more devastating earthquakes of the 20th century have occurred along this fault as the result of extrusion tectonics.

Transform Margins

Transform Margins or FAULTS are strike-slip faults that separate two lithospheric plates. Furthermore, they must connect two other types of plate margins. Therefore, there are three types of transform margins based upon the margins connected. There are:

1. divergent-divergent transform margins;
2. convergent-convergent transform margins; and
3. divergent-convergent transform margins.

Because they are plate tectonic margins, they are constantly active, potentially for very long periods of time. They are some of the most active faults on Earth. Luckily, more than 99 percent of transform faults are on the ocean floor. The vast majority of these are divergent-divergent transform margins that accommodate bends and offsets on the mid-ocean ridges. These faults are also called geofractures or fracture zones and can be seen on bathymetric maps of the ocean floor at high angles to the mid-ocean ridges. The offsets give them a dentate appearance with the high density of transform faults along the ridge. Convergent-convergent transform margins accommodate jogs and offsets in island arcs, but they are not common. Divergent-convergent transform margins are also uncommon and are usually associated with smaller plates such as the Caribbean plate.

Transform faults on continental crust are very dangerous because of their persistent activity. The best example is the SAN ANDREAS FAULT in CALIFORNIA. Every few years it produces an earthquake with a MAGNITUDE of 7.0 or greater, many of which cause great damage. Another example is the South Alpine Fault in NEW ZEALAND, which also produces numerous strong earthquakes.

ACKNOWLEDGMENTS

My thanks to the Department of Earth and Environmental Sciences at Rutgers University–Newark for support and resources; to the students in my natural disasters classes for leading me to new resources for earthquakes and volcanoes; and to reference librarian Veronica Calderhead at Dana Library for efforts above and beyond the call of duty in finding obscure references for me.

To Dr. Tobi Zausner for her excellent work in scanning slides and improving images, and to Colin Gates for producing several of the tables. To Dr. David Valentino at SUNY–Oswego for his suggestions to improve this volume; to executive editor Frank K. Darmstadt for his suggestions, encouragement, and patience; to Alana Braithwaite for her invaluable help in getting the manuscript into shape; and to literary agent Max Gartenberg for his sage advice.

I would also like to acknowledge the phenomenal resources of the U.S. Geological Survey and National Oceanographic and Atmospheric Administration, without which the production of this book would have been impossible.

INTRODUCTION

The volcano on the island of Thira (now Santorini) in the Aegean Sea exploded in about 1500 B.C. As a result, it sent a tsunami now estimated at over 200 feet (61 m) high across the eastern Mediterranean Sea that destroyed the dominant Minoan civilization on the island of Crete. The hill people of Greece thereby formed the dominant culture, eventually heralding the Golden Age of Greece. The huge tsunami caused an extensive retreat of seawater along the Mediterranean shoreline for up to 10 minutes before it came crashing back. It is possible that this opening of the sea allowed Moses and the Israelites to escape the Egyptians in 1500 B.C., as described in the Bible.

It is to this astounding degree that earthquakes and volcanoes have affected humans' history, culture, and civilization. In ancient times, earthquakes sometimes tipped the scales in power struggles, allowing one kingdom to defeat another. It is no wonder that earthquakes are sometimes said to be connected to punishment imposed by a displeased deity in ancient and even modern religions. Volcanoes have even been credited with altering the Earth's climate, thereby causing some of the major famines and even plagues in history. Earthquakes and volcanoes are the most powerful and destructive natural phenomena on the planet, fascinating people both young and old.

It was this fascination that inspired David Ritchie to write the first edition of the *Encyclopedia of Earthquakes and Volcanoes*. Untrained in science, he concentrated on historical accounts of the disasters. When I revised the first edition, I added the missing scientific aspect to the encyclopedia. I also expanded the coverage of volcanoes as well as updated information on recent geological phenomena. In addition, new photos and maps were added to show where the events took place, and diagrams were provided to illustrate the processes.

One would think that with all the human advances in monitoring technology and earthquake engineering that earthquakes and volcanoes would become less damaging and thus less important with every year. However, the still young 21st century is showing humans that this is not the case. The horrifying recent disasters associated with the Banda Aceh, Indonesia, tsunami and the Bam, Iran, and Muzaffarabad, Pakistan, earthquakes are among the worst in history. The burgeoning world population in geologically hazardous areas is clearly growing faster than the disaster-reduction technology.

These new events made it clear that the second edition of the encyclopedia lacked the resources to put these disasters into proper historical context. This third edition not only includes the new events but also many examples of historical earthquakes for contrast or comparison. The third edition has earthquake information for Italy, Greece, Egypt, Iran, Pakistan, China, India, and several other areas. It is especially timely, with greatly expanded coverage in the Middle East, where the world's focus has rested in recent years. Because many of these disasters involve destruction by landslides and avalanches, entries related to such events and processes are significantly expanded. Tsunami research and technology is also updated in the new edition to answer questions about the Banda Aceh disaster, and many other tsunamis are described for comparison. Finally, there are many tables to place the magnitude of these recent disasters into historical context.

The third edition of *Encyclopedia of Earthquakes and Volcanoes* provides a unique single source on historical earthquakes and volcanic eruptions for students or laypeople. Many of the sources included are obscure and difficult to obtain. With the new "Preface Essay on Plate Tectonics," this edition may serve as a stand-alone reference for introductory courses on earthquakes and volcanoes.

ENTRIES A–Z

A

aa A Hawaiian word (pronounced AH-ah), aa is a particular kind of LAVA FLOW with an irregular, jagged surface. Aa is very stiff and blocky because much of its mass is hardened LAVA. It flows slowly, with lava rubble tumbling down the advancing slope. It typically occurs far from the volcanic VENT at the leading edge of the flow.

See also PAHOEHOE.

acceleration The rate at which velocity of a body or particle is increased as compared with deceleration which is the rate at which velocity is decreased. When seismic waves pass through some material (soil, rock, or buildings), the shaking produces acceleration of the contained particles. The measure of this acceleration from shaking determines the amount of potential damage. Shaking is among the most deadly forces in an earthquake.

accelerogram The typically graphic recording of the ACCELERATION of the ground surface as surface waves arrive at an ACCELEROGRAPH station. The accelerograph produces the accelerogram.

accelerograph An instrument that measures the ACCELERATION of the ground surface at a given location. The record produced by an accelerograph is called an ACCELEROGRAM.

acidic An old term for FELSIC.

acoustics Various noises are associated with earthquakes and volcanic eruptions. Earthquakes are often accompanied by a deep, audible, rumbling noise. The noise often is compared to that of thunder or of heavy traffic or trucks passing on the streets. In one instance, noises associated with an area of occasional earthquake activity have become a tourist attraction. This case involves the "Moodus noises" in the state of CONNECTICUT. These are mysterious sounds similar to gunfire that have been reported in the vicinity of East Haddam. The name *Moodus* is derived from the Native American word *Mackimoodus,* meaning "meeting-place."

The acoustic effects of volcanic eruptions can be surprising. The noise that accompanied the explosion of the volcano KRAKATOA in 1883, for example, was heard some 3,000 miles (4,828 km) away on the island of Rodrigues in the Indian Ocean. This is said to be the greatest distance at which the noise of a natural event has been heard within historic times without the aid of electronic communications. In some eruptions, the noise may be audible hundreds of miles away and yet go unheard in areas much closer to the point of eruption. When Mount KATMAI in ALASKA erupted on June 6, 1912, for example, the sound of the eruption was heard some 800 miles (1,287 km) away but reportedly was not distinct at Kodiak, only about 100 miles (161 km) from the volcano.

active fault A FAULT that is actively moving. Each increment of movement, each jerk, produces earthquakes.

active volcano An active volcano is considered to be one that has shown activity within historical times or the past several thousand years. A historically active volcano, however, may be inactive at present and indeed may have shown no activity for hundreds of years. Approximately 500 volcanoes around the globe are thought to be active, but this figure may be a serious underestimate because of some submarine volcanoes whose activity has not been observed and reported.

Adana *earthquake, Turkey* On May 27, 1998, an earthquake of MAGNITUDE 6.2 occurred. It killed 145 people and injured 1,500. More than 17,000 houses were destroyed. Several major AFTERSHOCKS also occurred.

Adatara *volcano, Honshu, Japan* It is a STRATOVOLCANO that is located nine miles (15 km) from Fukushima City. It contains three cones (Adatara, Maegatake, and Osoyozan) that are andesitic (with minor BASALT) in composition. Adatara has experienced two historical eruptions, the last in 1990.

Advanced National Seismic System (ANSS) In response to the severe seismic risk that Americans are exposed to in many areas of the country, the U.S. government established the ANSS. This program provides accurate and timely data as well as information on seismic events, including the effects on buildings and other structures. The ANSS is a basic function of the National Earthquake Hazard Reduction Program. When the system is complete, it will have some 7,000 seismic stations, with dense concentrations at 26 designated high-risk urban centers. Functions of the ANSS include constant monitoring of seismicity, thorough analysis of seismic events, and automatic broadcasts of potential seismic hazards, all in real time.

Aeolian Islands *See* LIPARI ISLANDS.

Aeseput (Aeseput-weru) *See* TONDANO.

Afghanistan Although it is one of the more seismically active countries in the world, the lack of accurate records makes reporting on Afghanistan's historical earthquakes difficult. Recent earthquakes have been better reported, such as the ROSTAQ and MAZAR-E SHARIF earthquakes of 1998 and the series of earthquakes of 2002 during the U.S.–Al Qaeda conflict. Prior earthquakes are said to have killed as many as 20,000 people in a single event, but information is largely conflicting. The seismic activity results from the position of Afghanistan as a promontory of the EURASIAN CRUSTAL PLATE pushing southward into the Indian and ARABIAN CRUSTAL PLATES. The southern boundary of Afghanistan is marked by deep earthquakes associated with the SUBDUCTION of the Arabian plate beneath the MAKRAN COAST. Both the western boundary with IRAN and the eastern boundary with PAKISTAN have a long history of intense seismic activity. Fortunately, the central part of the country is less active. The exception is the active Chaman Fault, which shows a SEISMIC GAP near Kabul. If it produces a major earthquake, it could be devastating.

Africa The African continent has areas of strong seismic and volcanic activity. Much volcanism and earthquake activity is concentrated along Africa's GREAT RIFT VALLEY, which extends through the eastern portion of the continent and contains numerous volcanoes and CALDERAS. The East African Rift is the third arm of a TRIPLE JUNCTION that includes active arms of the Red Sea and the Gulf of Aden. Africa is also moving north relative to Europe. A SUBDUCTION ZONE in the MEDITERRANEAN SEA produces earthquakes and volcanoes that cause TSUNAMIS that affect the north coast of Africa. The Betic Zone largely lies in Spain but also affects the nearby African coast. Volcanoes and volcanic deposits of Africa include Asawa, the Barrier, Cameroon, Corbetti, Deriba, Fantale, Kilimanjaro, K'One, Land of Giant Craters, Ngorongoro, Nyamulagira, and Nyos.

aftershocks Earthquakes less intense (weaker) than the main (strongest) earthquake. Aftershocks can be determined only in retrospect. Typically, seismic events begin with FORESHOCKS, followed by the MAIN SHOCK and finally after- shocks. For example, if a MAGNITUDE 6.4 earthquake occurs after a series of foreshocks, it could be the main shock; all succeeding earthquakes would be aftershocks. If a magnitude 7.4 earthquake occurs three earthquakes later, then the 6.4 earthquake was merely a strong foreshock, and aftershocks only start after the 7.4 earthquake. Aftershocks may continue for weeks or months after the main shock and be nearly as powerful. They tend to be quite destructive to property, as they can topple structures left unstable by the main shock. In the NORTH RIDGE earthquake of 1994, strong aftershocks continued for a year after the main shock. They had a terrifying psychological effect on the populace of Los Angeles, California.

Agadir *earthquake, Morocco* The earthquake of February 29, 1960, struck the community of Agadir with a population of 33,000 at the foot of the Atlas Mountains at 11:45 P.M. It killed some 12,000 people and injured 12,000 others. Destruction of the old part of the city was complete, and some 70% of the new structures in the city were destroyed. The earthquake measured 6.25 on the RICHTER scale of MAGNITUDE and was preceded by two milder shocks. The earthquake was also accompanied by a TSUNAMI that reached almost a hundred yards inland from the sea. Effects of the earthquake included ruptured sewers, from which large numbers of rats were reportedly released into the city. The earthquake neutralized the city's fire-fighting capability, with the result that fires burned unchecked. The dome of a mosque collapsed on a group of praying Muslims, and the Jewish community of Agadir was devastated; of some 2,200 Jews in Agadir, approximately 1,500 were said to have died in the earthquake. Corpses were so numerous in Agadir after the earthquake that most of the 12,000 dead were simply buried in common graves, using a bulldozer.

The reasons that this earthquake was so devastating were threefold. First, the FOCUS of the earthquake was very shallow (less than two miles [3 km]). Typically, the energy from a deep earthquake is already spread out and somewhat diffuse by the time the waves reach the surface. In this case, the energy was still concentrated so it was more destructive than many earthquakes of higher magnitude. Second, the EPICENTER was right in the middle of the city. The zone of highest potential destruction was in the worst possible place. Third, the city was totally unprepared for the earthquake. There had been a very destructive earthquake in Agadir in 1751 (more than 200 years earlier) but only minor seismic events since.

agglomerate A chaotic jumble of mixed sizes and types of PYROCLASTIC material (EJECTA) that is lithified into a rock. An agglomerate is typically formed close to a volcanic VENT where the power of the eruption can pulverize the existing rock and dump it into a deposit. They indicate very high energy and must include a large component of BOMBS and BLOCKS.

Agnano *volcano, Italy* The Agnano volcano formed a crater in the PHLEGRAEAN FIELDS near NAPLES. The crater once was filled with a lake but later was drained and converted into a racetrack.

Agua de Pau *caldera, Azores* The STRATOVOLCANO Agua de Pau has a record of historical activity extending back to 1563, when an eruption of PUMICE reportedly covered the nearby island of São Miguel. Strong earthquakes preceded and continued during the eruption and destroyed most of the community of Ribeira Grande, several miles north of the Agua de Pau caldera. BASALT lava extruded from the volcano following the eruption from the main VENT. Another, less powerful eruption took place in the CALDERA the following year. In October 1952, destructive earthquakes preceded an eruption in which FISSURES opened at the foot of Agua de Pau. This eruption lasted one week and produced a LAVA FLOW and a small cone. Very small earthquakes occurred at Agua de Pau during the 1980s.

Agung *volcano, Bali, Indonesia* Agung is regarded as the "Navel of the Universe" and the home of the Supreme God by the Balinese. It is best known for its powerful eruption in 1963, which killed between 1,200 and 2,000 people and sent large amounts of ASH into the upper atmosphere. This airborne material is thought to have caused spectacular atmospheric effects in the following weeks, such as brilliant red sunsets and halos around the Moon and Sun. The high-altitude cloud from this eruption of Agung was also implicated in a sharp decrease in starlight as measured at observatories. Average temperatures at Earth's surface dropped measurably for three years after this eruption.

Many of the fatalities occurred at a religious festival that was in progress near the volcano at the time of the eruption. Clouds of lethal gas swept down from the volcano and killed large numbers of participants in the religious rites. Lava overwhelmed the villages of Sebih, Sebudi, and Sorgah. A combination of heat, ash, and poisonous gases is said to have killed animals for miles around the volcano. Huge boulders cast out from the volcano during the eruption landed in the village of Subagan.

See also CLIMATE, VOLCANOES AND.

Aira *caldera, Japan* The Aira lies a few miles north of the ATA caldera in southern Japan, in the region of Kagoshima Bay. The bay itself is thought to be a GRABEN, formed by volcanic and tectonic activity. Uplift of the bay floor has also occurred on occasion, and the Aira caldera has been cited to show that volcanically related uplift and subsidence can affect the whole area of a CALDERA even when an active volcano is located at the caldera's edge. The Aira caldera is famous for the violent eruptions of the SAKURAJIMA volcano (now known as On-take), although volcanic activity occurs at other points in the caldera as well.

An eruption from 1779 to 1781 began with a series of strong earthquakes in early September. Changes in water level (sometimes involving energetic spouting) were observed at water wells on Sakurajima on the morning of November 8, 1779, at about the same time clouds of steam started to rise from the summit of the volcano. On the afternoon of November 8, a major eruption started. After several days, small islands began to emerge from the waters near Sakurajima. These islands are thought to have been formed partly through underwater eruptions but also in part through uplift of the bay floor.

Changes in hot springs in the vicinity also were observed; two new hot springs emerged, and another stopped flowing. The area around Sakurajima appears to have subsided in the decades following this eruption because the waters encroached on low areas along the shore, flooding parts of the city of Kagoshima. In some areas, floods covered the land to a depth of perhaps 10 feet (3 m) or more. Local authorities tried building embankments to bar the rising waters but were unsuccessful. The rising waters wiped out some communities along the shore. Uplift affected other areas around Sakurajima about this time. A seacoast road on the southern shore of Sakurajima rose several feet until it lay more than 100 feet (30 m) inland from the waters. Uplift also affected the northwest shore of Kagoshima Bay so that trade in the harbor at one community had to be conducted using wagons rather than boats.

Another major eruption began in 1913, when earthquake activity to the north in May and June signaled the beginnings of renewed volcanism. The nearby Kirishima volcano erupted in late 1913 and early 1914. Strong earthquakes occurred near Kagoshima in late June. Dramatic changes also were observed in the activity of local hot springs. On the eastern shore of Sakurajima, hot springs stopped flowing in the spring of 1913, and shortly afterward, other hot springs on the south side of the volcano became too hot for bathing when the tide was low. Changes in the water table occurred early in 1914; a pond on the south side of Sakurajima dried up, killing the fish in it, while the water table dropped, and some of the island's water wells went dry. On the morning of January 12, 1914, a spring at a beach on the north side of Sakurajima emitted a gush of cool water while water spouted to a height of several feet from hot springs on the other side of the island. In addition, extremely hot water poured out from the ground at several other locations. On the same morning, a powerful eruption of Sakurajima began. This eruption was preceded by strong earthquake activity over more than 24 hours. Earthquakes were especially frequent on Sakurajima itself. A particularly strong earthquake occurred several hours after the eruption began. Since the 1913–15 eruptions, numerous small eruptions have been recorded. In 1935 earthquakes felt on the southern side of Sakurajima in the middle of the year were followed by eruptions of ASH beginning in September. Occurring only a few years after the violent events of 1913–15, these eruptions convinced several hundred residents of the area to evacuate. Ash was deposited to a depth of several inches on the south and east sides of the volcano, and some damage to crops occurred. An eruption may have occurred underwater on March 13, 1938, when waters about 1,000 feet (304 m) offshore rose abruptly while a roar was heard. This phenomenon was repeated soon afterward, some distance away. Sakurajima itself started to erupt again two weeks later. Minor eruptions took place between 1939 and 1942. In 1946, the volcano exhibited explosive activity and extruded LAVA. Minor explosions also occurred during the next eight years. Starting in 1955, explosions concentrated on the summit of the volcano. In the middle to late 1980s, explosive activity appeared to become more frequent after a comparatively quiet period. Earthquake activity at the Aira caldera has not always been related clearly to eruptions, although in some cases, eruptions plainly had earthquakes as precursors. The character of slippage along FAULTS has been seen to change during periods of eruptive activity. When the mountain is not erupting, earthquakes are characterized by STRIKE-SLIP (or predominantly horizontal) movement, which changes to oblique-slip normal faulting in the initial stages of eruptions and then to oblique-slip normal or REVERSE FAULTING when eruptive activity is at its height. As the eruption subsides, movement along faults returns to strike-slip or reverse.

As noted earlier, the Aira caldera is noted for the dramatic uplift and subsidence it has displayed on occasion. After the eruption of 1914, the caldera and adjacent areas displayed dramatic subsidence, almost 20 feet (6 m) in some locations. A few months later, uplift started again and has continued through the 1980s. Measurements of uplift at various points on and around Sakurajima indicate that a reservoir of MAGMA under the caldera has expanded at an average rate of perhaps 30 million cubic feet (850,000 m³) per year, at an estimated depth of perhaps four miles (6 km). The caldera does not appear to show DEFORMATION in a uniform pattern; some scientists have suggested that there is more than one source of uplift within Aira. In addition to deformation patterns observed in the caldera as a whole, Sakurajima may exhibit comparatively brief and shallow deformations.

In the 20th century, the Aira caldera has shown some curious phenomena related to heat flow. After the major eruption in 1914, the temperature of the SOIL began to rise near the northwest shore of Sakurajima. Fruit trees and other flora died. Eventually, trees were killed within an area some 500 feet (152 m) wide, and benzene and chlorine fumes were detected there. Some months earlier, similar emissions of fumes at a spot on the mainland, near a line of vents passing through the summit of Sakurajima, reportedly killed an ox and made several humans sick. By the spring of 1915, soil dug up at the heat-affected area was too hot to hold in one's hands. After 1915, this unusual concentration of heat diminished.

Akan *caldera, Japan* The Akan CALDERA is located in northern JAPAN near the south end of the Kuril Islands. Lake Akan occupies part of the caldera. Several cones are also found in the caldera; these include Furebetsu, Fuppushi, 0-akan and Me-akan, the last of which has been active within historical times. Strong earthquakes were felt in the vicinity of Akan caldera in the late 1920s and early 1930s, and in late 1937, a cloud of vapor was seen rising from the foot of Me-akan. Marked seismic activity increased in the early 1950s, and small quantities of ASH may have been released during this period, although it is not known if there was any direct relationship between eruptive and seismic activity at that time. An eruption began in November 1955, and the following year, observations of earthquakes showed a rise in activity of EARTHQUAKE SWARMS several days before an explosion on June 15. Seismicity increased for approximately three weeks preceding an explosion in 1959. Earthquakes accompanied eruptions of ash in 1988. The Akan caldera has been studied intensively to examine the relationship between earthquakes and a magmatic system. TECTONIC (as opposed to volcanic)

earthquakes have also been studied for their relationship to earthquake swarms at Akan. On one occasion, a major tectonic earthquake followed changes in temperature at hot springs in Akan.

Akutan *volcano, Aleutian Islands, Alaska, United States* There is a STRATOVOLCANO on Akutan Island in the eastern ALEUTIAN ISLANDS. Akutan has a summit CRATER with a lake and a cinder cone. It has erupted at least 27 times since its discovery in 1790. The most recent full eruption was in 1992. In March 1996, earthquake activity intensified with many swarms and magnitudes up to 5.1.

Alabama *United States* The state of Alabama varies geographically in its degree of seismic risk. The southern portion of the state is characterized by low seismic risk, whereas the degree of risk generally increases northward toward the TENNESSEE border. There have been several notable earthquakes in the history of Alabama, including the earthquakes of February 4 and 13, 1886, in Sumter and Marengo Counties, where perceptible movement of Earth was reported along the Tombigbee River. An earthquake on May 5, 1931, in northern Alabama was felt in Birmingham and caused minor damage at Cullman; the MERCALLI intensity was V–VI, and the affected area was about 6,500 square miles (17,000 km²). On April 23, 1957, an earthquake in the area of Birmingham, estimated at INTENSITY VI on the Mercalli scale and affecting an area of about 2,800 square miles (7,000 km²), caused minor damage in Birmingham; loud noises were associated with this earthquake in some locations. The August 12, 1959, earthquake along the border of Alabama and Tennessee caused minor damage and was estimated at intensity V; the earthquake affected an area of some 2,800 square miles. An earthquake on February 18, 1964, along the border of Alabama and GEORGIA measured Mercalli intensity V and RICHTER magnitude 4.4.

Alaska *United States* The largest and northernmost state of the UNITED STATES, Alaska also is one of the most seismically and volcanically active parts of the country. Earthquakes in Alaska are concentrated in two belts, one extending along the southeast coast and another reaching from the interior near Fairbanks southwestward along the ALEUTIAN ISLANDS. The 1964 GOOD FRIDAY EARTHQUAKE, one of the most powerful and destructive earthquakes of the 20th century, occurred along the southern coast of Alaska and, along with the TSUNAMI associated with it, caused extensive destruction as far south as CRESCENT CITY, CALIFORNIA.

Volcanism in Alaska has been both frequent and destructive throughout history. A familiar case in point is the eruption of KATMAI in 1912. This eruption created a CALDERA some three miles wide and laid down a plain of FUMAROLES later named the VALLEY OF TEN THOUSAND SMOKES. The volcanic arc in Alaska extends more than 1,000 miles (1,600 km), from Cook Inlet in the east to Buldir Island near the tip of the Aleutian chain in the west. More than 70 volcanoes exist in the Aleutian Islands and on the Alaska Peninsula. The Alaskan volcanoes and earthquakes are an expression of activity along a SUBDUCTION ZONE marked by the Aleu-

As a result of the 1964 Good Friday earthquake in Alaska, a rockslide-debris avalanche (dark area) was generated in Shattered Peak and spilled over Sherman Glacier (white). *(Courtesy of the USGS)*

tian Trench south of the Alaska Peninsula and the Aleutian Islands. The Aleutian Trench reaches depths greater than 20,000 feet (6,000 m). North of the Aleutian Islands, in the Bering Sea, lie the Pribilof Islands, which were formed by eruptive activity but do not constitute part of the Aleutians. A history of earthquakes and volcanism in Alaska would occupy an entire volume, and all an article of this length can do is present a few examples.

A very strong earthquake accompanied the eruption of PAVLOF volcano on the Alaska Peninsula in 1786. A tsunami, or seismic sea wave, reportedly flooded land on Sanak Island, the Shumagin Islands, and the Alaska Peninsula on July 27, 1788, with considerable loss of human life and of livestock. In May 1796, an earthquake with frightening noises affected Unalaska Island, and BOGOSLOF volcano cast out rocks as far away as Umnak Island. In 1812, powerful earthquakes accompanied an eruption of a Sarycheff volcano on Atka Island. Umnak Island underwent a strong earthquake in April 1817, when Yunaska volcano erupted. Sometime in 1818, an earthquake near Makushin volcano and Unalaska Island is said to have caused great alterations in the landscape. Unalaska Island experienced two earthquakes in June 1826, but details are unavailable. An earthquake described as "severe" struck the Pribilof Islands on April 14, 1835; and in April 1836, the Pribilofs were subjected to shocks so powerful that they knocked people off their feet. An earthquake on September 8, 1857, was very powerful but apparently caused no damage. A minor earthquake on May 3, 1861, at St. George Island in the Pribilof Islands was accompanied

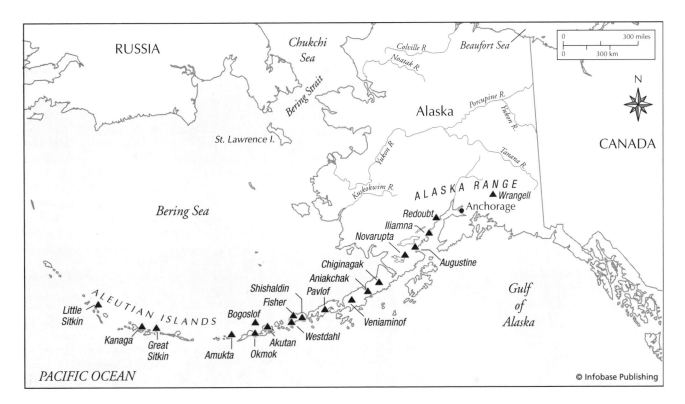

Locations of many active volcanoes of the Aleutian arc, Alaska

by noise from underground. On August 29, 1878, an entire town on Unalaska Island appears to have been destroyed by a tsunami and earthquake.

AUGUSTINE volcano erupted on October 6, 1883; a very powerful earthquake and a tsunami occurred in connection with this eruption. An earthquake in the area of Prince William Sound in May 1896 was so violent that people who were standing had trouble remaining on their feet. The Yakutat Bay earthquakes of September 3 and 10, 1899, were estimated at MERCALLI intensity XI and at RICHTER magnitudes 8.3 and 8.6, respectively. The EPICENTER was located near Cape Yakataga. The first of these earthquakes was felt with tremendous violence at Cape Yakataga, but the second earthquake was the one that caused major changes in topography. A U.S. Geological Survey expedition to the region six years after the earthquakes found widespread evidence of topographic changes. Beaches had been raised, and barnacles and other aquatic organisms were lifted out of the water. On the west shore of Disenchantment Bay, an uplift of more than 47 feet (14 m) was measured—approximately the height of a five-story building. Over a wide area, uplift of 17 feet (6 m) or more was observed. In some areas, depressions of several feet occurred. A tsunami thought to have been perhaps 35 feet (11 m) high occurred in Yakutat Bay, and tsunamis were reported at other locations along the coast of Alaska as well. There were reports of volcanic eruptions associated with these earthquakes, but the "eruptions" are presumed to have been merely large clouds of snow released in slides caused by the earthquakes. Strong AFTERSHOCKS occurred over several months following these earthquakes. No loss of life was attributed to the earthquakes because the area was not yet settled; a small number of Native Americans and prospectors, however, witnessed the earthquakes firsthand.

On September 21, 1911, an earthquake of Richter magnitude 6.9 on the Kenai Peninsula and Prince William Sound broke cables, caused great rockslides, and killed large numbers of fish; water at Wells Bay was reportedly disturbed greatly. Cables broke also in another earthquake on January 31, 1912, in the vicinity of Prince William Sound; this earthquake, which was measured at Richter magnitude 7.25, appears to have been centered west of Valdez and was felt in Fairbanks. Very strong shocks occurred at Kanatak, Nushagak, and Uyak on June 4–5, 1912, and were felt more than 100 miles (161 km) away from Mount Katmai, although the earthquakes may have been unaffiliated with the June 6 eruption of Mount Katmai. An earthquake of Richter magnitude 6.4 at Cook Inlet on June 6, 1912, coincided with a bright display of light from Katmai, and the shock was recorded at many distant locations, including Ottawa, Ontario, and Irkutsk in Russia. Very strong earthquakes were reported on the night of June 6 at Kodiak, and on June 7, a strong earthquake struck Kanatak, together with ROCKSLIDES and a powerful rumbling noise.

An earthquake near Seward on January 3, 1933, measured at Richter magnitude 6.25, was felt very strongly at Anchorage, and caused alarm at Seward; the ground cracked in numerous places in the vicinity of Seward, notably for a distance of 20 miles (33 km) along a road extending north from the city. On April 26, 1933, an earthquake northwest of Anchorage severed telegraph lines and broke plate-glass windows and was felt also in Fairbanks and in the Aleutian Islands. Houses were displaced from their foundations at Old Tyonek. The principal shock measured Richter magnitude 7.0. Old Tyonek experienced further damage several weeks later when an earthquake of magnitude 6.25 occurred there on June 13, 1933. The May 14, 1934, earthquake on Kodiak Island measured magnitude 6.5 and was felt strongly on Whale and Kodiak Islands; plaster was cracked, and roads were blocked by LANDSLIDES. An earthquake of magnitude 6.75 in south central Alaska was strong enough to break plate glass in Anchorage on August 1, 1934.

Tsunami damage was remarkable in the magnitude 7.4 earthquake of April 1, 1946. Centered about 90 miles (145 km) southeast of Scotch Cap Lighthouse, the earthquake produced a tsunami that demolished the lighthouse and caused damage at widely separated locations in and around the Pacific Basin, along the Pacific coasts of North and SOUTH AMERICA, in the Aleutian Islands, and in the HAWAIIAN ISLANDS, where 173 persons drowned and property damage was estimated at $25 million.

The earthquake of March 9, 1957, measured Richter magnitude 8.3 and was one of the greatest natural calamities in Alaskan history. The earthquake, which involved hundreds of aftershocks and affected an area approximately 700 miles (1,127 km) in length along the southern border of the Aleutian Islands between Amchitka Pass and Unimak Island, was accompanied by a tsunami 40 feet (12 m) high that struck the shore at Scotch Cap, and a 26-foot (8-m)-high tsunami that caused extensive damage at Sand Bay. On the islands of Kauai and Oahu in Hawaii, the waves destroyed two villages and caused several million dollars in damage. The tsunami was 10 feet (3 m) high along the coast of JAPAN, and a wave six feet (2 m) high was reported in CHILE.

The earthquake of July 9, 1958, is famous for the dramatic effect it produced at LITUYA BAY, on the Gulf of Alaska in the southeastern part of the state. A tremendous rockslide at the head of the bay produced a giant wave (SEICHE)—more than 1,700 feet (518 m) high—that swept outward through the mouth of the bay and is thought to have killed two people who were caught in the wave. A fishing boat with two occupants was carried out of the bay by the wave front and reportedly cleared the spit at the mouth of the bay by at least 100 feet (30 m). The wave also wiped the rim of the bay clean of trees. Otherwise, little damage was reported from this earthquake, except that underwater communication cables were broken in the vicinity of Skagway, and Yakutat experienced damage to bridges, a dock, and oil lines. Great landslides reportedly occurred in the mountains, and FISSURES and sand blows were reportedly widespread on the coastal plain near Yakutat.

The Good Friday earthquake of March 27, 1964, is covered in detail elsewhere in this volume.

Among the volcanoes of Alaska are Katmai, Augustine, Pavlof, REDOUBT, ILIAMNA, and SHISHALDIN. Numerous calderas, indicative of eruptive activity followed by collapse, are found at locations such as ANIAKCHAK, EMMONS LAKE, FISHER, LITTLE SITKIN, OKMOK, SEMISOPOCHNOI, VENIAMINOF, and the WRANGELL MOUNTAINS.

Alban Hills *volcanic structures, Italy* The Alban Hills are located near Rome and are believed to have originated through a combination of explosive and effusive eruptions. A period of predominantly effusive eruptive activity is thought to have produced a STRATOVOLCANO that developed a collapsed CALDERA from which a new central cone arose later. The record of activity at the Alban Hills in historical times is uncertain. Eruptions are reported in 114 B.C. and possibly several centuries earlier, but there is some question whether these events were volcanic in nature or represent other natural phenomena such as fires and falls of hail. An ASHFALL was reported in nearby Rome in 540 B.C. The volcano Albano is located in the Alban Hills area, and Lago Albano occupies an eccentric CRATER just west of the rim of the inner caldera. The Via Appia Nuova, Via Appia Antica, and Via Tuscolana traverse the Alban Hills.

Alcedo *volcano, Galápagos Islands, Ecuador* Alcedo is one of several volcanoes on Isabela Island in the Galápagos. A CALDERA is present. A LAVA FLOW, identified from aerial photos, appears to have occurred on the southeast side of the volcano between 1945 and 1961. Radial FISSURES account for many of the lava flows on the volcano. Volcanic activity is suspected (although this has not been proven) as the cause of an uplift of a short length of shoreline on the west side of the island, possibly in 1954. A large amount of coral reef was lifted above sea level, probably at the same time.

Aleppo *earthquake, Syria* This ancient city in Syria is now called Halab. It is credited with having experienced one of the greatest earthquake disasters of all time. On September 8, 1138, a devastating earthquake struck the city. The damage was estimated from historical records to have been XI on the modified MERCALLI scale, with virtually all buildings destroyed by the intense ground shaking. The estimated DEATH TOLL was a staggering 230,000, though other sources place it at 100,000. Aleppo was the regional capital at the time, so the population may have been higher than usual, but the city of Halab only had a population of about 200,000 in the last census. In addition, historical records are poor and incomplete, suggesting that this number could be in error. Aleppo was struck by another major earthquake on September 5, 1822. Damage from this event was estimated at X on the modified Mercalli scale; details of the hazards, however, are again poor. This earthquake is historically well known because of a report by the American missionary Benjamin Barker. The report, "Earthquake at Aleppo," was one of the first examples of newspaper-style reporting in the United States. It described well the human suffering and the religious context but contained no information on hazards. The death toll from this event was 22,000 people, a more reasonable number than that for the 1138 event.

Aleutian Islands *Alaska, United States* The volcanic Aleutian Island chain extends westward from the south shores of ALASKA. The islands are part of a range of volcanic mountains, the Aleutian Range, extending more than 1,600 miles (2,600 km), from the Alaska Peninsula to a point just east of the International Date Line. The Aleutian Range contains dozens of recently active volcanoes, including AUGUSTINE, BOGOSLOF, KATMAI, Novarupta, PAVLOF, REDOUBT, and TRIDENT. The tallest volcanoes (up to 11,000 feet [3,400 m]) occur at the northeast end of the range. Summit elevations generally diminish southwestward along the Aleutians. Several kinds of volcanoes occur in the Aleutian Range. Some are SHIELD VOLCANOES made up of numerous thin flows of LAVA. Other Aleutian volcanoes are composite volcanoes with steep sides. In some places, these composite cones occur atop older, shield volcanoes, resulting in a structure much like that of the CASCADE MOUNTAINS in the northwest UNITED STATES and the Canadian province of British Columbia. Volcanic domes may also be seen where viscous lava has emerged. A notable CALDERA formed from the collapse of Katmai during its 1912 eruption. The Aleutian volcanoes are associated with an offshore subduction zone marked by the presence of the Aleutian trench, a deep undersea trough located to the south of the Aleutians and the Alaska Peninsula. The trench is shallower and eventually vanishes toward the mainland. The progressive shallowing of the trough is thought to be due to a buildup of sediment.

The Aleutian Range has been the site of powerful earthquakes, such as the GOOD FRIDAY EARTHQUAKE of 1964, which caused great destruction in the vicinity of Anchorage. Ground subsidence destroyed much of Anchorage's main street. Approximately 75 homes in a residential neighborhood on Turnagain Bluff were wrecked when the land on which they rested underwent a sudden slump. The earthquake also demolished the airport control tower and killed the controller on duty when the structure collapsed. Alaskan earthquakes have been accompanied by powerful TSUNAMIS on several occasions in this century. The tsunami that accompanied the Good Friday earthquake, for example, caused tremendous damage along the south coast of Alaska, wiped out much of the state's commercial fishing fleet, and carried destruction as far south as CRESCENT CITY, CALIFORNIA. More than 200 persons were killed as a result of tsunamis originating in Alaskan waters between 1946 and the Good Friday earthquake.

In recent years, areas along the Aleutian Range have emerged as cause for concern as potential sites of major future earthquakes. One of these so-called gaps is the Commander gap, an area near the west end of the Aleutian chain where no major earthquake has occurred since the mid-19th century. The Shumagin gap near the west tip of the Alaska Peninsula also has been identified as a prospective source of powerful earthquakes because others have occurred there in 1788 and 1946. (Another strong earthquake in 1903 may have originated in this area.) The potential for destructive tsunamis from earthquakes in the Shumagin gap is also considerable. A third "gap" along the southern Alaska coast, the Yakataga gap, lies near the north tip of the Alaska panhandle. It has been an area of concern for the U.S. Geological Survey, which expects that strain in the Yakataga gap may manifest itself in the near future in the form of earthquakes of magnitude 8.0 or stronger.

Alexandria *earthquake, Egypt* An earthquake on July 21 in A.D. 365 shook much of the MEDITERRANEAN basin and appears to have caused widespread destruction. Among the most notable casualties of this earthquake was reportedly the great lighthouse at Alexandria in EGYPT. Said to have been some 600 feet (183 m) high, the lighthouse was reduced to a ruin that remained in place for the next five centuries. More than 50,000 people in Alexandria were reportedly killed in this earthquake, which was accompanied by TSUNAMIS. Edward Gibbon, in his history *The Decline and Fall of the Roman Empire,* describes the effects of this earthquake on the shores of the Mediterranean:

> In the second year of the reign of Valentinian and Valens, on the morning of the twenty-first day of July, a violent and destructive earthquake shook the greatest part of the Roman world. The impression was communicated to the waters; the shores of the Mediterranean were left dry by the sudden retreat of the sea; great quantities of fish were caught with the hand; large vessels were stranded on the mud; and a curious spectator [evidently the historian Ammanius, whose accuracy Gibbon questions in a footnote to the work] amused his eye, or rather his fancy, by contemplating the various appearance of valleys and mountains which had never, since the formation of the globe, been exposed to the sun. But the tide soon returned with the weight of an immense and irresistible deluge, which was severely felt on the coasts of Sicily, of Dalmatia, of Greece, and of Egypt; large boats were transported and lodged on the roofs of houses, or at the distance of two miles from the shore; the people, with their habitations, were swept away by the waters; and the community of Alexandria annually commemorated the fatal day on which fifty thousand persons had lost their lives in the inundation.

The psychological impact of this earthquake on the Romans appears to have been considerable. Gibbon continues:

> This calamity, the report of which was magnified from one province to another, astonished and terrified the subjects of Rome, and their affrighted imagination enlarged the real extent of a momentary evil. They recollected the preceding earthquakes, which had subverted the cities of Palestine and Bithynia; they considered these alarming strokes as the prelude only of still more dreadful calamities; and their fearful vanity was disposed to confound the symptoms of a declining empire and a sinking world.

Algeria Algeria lies along the western Mediterranean Sea, which experiences moderate seismic activity as a result of its position at the northern margin of the African plate. It is the grinding between the African and EURASIAN CRUSTAL PLATES that produces the seismicity. The most destructive events have their sources in the Tellian Atlas Mountains in northern Algeria. This area is dominated by northeast-southwest-oriented REVERSE FAULTS, but other areas contain STRIKE-SLIP and normal faults as well. The southern region of Algeria contains the large east-west Sahel Fault, which is also a reverse fault. The more destructive earthquakes in Algeria include those in Algiers, 1916 (IX); ORAN, 1790 (XI); Mascara, 1889 (IX); El ASNAM, 1980 (Ms=7.3); Constantine, 1985 (Ms=6.0); Tipasa, 1989 (Ms=6.0); Mascara, 1994 (Ms=6.0); and ALGIERS, 1996 (Ms=5.7) and 2003 (Ms=6.8).

Algiers *earthquake, Algeria* At 7:44 A.M. local time on Wednesday May 21, 2003, a major earthquake struck northern Algeria. Its EPICENTER was located 40 miles (65 km) east-northeast of Algeria, and the FOCUS was a shallow six miles (10 km) in depth. The RICHTER magnitude of the event was 6.8, and damage reached X on the modified MERCALLI scale. More than 1,243 buildings were damaged or destroyed, leaving some 150,000 homeless. A TSUNAMI with a maximum RUN-UP height of 6.5 feet (2 m) was generated in the MEDITERRANEAN SEA and damaged boats along the coast. Total damage was estimated at $100 million. The DEATH TOLL for this event was 2,266, and there were 10,261 people injured.

Alkalic basalt BASALT (LAVA) that is unusually rich in alkali elements (K, Na, etc.). These lavas are common in continental RIFTing but only in the early stages or at the end of a volcanic period. They occur in such HOT SPOTS as HAWAII under the same conditions. There is no visual way to tell an Alkalic basalt from a regular basalt; chemical analysis is necessary.

Allahbund (Gujarat) *earthquake, India* The Rann of Kutchh was shaken by a violent earthquake on June 16, 1819, at about 7:00 P.M. A 55-mile (90-km) stretch of land was elevated by up to 14.3 feet (4.3 m) during this event and named the Allah Bund which translates to "mound of God." Trenching of the feature indicated that it was a fold rather than a fault, and it had been elevated at least twice prior to the 1819 earthquake. The fault that generated the earthquake was therefore a BLIND FAULT and appeared to have been a northwest-oriented REVERSE FAULT. To the south of this area, subsidence in the range of 16.5 to 20 feet (5 to 6 m) flooded the Fort of Sindri. The RICHTER magnitude of this earthquake was estimated to have been between 7.5 and 8.3, but the lower number is more probable. Shaking appeared to have lasted almost two minutes.

Records of casualties from the earthquake are inconsistent, but there were at least 1,550 people killed. The majority of the damage occurred within a 49.7-mile (80-km) radius of the epicenter, although excessive shaking was reported from more than 186.4 miles (300 km) away. LIQUEFACTION was one of the main surface effects. Many buildings tilted and fell over, and mud volcanoes were 12 to 20 feet (3.7 to 6 m) in diameter and active for two to three days. The tributaries of the Indus River were also strongly affected. Flow stopped for three days in the Fullalee River, and the Nara River was blocked by a landslide, forming a large pool, while the downstream portion dried up. AFTERSHOCKS continued for several months, with the strongest on June 17 and July 15.

Aluto *See* ASAWA.

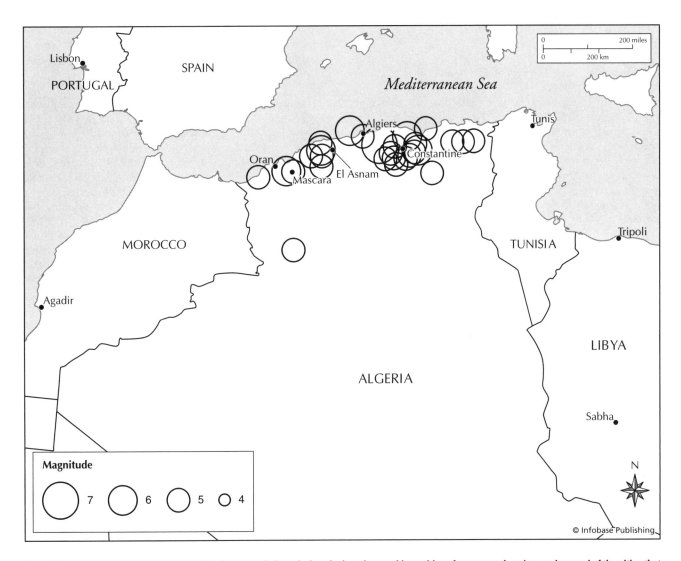

Map of Algeria and neighboring countries showing tectonic boundaries, the locations and intensities of recent earthquakes, and several of the cities that experienced serious historical earthquakes

Amatitlán *caldera, Guatemala* The Amatitlán CALDERA is located several miles south of Guatemala City and includes the volcano PACAYA, which has a long history of unrest within historical times.

Ambrim *volcano, Vanuatu* Situated at the intersection of the Vanuatu archipelago and the D'Entrecasteaux fracture zone near the Loyalty Islands, the volcanic island Ambrim has exhibited numerous explosive eruptions and LAVA FLOWS during the past two centuries. Roughly triangular in shape, Ambrim is approximately 30 miles (48 km) long and 20 miles (32 km) wide at its broadest point. A large CALDERA occupies the summit. Two cones inside this caldera, Mount Marum and Mount Benbow, show nearly constant activity. Several small volcanoes (Rahoum, Tower Peak, and Tuvio) also are found on the island. The historical record of activity on Ambrim is brief but colorful. It formed about A.D. 50 and has had 48 eruptions since then. Emissions resembling smoke

were reported in 1774, and in 1888 a flow of LAVA was observed from a rift on the southeastern side of the island. Large numbers of earthquakes characterized an eruption in 1894. Dates of eruptive activity between 1912 and 1915 are not entirely certain, but Mount Benbow and various other sites on the island appear to have shown eruptive activity. Strong earthquakes accompanied eruptions in 1913 along FISSURE lines lying east to west across the island, and reports mention a large eruption cloud and emissions of flames. A hospital was destroyed in this set of eruptions. Strong earthquakes accompanied another eruption in late March 1937, and marked seismic activity was noted before the lengthy eruption of 1950 through 1954. An explosion in 1972 did not show any precursor earthquake activity. Ambrim remained in nearly continuous eruption between 1964 and 1980. Acid rain generated by emissions of sulfur dioxide harmed crops in February 1979. Eruptions are thought to have occurred around the end of 1985 and in early 1986. In February 1988, the crew of an aircraft flying near the island observed an eruption from Mount Benbow.

amplification Amplification is any process that increases the amplitude of seismic waves. Amplification may be the enhancement of the strength of a signal through electronic means, so that even small earthquakes can be analyzed. The seismic energy may also be increased locally by focusing it through geologic means. This focusing is typically through the geometry of the structure of a basin, the topography of a basin, or the sediment (stratigraphic) velocity structure. The result is great damage in specific areas distant from the earthquake EPICENTER and with much less damage in all surrounding areas.

amplitude Physical distances of all waves, whether ocean, radio, or seismic (earthquake), define certain quantities. One-half height of a wave from crest to trough is the amplitude.

Amukta *volcano, Alaska, United States* It is a poorly known, seismically unmonitored, and uninhabited island STRATOVOLCANO. It has erupted at least five times since its discovery in 1760. Its last major eruption was in 1987, but it underwent a minor eruption in September 1996.

amygdaloidal VESICLES in volcanic rocks are holes that look like worms made them. Mineral-bearing waters flow through these holes and deposit minerals in them. When the holes are filled with mineral, they are called amygdules. *Amygdaloidal* refers to a rock that contains amygdules. Minerals common to amygdules include calcite, QUARTZ (including amethyst), and prehnite.

Anak Krakatoa *See* KRAKATOA.

Anatolia *earthquake, Turkey* On October 16, 1883, an earthquake at Anatolia in Asia Minor, now part of TURKEY, killed perhaps 1,000 people and left some 20,000 homeless. Great FISSURES are said to have opened and shut in the earth during this earthquake. Starvation and cold temperatures reportedly killed several hundred more residents of Anatolia before assistance could arrive.

Ancash *earthquake, Peru* On November 10, 1946, at 12:40 P.M., a devastating earthquake struck the mountainous region of Peru and, in particular, the villages of Ancash and Quiches. The quake had a RICHTER magnitude of 7.4 and was felt 400 miles (680 km) to the north in Guayaquil, Ecuador, and 245 miles (410 km) to the south in Lima, Peru. This means that it affected an area of about 175,000 square miles (450,000 km²). Although the FOCUS of this earthquake was 18–24 miles (30–40 km) deep, making it SUBDUCTION ZONE–related, there was significant surface rupturing, leaving a SCARP along the Quiches fault for some three miles (5 km) in length. Whether this was a related rupture through an AFTERSHOCK or a FORESHOCK is not clear.

An estimated 1,400 people lost their lives in the Ancash earthquake. The reason for the relatively high death toll in such a sparsely populated area was the mass movements. The earthquake shook loose multiple LANDSLIDES on many scales, including several AVALANCHES, ROCKFALLS, and ROCKSLIDES. These mass movements slid down the steep slopes, destroying the villages in the valleys below.

Andaman Islands *earthquake, India* A devastating earthquake struck the Andaman Islands INDIA, on June 26, 1941. It had a RICHTER magnitude of 7.7, with a duration of four minutes, making it one of the strongest in the area. The FOCUS of the earthquake was 33 miles (55 km) deep, making it SUBDUCTION ZONE–related. Tremors were felt strongly in the area around CALCUTTA, India, but also as far away as Sri Lanka. There were many AFTERSHOCKS, including two of MAGNITUDE 6.0 within 24 hours of the MAIN SHOCK and 14 of magnitude 6.0 through January 1942. A TSUNAMI was generated in the Bay of Bengal by this event. The height of the waves was less than 6.6 feet (2 m) in all cases, but they nonetheless devastated the Indian coast, causing most of the loss of life. The death toll for the entire event was in excess of 5,000 people, but records are poor, and the numbers could be much higher.

andesite One of the most common volcanic rocks, andesite is widely distributed around the Pacific basin (the "RING OF FIRE"), where chains of andesitic volcanoes form the "andesite line" that has been used to mark the boundary of the Pacific basin. Andesitic volcanic rocks also are found in other regions of the world, including Europe's Carpathian Mountains, the Himalaya Mountains, the Zagros Mountains, some of the active volcanoes in the MEDITERRANEAN, the CARIBBEAN Islands, and the South Sandwich Islands. Andesite varies in composition but is generally characterized as a gray rock that is lighter in color than BASALT but most commonly darker than RHYOLITE. It is also intermediate between basalt and rhyolite in terms of silica content. The composition of andesite indicates that it derives its chemical makeup from conditions in the mantle below the continents. Experimental studies reveal that the melting of the subducting wet ocean crust or melting of MANTLE that has been injected with water from the subducting OCEANIC CRUST can account for the composition. However, in cases of MAGMATIC ARCS where MAGMA must pass through thick continental crust, andesite contains commonly ASSIMILATED components of crustal rock as it rises to the surface. The characteristics of andesite output vary greatly from one volcano or cluster of volcanoes to another. In some areas, a group of volcanoes may emit andesite of remarkably consistent composition, whereas a single volcano elsewhere may put out a variety of types. Many volcanoes release two main types of rocks, a principal andesitic series and another group made up largely of basalt with either rhyolite or dacite mixed in. Andesite volcanoes also emit large quantities of ASH and other EJECTA. Composite andesitic volcanoes, widely found around the PACIFIC OCEAN basin, are composed of TEPHRA and flows of andesite and rhyolite and commonly feature CALDERAS formed by explosive eruption and the collapse of a cone into a depleted MAGMA CHAMBER beneath the mountain, as in the case of CRATER LAKE in OREGON.

andesite line *See* ANDESITE.

Andes Mountains The Andes mountain range is part of the Andean cordillera, which extends along the western edge of SOUTH AMERICA. The Andes are formed by an ongoing collision between the South America plate and primarily the NAZCA CRUSTAL PLATE (Antarctic and Cocos as well) that underlies the southeast PACIFIC OCEAN. This SUBDUCTION ZONE is considered the model for CONVERGENT PLATE BOUNDARY in which oceanic crust is subducted beneath continental crust. All other such geometries are referred to as an Andean Margin. The Andes Mountains form a MAGMATIC ARC with extensive plutonic and volcanic activity. Volcanic and HYDROTHERMAL ACTIVITY along the Andean cordillera has generated hot springs and numerous commercially viable deposits of ores of various metals, among them copper, silver, and gold. TSUNAMI activity also has been associated with the numerous and regular strong earthquakes along the Andean cordillera. The strongest recorded earthquake in the world occurred in CHILE in 1960.

A curious magnetic anomaly has been reported along the Andean cordillera. A reversal of change in the vertical geomagnetic field has been accompanied by an intensified change in the horizontal magnetic field. This pattern indicates the existence of internal induction currents at a depth of perhaps 40 miles (70 km), along a high-conductivity zone beneath the mountains.

Earthquakes occur frequently in and near the Andean cordillera. An earthquake in Chile in 1822, for example, reportedly killed some 10,000 persons and raised the shore-line by several feet. Along the Pacific shore, earthquakes are sometimes accompanied by tsunamis. The volcanoes and volcanic deposits in the Andes include CERRO RICO, COPA-HUÉ, CORDILLERA NEVADA, DIAMANTE, NEVADO DEL RUIZ, NEVADOS DE CHILIAN, and PURACÉ.

See also BOLIVIA; CONCEPCIÓN; PERU; PLATE TECTONICS.

Andijan *earthquake, Turkestan (Uzbekistan)* An earthquake of RICHTER magnitude 6.4 struck the town of Andijan in Turkestan (now Uzbekistan) on December 16, 1902, at 5:07 A.M. The FOCUS of the earthquake was at a depth of 5.5 miles (9 km). The damage was extreme, and freezing temperatures and continuing AFTERSHOCKS exacerbated the disaster. The DEATH TOLL was at least 4,500, but some reports placed it as high as 10,000 people. Some 15,000 houses were destroyed, and the cost of the damage was upward of $6 million.

angle of repose The angle of repose is the maximum angle that a slope of a certain material can exist at under a certain set of conditions before it fails. If a person slowly dumps a bucket of dry sand on the floor, it will form a cone-shaped pile with a fixed angle of the slopes regardless of where it is measured. To make the pile higher, the person adds more sand, but the pile does not grow straight up. Instead, most of the sand slides down the slopes (slope failure), and the pile grows wider as it gets higher. The angle of the slopes of the bigger pile will be exactly the same as that for the smaller

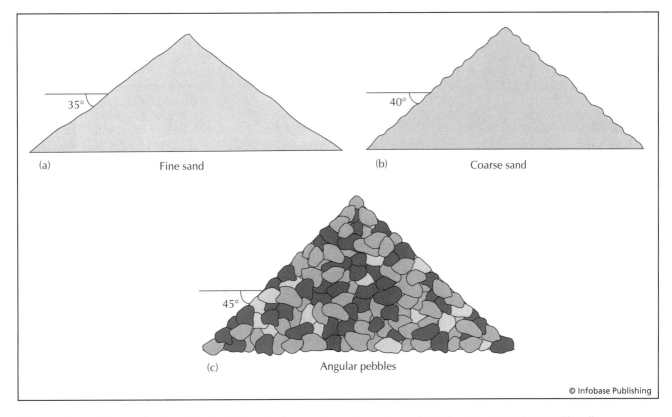

Piles of various particle types (sediment) showing their angle of repose, the maximum slope angle that they can attain before sloughing off.

pile. Different materials (gravel, clay, boulders, etc.) form different slope angles; typically, the more friction among particles, the steeper the slope.

The condition of the material may also control the angle of repose. For example, the angle of repose of damp sand is very high. The moisture has surface tension that produces capillary action among the grains and holds them tightly together. That is why people can make sand castles on the beach. On the other hand, if a lot of water is added to a material, it turns into a slurry and has a very low angle of repose. Soupy mud lies almost flat when spilled onto the floor. If a slope is frozen, it can maintain a much higher angle of repose than one that is simply wet. Temperature may also play a role in some cases, as can wind, which may blow down slopes in some cases, or earthquakes and other vibrations, which tend to shake slopes down.

Vegetation can also enhance the angle of repose for large slopes outside. Trees have very deep roots and tend to hold the soil together, thus greatly enhancing the angle of repose. Grass can also hold the slope together a little but not as well as trees. People also affect the angle of repose for larger slopes through a variety of methods, such as cement, retaining walls, gravel or riprap, netting, and drainage ditches and culverts, among others. Typically, people attempt to increase the angle of repose with varying degrees of success. The LANDSLIDES and MUDFLOWS in CALIFORNIA that are reported by the media on occasion are examples of poor results.

Aniakchak *caldera, Aleutian Islands, Alaska, United States* The Aniakchak CALDERA is located in the eastern Aleutian ISLAND ARC near Bristol Bay. Surprise Lake occupies part of the CRATER. Several cones and necks are found on the floor of the caldera, including Vent Mountain. Aniakchak formed about 3,400 years ago and has had about 10 eruptions since. Powerful explosions occurred at Aniakchak in 1931, possibly from a cinder cone. A dome formed in the vent late in the eruption.

Ansei Tokai *earthquake, Japan* A strong earthquake struck the island of Honshu, JAPAN, on December 23, 1854, in the Ansei-Tokai area. The MAIN SHOCK was estimated at 8.4 on the RICHTER scale, but a strong AFTERSHOCK with estimated MAGNITUDE of 7–7.5 occurred less than 24 hours later, at 8 A.M. on December 24, 1854. The FOCUS was on the Tokai Fault, which extends from the east coast of the Kii Peninsula to the Suruga Bay. It was generated at a shallow level in the crust, unlike the typical SUBDUCTION ZONE earthquakes in Japan.

The thick, soft sediment in this part of Japan enhanced the SURFACE WAVES, which led to excessive ground movement. LIQUEFACTION produced large water spouts that were even seen being emitted from Suruga Bay. One of the most devastating effects of both strong shocks were the TSUNAMIS they produced. These waves ranged from 6.5 to 23 feet (2 to 7 m) in height and reportedly battered and beached some 200 boats in Osaka Bay alone. The reported DEATH TOLLS for this combined event were conflicting. The most reliable figure was in excess of 5,000 people, but some reports were as high as 31,000 people. One of the most ominous aspects of this event is the lack of subsequent large earthquakes in the area. With a RECURRENCE INTERVAL of 100–150 years, the Tokai area is overdue. The predicted casualties in this now highly populated area from a magnitude 7 earthquake are approximately 6,000 deaths and 19,000 injuries. Another 8.4 earthquake would be even more disastrous.

Antarctica Although the historical record of seismic and volcanic activity in Antarctica is not as extensive as the record for more densely settled portions of the world, much is known about earthquake and volcanic activity on the Antarctic continent. The volcano Mount EREBUS was discovered by Captain James Clark Ross of Britain on an expedition to reach the south magnetic pole. Erebus reportedly was erupting at the time of Ross's visit. The volcano and a nearby CRATER were named Erebus and Terror respectively after the two ships on Ross's expedition. A later expedition under the command of Ernest Shackleton climbed Erebus in the first ascent of a mountain in Antarctica. However, Antarctica is considered tectonically quiet, and earthquakes and volcanoes are few and are mostly located along the edges of the continent and offshore. Other sites of volcanic activity on or near Antarctica include DECEPTION ISLAND, Hampton/Whitney, Takahe, Thule Island, and Waesche.

Antioch *earthquake, Syria* On the evening of May 29, A.D. 526, one of the world's great natural disasters struck the city of Antioch, Syria, which is now Antakya, TURKEY. The city of Antioch was founded about 300 B.C. by Syrian emperor Seleucus I but was captured by Rome in A.D. 25. Saint Paul selected it as the center of his work in Galatia around A.D. 50. Antioch thus became a prominent city for both trade and Christianity. During the sixth century A.D., the Feast of the Ascension, 40 days after Easter, became one of the most prominent Christian festivals. In A.D. 526, the holiday fell on May 30, and hordes of people flooded Antioch in the preceding days in anticipation. When the huge earthquake, estimated at a 9.0 on the RICHTER scale, struck the city, it was devastating. It was reported that most buildings simply collapsed, killing virtually all inside, when the MAIN SHOCK struck. Fire swept through the remaining buildings, and AFTERSHOCKS toppled remaining walls and buildings on escaping survivors. In all, a shocking 250,000 people were said to have perished. The survivors fled from the city and were accosted and even killed by people in the surrounding countryside. Sensing an opportunity to advance their standing, bands of these country people came back into Antioch to loot the city. They were reported to have stripped rotting corpses of jewelry and other valuables. Rescue and relief efforts appeared to have been slow and only undertaken by a small group of survivors. Fortunately, Emperor Justin I (A.D. 518–527) made a strong commitment to rebuilding the city, which was carried on by his successor, Emperor Justinian I (A.D. 527–565). The city never regained its former splendor or prominence.

Aoba *volcano, Vanuatu* A SHIELD VOLCANO, Aoba is BASALTIC and is situated along a FISSURE system that has given the island an elongated shape. Although the volcano appears to have grown through outpourings of fluid LAVA, there is evidence of explosive activity in the island's history as well, signified by PYROCLASTIC materials. Two nested CALDERAS occupy the summit of the island, and a line of SPATTER CONES accompanies the fissure system along its trace from southwest to northeast. Several CRATERS less than a mile in diameter are located at the extreme northeast and southwest ends of the island. Dates of recent eruptions are inexact, but effusive and explosive eruptions are thought to have occurred several hundred years ago, possibly involving the inner caldera's collapse. An explosive eruption approximately a century ago cast out large quantities of ash, and LAHARS reportedly wiped out villages on the southeastern side of the island. Emissions of steam from the SUMMIT caldera increased and then subsided in 1971, and FUMAROLE activity may have been responsible for discoloring a lake on the summit in 1971. There was a large steam explosion and increased and unusual earthquake activity in 1995. Aoba was classified as the potentially most dangerous volcano in Vanuatu as a result.

Apache tears OBSIDIAN is volcanic glass and commonly black or brown. Obsidian nodules (LAPILLI) from the southwest UNITED STATES that are teardrop-shaped as the result of being shot out of a volcano are termed apache tears.

aphanitic The fine-grain size of volcanic rock. When LAVA comes out of a volcano, it goes from a hot area underground to cool air or water at surface conditions. When lava or MAGMA cools quickly, the minerals do not have time to grow large and are very fine-grained.

Apoyo *caldera, Nicaragua* The Apoyo CALDERA is located in the Nicaraguan Depression near the town of Granada. Lake Apoyo occupies much of the caldera. The collapse of the volcano, forming the caldera, is thought to have occurred following great eruptions that expelled perhaps a third of a cubic mile (0.8 km³) of MAGMA. Apoyo caldera is noted for a long history of earthquake activity that appears to have started in the 16th century, although some of this earthquake activity may have been TECTONIC rather than volcanic in origin and involved the whole region, not merely this caldera. Although no actual eruptions have been observed at Apoyo, EARTHQUAKE SWARMS, along with changes in the temperature of Lake Apoyo and in its sulfate content, indicate that some of the disturbances at Apoyo are due to volcanic processes. Several domes (El Cerrito, Lomo Poisentepe, and Apoyoito) are located near the caldera, as is a line of cinder cones along a FAULT roughly along a north-south line immediately to the east of the caldera between Apoyo and the shore of nearby Lake Nicaragua.

Apoyoito *See* APOYO.

applied seismology The use of seismic waves for exploration purposes. When an earthquake occurs, it sends out seismic waves in all directions. The speed and path they travel depends on the rocks and SOIL they pass through. In applied seismology, seismic waves are produced synthetically using an explosive or falling weight (hammer). Using seismographs, these waves are recorded and a kind of sonogram (like an X-ray) of the underground rock and soil layers is produced. Geologists can then tell where oil, gas, precious metals, or buried environmental hazards are.

Arabian crustal plate A plate of the CRUST adjacent to the African plate, the Arabian plate is separated from AFRICA by the Red Sea and also borders on the Asian plate to the north. The Arabian plate used to be part of the African plate. However, about 25 million years ago, a TRIPLE JUNCTION formed at the intersection of the Gulf of Aden, Red Sea, and East African rift at the northern end of Ethiopia and Eritrea. The Red Sea and Gulf of Aden formed an active DIVERGENT BOUNDARY in which Arabia pulled away from Africa northward into Asia. It quickly closed up the remains of a once great sea called Tethys and collided with Asia forming the Zagros Mountains. That collision continues today. The Zagros Mountains of Iraq, IRAN, and TURKEY grow ever taller and wider. They are expanding southward and will eventually force the Persian Gulf to dry out and become land as it is uplifted. The collision is also responsible for forcing Turkey westward and into the MEDITERRANEAN SEA. Turkey is being squirted out like a watermelon seed being squeezed between thumb and forefinger. This movement causes the devastating earthquakes that plague Turkey.

See also PLATE TECTONICS.

Ardabil *earthquake, Iran-Armenia* In the year A.D. 893, a massive earthquake struck an area from IRAN to Armenia, causing great destruction and loss of life. The U.S. Geological Survey lists the date as March 23, but several other sources list the date as between December 14 and January 11, A.D. 894, ostensibly as recording a MAIN SHOCK and a group of strong AFTERSHOCKS. Either the main shock or a very strong aftershock was reported to have struck the city of Tovin, Armenia, but also Ardabil. Both areas were demolished, and reports of drifting gray and black ash indicated that the earthquake may have been accompanied by volcanic activity. The DEATH TOLL from this event or sequence of events is debatable, with reports of 82,000 for Tovin alone but 100,000 to 180,000 for the entire event. The U.S. Geological Survey lists the death toll at 150,000, placing it in the top 10 for causing the greatest loss of life for an earthquake. Instability apparently continued into early A.D. 894 as another earthquake in the same area (town of Davin) was said to have killed another 20,000 people. It is strange that this earthquake was approximately the same time as the DEBAL, PAKISTAN, earthquake, which was also reported to have killed 150,000 people.

On February 28, 1997, an earthquake of magnitude 6.0 occurred. At least 965 people were killed and 2,600 were injured. More than 36,000 people were left homeless and 12,000 homes were destroyed. Some 160,000 livestock were also killed. Transportation, communication, and other services were severely disrupted in the area.

Arenal *volcano, Costa Rica* Arenal was a dormant STRATOVOLCANO that had not erupted in historic times (since A.D. 1500). In July 1968, however, the giant awoke. It produced glowing AVALANCHES and PYROCLASTICS that destroyed the west flank of the volcano. It destroyed the town of Pueblo

Nuevo and killed 78 people. It has produced loud explosions, gas clouds, ASHFALLS, and LAVA FLOWS discontinuously since then. It had major eruptions in 1993 and again in May 1998. In the latest eruption, lava flows and flying EJECTA at reported speeds of 120 miles (200 km) per hour forced the evacuation of 450 people. The volcano is relatively quiet once again and serves as a tourist attraction.

Arica *earthquake and tsunamis, Chile* Sometime between August 13 and 15, 1868, an earthquake and subsequent TSUNAMIS struck the west coast of SOUTH AMERICA from COLOMBIA to CHILE. The EPICENTER of the earthquake was beneath the city of Arica, Chile, near the border with Peru. The earthquake was generated by movement on the Andean SUBDUCTION ZONE at great depth. The earthquake destroyed Arica in seconds and enveloped it in a cloud of dust. Many of the survivors rushed to the shoreline, where the USS *Wateree*, which was moored in the harbor, launched a rescue boat to save them. Just as the boat reached the survivors, the first tsunami came barreling into the harbor and drowned the survivors along with 13 crewmen. Several minutes later, the water receded out of the harbor, leaving the *Wateree* stuck on wet sand of the seafloor. The second tsunami was much larger than the first and rolled over the city. It unearthed hundreds of tombs on a mountain on the outskirts of the city. The dead had been buried upright and wound up standing in ranks before being dragged out to sea. A third tsunami struck the city at nightfall, sealing the fate of the devastated town and carrying the battered *Wateree* over one mile inland. In all, more than 25,000 people were reported killed, but the DEATH TOLL was an approximate number.

Arizona *United States* Located in a region of moderate seismic risk, the state of Arizona experiences earthquakes that originate within its own territory as well as vibrations from earthquakes centered in neighboring states, notably CALIFORNIA. A very powerful earthquake, estimated at MERCALLI intensity VIII–IX, occurred near Fort Yuma on November 9, 1852; FISSURES opened in the desert along the Colorado River, and the earthquake knocked down parts of Chimney Peak. Shocks were reported on almost a daily basis for months.

There is abundant evidence of volcanic activity in Arizona. One interesting example is Vulcan's Throne, a cinder cone on the northern rim of the Grand Canyon. Output from its eruptions is thought to have blocked the canyon time and again, but on each occasion, the river formed a new channel. The Kitt Peak Observatory near Tucson is built atop a mountain that formed as intrusive igneous rock, and hydrothermal activity in Arizona deposited ores that made the state a major source of copper. The spectacular San Francisco Peaks near Flagstaff are also volcanic in origin, as is nearby Sunset Crater, which is thought to have erupted in the 11th century. Arizona also has one of the most famous IMPACT STRUCTURES on Earth, namely Meteor Crater, an impact crater near Flagstaff.

Arkansas *United States* Although the state of Arkansas has seldom experienced powerful earthquakes, one series of such earthquakes was the strongest in UNITED STATES history: the NEW MADRID, Missouri earthquakes of 1811–12, which

altered the topography of northeastern Arkansas considerably. A less destructive, but nonetheless powerful, earthquake occurred in Arkansas on October 22, 1882; this earthquake was estimated at MERCALLI magnitude VI–VII and affected an area of some 135,000 square miles (350,000 km²), although the EPICENTER was difficult to ascertain because reports from the affected area were so few. The October 28, 1923, earthquake at Marked Tree was remarkably strong (Mercalli INTENSITY VII) and affected some 40,000 square miles (104,000 km²); the earthquake was felt in Arkansas and in nearby states, caused considerable damage to buildings, and disturbed the surface of the St. Francis River. On November 16, 1970, an earthquake of Mercalli intensity VI and Richter magnitude 3.6 in northeastern Arkansas was felt over some 30,000 square miles (78,000 km²) and resulted in minor damage.

Armenia (1) *earthquake, Colombia* On January 25, 1999, an earthquake of MAGNITUDE 6.2 occurred. Approximately 1,885 people were killed and more than 4,750 were injured. It left 250,000 people homeless. Severe LANDSLIDES resulted from the earthquake and blocked many of the major roads.

Armenia (2) *earthquake, formerly a republic of the Soviet Union* On December 7, 1988, a devastating earthquake of MAGNITUDE 6.9 struck northwestern Armenia. The MAIN SHOCK was followed by an AFTERSHOCK of magnitude 5.8 just four minutes later. More than 25,000 people were killed, and some 15,000 were injured. Economic losses were estimated at $14.2 billion. The epicentral area encompassed the towns of Leninakan, Spitak, Stepanovan, and Kirovakan. The main cause of death was collapsing buildings. The streets were impassible as the result of all of the rubble. It took years to clean out these towns. This earthquake provides a sharp contrast to the LOMA PRIETA earthquake of CALIFORNIA of 10 months later that was even more powerful. Yet the Loma Prieta earthquake only caused 67 fatalities, most of which came from the collapse of a single freeway. The reason for the sharp contrast in the DEATH TOLL is that California has building codes to minimize damage from earthquakes, whereas Armenia does not. This contrast is a tribute to advances in earthquake engineering.

Arunachal Pradesh *earthquake, India* A huge earthquake struck Arunachal Pradesh in far northern INDIA at 7:40 P.M. on August 15, 1950. It registered 8.6 on the RICHTER scale and was reportedly felt over an area of 4.5 million square miles (12.5 million km²) and the rumble heard over 750 miles (1,200 km) away. Incredibly, the shock lasted over four minutes. AFTERSHOCKS with magnitudes up to 6 continued for years after the MAIN SHOCK. Luckily, the area around the EPICENTER was sparsely populated and only 1,526 people died. Most of these deaths resulted from aftereffects of the earthquake rather than the earthquake itself. LANDSLIDES shaken loose from the mountains resulted in 156 deaths and the destruction of 70 villages. Some landslides dammed the tributaries of the Bramaputra River, which flooded, drowning many victims. A dike along the river at Subansiri broke eight days after the earthquake, and a 23-foot (7-m)-high wave swept over many villages, killing 532 people. Heavy LIQUEFACTION, SLUMPING, and FISSURES were also rampant in the area. Combined with the direct effects of shaking, over 2,000 buildings were destroyed.

Asamayama *volcano, Japan* The volcano Asamayama on Honshu, the central and largest island of JAPAN. This extremely active STRATOVOLCANO has had at least 121 eruptions in historic times. The last eruption was in 1990. Most of these eruptions are of the vulcanian type. Asamayama is composed of a young stratovolcano with two CRATERS all on a SHIELD VOLCANO and in turn on an older stratovolcano. The older stratovolcano is called Kurohuyama and the two craters and Maekakeyama and Kamayama. The shield volcano is called Hotoke-iwa.

Asamayama underwent its most famous eruption in 1783, when the volcano cast out large numbers of hot rocks that landed on nearby communities. The eruption is said to have killed about 5,000 people. One rock expelled in this eruption reportedly measured 120 feet (37 m) by more than 260 feet (79 m) and formed an island where it landed. Together with the ICELANDIC volcano SKAPTARJÖKULL, Asamayama was implicated by Benjamin Franklin in the unusually low temperatures that affected the Northern Hemisphere that year and in a curious "dry fog" seen hanging over the land. ASH cast out by the two volcanoes may have been responsible for the apparent fog and the drop in temperature.

Asawa *volcanic complex, Ethiopia* The Asawa volcanic complex is located in the central part of the main Ethiopian rift valley near Lakes Abaya and Shalla. Also known as Asawa/Corbetti, the Asawa complex includes the CALDERAS Aluto, Awasa, Corbetti, Duguna, Gadamsa, Gademota, Hobicha, Shalla, and Wonchi and the Wagebeta caldera complex. The area is characterized by flows of BASALT, OBSIDIAN, and SCORIA, as well as by layered PUMICE and lava domes. The Asawa caldera adjoins the Corbetti caldera, which is thought to have formed through the eruption of large amounts of material from fissure VENTS. The historical record of activity at Asawa/Corbetti is brief. There are reports of eruptions dating back to the early 20th century, but the accuracy of these reports is questioned. Most of the volcanoes are flooded calderas that have been inactive for thousands of years. In recent years, FUMAROLES have been active here. In 1984, earthquakes damaged buildings and caused the evacuation of a school.

Ascension Island Ascension Island is the summit of a volcanic mountain along the Mid-Atlantic Ridge approximately midway between AFRICA and SOUTH AMERICA.

aseismic An area in which there is no earthquake activity. Used as an adjective, it might also be a structural or topographic feature that might be expected to produce earthquakes but which does not. For example, an aseismic ridge is an oceanic ridge that does not produce earthquakes, whereas most are seismically active.

ash Ash is fine solid material ejected from a volcanic eruption. Volcanic ash differs in color and composition but is

This regular snowplow is plowing the road on May 18, 1980, in Washington. However, it is not plowing snow, it is plowing volcanic ash from the eruption from Mount Saint Helens. Ash blanketed the area around the volcano so deeply that it looked like there had been a gray snowstorm. The problem is that ash is much heavier than snow and thus caused many roofs to cave. *(Courtesy of the USGS)*

usually gray. Four eruption processes give rise to volcanic ash. One is magmatic. In this process, gas bubbles form in MAGMA as pressure on the molten rock diminishes on its way to the surface. The bubble-filled magma then fragments in the VENT of the volcano and is expelled as finely divided solid material. In the second process, the hydrovolcanic process, magma mixes with groundwater or surface water in an explosive manner. The PHREATIC process involves fast expansion of steam and/or hot water and fragments of COUNTRY ROCK. The fourth process is abrasion, which occurs when grains of ash collide with each other. The shape of ash particles depends on the conditions in which they were formed. Where large bubbles form in the vent of the volcano during decompression of magma, for example, resulting bits of ash may occur as thin sheets of glass formed when the bubbles solidified and broke apart. Hollow "needles" may be found in ash where the flow of magma within the vent elongated gas bubbles in the molten rock.

Many ash particles fall out of the air soon after being ejected from the volcano that gave rise to them, but extremely fine ash may rise into the upper atmosphere and block incoming sunlight, causing a drop in surface temperatures. Another observed effect of volcanic ash in the upper atmosphere

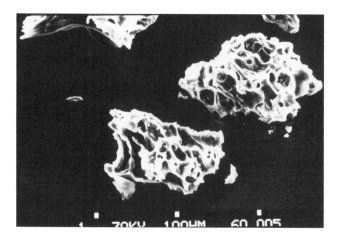

Scanning electron microscope image of minute ash fragments showing that they are composed of pumice *(Courtesy of the USGS)*

is extremely colorful sunsets, caused by the high-level ash clouds' tendency to block short wavelengths of solar radiation and let through only the longer wavelengths, notably orange and red. Such vivid sunsets followed the 1883 eruption of the volcano KRAKATOA, for example.

Ash in the upper atmosphere may remain there for years. In the lower atmosphere, ash from eruptions may fall and cover the land or sea in layers many feet thick. One of the most famous ashfalls, from VESUVIUS in A.D. 79, preserved the entire cities of POMPEII AND HERCULANEUM and hid them from discovery for some 18 centuries. Archaeologists exploring the buried cities found curious cavities, or lacunae, in the ash. These lacunae turned out to be the preserved outlines of persons killed by the eruption. The ash buried them where they fell, and the lacunae remained after their bodies decayed. Plaster casts have been made of some of these lacunae and offer a striking glimpse of the human aspect of the two cities destruction. The 1991 eruption of Mount PINATUBO in the PHILIPPINE ISLANDS deposited so much ash on nearby Clark Air Base that buildings collapsed. During an eruption, ash clouds may interfere with the navigation of aircraft. Pilots flying near Mount SAINT HELENS in WASHINGTON State during its 1980 eruptions reported that very fine ash from the volcano made its way into their aircraft. Airborne ash also caused severe damage to an airliner flying through the cloud from an eruption of ALASKA's REDOUBT volcano in 1989.

See also AVIATION AND VOLCANOES; ISOLATION; TEPHRA; "YEAR WITHOUT A SUMMER."

ashfall ASH that rains down after a volcanic eruption. In a summit volcanic eruption, huge clouds of ash are shot miles into the atmosphere. This meteoric ash then rains back down to Earth like snow, blanketing houses, lawns, forests, and roads. The thickest deposits of ash are closest to the volcano, but ash can stay suspended in the atmosphere and travel all around the world.

ashfall tuff An ASHFALL deposit. As layers of ASH from an ashfall are buried and compressed, they are lithified into a rock. This burial might be from LAVA FLOWS, other ashfalls, ash flows, or even sedimentary deposits. Heat may therefore be involved in the lithification process (WELDED TUFF). The rock is fine grained, banded, and porous. It is typically light gray to tan, but weathering and interaction with groundwater may make it red to orange.

ash-flow eruption In an ash-flow eruption, a *NUÉE ARDENTE* or similar phenomenon lays down a deposit of very hot ASH, which may range in thickness from only several feet to perhaps a thousand feet locally. In portions of the western UNITED STATES, particular ash-flow deposits may extend for 100 miles (161 km). Material from these eruptions covers large areas of the western United States, MEXICO, NEW ZEALAND, and other parts of the world. Deeper levels of such a deposit may become denser and resemble lavalike rock, which may include pieces of OBSIDIAN. This increase in density is thought to result from intense heat acting on the deeply buried layers of material. The heat fuses the ash particles together and in some places turns them into WELDED TUFF.

Ash-flow eruptions commonly lay down deposits very quickly, even over wide areas. The rapidity of this process is reflected in the fusing of ash particles, which were deposited so quickly that the initial heat had little or no chance to dissipate. In ALASKA, an eruption of Mount KATMAI produced in less than 24 hours the VALLEY OF TEN THOUSAND SMOKES, a famous ash-flow deposit that endured for decades as a plain of FUMAROLE. A similar phenomenon occurred during the violent eruption of BEZYMIANNY volcano in Russia. That eruption produced an ash-flow deposit later named Valley of Ten Thousand Smokes of Kamchatka.

ash-flow tuff A lithified ash-flow deposit. A PYROCLASTIC FLOW leaves a deposit of EJECTA with mixed grain sizes (ASH, LAPILLI, etc.). The sizes of grains are all jumbled within the layer. The deposit is lithified into a porous rock that is tan to gray.

Ashgabat *earthquake, Turkmenistan (USSR)* On October 6, 1948, at 3:50 A.M., a deadly earthquake rocked Asgabat, the capital city of Turkmenistan in the then USSR. The powerful quake had a RICHTER magnitude of 7.3, and the INTENSITY achieved a rating of X on the modified MERCALLI scale at Ashgabat. It had a FOCUS at 11 miles (18 km) depth, and the duration was reported as two minutes. EARTHQUAKE LIGHT was reported to have occurred during the event. Troops were deployed to keep out plunderers, and order was restored quickly. The DEATH TOLL for this event was a subject of wild debate. The Stalin regime downplayed the disaster, claiming 25,000 to 30,000 casualties and finally releasing an official tally of 19,800 dead. The U.S. Geological Survey lists the death toll at 110,000 and claims it was one of the most destructive earthquakes ever. Local Turkmenistan authorities still claim that 174,000 people perished in the event. Recent compilations by external, reportedly unbiased, accounts placed the death toll between 60,000 and 70,000 people, still a significant number. All accounts agreed that shoddy building and house construction was the reason for the magnitude of the disaster.

Askja *caldera, Iceland*   Prehistoric in origin and located in the rift zone through the middle of ICELAND, the Askja CALDERA was the site of an eruption in 1875 that created the Öskjuvatn caldera. The Askja caldera itself appears to have formed when MAGMA moved underground into a nearby FISSURE swarm. A comparable set of circumstances is thought to have produced the Öskjuvatn caldera, which is notable for having emerged during an episode of RIFTING. A major volcanic eruption took place in 1874 and 1875 along a FRACTURE ZONE about 60 miles (97 km) north-northwest of the Askja caldera. These events were preceded and accompanied by earthquakes, including one episode that shook all of northern Iceland continually for the last week of December 1874. Eruptions (both effusive and explosive) began at Askja on January 1, 1875. A violent eruption took place at Askja on March 28 and 29. Eruptions continued through April at Askja and through October at the fissure swarm to the north. Magma moved underground from the vicinity of Askja toward the fissure swarm, and the collapse of Öskjuvatn caldera started in February 1875, before the great eruption of March 28–29. The collapse was largely completed by July.

There may have been a minor eruption in 1919, but the only sign of it is a layer of TEPHRA in an ice core. LAVA flowed from the rim of Öskjuvatn between 1921 and 1926, and around 1929 (the exact date has not been determined), a FISSURE ERUPTION took place to the south of Askja. An eruption in 1961 was heralded by earthquakes and began with emergence of GEYSERS and SOLFATARAS at Askja in early October, along with numerous fissures on the floor of the caldera. Hot water flowed from the main fissure. Very energetic geyser activity occurred for several days in mid-October. In an eruption beginning on October 26, BASALTIC lava emanated from a fissure lying east to west across the floor of Askja caldera. The water level in a caldera lake dropped several feet, and fissures formed in the walls of the caldera.

Asnam, El *earthquake, Algeria*   A large earthquake struck the northern Algerian city of El Asnam at 12:25 A.M. on October 10, 1980. The shock registered a 7.3 on the RICHTER scale and was followed closely by another that registered 6.5. The modified MERCALLI scale damage was X at the EPICENTER, and the FOCUS was six miles (10 km) in depth. FAULTing produced a steep SCARP with up to 10 feet (3 m) vertical offset that crossed the landscape for several kilometers.

Damage was extensive in El Asnam, with upward of 80% of the city in ruins. The estimates of economic loss ranged from $3 billion to $5.2 billion. The DEATH TOLL from this event is debatable but ranged from 5,000 to 11,000 people. (It is probably close to the lower number.) The damage displaced some 250,000 people.

Building destroyed by surface waves from the 1980 El Asnam earthquake *(Courtesy of the USGS)*

Aso *caldera, Japan* The Aso CALDERA is located on the southern Japanese island of Kyūshū and is situated on the Oita-Kumamoto Fault Zone, which extends northeast to southwest through the center of Kyūshū. Several great eruptions from Aso are thought to have occurred in prehistoric times, and one of these eruptions appears to have left a layer of ASH all over JAPAN. It has been labeled as having produced more explosive eruptions than any volcano in the world. The Nakadake volcano within the caldera has erupted 167 times since its first recorded eruption in A.D. 553. Activity at Nakadake has not been confined to one area of the CRATER but rather has migrated during the last few decades. LAVA emanating from Nakadake is basaltic ANDESITE. Several vents occupy the caldera. Hot springs also occur there.

Aso caldera is noted for the earthquake activity associated with eruptions there. There is evidence from studies of s-WAVES and P-WAVES that MAGMA is present in large amounts under the eastern and central portions of the caldera, several miles down.

The Aso caldera has exhibited frequent seismic activity in the latter half of the 20th century. Several types of earthquakes have been observed and associated with eruptive activity at the caldera. One type is characterized by SURFACE WAVES with periods of about one second and is thought to be produced by internal volcanic gas explosions. A second type of earthquake is characterized by surface waves lasting for slightly longer periods (up to eight seconds), resulting from vibration by the MAGMA CHAMBER and possibly also from explosions. The third type consists of BODY WAVES averaging about a half-second long and produced by explosions in the magma during eruptions. The fourth type, associated with periods of eruptive activity, is characterized by body waves with a period of approximately one-fifth of a second. (During one eruption of Nakadake in 1958, vibrations lasting a much longer period, 40 seconds or longer, were noted for two days before the eruption but ceased about an hour before the eruption began.)

Tilt or inclination measurements of the ground conducted since 1931 at Aso caldera have provided information that has helped scientists understand the relationship between tilt and eruptive activity. Measurements made in the 1950s revealed inflation for some months before major eruptions, and deflation following eruptions. Tilt measurements made in the middle 1960s also showed inflation occurring during a period of eruptive activity. Temperature changes and gas emissions have also been studied as possible precursors of eruptions in the Aso caldera. Temperatures in the crater of Nakadake, monitored in the late 1950s, went up sharply in the days just prior to an eruption, and the level of dissolved carbon dioxide in hot springs in the vicinity rose for several months before an eruption in 1979. Temperature measurements and fluctuations in water level in a pond in Nakadake's crater were used to anticipate explosive activity and emissions of ash that began in late 1984 and continued through early 1985. Its last eruption was in 1993. There also appears to be a relationship between emissions of SULFUR DIOXIDE and eruptive activity at Aso caldera: Sulfur dioxide output rises before and during periods of explosive activity but then diminishes as explosions subside.

See also SEISMOLOGY.

asperity An asperity is a sticking point in a FAULT plane that is usually the result of an irregularity on the fault surface. The irregularity is commonly a protrusion that sticks across from one side of the fault into the other. It grinds across the fault and can lock it up for an extended period of time, creating a stress concentration. When the STRESS finally overcomes this hindrance, the resulting earthquake may be much more powerful than usual for the particular fault zone.

Assam *earthquake, India* A huge earthquake struck the Assam area of INDIA at 8:46 A.M. Indian Standard Time on June 12, 1897. The EPICENTER of the earthquake was located near Sangsik, eastern India, and the Richter magnitude was 8.0. Ground ACCELERATION was recorded as high as 1 g. The phenomenal aspect of this earthquake was the amount of movement. Movement occurred on the east-west-oriented Oldham Fault and was in a STRIKE-SLIP sense; offset reached up to 53 feet (16 m). This is among the greatest for any known earthquake. It has now been determined that the earthquake was on a BLIND FAULT, with movement from 5.5 to 21 miles (9 to 35 km) depth. The surface rupturing that accompanied this event was from associated movement on subsidiary faults. Telegraph poles were offset 10–12 feet (3–3.5 m) laterally, and one railroad segment shifted seven feet (2 m). Other surface effects were equally impressive. Fountains of water from LIQUEFACTION were reported up to four feet (1.1 m) high. Ten-foot (3-m) waves were reported on the Bramaputra River, which rose 25 feet (7.6 m) in one area and reversed its flow. The Chedrang River developed sag ponds and waterfalls as it crossed the fault SCARPS. Frozen earthwaves were described in a rice field, where the crest to trough height was 6.5–10 feet (2–3 m). Stones on one road were said to have "vibrated like peas on a drum" during the earthquake.

This massive earthquake killed 1,542 people and injured thousands. LANDSLIDES destroyed many of the hill towns in the area. FISSURES opened across northeastern India and PAKISTAN. AFTERSHOCKS continued to rattle the area for years. Particularly strong events occurred on June 13 at 1:30 A.M., 1:00 P.M. and 10:40 P.M.; on June 14 at 12:47 A.M.; on June 22 at 7:24 A.M.; on June 29 at 10:19 A.M.; and October 2 at 8:58 P.M. The last reported aftershock of any size was on October 9, 1897 at 1:40 A.M.

The following were observations by R. D. Oldham, a British scientist, at Shillong:

> I was out for a walk at the time. At 5:15 a deep rumbling sound, like near thunder commenced . . . followed immediately by the shock . . . The ground began to rock violently, and in a few seconds it was impossible to stand upright, and I had to sit down suddenly on the road. . . . The feeling was as if the ground was being violently jerked backwards and forwards very rapidly, every third or fourth jerk being of greater scope than the intermediate ones.
>
> The surface of the ground vibrated visibly in every direction, as if it was made of soft jelly; and long cracks appeared at once along the road. The sloping earth-bank around the water tank, which

was some ten feet high, began to shake down, and at one point cracked and opened bodily. The road is bounded here and there by low banks of earth, about two feet high, and these were all shaken down quite. The school building, which was in sight, began to shake at the first shock, and large slabs of plaster fell from the walls at once. A few moments afterwards the whole building was lying bent and broken on the ground. A pink cloud of plaster and dust was seen hanging over every house in Shillong at the end of the shock.

My impression at the end of the shock was that its duration was certainly under one minute. . . . Subsequent tremors lasted some time. . . . The whole of the damage done was completed in the first ten or fifteen seconds. . . .

assimilated   Existing rock (COUNTRY ROCK) can be broken off and pulled into MAGMA or LAVA as it passes through. The pulled-in pieces of rock can be melted and mixed with the molten rock. The process of assimilation changes the composition of the molten rock. The pieces of rock that are melted are said to be assimilated.

asthenosphere   The layer of Earth's MANTLE, soft and gumlike in contrast to the typically rigid rock that comprises the rest of the mantle. The rocks in the asthenosphere have the same composition as the rest of the mantle. The pressure and temperature conditions, however, dictate that the rock behaves in a DUCTILE manner in the asthenosphere. Therefore, the layer is really defined by the behavior of the material. It is more like the contrast between a candle left out in the hot sun versus one in the freezer. The warm wax will bend easily whereas the cold wax will only break. The plates, which are made of a layer of CRUST adhered to a layer of rigid mantle underneath, float on the soft asthenosphere. The movement of the asthenosphere is what drives PLATE TECTONICS.

See also EARTH, INTERNAL STRUCTURE OF.

Ata *caldera, Japan*   The Ata CALDERA is located in southern JAPAN near the end of the Satsuma Peninsula and under the waters of Kagoshima Bay. The caldera extends from the tip of the Satsuma Peninsula to the tip of the Osumi Peninsula several miles across the water. This has been an area of intense volcanic activity in recent centuries, notably at SAKURAJIMA volcano in the AIRA caldera and at Kaimondake in the nearby Ibusuki volcanic field. The small Ikeda and Yamakawa calderas are thought to be nested inside Ata. Strong eruptions occurred in the first and ninth centuries. Pronounced earthquake activity occurred in the vicinity of Ata caldera in the late 1960s and in 1970.

Atitlán *caldera, Guatemala*   The Atitlán CALDERA is believed to have formed during an eruption of several cubic miles of LAVA about 84,000 years ago. Lake Atitlán occupies a portion of the caldera. Three STRATOVOLCANOES—Atitlán, San Pedro, and Toliman—have formed in the caldera since

its origin, along the caldera's southern edge. Atitlán volcano reportedly erupted in 1469 and then again in 1717 and 1721. Eruptions continued at intervals of several years through the first half of the 19th century. Of these eruptions, only one was strong, in May 1853. Activity at Atitlán volcano subsided following an eruption in 1856.

Atlantic Ocean   Although earthquake and volcanic activity is less pronounced around the margins of the Atlantic Ocean basin than in and around the PACIFIC OCEAN, the Atlantic contains numerous features with earthquakes and volcanism. Most prominent on a physiographic map of the Atlantic basin is the MID-OCEAN RIDGE, which lies along the middle of the Atlantic Ocean, roughly equidistant from the Americas to Europe and AFRICA. ICELAND is an active volcanic island on the Mid-Atlantic Ridge. SURTSEY is a famous volcano of Iceland. Exploration of the Mid-Atlantic Ridge in the late 1970s revealed the existence of hydrothermal vents and large colonies of animals living in the vicinity of the vents. Among the animals found there were mussels with shells approximately one foot long and polychaete worms up to six feet (2 m) long. The AZORES, a group of islands off the western coast of Africa, consist of several mountains along the flank of this ridge, which was discovered during World War II by military ships making depth measurements. The Azores are produced by a HOT SPOT. There are two ISLAND ARCS in the Atlantic Ocean basin. The CARIBBEAN Islands form an active island arc with several dangerous volcanoes. Mount PELÉE killed some 29,000 people in 1902 with a large *NUÉE ARDENTE*. SOUFRIÈRE HILLS is a volcano that closed the island resort on Monserrat in 1998. The South Sandwich Islands form an island arc in the South Atlantic. Numerous SEAMOUNTS are found in the Atlantic Ocean. Bermuda occupies the summit of one seamount. There are several major fracture zones in the Atlantic Ocean, namely the Romanche and the Charlie Gibbs. These are large TRANSFORM FAULTS that are seismically active. There are hundreds of other smaller transform faults that are also seismically active.

Atlantis   An enormous volcanic eruption in the MEDITERRANEAN may have contributed to the origin of the Atlantis legend. The story of Atlantis, as related by Plato in writings dated around 400 B.C., concerns a CONTINENT that disappeared beneath the sea within 24 hours, taking with it an advanced civilization. Plato attributes the story to Critas, an Athenian politician. Critas in turn heard the story from his father, who was a friend of Solon, who lived in the sixth and seventh centuries B.C. and is considered the founder of democracy in Athens. In a period of exile, when his political career was at a low, Solon visited EGYPT and heard there the story of a gigantic island, located somewhere west of the Straits of Gibraltar. The island was called Atlantis, and (according to legend) it sank beneath the sea in a single day and night after an alliance led by the Athenians had defeated the Atlanteans in battle some 9,000 years ago. The legend of Atlantis has exerted a powerful fascination on the Western imagination, and there have been efforts to locate an actual land or natural catastrophe that might account for the Atlantis story. Eventually, speculation focused on the island of SANTORINI (THIRA)

This reef-rimmed tropical volcanic island is an atoll from the Pacific Ocean. *(Courtesy of NOAA)*

in the Aegean Sea. Santorini once had been the hub of the advanced Minoan culture, which vanished suddenly and mysteriously from the Mediterranean in approximately 1400 B.C. It is now widely presumed that the Minoan culture perished in an eruption that resulted in the collapse of the volcano and the formation of a CALDERA, accompanied by a 200 foot high (61 m) TSUNAMI that emanated from the island and spread destruction through nearby portions of the Mediterranean basin. The destruction of Thira and the Minoan civilization appears to have been recorded, in an exaggerated form, as the legend of Atlantis.

atoll By definition, an atoll is a circular-to-elliptical ring-shaped island in the ocean that is composed of coral reef but with no land in the middle. Atolls range in diameter from less than one mile (1 km) to 100 miles (130 km). Although they are common in the tropics of the western PACIFIC OCEAN, the name derives from the Maldive Islands, so they also exist elsewhere. They are most commonly formed as a fringe coral reef around an extinct ocean island volcano. The volcanic rocks are quickly weathered in the tropical climate, and the volcano disappears beneath the waves, leaving only the reef. The Bikini Atoll is famous as the location for the early nuclear bomb tests.

attenuation Reduction in seismic wave energy as they travel away from the earthquake source. Waves from an earthquake travel away like ripples on a pond when a rock is thrown in. Because the circles grow larger as they move away, the energy of the ripple is spread out farther and consequently the ripple height becomes smaller. The same is true of seismic waves. In addition, certain rock and SOIL transmits seismic waves well and others do not. Instead, they absorb the energy and the waves are reduced. Rocks can also attenuate seismic energy.

Augustine *volcano, Alaska, United States* An island STRATOVOLCANO in the Cook Inlet area of south ALASKA, Augustine has undergone highly explosive eruptions in 1812, 1883, 1935, 1963–64, 1976, and 1986. DEBRIS FLOW avalanches are common in this volcano. During the 1883 eruption, a DEBRIS AVALANCHE produced a 30-feet (9-m)-high TSUNAMI. Eruptions commonly end with the formation of a LAVA DOME.

Australia Unlike its neighbor NEW ZEALAND, Australia is not especially noted for earthquake activity in modern times, although several powerful earthquakes have occurred there in the past century, including quakes in Adelaide in 1897 and 1954, southeast Queensland in 1918, New South Wales in 1961, and Victoria in 1966. On its north coast, Australia is colliding with Indonesia. It forms a SUBDUCTION ZONE beneath INDONESIA producing the earthquakes and volcanoes there.

Although there are no active volcanoes in Australia, it was highly volcanically active in the relatively recent geologic past. This activity is interpreted to result from movement of Australia over a HOT SPOT. Unlike HAWAII, where there is a single spot of concentrated igneous activity, volcanism in Australia was diffuse. It resulted in a whole group of mostly volcanic chains around the east coast of Australia. These chains are progressively younger toward the south as the result of the northward movement of Australia. Each volcanic field and center began with BASALT and progressed to SILICA-rich TRACHYTE and RHYOLITE in most areas. However, in the central area, there was a chain of volcanoes that compositionally moved toward very uncommon rocks that are unusually silica-poor. The volcanic rocks are mainly subdivided into provinces. The Monaro Volcanic Province in southeastern New South Wales is 57.5 to 34 million years old and covers 1,650 square miles (4,200 km²). The Glass House Mountains are rhyolitic and 27 to 25 million years old. The West Kimberly Province is 22 to 20 million years old and has more than 100 intrusive and volcanic centers including the Ellendale sequence. The Nulla Volcanic Province has 45 volcanic centers. The most famous part of the province is the 13,000-year-old Toomba lava flow that extends for 75 miles (121 km). It is the longest documented flow on Earth. The McBride Volcanic Province contains 160 volcanoes, the youngest of which in Mount Kinara at approximately 20,000 years. The younger volcanoes approach historical times. The Newer Volcanic Province has more than 400 vents and covers some 600 square miles (1,600 km²). Included in this field are the 7,290-year-old Mount Napier and 4,900-year-old Mount Gambier. A volcano not part of a province, Mount Schank, is also very young at 5,000 years old.

auxiliary fault plane   When a FAULT slips to create an earthquake, there is another plane at a high angle to it and with a different sense of movement that is also favored to move by the specific geometry of forces. For example, if a DEXTRAL strike-slip fault that lies in a northeast direction slips and produces an earthquake, the same stress field also could have produced a sinistral strike-slip fault that lies in a northwest direction. These two faults are termed *conjugate* and, commonly, both will move sooner or later. Auxiliary fault planes are particularly important in FAULT PLANE SOLUTIONS.

avalanche   A large mass of unconsolidated material, such as rock, soil, ice, or snow, falls at high speed under the influence of gravity. Earthquakes may set off avalanches, which are especially hazardous in seismically active areas where population centers are located close to mountains. Some of these avalanches can travel at speeds up to 270 miles (435 km) per hour. In NEVADOS HUASCARÁN, Peru, 1970, a DEBRIS FLOW avalanche killed some 18,000 people in such a fall.

Avezzano *earthquake, Italy*   A disastrous earthquake struck the central Italian town of Avezzano at 6:52 A.M. on the morning of January 13, 1915. The quake registered a 6.9 on the RICHTER scale, and local damage was estimated at XI on the modified MERCALLI scale. The EPICENTER of the quake was located between the small towns of Gioia dei Marsi and Ortucchio, and the FOCUS was at five miles (8 km) depth, with the causative rupture some 24 miles (40 km) in length. It was on the Avezzano-Celano Fault and was classified as an Appenine type event. Faulting produced surface ruptures, and a large SCARP was formed on the east side of the drained Fucino Lake basin that was three miles (5 km) long and up to 10–13 feet (3–4 m) high. There was also significant subsidence in the area. Because the city sits on soft lake sediments, SURFACE WAVES were strongly amplified during this event. Thus, damage to buildings was much more intense than if Avezzano had been situated on BEDROCK.

The DEATH TOLL from the earthquake was best estimated at 29,800 people, but up to 35,000 people, making it one of the worst in Italian history. It was especially impressive considering that there are only 40,000 people living in Avezzano today. The intense damage zone was about 500 square miles (1,300 km²). Damage to the entire area was estimated at US$600,000 (in 1915 dollars). This damage extended all the way to Rome, where 22 churches, 20 palazzos, and the ancient aqueduct were affected. Avezzano was rebuilt with large, straight roads and wide, green areas—only to be destroyed again during World War II.

aviation and volcanoes   Although volcanic eruptions are infrequent compared to other phenomena that pose dangers to aircraft, such as thunderstorms, clouds of ASH from volcanoes have the potential to do tremendous damage both to aircraft on the ground and to those in flight. This danger is especially great to aircraft in flight because clouds of volcanic ash do not show up on airplane radar—a result of the limitations on the sensitivity and power/aperture of these radars. The fine airborne ash can cause a wide variety of damage to aircraft that encounter it in flight. Abrasive action from the ash can damage engines, landing lights, control surfaces and windows and windshields. The windshield of a jet aircraft passing through an ash cloud may become opaque. Jet engines may cease operating, leaving an aircraft in a powerless descent. Damage to a single aircraft may amount to tens of millions of dollars and leave the airplane unusable without extensive repair.

SULFUR emissions present a secondary hazard that typically accompanies the ash. Sulfur oxides form aerosols that may actually contribute more to the stalling of engines than ash. The sulfur oxides are taken into the engines instead of the oxygen required for combustion and they stall. The hot sulfuric compounds form strong acid that can eat away at the engines as they pass through.

A well-documented case of aircraft damage from a volcanic ash-sulfur cloud occurred during the 1989 eruption of REDOUBT volcano in ALASKA. A Boeing 747–400 aircraft entered the cloud from Redoubt at approximately 26,000 feet (8,000 m) while descending for landing at Anchorage. The aircrew tried at once to gain altitude and escape the cloud,

A World Airways DC-10 jet at Cubi Point Naval Air Station is covered with volcanic ash after the eruption of Mount Pinatubo, Philippines, June 17, 1991. Ash is very abrasive and can cause engines to stall or fail if an airplane flies through an eruption cloud. The eruption of Mount Pinatubo caused 11 commercial aircraft emergencies. *(Courtesy of the USGS)*

but all four engines died after climbing only some 3,000 feet (900 m). Compressor erosion and other damage were considerable. Ash that melted and resolidified on the stage-one turbine nozzle guide vanes was found later to be the primary cause of loss of engine thrust. Eight minutes after the engines stalled, after the airplane had descended approximately 13,000 feet (4,000 m), the aircrew succeeded in restarting the engines. All engines, nose cowls and thrust reverses had to be replaced. The sandblasting effect of passing through the ash cloud caused heavy damage to the pilots' windshields, some cabin windows, and landing light covers and required their removal and replacement. Sandblasting also forced the removal and replacement of leading edges on the wing, stabilizer, and vertical fin. Ash contaminated the whole interior of the aircraft, so that all seats, side walls, and other interior furnishings had to be taken out and either cleaned or replaced. Contamination from ash extended to the electronic equipment. The fuel, oil, hydraulic, and drinking-water systems also were contaminated and required draining and cleaning.

Total damage to the aircraft exceeded $80 million. Before this incident, other aircraft had encountered clouds of volcanic ash over Alaska on several occasions and undergone various degrees of damage. The 1986 eruption of AUGUSTINE volcano had extensive effects on aircraft operation in the vicinity of Cook Inlet. Air carrier service at Anchorage all but ceased during the eruption, and the U.S. Air Force moved most of its aircraft out of Anchorage for several days.

No disabling incidents involving aircraft were reported during this eruption. During an earlier eruption of Augustine, in 1976, two F-4E Phantom jets passed through the volcano's ash plume and had their canopies scoured. Sandblasting removed some paint from the wings, and fine material penetrated many portions of the planes' interiors. A DC-8 passing through the ash cloud on the way to TOKYO had its center windshield scoured, and the windshield had to be replaced. Large amounts of ash stuck to the plane, and some abrasion was reported on landing gear and other external parts of the aircraft. Two other passenger aircraft had ash adhering to them, but damage was less extensive than to the DC-8. A brief eruption of Mount Spurr near Anchorage in 1953 resulted in sandblasting damage to three aircraft that flew through the ash plume. In this eruption, the Air Force's 5039th Air Transport Squadron evacuated more than 20 of its aircraft to Laird and Eielson Air Force Bases. Three big military cargo aircraft that remained on the ground, exposed to the ASHFALL, needed 10 days to be cleaned of ash.

Other eruptive activity in various parts of the world has provided further information on airborne ash and its effect on aircraft. Eruptions of the volcano GALUNG GUNG in INDONESIA in 1982 affected two aircraft that flew above the volcano on June 24 and July 13. The planes had to make emergency landings at Jakarta. The 1986 eruption of the Lascar volcano in CHILE also had a bearing on aviation safety, although no incidents involving aircraft were reported. The eruption of Lascar occurred at a remote location and lasted only several minutes, but its ash plume traveled quickly over populated areas served by commercial air travel and illustrated how even a largely ignored volcano such as Lascar (it was not considered dangerous before its eruption) could pose hazards to air travel.

The Boeing Commercial Airplane Group has reported that flights have observed many peculiar conditions during flights through volcanic ash, including heavy discharges of static, a glow in the engine inlets, and false cargo fire warnings. Ash also may enter the cockpit, accompanied by an odor of acid.

Azores The Azores island group off the western coast of AFRICA in the ATLANTIC OCEAN includes several volcanoes, notably AGUA DE PAU, FURNAS, and SETE CIDADES. Five of the volcanoes have undergone eruptions within historical times starting from 1563, and submarine eruptions have been numerous. Minor eruptions occurred at the east end of the Azores in 1638 and 1881 but did not form any lasting additions to the island chain. Other volcanoes include Fayal, Graciosa, Monaco Bank, and Pico.

See also HOT SPOT.

B

backarc The area directly adjacent to an ISLAND ARC on the opposite side of the SUBDUCTION ZONE. Typically, this area is on the landward side of the island arc. In many cases, secondary convection develops in the MANTLE above the subducting OCEANIC CRUST below this area. The convection causes the backarc area to undergo extension and begin to RIFT, creating a backarc basin. The basin is active with earthquake-producing normal faults and even minor MAFIC igneous activity, including volcanic eruptions. Unlike the volcanoes of the island arc, they are typically not particularly dangerous.

Baikal, Lake *Russia* Lake Baikal, in eastern Siberia, Russia, has been the site of many notable earthquakes, including events recorded in 1828, 1839, 1862, 1869, 1871, and 1959. Many extinct volcanoes are located in the vicinity of the lake. A RIFT passing through Lake Baikal has been linked tentatively to a nascent MID-OCEAN RIDGE within the continental landmass of eastern Asia. Earthquakes around Lake Baikal are mostly restricted to a narrow zone around the rift. normal faults and STRIKE-SLIP FAULTS are common in the Lake Baikal rift.

Baitoushan *volcano, China* Situated on the China-Korea border, the Baitoushan STRATOVOLCANO in the TIEN-CHI caldera has erupted five times in the last millennium. Its most recent eruption was in 1702. Its most impressive eruption was RHYOLITIC and occurred at about A.D. 1050. This eruption had a VEI = 7 and produced 32 cubic miles (150 km^3) of PUM-ICE. There was a thick circular apron of pumice around the volcano of radius at least 25 miles (40 km) wide. These numbers make it one of the largest eruptions of the last 10,000 years.

Baja California *Mexico* The elongate strip of land that lies approximately north to south along the west coast of MEXICO and is separated from continental Mexico by the Gulf of California. Baja California is a small piece of land that has RIFTED away from NORTH AMERICA. The Gulf of

California contains a MID-OCEAN RIDGE (divergent margin) that connects to the SAN ANDREAS FAULT of California. These tectonics cause Baja California to push northward into California. This movement is part of the tectonic environment that gives rise to frequent strong earthquakes in southern California, as the northward-moving plate strikes and grinds against the deeply rooted mass of the SIERRA NEVADA.

See also LOS ANGELES.

Baker, Mount *Washington, United States* One of the peaks of the volcanic CASCADE MOUNTAINS, Mount Baker has been active on numerous occasions over the past few centuries. Many eruptions were observed in the last 200 years, notably 1792, 1843, and 1880. The 1843 eruption deposited ash over a large area. Although Mount Baker appears to have emitted LAVA in the recent geologic past, it is considered more likely to expel ASH and fragments of rock. Mount Baker stands just north of Seattle and is heavily glaciated. It was the site of an interesting phenomenon in 1975: the rapid formation of a crater lake. The lake appeared as a result of increased activity at Sherman Crater, a vent several hundred feet from the mountain's summit. Ice in the CRATER broke apart and melted, forming a lake more than 150 feet (46 m) wide and 200 feet (61 m) long. The lake is thought to have been buried under an AVALANCHE of ice in 1977.

Bam *earthquake, Iran* The third most devastating earthquake of the 21st century (as of March 2006) struck the oasis city of Bam, in southeastern IRAN, at 5:26 A.M. local time on December 26, 2003. Sadly, one year later, this earthquake would be upstaged by the BANDA ACEH earthquake as the most devastating of the century. The magnitude of the Bam earthquake was 6.7, and the shaking lasted 20 seconds. Local INTENSITY of the quake reached IX on the modified MERCALLI scale, and activity on the Bam Fault was responsible for the event.

The earthquake killed about 43,200 people and left over 90,000 homeless. More than 200,000 people lost their livelihood. This is especially tragic because Bam was a prosperous city, with a wealthy middle class based on agriculture and enterprise. Its ancient walled citadel had been a popular tourist destination but was flattened by the earthquake. Electric and water lines were also severed, destroying the city's agriculture. Besides internal aid from the government of Iran, the United Nations and other world organizations provided aid. Unfortunately, relief came slowly, and riots broke out as a result, causing further distress to an already beleaguered population.

Banda Aceh *earthquake and tsunamis, Indonesia* At 7:58 A.M. on December 26, 2004, a huge SUBMARINE EARTHQUAKE of magnitude 9.0 occurred about 100 miles (160 km) off the west coast of northern Sumatra near Banda Aceh. At 9.0, it was tied for the fourth-strongest earthquake since measurements began in 1899. Not since the 1964 GOOD FRIDAY EARTHQUAKE, 40 years earlier, had a quake been so strong. The energy released by this earthquake was equivalent to

23,000 Hiroshima-type atomic bombs. The FOCUS was shallow, at six miles (10 km) depth.

The FAULT that generated the earthquake is part of the SUBDUCTION ZONE that forms the Sunda trench where the INDIA and Burma plates collide. This fault is considered a MEGATHRUST by its form and size. During the earthquake, a 720-mile (1,200-km) stretch of this plate boundary fault, roughly the length of California, slipped with up to 50 feet (15 m) of DISPLACEMENT, though most of the movement was concentrated to a 240-mile (400-km) section. This displacement uplifted the seafloor and in turn the ocean water above it. It was the redistribution of the ocean surface that formed the TELETSUNAMI that devastated the Indian Ocean basin.

GEOPHYSICISTS from the U.S. Geological Survey realized the severity of this earthquake as the arrivals began to send the SEISMOGRAPHS wild. They located the EPICENTER and immediately suspected that a TSUNAMI was possible. The problem was that unlike the Pacific Ocean, which has a real-

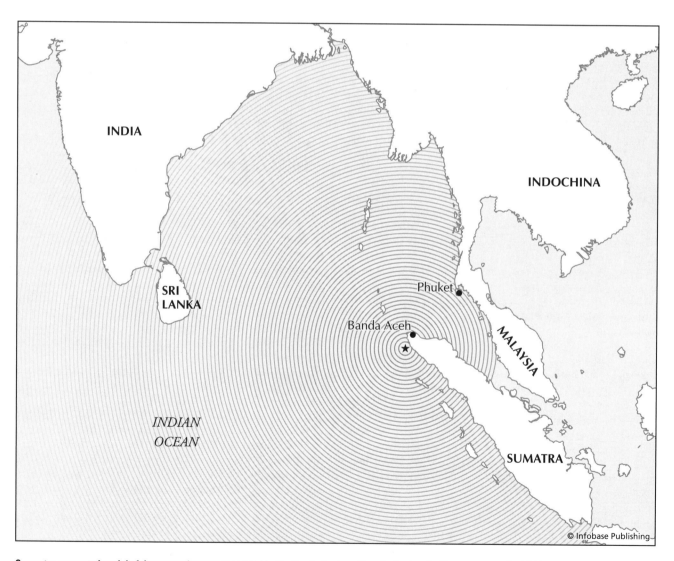

Computer-generated model of the tsunami waves generated by the Banda Aceh earthquake just as the first waves reached Sri Lanka. The star shows the location of the epicenter.

time tsunami monitoring system (PACIFIC TSUNAMI WARNING SYSTEM), the Indian Ocean does not. Further, there was no system to dispatch warnings around the ocean basin, so the geophysicists could not send out a widespread alert. As a result, the Indonesian coastal resorts, overbooked because of the holidays, were struck blind. It was a bad event at a bad time.

A tsunami struck the Indonesian coast less than one hour later with waves ranging from 50 feet (15 m) up to 90 feet (27 m) in height. The waves penetrated inland one-half mile to nearly one mile, depending upon topography. The DEATH TOLL was phenomenal. The waves made their way around the basin rim for the rest of the morning, slowing and losing height with time but still causing death and destruction everywhere they touched. Resorts in Thailand, especially Phuket, were literally demolished as the now 20-foot (6-m) waves came ashore. In typical fashion, as the waves approached, they drew the water away from the coast, exposing the seabed for several minutes before crashing back, sweeping away tourists and inhabitants alike. Two hours later, 40-foot (12-m) waves hit India and Sri Lanka, continuing inland one-half mile (0.8 km). There was a story of an eight-year-old girl in India who had done a report on tsunamis in school the previous semester. When she saw the water retreat, she sounded an alarm and saved the lives of 100 people who had been on the beach. The waves devastated all of the ocean islands (Maldives, for example) and even caused death and destruction on the east coast of AFRICA, with smaller but still powerful waves.

The death toll for the total event was staggering. Estimates continue to be revised, but as of 2006, 283,100 people were believed to have perished, making the earthquake the second or third worst earthquake disaster in history. There are still 14,100 people listed as missing, and a staggering 1,126,900 people are considered homeless. The worst damage occurred in Indonesia, where there were 108,000 confirmed dead and 127,700 listed as missing and believed dead. Other losses included 30,900 dead in Sri Lanka, 10,700 in India, 5,300 in Thailand, and even 150 in Somalia, Africa.

Media coverage of the disaster was unparalleled, with television and Internet news continuing for months. Amateur videos taken by tourists during the tsunami and stories of rescued people who had been clinging to trees or ocean debris for days tugged at the world's heartstrings prompting an outpouring of aid. In addition to many countries delivering record amounts of aid, an extraordinary amount of private donations came pouring in. The U.S. military conducted extensive rescue and relief operations, and there were fundraising concerts and campaigns.

The numerous aftershocks on the days subsequent to the MAIN SHOCK numbered in the hundreds but died down in a few days. The FAULT, however was not finished. The biggest aftershock ever on record and among the top 10 earthquakes ever occurred on March 28, 2005, at 11:09 P.M. It had a phenomenal 8.7 MAGNITUDE and also generated tsunamis, though of much smaller size. (Approximately 1,000 people were killed in Indonesia.) There was an aftershock of magnitude 6.7 on April 10, 2005; a 6.8 on May 14, 2005; and a 7.2 on July 22, 2005. This event does not seem to end.

Banda Api *caldera, Banda Sea, Indonesia* Banda Api is located in the Banda Sea near the islands of Buru and Ceram and is part of the Banda group, which includes 10 islands. Two of these islands (Banda Besar and Pisang) are situated along the rim of the Banda Api CALDERA. Another island, Neira, stands on the eastern rim of a smaller caldera nested inside the one marked by Banda Besar and Pisang. Banda Api appears to have formed in much the same pattern as KRAKATOA, with two or more episodes of caldera-building, each one involving the growth and destruction of a STRATOVOLCANO.

Banda Api has been active frequently in the 19th and 20th centuries. One frightening eruption occurred without warning in June 1820 and scared the inhabitants into fleeing by boat. Clouds of ASH and smoke emanated from the volcano in 1824 during the formation of a new CRATER. SULFUR vapor from the volcano caused a thick mist to form in late 1835, and a subsequent strong earthquake was followed by still more shocks that continued for several days.

Less powerful earthquakes occurred in late 1853, and the volcano put out unusually large amounts of steam. Noises from underground were heard in December 1855, and a minor earthquake occurred in January 1856. More subterranean noises were heard later that month. The volcano reportedly emitted large amounts of smoke during these times of rumbling. Five earthquakes, one of them accompanied by a TSUNAMI, occurred in June and July 1859. Earthquakes continued in August. One especially strong earthquake on September 25 was associated with a dramatic rise in the sea along the southern coast of Neira, although this disturbance of the sea appears to have caused no significant damage. More strong earthquakes occurred on October 18, November 7–8, and December 29. In 1859 and 1860, Banda Api emitted unusually large amounts of smoke.

Several notable earthquakes occurred in 1860, including one that was accompanied by a strange underground roaring noise that reportedly appeared to move from southeast to northeast. Neira experienced frequent earthquakes between early January and late June of 1877, although the volcano did not erupt. A fairly strong earthquake in 1887 that appeared to be centered near Banda Api was followed by an increase in steam from the volcano, although no such rise in activity was observed in connection with other, strong shocks that followed several months later. A very strong earthquake with pronounced subterranean rumbling shook Banda Api on August 12, 1890, followed by an even stronger earthquake on November 23. AFTERSHOCKS continued for days. Damage to buildings was widespread. A fresh active crater with three FISSURES emanating from it appeared at this time.

Earthquakes and detonations occurred on May 18, 1901, and a glow was seen near the summit of Gunungapi. A minor earthquake occurred in March 1902, and more subterranean rumbling was heard. A strong earthquake in June 1987, approximately 90 miles (145 km) southeast of Banda Api, preceded the release of a steam plume from the volcano. Earthquake activity increased dramatically in early May before an eruption on May 9.

Bandai-San *volcano, Japan* The July 15, 1888, eruption of Bandai-San is one of the worst volcanic disasters in Japanese history. Although the mountain had displayed no signs of activity for perhaps a thousand years beforehand, Bandai-San blew up in a series of approximately 20 steam-blast explosions in a single day. The eruption had a VEI = 4. Most of the destruction occurred within several minutes, when the most powerful explosion set off a huge LANDSLIDE that sent the entire north slope of the mountain rolling into the Nagase Valley. More than 461 people were reported killed, but only 116 bodies were found. According to one estimate, the eruption expelled some 1 billion cubic yards (750,000,000 m³) of solid material into the valley below the volcano. One VOLCANOLOGIST put this volume of material into more easily understandable terms when he explained that if the material from this eruption were formed into ships of 15,000 tons each, the resulting line of ships would span the PACIFIC OCEAN from JAPAN to CALIFORNIA. This eruption was one of several (including BOGOSLOF and VULCANO) that occurred about the same time and were implicated in a sharp drop in the amount of solar radiation reaching Earth in 1890–91. (*See* CLIMATE, VOLCANOES AND.) In total, Bandai-San volcano has erupted four times between 806 and 1888. Besides the 1888 eruption, most eruptions are moderate in size (VEI = 2–3).

Bandelier Tuff *southwestern United States* The Bandelier Tuff in NEW MEXICO is made up of volcanic ash deposited over a wide area by an eruption of the VALLES caldera.

Bantul *earthquake, Indonesia* On Saturday, May 27, 2006, at 5:54 A.M. local time, a strong earthquake struck the Bantul-Yogyakarta area of INDONESIA. The earthquake had a MAGNITUDE of 6.3, a maximum INTENSITY of VII–IX on the modified MERCALLI scale, and a focal depth of 6.2 miles (10 km). The DEATH TOLL for this event was more than 6,000 victims, with more that 30,000 injured. Damage to Bantul was extreme, with more than 80 percent of homes (135,000 houses) destroyed, leaving more than 650,000 people displaced.

The earthquake occurred just weeks after renewed eruptive activity on Mount MERAPI volcano just 20–25 miles (32–40 km) to the north. It is possible that the eruption and earthquake were related, and fear that the earthquake would cause much stronger eruptions was widespread. Indonesia lies above a SUBDUCTION ZONE between the Australian plate to the south and the EURASIAN CRUSTAL PLATE. The collision is not orthogonal (head-on) but is oblique, with Australia moving northwestward. Instead of Australia subducting at an oblique angle under Indonesia, it is partitioned into orthogonal at the trench, and all westward motion is taken up on a large SAN ANDREAS–like strike-slip fault that produces destructive shallow earthquakes as that in Batul.

Barcena *Mexico* Barcena is a volcano on San Benedicto Island south of BAJA CALIFORNIA, and west of MEXICO CITY in the PACIFIC OCEAN. The only historic eruption of Barcena was in 1952. It produced a LAVA FLOW and a glowing AVALANCHE during this eruption.

Bardarbunga *volcano, Iceland* Buried under the ice of the Vatnajökull, Bardarbunga, and its surrounding area are thought to have been the site of several notable eruptions since the late ninth century. A large eruption occurred along FISSURES to the southwest of Bardarbunga around 900 and again around 1480. Layers of TEPHRA have been associated with eruptions in 1477 and 1717. An eruption from a fissure some 20 to 25 miles (32–40 km) southwest of Bardarbunga in the 1860s released a substantial LAVA FLOW and apparently some tephra as well. Earthquake activity around Bardarbunga showed a marked increase in the mid-1970s. Between 1974 and 1986, several moderately strong earthquakes occurred. Activity at Bardarbunga may be related to activity at KRAFLA, some 70 miles (113 km) north of Bardarbunga. It has been suggested that inflation at Krafla accompanies deflation at Bardarbunga and that changes in pressure are transmitted through a layer of partly molten crustal rock.

barranco Steep-sided drainage channels in a volcanic cone formed by coalescence of smaller channels. They typically form a radial pattern around the volcanic cone.

Barren Island *volcano, Andaman Islands, India* Barren Island is a volcano with a .99-mile (1.6-km)-wide crater that contains a central PYROCLASTIC cone. It erupted in 1787, 1789, 1795, 1803, 1852, 1991, and 1995. The 1803 eruption was the most intense of the older group and produced LAVA FLOWS that reached the coast. The 1991 eruption produced basaltic ANDESITE flows that covered 17,000 square feet (1,600 m²) and reached the northwest coast. The 1995 eruption was a large STROMBOLIAN type that produced extensive ASH and EJECTA eruptions with volumetrically lesser amounts of lava.

barrier (fault) Part of a FAULT surface that resists movement because of its geometry or some geologic body or change. A bend or step in the fault surface or another rock body, such as an igneous intrusion, can stop a fault from moving in that area. Stress can build to excessive levels in this area, and if it finally ruptures, the resultant earthquake can be very strong.

Barrier, The *caldera, Kenya* The Barrier is located south of Lake Turkana and separates the lake from Lake Logipi and the Sugata Valley to the south. The Barrier has an outer and an inner, younger CALDERA. Although there are reports of eruptions in the mid- to late 19th century and in 1921–22, details are unavailable.

basalt A dark volcanic rock, basalt is the most abundant volcanic rock on Earth's surface. Basalt is composed of PYROXENE, PLAGIOCLASE, and commonly OLIVINE. It occurs in several different forms. The ocean floor from all ocean basins is composed of basalt covered by ocean sediments. The basalt is produced at the MID-OCEAN RIDGE as part of the new CRUST. When basalt is extruded underwater, it comes out as small spherical blobs that settle to the ocean floor and form PILLOW LAVA (BASALT) because of their shape. "Plateau basalts," or FLOOD BASALTs, are found where large quantities of basalt have poured from FISSURES, sweeping over the surface and overwhelming landforms in their path. Plateau basalts may contain hundreds of thousands of cubic miles of rock. Examples of plateau basalts are found in the Pacific Northwest of the UNITED STATES, specifically the COLUMBIA PLATEAU and the SNAKE RIVER PLAIN. Other examples of plateau basalts are found in India's Deccan Plateau and along Manchuria's borders with Korea and Russia, notably around the Baekdoo-San volcano on the Korean border. Yet another plateau basalt, possibly made up of several plateaus, stretches from Scotland to Greenland. The Karoo basalts of AFRICA, the Pirana basalts of SOUTH AMERICA, and the WATCHUNG basalts of NEW JERSEY are other examples. Plateau basalts are noted for their absence of pyroclastic material. Individual lava flows in a plateau may be up to approximately 150 feet (46 m) thick and may extend for 10 miles (16 km) or more. The geological record indicates these flows were not accompanied by explosive eruptive activity. Basalt sometimes exhibits a phenomenon called COLUMNAR JOINTING, in which the rock forms a vertical prismatic structure reminiscent in some ways of a honeycomb. A spectacular example of a basalt formation exhibiting columnar jointing is Giant's Causeway in the British Isles. It occurs only in basalt that has flowed to Earth's surface and cooled in air. Basalt is believed to form from MAGMA originating in the UPPER MANTLE. The process that causes the melting solid rock in the upper mantle to produce basaltic MAGMA is called decompression melting.

basal wreck The bowl-like structure formed when the peak of a volcano is destroyed in an eruption. A basal wreck is characterized by a roughly circular shape, a raised rim, and a central depression.

basic An old term for MAFIC.

Basin and Range Province *western United States* The Basin and Range Province in the UNITED STATES extends from the vicinity of Klamath Falls, OREGON, and Bear Lake, IDAHO, southward through NEVADA and UTAH, and occupies large portions of southeastern CALIFORNIA, southern ARIZONA and southwestern and central NEW MEXICO. The province covers some 300,000 square miles (777,000 km²), or roughly 8% of the United States. More than 200 mountain ranges occupy the Basin and Range Province, which shows widespread evidence of recent faulting (notably along the Wasatch Mountains in Utah, at the base of which contains fresh exposures of unweathered rock indicating movement along a FAULT there in the very recent geologic past).

The Basin and Range Province consists of a series of parallel mountain ranges that trend north-south. The ranges enclose and are separated by long depressions. The Basin and Range Province represents a region of thinning and RIFTING of Earth's crust, where crustal rock has split into many blocks that are separated by normal faults. The blocks that sank became depressions, or GRABENS. The blocks that remained elevated became the ranges, or HORSTS. The Basin and Range Province represents the early stages of a DIVERGENT BOUNDARY.

The Basin and Range Province is noted for strong earthquake activity. A recent example was the Dixie Valley, Nevada, earthquake of 1952 (MAGNITUDE 7.1). One of the greatest earthquakes recorded in the Basin and Range Prov-

ince occurred in 1872 and involved more than 20 feet (6 m) of vertical displacement between Owens Valley and the SIERRA NEVADA. Seismic activity is generally most intense along the eastern and western edges of the province. The powerful KERN COUNTY, California, earthquake of 1952 took place just beyond the western boundary of the Basin and Range Province, and the Owens Valley earthquake of 1872 took place on that boundary at the Southern Sierra Gap. On the opposite boundary, a 1959 earthquake near HEBGEN LAKE, MONTANA, had a magnitude of 7.1 and was felt over an area of more than a half-million square miles (1.4 million km²). This earthquake set off an AVALANCHE that dammed the Madison River, formed a lake, and killed more than 20 persons. Another earthquake of approximately the same magnitude struck Idaho in 1983, caused two deaths, and resulted in more than $12 million in damage to property. Other strong earthquakes in the Basin and Range Province affected Nevada in 1869, 1903, 1915, 1932, 1934, and 1954. The GREAT BASIN, which includes DEATH VALLEY, is part of the Basin and Range Province. Volcanic activity also occurs along the boundaries of the province, a notable example being the LONG VALLEY caldera.

Evidence of volcanism is also abundant within the Basin and Range Province. LAVAS and other products of eruptions have buried older rock along the northern edge of the Great Basin. Volcanic rocks extending across the Great Basin include huge volumes of TUFF, or pyroclastic material welded together. Large flows of basalt are found on the northwestern and southeastern edges of the Great Basin, along the COLUMBIA PLATEAU and the SNAKE RIVER PLAIN and the southern edge of the Colorado Plateau respectively. Ubehebe Crater is a famous product of relatively recent volcanic activity in the Basin and Range Province.

Mines and mining have played an important part in the development of the Basin and Range Province during the past two centuries. Copper, gold, silver, lead, zinc, antimony, iron, mercury, magnesium, manganese, and beryllium have all been mined in large quantities from deposits in the Basin and Range Province.

See also NORTH AMERICA; NORTH AMERICAN CRUSTAL PLATE.

batholith A large PLUTON or body of intrusive IGNEOUS ROCK that has a surface (exposed) area of greater than 39 square miles (100 km²). Less than that and the body is a STOCK. Batholiths are prominent in the landscapes of the western UNITED STATES and CANADA. The northern Rocky Mountains contain tremendous batholiths, but it is uncertain what combination of processes created them. The IDAHO Batholith, roughly 250 miles (402 km) long and 100 miles (161 km) across at its broadest point, contains the Salmon River and Clearwater Mountains. The Bitterroot Range along the eastern edge of the batholith lies along the border between MONTANA and Idaho. Batholiths may have great economic importance if associated with valuable ore deposits. A break along the northern part of the Idaho Batholith contains the Coeur d'Alene district, which is famous for its lead, zinc, and silver mines. The nearby Boulder Batholith in Montana is also a rich source of metals and has yielded about a third of the copper mined in the United States. Granite batholiths form large portions of the northern CASCADE MOUNTAINS, the Coast Mountains in British Columbia, southern ALASKA, and the SIERRA NEVADA in CALIFORNIA. The GRANITE in these batholiths may be very coarse-grained, indicating that the MAGMA that formed the rock cooled slowly and at great depth. (Fine-grained igneous rocks generally cooled more rapidly, at or near the surface, so that coarse grains did not have an opportunity to develop.) The chemical composition of these granitic rocks may vary. Sections of COUNTRY ROCK may separate individual plutons, and entire blocks of country rock may be included with plutons, having become separated from the walls or roof of a pluton and fixed in the intruding magma as an INCLUSION.

Batholiths are most commonly FELSIC or INTERMEDIATE in composition, but some notable MAFIC varieties exist. Batholiths form the roots of large MAGMATIC ARCS, so they are common in areas of such ocean-continent convergence as the ANDES. They may also form from HOT SPOTS on land, such as those in AUSTRALIA and even along RIFTS in the early stages of a DIVERGENT BOUNDARY.

Batur *caldera, Bali, Indonesia* Batur is situated on the island of Bali and is one of two active volcanoes, the other being the very dangerous AGUNG. Batur undergoes frequent explosive eruptions but is less dangerous than Agung. The Batur CALDERA began its formation about 500,000 years ago. About 50,000 years ago, it underwent a huge explosion that produced about 31 cubic miles (140 km³) of EJECTA (VEI = 7). About 20,000 years ago, another similarly massive eruption produced the large caldera that exists today.

Two calderas, one inner and one outer, have been described at Batur within the larger old caldera. Inside the inner caldera, a small STRATOVOLCANO (also called Batur) has arisen and is active. There have been 23 DACITE and BASALT eruptions from 1824 through 1994. LAVA FLOWS occurred in 1849, 1888, 1904, 1905, 1921, 1926, 1968, and 1974. In 1888, numerous earthquakes were felt in Bali in late May, days before an eruption of Batur. The Batur volcano also may have been involved in detonations and earthquakes perceived from October 9 to October 11, 1902. A lava flow from Batur invaded a local village in 1904 or 1905. Although strong earthquakes occurred in the vicinity of Batur in 1917, these are thought to have been TECTONIC rather than volcanic in origin. Batur erupted again from 1921 to 1926. (It is not certain whether this resumption of eruptive activity was related to the earthquakes of 1917.) The volcano expelled incandescent pieces of rock, and detonations shook the ground. There was a brief diminution in eruptions at the summit of the volcano as LAVA started to flow on April 9 and 10. Powerful explosions began again, however, in mid-April. Lesser eruptions took place between 1922 and 1925, followed by a more powerful eruption with accompanying earthquakes in 1926. It was probably the largest historical eruption of Batur. Lava emerged from a 1.3-kilometer-long FISSURE on the southwestern side of Batur, and soon afterward another fissure opened on the southeast side of the volcano, releasing a lava flow that overran the nearby village of Batur. Lava tubes formed during the 1926 eruption. In 1959 emissions of hydrogen sulfide increased from the northern part of the caldera lake.

Small earthquakes were observed on the rim of the caldera during the eruption of Agung volcano, several miles southeast of Batur, on September 5, 1963. Batur started erupting later that day. Lava flowed from several VENTS on the west side of Batur during this eruption. Minor explosions occurred every few days at Batur between August and December 1965. Explosions occurred again on a frequent basis in early 1968, and lava flowed from Batur. Bubbles of gas, smelling of SULFUR, rose from the lake in 1969, and small explosions took place in 1970. The latest 1994 eruption produced only ASH.

Bayonnaise *caldera, Japan* A mostly submerged CALDERA in southern Japan, Bayonnaise has a record of SUBMARINE ERUPTIONS dating back to 1896. An eruption of the volcano Myozin-Syo, a lava dome in the middle of the caldera, in 1952 destroyed a research ship that crossed over a vent and killed more than 30 people on board. Myozin-Syo has produced some spectacular PHREATIC explosions.

bedrock All solid rock that underlies the sediment and SOIL that typically covers the Earth. Besides the thin veneer of soft material that covers the planet, the entire Earth is made of rock. If a person were to dig down and hit that rock, he or she would have struck bedrock. Where bedrock reaches the surface, it is called an outcrop or exposure.

Benbow, Mount *See* AMBRIM.

Bengkulu *earthquake, Indonesia* On June 4, 2000, a strong earthquake occurred 21 miles (33 km) beneath the Indian Ocean just west of Bengkulu, whose population is more than 1.2 million. The quake measured 7.9 on the RICHTER scale and was followed 11 minutes later by an AFTERSHOCK of MAGNITUDE 6.0. During the next six hours, some 53 smaller aftershocks were recorded. Many LANDSLIDES occurred from the shaking. More than 97 people were killed, more than 1,900 were injured, and many thousands were left homeless. Electricity, water supplies, and telephone lines were all interrupted and many roads were left impassable. The police reported that virtually every building in the city was damaged, and many were destroyed.

Benioff zone A zone of earthquake activity that tracks the downgoing OCEANIC CRUST within a SUBDUCTION ZONE. The Benioff zone starts near the surface at the TRENCH and progressively extends perhaps 400 miles (644 km) below the surface proportional to the distance into the ISLAND ARC or MAGMATIC ARC. If the foci of the earthquakes are plotted on a map and cross-section of an arc, the top of the downgoing crust can be imaged and a relatively accurate picture of what the subduction zone looks like, including the angle of descent and how much the crust is bent, can be made. This discovery was one of the key elements in developing the PLATE TECTONIC model. It is named after Hugo Benioff (1899–1968), an American seismologist who mapped such zones in the 1940s and 1950s.

bentonite An ASH bed from an ASHFALL within a sequence of predominantly sedimentary rock layers. The ash is converted to a layer of clay with a distinctive composition.

Bermuda The island of Bermuda occupies the summit of a volcanic SEAMOUNT that rises some 6,000 feet (1,800 m) from the ocean floor. At one time, waters covered the peak, and coral formed on the summit. The coral now constitutes the surface of the island. A well-drilling project showed that almost 400 feet (130 m) of limestone overlies the BASALTIC rocks that make up the volcanic portion of the seamount.

Bezymianny *volcano, Kamchatka, Russia* Part of the KLIUCHEVSKOI complex of volcanoes, Bezymianny is a STRATOVOLCANO whose name means "no name" because it is small compared to its larger neighbors like Kliuchevskoi and because it had no historical eruptions. This situation changed when it erupted with great violence in 1955–56. In many ways, this eruption closely resembled that of Mount SAINT HELENS in WASHINGTON in the UNITED STATES in 1980, except that it was much bigger. On March 30, 1956, a tremendous explosion destroyed the summit of Bezymianny and produced a CRATER one mile (1.6 km) × 1¼ miles (2 km) wide. The eastern side of the mountain was blown open. The blast leveled trees one foot (0.3 m) in diameter at a distance of 15 miles (24 km). Fires were started at 18 miles (29 km) distance by the heat of the gas cloud. Glowing AVALANCHES and LAHARS swept down the volcano with great force. A LAVA DOME formed in the crater. The March 30 eruption alone is believed to have cast out about 2.4 billion tons of rock. The ERUPTION column was 24 miles (39 km) high. An ash flow laid down by the volcano produced an area of FUMAROLES similar to the VALLEY OF TEN THOUSAND SMOKES created by the eruption of Mount KATMAI in ALASKA; this site came to be known as the Valley of Ten Thousand Smokes of KAMCHATKA.

There have been 34 eruptions of Bezymianny since 1955, making it one of the most active volcanoes on Earth. Its most recent eruptions were in 1994, 1997, 1999, and 2000. The 1997 eruption produced a 40,000- to 45,000-foot (12,192- to 13,716-m) eruptive column and a plume that spread some 260 miles (418 km) to the east-northeast. It was accompanied by significant seismic activity. The 1999 and 2000 eruptions were similar in character but smaller.

Bhuj *earthquake, India* One of the major earthquakes of the still young 21st century struck the Gujarat area of westernmost India at 8:46 A.M. on January 26, 2001. (Notably, the quake struck virtually the same area as the ALLAHBUND earthquake of 1819.) The Bhuj earthquake has also been unofficially called the Gujarat earthquake and the "Republic Day" earthquake. Its MOMENT MAGNITUDE was 7.6, with a duration of 85–90 seconds. The FOCUS of the earthquake was on the newly discovered east-west-oriented North Wagad Fault, which is a REVERSE FAULT. Slip is believed to have been three to 13 feet (1 to 4 m) on the FAULT, and surface expression included a line of minor cracking and DEFORMATION in the form of a bulge similar to that in the 1819 earthquake about 1.2 miles (2 km) long. LIQUEFACTION was common, with MUD VOLCANOES hundreds of feet to even three miles (5 km) across in one case. In some cases, GROUNDWATER was driven to the surface in such quantity to activate desert rivers that had been dry for more than 100 years. AFTERSHOCKS continued for months, with the stron-

gest reported on January 28 with a MAGNITUDE of 5.8 and a duration of 30 seconds.

The official DEATH TOLL for this event was 19,727 although some sources place it at 13,805. Over 166,000 were injured, and more than 600,000 were left homeless. There were 348,000 houses destroyed and 844,000 damaged. More than 20,000 cattle were killed. The Indian government estimated that 15.9 million people were directly or indirectly affected by the quake, and the economic losses exceeded $1.3 billion, although some estimates place it as high as $5 billion.

Bible, earthquakes in the The Bible contains numerous references to earthquakes, in both the Old and New Testaments. In the first book of Kings (I Kings 19:11–12), the prophet Elijah experienced an earthquake:

> And, behold, the Lord passed by, and a great and strong wind rent the mountains, and brake in pieces the rocks before the Lord; but the Lord was not in the wind; and after the wind an earthquake; but the Lord was not in the earthquake: and after the earthquake a fire; but the Lord was not in the fire.

Isaiah also described God's power in terms of an earthquake (Isaiah 9:6):

> Thou shalt be visited of the LORD of hosts with thunder, and with earthquake, and great noise, with storm and tempest, and the flame of devouring fire.

Amos, who lived around 750 B.C., used a famous earthquake to set the date of his prophecies (Amos 1:1):

> The words of Amos, who was among the herdmen of Tekoa, which he saw concerning Israel in the days of Uzziah king of Judah, and in the days of Jeroboam the son of Joash king of Israel, two years before the earthquake.

This same earthquake, apparently, is mentioned in Zechariah 14:5:

> And ye shall flee to the valley of the mountains; for the valley of the mountains shall reach unto Azal; yea, ye shall flee, like as ye fled from before the earthquake in the days of Uzziah king of Judah: and the Lord my God shall come, and all the saints with thee.

Matthew wrote that Christ predicted His return would be preceded by earthquakes (Matthew 24:7):

> For nation shall rise up against nation, and kingdom against kingdom: and there shall be famines, and pestilences, and earthquakes, in divers places.

Mark repeated this same warning in similar language (Mark 13:8):

For nation shall rise against nation, and kingdom against kingdom: and there shall be earthquakes in divers places, and there shall be famines and troubles: these are the beginnings of sorrows.

Luke's version of this prophecy (Luke 21:11) reads as follows:

And great earthquakes shall be in divers places, and famines, and pestilences; and fearful sights and great signs shall there be from heaven.

Matthew also reported that an earthquake accompanied the crucifixion of Christ and greatly impressed onlookers (Matthew 27:54):

Now when the centurion, and they that were with him, watching Jesus, saw the earthquake, and those things that were done, they feared greatly, saying, Truly this was the Son of God.

Matthew added (Matthew 28:2) that another earthquake occurred when the stone was rolled away, by angelic action, from Christ's tomb:

And behold, there was a great earthquake: for the angel of the Lord descended from heaven, and came and rolled back the stone from the door, and sat upon it.

In Acts 16:26, an earthquake released the apostles Paul and Silas from prison:

And suddenly there was a great earthquake, so that the foundations of the prison were shaken: and immediately all the doors were opened, and every one's hands were loosed.

Revelation, the final book of the New Testament, abounds in references to earthquakes in connection with prophecies of the final days before Christ's return. The first-person narrator was the author, the apostle John:

And I beheld when he had opened the sixth seal, and, lo, there was a great earthquake; and the sun became black as sackcloth of hair, and the moon became as blood. [6:12]
    And the angel took the censer, and filled it with the fire of the altar, and cast it into the earth: and there were voices, and thunderings, and lightnings, and an earthquake. [8:5]
    And the same hour was there a great earthquake, and the tenth part of the city fell, and in the earthquake were slain of men seven thousand; and the remnant were affrighted, and gave glory to the God of heaven. [11:13]
    And the temple of God was opened in heaven, and there was seen in his temple the ark of his testament; and there were lightnings, and voices, and thunderings, and an earthquake, and great hail. [11:19]

And there were voices, and thunders, and lightnings; and there was a great earthquake, such as was not since men were upon the earth, so mighty an earthquake, and so great. [16:18]

Bihar *earthquake, India*   A deadly earthquake struck INDIA near the Nepal border and devastated the Bihar area at 2:21 P.M. local time on January 15, 1934, registering an 8.0 on the RICHTER scale. The FAULT that generated the earthquake experienced LEFT-LATERAL strike-slip motion, although reports of normal fault movement also exist.

The DEATH TOLL from the Bihar earthquake was 10,653, making it one of the deadliest in India. Damage was primarily the result of intense LIQUEFACTION, SLUMPing, and intense ground shaking. FISSURES opened in the center of towns and were reported as being up to 240 feet (72 m) long and eight feet (2.5 m) wide. Subsidence occurred as a result of slumping and lateral spreading, with the ground surface dropping up to 10 feet (3 m) in several cases. Liquefaction caused the greatest damage, leaving many buildings in various states of distress. It was accompanied by great sand blows and MUD VOLCANOes. In all, sand and water were emitted over an area of about 18,000 square miles (46,600 km^2). AFTERSHOCKS continued to affect the area for years after the MAIN SHOCK. Especially strong aftershocks (MAGNITUDEs between 5 and 6) occurred on January 19, 1934; June 2, 1934; and February 11, 1936—causing damage and loss of life in some cases.

biotite   Otherwise known as black mica, biotite is a common mineral especially in INTERMEDIATE rocks but also in many FELSIC rocks. Biotite occurs in shiny black booklike sheets or flakes.

Birjand-Qā-yen *earthquake, northern Iran*   On May 10, 1997, an earthquake of RICHTER magnitude of 7.2 occurred. At least 1,567 people were killed, 2,300 people were injured, and 50,000 were left homeless. 10,533 homes were destroyed either directly from the earthquake or from the LANDSLIDES it triggered. The earthquake occurred on the Abiz Fault, which underwent left-lateral STRIKE-SLIP movement. The Abiz Fault produced the Dasht-e-Bayez earthquake of magnitude 7.3 in 1968. It killed 12,000 to 20,000 people.

Bishop Tuff *western United States*   A PYROCLASTIC FLOW, the Bishop Tuff covers almost 600 square miles (1,600 km^2) of NEVADA and central CALIFORNIA. The TUFF is thought to have been laid down in the catastrophic eruptions associated with the formation of the LONG VALLEY caldera in California about 730,000 years ago. This eruption involved the collapse of the roof of a MAGMA CHAMBER. This collapse forced 150 cubic miles (2,300 km^3) of RHYOLITIC magma into PLINIAN ash columns and associated ASHFALLS and ash flows. Compare the volume of ash to that of Mount SAINT HELENS, which produced one cubic mile of ash. The Bishop Tuff extends through ARIZONA, COLORADO, and UTAH and has been identified as far east as KANSAS.
    See also TEPHRA.

black smoker   *See* MID-OCEAN RIDGE.

blind fault A FAULT that does not reach the ground surface. Faults may grow off of a larger fault. The fault will grow with continued stress if it is great enough. The fault will produce earthquakes as it grows but it will form no surface expression. Because they form no surface expression, geologists may not know that the faults exist until they produce an earthquake. Therefore they are hidden and will take people by surprise when they rupture. The NORTH RIDGE earthquake of 1994 near LOS ANGELES was produced by a blind fault. Because the area is riddled with active faults, a major earthquake is not unexpected. In other areas, where earthquake activity is less, sudden activity from a blind fault may be devastating.

block The largest type of volcanic EJECTA, blocks are defined as being larger than 10 inches (25.6 cm). Blocks are larger than BOMBS but are similar in appearance.

body wave In contrast to SURFACE WAVES, which only travel along the Earth's surface, body waves are seismic waves that travel throughout the volume of the planet. Body waves are P-WAVES and S-WAVES and are much faster than surface waves but less destructive. Body waves are the first arrivals on the seismic records for an earthquake event.

body wave magnitude The body wave magnitude m_b is a determination of the magnitude of an earthquake based purely upon the AMPLITUDE of the initial P-WAVE, which is a BODY WAVE.

Bogoslof *volcano, Aleutian Islands, Alaska, United States* Bogoslof volcano is one of the most famous in the world because of its record of self-destruction and reemergence over the past few centuries. It is visible as an island that is the summit of the volcano, based on the seabed some 6,000 feet (1,900 m) below. The summit of the volcano keeps building itself up and tearing itself apart in explosive eruptions or continues being destroyed by wave activity. The island actually is a set of LAVA DOMES that have appeared at various locations. An island at the location of Bogoslof was reported first in 1768 and was named Ship Rock. Castle Rock, another island, appeared in 1796 during a loud eruption that frightened natives on a neighboring island. By 1826, Castle Rock was about two miles (3 km) long, less than a mile wide, and more than 300 feet (91 m) high. A great mass of volcanic rock arose from the waters near Ship Rock, on the opposite side from Castle Rock, in 1883. This new island was called New Bogoslof. A bar of clastic, or loose and fragmented, material linked New Bogoslof with Ship Rock and Castle Rock. Wave activity soon produced a channel, however, and thus cut the temporary, single island into two. Two new domes, Metcalf Cone and McCulloch Peak, formed in the waters between the two islands in 1906 and 1907. Metcalf Cone appeared first

A wave-modified volcanic island, Bogoslof Island, Aleutian Islands, Alaska. It started as a circular volcanic island in the late 1700s. Wave action has turned the island elongate. This evolution is one step in the destruction of a volcanic island. *(Courtesy of the USGS)*

but was partially destroyed by an explosive eruption before McCulloch Peak formed. McCulloch Peak was itself short-lived; only 10 months after it appeared, the peak was demolished by an explosion, leaving behind a heated lagoon almost a mile wide. This eruption is said to have been spectacular, with a thick black cloud and lightning reported. Further eruptions were reported in 1910, 1926, and 1931. The 1926 eruption produced a new dome. An eruption of Bogoslof in the 1890s was one of several eruptions that occurred around the same time (BANDAI-SAN in 1888 and VULCANO in 1888–90).

Bolivia Located in the seismically and volcanically active "RING OF FIRE" that encircles the PACIFIC OCEAN basin, Bolivia is situated in northwest SOUTH AMERICA. A volcanic mountain in Bolivia, CERRO RICO, played an important part in the economic exploitation of the Inca by the Spanish Empire. The mountain contained rich veins of silver ore and yielded great quantities of silver to be shipped to Europe. Volcanoes in Bolivia include Illimani, Parinacota, and Sajama.

On June 9, 1994, a massive earthquake of magnitude 8.2 occurred in a remote area of northern Bolivia. It triggered many LANDSLIDES and ROCKFALLS, but damage to buildings and roads was minor. The importance of this earthquake was that it was felt throughout the UNITED STATES and even in parts of CANADA. It is the first recorded incidence of an earthquake from this part of SOUTH AMERICA being felt in NORTH AMERICA. It is also the highest magnitude ever recorded in this part of South America.

See also ANDES MOUNTAINS.

Bolu-Duzce *Turkey* On November 12, 1999, an earthquake of magnitude 6.0 occurred. The earthquake resulted in 834 deaths and 4,566 injuries. It triggered several major LANDSLIDES that blocked roadways. This earthquake was the largest in a complex swarm of earthquakes that affected this part of TURKEY at the time.

bomb Fluid to blocky mass of LAVA that is shot out of a volcano and solidifies or cools in midair before landing on the ground. In general, volcanic bombs tend to be streamlined but may occur in many different shapes. If they are still fluid, when they land they flatten. If solid, they tend to crack and break apart. There are several different names for bombs depending upon their shape, including BREAD-CRUST BOMBS and COW-PIE BOMBS.

Bonneville Slide *Columbia River Gorge, Oregon and Washington, United States* The Bonneville Slide is a huge quantity of SOIL and rock that fell from the sides of the Columbia River gorge near Crown Point, east of Portland, OREGON, some centuries ago, possibly around A.D. 1100. Almost a half cubic mile of solid material was involved in the Bonneville Slide, which may have occurred as a result of an earthquake engendered by the CASCADIA SUBDUCTION ZONE offshore to the west. The northern side of the gorge is especially susceptible to LANDSLIDES because the uplift of the CASCADE MOUNTAINS has tilted the gorge toward the south and because easily eroded sedimentary material underlying LAVA FLOWS along the northern side of the gorge undermines the

Breadcrust volcanic bomb named for the breadcrust-like cracked surface of the bomb. *(Courtesy of the USGS)*

land. Landslides along the gorge are believed to have buried more than 10 square miles (26 km^2) of the gorge and created a huge natural dam perhaps 200 feet (61 m) high and more than 5,000 feet (1,500 m) wide. Other slides in the vicinity of the Bonneville Slide include the Fountain Slide, Ruckel Slide, Skamania Slide, Wind Mountain Slide, and Washougal Slide.

Boquerón *volcano, El Salvador* Located near the city of San Salvador, Boquerón erupted explosively on June 6, 1917. Though brief, the eruption destroyed most of San Salvador and carried destruction over an area of some 20 miles (32 km) around the volcano. A lake that had formed inside the CRATER of Boquerón spilled out in the form of a hot flood that overwhelmed more than a dozen communities. LAVA from this eruption collected to depths of 160 feet (49 m) in some locations, and ash accumulated in layers several feet in depth. One curious effect of this eruption was hair loss among residents of the area; evidently acid in an ASHFALL was responsible for this occurrence. More than 400 people were reported killed in this eruption of Boquerón. On more than a dozen previous occasions, eruptions had destroyed the city of San Salvador.

bore Among the many meanings of the word *bore*, it also describes a single wave in the ocean with a near vertical front. These waves can result from seismic activity.

Bowens Reaction Series A theoretical model for the order of crystallization of minerals in IGNEOUS ROCKS that explains most rocks quite well. The diagram for crystallization is Y-shaped with temperature decreasing downward. The left arm of the Y represents the discontinuous reaction series (different minerals) and the right arm represents the continuous reaction series (one mineral). At about 1,400°C, OLIVINE begins to crystallize from the melt on the discontinuous side. As temperature decreases, PYROXENE, HORNBLENDE, BIOTITE, and MUSCOVITE will crystallize in progression, depending upon composition of the melt. These minerals are largely FERRO-MAGNESIAN and dark in color. At 1,400°C, PLAGIOCLASE also begins to crystallize on the continuous side and it crystallizes all the way down with decreasing temperature. It is calcium-rich at high temperature and sodium-rich at low temperature. The ratio of calcium to sodium depends upon the temperature. At the stem of the Y are the lowest temperature minerals K-FELDSPAR and finally QUARTZ. They crystallize at about 750°C. This model also shows the MAFIC rocks crystallize (and erupt) at the highest temperature (1,400°C) because they are composed of olivine, pyroxene, and Ca-plagioclase. INTERMEDIATE rocks are composed of hornblende, biotite, and Ca-Na plagioclase and form at intermediate temperatures (1,000°C). FELSIC rocks are composed of quartz and feldspar and form at the lowest temperatures (750°C).

bradyseism   A long-continued, extremely slow vertical instability in the crust near a volcano as in the volcanic district west of NAPLES, Italy. The PHLEGRAEAN bradyseism involves movements from 20 feet (6 m) above to 20 feet (6 m) below sea level during the last 2,000 years.

branching fault   Branch off of a main FAULT. Also known as a SPLAY FAULT. These faults may take up some of the movement that would have otherwise occurred on a main fault. They may also relieve STRESSES that have built up but which cannot be relieved by movement on a main fault. Most branching faults are smaller than the main fault, but in some cases the major movement can shift over to the branching fault to the point of leaving parts of the main fault inactive.

bread-crust bomb   Volcanic BOMB with a surface that looks like bread crust. The bomb has a rough surface that develops cracks as it cools. It looks like old-fashioned or bakery bread in which the crust is cracked as the result of expansion during baking.

breccia   A coarse-grained SEDIMENTARY ROCK that is made up of angular clasts, or particles. They typically indicate that the source of the rock fragments is nearby. There are also volcanic breccias. In the case of volcanoes, a breccia is a rock made up of angular clasts that are fragmented from explosions at the vent of a volcano. They look similar to sedimentary breccia but all of the fragments are volcanic rock and there can be distinctive cementing material. There is also FAULT breccia. Fault breccia is made of angular clasts that were fragmented by fault movement during an earthquake. The size of the clasts in fault breccia can vary widely.

British Columbia *Canada*   The Canadian province of British Columbia is located on the Pacific coast within the circum-Pacific "RING OF FIRE," the belt of earthquake and volcanic activity that encircles the Pacific basin. Although not as noted for earthquake and volcanic activity as the coastal UNITED STATES (CALIFORNIA, OREGON, and WASHINGTON) to its south, British Columbia has a history of considerable seismic and volcanic activity. This activity is due to the

Volcanic vent breccia produced during an explosive eruption. The white broken fragments are encased in lava. *(Courtesy of the USGS)*

The 1977 earthquake in Bucharest caused the partial collapse of the reinforced concrete frames and masonry walls in office and apartment buildings. *(Courtesy of the USGS)*

subduction of the JUAN DE FUCA CRUSTAL PLATE along the CASCADIA SUBDUCTION ZONE beneath the PACIFIC OCEAN to the west of the province. This subduction zone produces the volcanoes of the CASCADE MOUNTAINS, which reach into southern British Columbia and include Mount GARIBALDI, a volcano north of Vancouver.

brittle   A brittle response of a material to STRESS is to crack and shatter like glass. Brittle rocks break into unconnected pieces that are internally unchanged by the stress. For them to break, stress must increase until it exceeds the strength of the rock. The rock snaps, releasing sound and seismic waves. The snapping is an earthquake. Therefore, all shallow FAULTS are brittle. In contrast, ductile DEFORMATION involves the slow and steady shape change as would occur in squeezing gum.

brittle-ductile boundary   In the shallow CRUST, rocks deform in a BRITTLE manner. They crack upon failure and produce earthquakes. Deeper in the crust, rocks behave in a DUCTILE manner, distorting but not breaking. These rocks typically do not produce earthquakes. There is a diffuse boundary between rocks exhibiting each of these mechanical behaviors that depends upon various factors. Typically, this boundary lies at a certain depth below the surface as a function of increasing temperatures and pressures. For the dominant quartz-feldspar-rich rock, the transition occurs between six and nine miles (10 and 15 km) of depth. Under most crustal conditions, earthquakes are generated at depths of less than six to nine miles (10–15 km). Rock type is the primary determining factor. Clay is ductile at the surface of the Earth, whereas ULTRAMAFIC and MAFIC igneous rocks can be brittle through the entire thickness of the crust. MANTLE rocks are ultramafic, so they can be brittle to much greater depths. Earthquakes can be generated at depths of 240 miles (400 km) or more in a SUBDUCTION ZONE because the rocks are driven to great depths before they can heat up. Normally, the transition from the LITHOSPHERE to the ASTHENOSPHERE marks the brittle-ductile transition in the mantle.

Bucharest *earthquake, Romania*   A major earthquake struck Bucharest, the capital city of Romania, at 9:20 P.M. on March 4, 1977. The earthquake had a RICHTER magnitude of 7.4

and was felt throughout the Balkans. Earthquakes in this area, termed the *Vrancea Zone*, are typically from deep foci. The RECURRENCE INTERVALS for seismic events are 30 years for MAGNITUDE 7, and 120 years for magnitude 7.5. This earthquake caused between 1,570 and 2,000 deaths. Damage was extensive. Some 30,000 apartments, 32 eight-to-12-story buildings and 150 older six-to-eight-story buildings were destroyed, leaving more than 35,000 families homeless. The total damage was estimated at $2 billion.

Bulusan *volcano, Philippine Islands*   Also known as Irosin, the STRATOVOLCANO Bulusan is located in the southern part of Luzon Island. Recorded explosive eruptions date back to 1852 and have occurred every few years since then. It erupted at least 13 times between 1886 and 1988. Seismic records from 1978 to 1983 indicate that earthquake activity increased before eruptions, and temperatures at hot springs in the vicinity during this period have shown marked fluctuations. EARTHQUAKE SWARMS, however, have not always been followed by eruptions. Earthquakes and hot-spring activity did not provide any warning of a small eruption of ASH in 1988, although an earthquake swarm was noticed some hours following the eruption.

butte   A geomorphic feature with steep sides and a flat top. Buttes must be higher than they are wide. Otherwise, they would be classified as a mesa or a plateau. Old weathered-away volcanic necks like DEVILS TOWER form buttes.

Buyin-Zahra *earthquake, Iran*   The earthquake-ravaged country of IRAN was once again struck by a deadly temblor in the Buyin-Zahra region in the northwestern part of the country on September 1, 1962. The quake registered a 7.3 on the RICHTER scale, and the FOCUS was determined to be at 12 miles (20 km) of depth. Damage to buildings occurred in Tehran, more than 100 miles (160 km) from the EPICENTER.

Some 12,220 people lost their lives in this disaster, and more than 30,000 people were made homeless. More than 21,000 houses were destroyed, and heavy damage extended 60 miles (100 km) from the epicenter. Total property damage was estimated at US$30 million in 1962 money, which made it the most expensive disaster in Iran up to that point. It shut down the Trans-Iranian railway for several days, stalling the relief effort. The UNITED STATES provided significant amounts of aid that was distributed by the Iranian Red Cross.

b-value   The b-value for an area is a calculated factor that describes the ratio of small to large earthquakes within a given time period. The b-value is commonly constant over a large range of earthquake magnitudes. Mathematically, it is the slope of the curve in the Guttenberg-Richter earthquake recurrence relationship.

C

Cairo *earthquake, Egypt* Approximately two-thirds of the great city of Cairo, EGYPT, was reportedly destroyed by a devastating earthquake on September 10, 1754. The estimated RICHTER magnitude for this event was only 5.4, but the EPICENTER was right under the city and the FOCUS was at a shallow depth. The damage rated X on the modified MERCALLI scale. The final DEATH TOLL was estimated at 40,000, making it one of the most destructive in Egyptian history.

Calabria *earthquakes, Italy* The earthquakes that began in Calabria in southern ITALY in early February 1783 and continued for the rest of that year were felt over a wide area of the Italian Peninsula and Sicily. The earthquakes began with a powerful shock on February 5 and continued with another strong event on March 28. The earthquakes appear to have been centered near the town of Oppido and to have caused extensive destruction for several miles around. Scarcely any structure in Oppido was said to remain standing after the first earthquake. Large chasms opened in the ground and reportedly swallowed numerous houses. One chasm opened in these disturbances measured some 500 feet (152 m) long and 200 feet (61 m) deep. LANDSLIDES occurred along the shores of the Straits of Messina and allegedly destroyed numerous villas.

Following the February 5 earthquake, residents of the area recognized the danger of remaining on the coast and evacuated the shoreline, either moving slightly inland to elevated ground or putting out to sea on fishing boats. These safety measures proved ineffective. The night after the evacuation, a strong earthquake dislodged a large mass of earth from Mount Jaci; part of the falling material landed in the sea and generated a disturbance that sent a wave rolling over the ground on which the refugees had gathered. The wave is also said to have caused the destruction of the fleet of boats. More than 1,400 people were reported killed. Much of the city of Messina was destroyed in the earthquakes by the shocks themselves and by subsequent fires.

The earthquakes in February and March are believed to have killed some 40,000 people, not counting an estimated 20,000 who died later from exposure, disease, and other causes. Reportedly, thermal springs burst forth from FISSURES in the ground. One spectacular account says that some individuals fell into these fissures and then were cast out again in outbursts of boiling water; although survivors of these ordeals were reportedly rare, some did live through them but were said to have been crippled by burns for the remainder of their lives.

In another great earthquake at Calabria on December 16, 1857, whole villages reportedly fell into huge fissures in the earth, and more than 10,000 people are thought to have been killed. According to one estimate, some 111,000 people in the vicinity of Calabria were killed in earthquakes between the years 1783 and 1857—an average of 1,500 deaths per year.

Calaveras Fault *California* One large SPLAY fault off the main SAN ANDREAS FAULT. It is an active fault that produces major earthquakes.

Calbuco *volcano, Chile* The STRATOVOLCANO Calbuco has erupted several times since 1837. A 1929 eruption buried forests in the vicinity under thick layers of ASH.

calc-alkaline series A chemical index for volcanic rocks that are produced in a SUBDUCTION ZONE. The series usually begins with BASALT of a composition distinct from that produced in the MID-OCEAN RIDGE or in a HOT SPOT. Later volcanic rocks are ANDESITE and possibly DACITE. This succession contrasts sharply with that produced in a mid-ocean ridge, which is referred to as a THOLEIITIC series.

Calcutta *earthquake?, India* The 1737 Calcutta earthquake was estimated to have caused some 300,000 deaths, placing it among the top 3 or 4 earthquakes of all time. Recent analyses of the event, however, not only cast grave doubt on the number of casualties but also whether there was an earthquake at all. The problem is that on the day of the supposed earthquake, October 11, 1737, a major cyclone struck the Calcutta area. It was such an intense storm that it scuttled and sank eight of the nine British ships moored in the Ganges River and virtually all of the boats operated by the natives. Certainly, many deaths were attributed to this storm. The British East India Company, however, did not compile an official report of the disaster until the survivors reached headquarters in London some six months later. Their official report listed 3,000 fatalities, with no mention of an earthquake at all. In the following year, *Gentlemen's Magazine* and *London Magazine* featured stories that claimed 300,000 deaths as the result of an earthquake. Thus began the confusion about the event. Historical records indicate that the urban population of Calcutta in 1737 was probably less than 20,000. There were other settlements outside of Calcutta, but it is doubtful that there were 300,000 in the whole area, much less that many killed. Although there were numerous small to moderate earthquakes in the Calcutta area, they were not the huge temblors that struck the plate boundary area to the north. It is unlikely, but not impossible, that such a huge earthquake could even strike Calcutta. It is most likely that the 1737 Calcutta earthquake is the earthquake that never happened.

caldera   A large, roughly circular depression commonly formed by the explosion or collapse of a volcanic cone at any elevation. A caldera may be miles in diameter and harbor lakes containing volcanic islands. A famous example of a caldera is CRATER LAKE in OREGON. A caldera is much wider than the volcanic VENT(s) involved in its formation. A caldera complex consists of the rock assemblage underlying a caldera and may involve a wide variety of constituents, including, but not limited to, TUFF, BRECCIA, DIKES, STOCKS, and SILLS. A "resurgent" caldera is one in which volcanic activity resumes after the caldera has formed. Crater Lake again is a familiar example of a resurgent caldera. Others include Creede Caldera in COLORADO and VALLES caldera in NEW MEXICO.

The formation of a caldera is a spectacular event. A tremendous explosion or series of explosions may be involved, as in the case of the destruction of KRAKATOA, where explosions demolished much of the island and left behind a fresh caldera. Such an event may affect areas hundreds or even

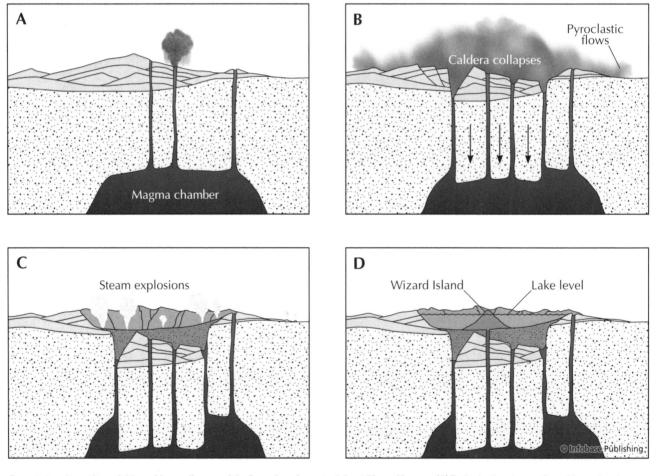

Four-stage schematic model for caldera collapse and the formation of a crater lake at Mount Mazama. (A) Typical volcanic eruption with overlapping stratovolcanoes. (B) Collapse of the caldera by gravity as magma is drained from the magma chamber. This results in extensive pyroclastic flows. (C) Post-eruption fumarole activity. (D) A small cinder cone is built as water fills the collapsed caldera.

thousands of miles away, through mechanisms such as ASH-FALLS and TSUNAMIS. The opposite process, collapse, rather than explosion, also may be responsible for forming calderas. Collapse may occur when MAGMA underneath a volcanic peak drains away through underground conduits or perhaps through a parasitic vent on the flank of the volcano; as the molten rock below it is removed, the peak loses its underpinnings, so to speak, and falls inward upon itself. This process appears to have created Crater Lake in Oregon, one of the most accessible and scenic calderas in the world. Calderas are not always so easy to identify as those at Krakatoa and Crater Lake; subsequent volcanic activity and/or erosion may cover or erase many traces of a caldera. Over many years, several calderas may form at a given location and form a complex pattern of overlapping, roughly circular features.

Activity at calderas may take many forms. Ground DEFORMATION may occur. In some cases, ground deformation is so slight that it can be measured only by using special, sensitive instruments. In other cases, ground deformation may be dramatic, measured in inches per day. In rare cases, ground deformation has lifted structures at harbors many feet out of the water. Giant SHIELD VOLCANOes, such as those in HAWAII, exhibit a special pattern of deformation. The entire summit inflates until lava emanates from the volcano. Sometimes uplift at calderas can elevate the surface more than 1,000 feet (305 m), as happened during an eruption at TOYA caldera in JAPAN. A caldera floor may also drop suddenly by hundreds of feet under the appropriate conditions. Perhaps the most famous case in point occurred at FERNANDINA caldera in the GALÁPAGOS ISLANDS in 1968, when a portion of caldera floor subsided up to 1,000 feet (305 m) or more in little more than a week. As a rule, however, uplift and subsidence at calderas tend to be more gradual.

Temperature changes are commonly seen when activity occurs at calderas. FUMAROLES, lakes, hot springs, and the SOIL itself may show marked increases in temperature. The water level in crater lakes may change, as may the colors of lakes. Chemical changes in groundwater, hot springs, and fumaroles also may be observed. Another phenomenon observed in and near calderas in connection with eruptions is change in Earth's magnetic field. Little is understood about these changes, although they are thought to be linked to tectonic forces or to demagnetization of COUNTRY ROCK. Changes in gravity have been noted at calderas and have been attributed to uplift or subsidence, changes in density of rock beneath the surface, and movement of magma or groundwater.

Unrest at a caldera may be brief or prolonged, depending on circumstances at the site. Some calderas have been active almost continually in the recent past, whereas others have been quiet for many centuries. There is also a great range of intensity in caldera unrest. Activity at some calderas has been extremely violent, whereas other calderas have a history of only mild activity. The most evident unrest may be confined to the area of an active vent, although it is commonplace for unrest at a caldera to involve the entire caldera in one way or another; deformation and other signs of unrest may occur all through a caldera during an episode of unrest.

Among the more exotic factors that appear to influence activity at calderas are *ocean loading* and *unloading*, mean-

A large caldera on Kilauea, Hawaii. The white steam comes from boiling groundwater. *(Courtesy of the USGS)*

ing the increase and decrease respectively of the weight from water piled atop the area of a caldera. Increased ocean loading is thought to contribute to eruptions in some situations by exerting pressure on a magma reservoir underground and forcing magma toward the surface. This process is believed to operate at PAVLOF volcano in Alaska from time to time. On the other hand, ocean unloading also is suspected of creating conditions favorable to eruptions; a case in point is RABAUL, where an eruption in 1983 and 1984 is believed to have been connected with ocean unloading.

As a rule, earthquakes in or near calderas are small, although there have been notable exceptions to this pattern, such as the powerful earthquakes, greater than magnitude 7 on the RICHTER scale, that have occurred in the vicinity of KILAUEA in Hawaii on occasion. Earthquakes in a caldera setting have various causes. Some are TECTONIC in origin, meaning they result from the readjustment of shifting blocks of Earth's crust rather than directly from volcanic processes. (A very strong earthquake in the vicinity of a caldera stands a good chance of being tectonic rather than volcanic in origin.) Some earthquakes at calderas are associated with intrusion of molten rock underground, whereas others may result from processes including tectonic activity and the draining away of molten rock from beneath the caldera. *Volcanic tremor* is the expression for earthquakes that are thought to result from flow of magma and/or gases through narrow space underground or other mechanisms such as the collapse of bubbles in magma. There is no single pattern of seismic activity associated with eruptions at calderas; although, one pattern often observed is for an eruption to occur as earthquakes reach a maximum of energy release after a quick increase. Another pattern sometimes seen is for an eruption to begin in the lull after seismic energy has increased and then diminished sharply. There appears to be no connection between the intensity of earthquake activity before an eruption at a caldera and the power of the eruption itself. Eruptions at calderas may be classified as normal explosions, involving magma and gases released from it, and phreatic explosions, involving groundwater turned into vapor. Phreato-magmatic explosions, as they are sometimes called, are combinations of the previous two varieties. Eruptive phenomena may include ashfalls, LAVA FLOWS, LAVA LAKES, domes, crypto-domes, LANDSLIDES, cinder cones, *NUÉE ARDENTES*, AVALANCHES, and SUBMARINE and subglacial eruptions.

caldera complex A group of multiple calderas on a single or even small group of volcanoes. Caldera complexes are common on SHIELD VOLCANOES. There can be several calderas that are concurrently active, or they can be intermittently active. Typically, one of the calderas will be dominant both in size and activity. In addition, new calderas may develop while others may become inactive. A good example of a caldera complex is KILAUEA on HAWAII. Several calderas on the same shield volcano have been active through recorded history. Each is named.

California *United States* California is known as the earthquake state because it has been the site of some of America's most powerful and destructive earthquakes, including the famous SAN FRANCISCO earthquake of 1906. The great SAN ANDREAS FAULT lies along the western shore of California, but it is not the only active FAULT in the state. Numerous faults, inactive and active, are found in California, and no part of the state is completely free from seismic disturbances. Well-known faults in California include the HAYWARD FAULT, CALAVERAS FAULT Zone, Imperial Fault, Manix Fault, and Garlock Fault. California is so active from a seismic standpoint that a complete list of all earthquakes in its history is impossible in the space available here. What follows is a partial list of major California earthquakes during the past two centuries:

EARTHQUAKE HISTORY OF CALIFORNIA

Early Earthquakes

An earthquake on July 28, 1769, in the region of LOS ANGELES involved four very strong shocks followed by many others over the next few days and possibly into the following year. In October 1800, an earthquake at San Juan Bautista left every dwelling there uninhabitable, and carts outdoors were used for sleeping; the date of the principal shock is not known, but it reportedly was accompanied by an extremely loud noise. An earthquake in June 1838 in the area of San Francisco is thought to have been comparable in power to the catastrophe that destroyed the city in 1906; the earthquake generated effects of MERCALLI intensity X, and large displacements were reported along the San Andreas Fault. The shock was especially strong at Yerba Buena, and heavy damage was reported at the missions in San Francisco, San Jose, and Santa Clara, as well as at the Presidio in San Francisco.

Hayward, 1868

The October 21, 1868, earthquake at Hayward caused maximum damage there and at other locations along the Hayward Fault. Virtually every building in Hayward experienced heavy damage, and some were ruined completely. Civic buildings were destroyed in San Leandro. San Francisco again experienced damage to buildings on landfill, and property damage appears to have exceeded $350,000. Some 30 persons are thought to have been killed in this earthquake, which was followed by AFTERSHOCKS continuing into the following month. The great San Francisco earthquake of April 18, 1906, is covered in detail elsewhere in this volume.

Santa Barbara, 1925

The Santa Barbara earthquake of June 29, 1925, caused an estimated $8 million in damage in Santa Barbara, where most buildings on the principal street in the commercial district were damaged. Thirteen people were killed in the earthquake, and movement of approximately one foot was seen along a boulevard built on a beach. An earth dam at a reservoir failed, but the freed water did little harm. Much of the exterior of a hotel collapsed. An office building failed, and the roof of a church collapsed. This earthquake served as a reminder that sound design and construction can preserve a

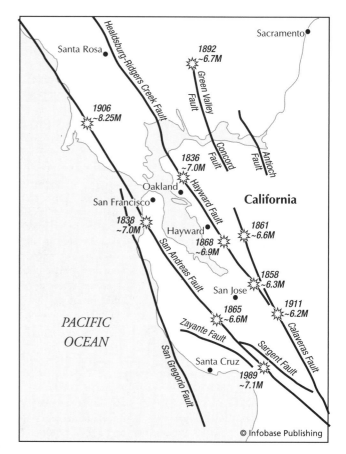

Fault and major recent earthquake map of northern California, San Francisco Bay area. The San Andreas Fault is a transform margin whose actions control the activity of the other faults in the area. They are either splays or compensation features. The star symbols are the locations of major earthquake epicenters with the year of the event and the Richter magnitude followed by an M.

building even in earthquake-prone territory and on less than stable SOIL. For example, one concrete building, well designed and given a good foundation, went through the earthquake with little or no damage, even though the building was constructed in a marshy area. Reinforced concrete proved resistant to the earthquake, except where construction was faulty. Strong aftershocks were reported on July 3. Another earthquake at Santa Barbara on June 29, 1926, killed a child and knocked down some chimneys.

Long Beach, 1933

The Long Beach earthquake of March 10, 1933, was not one of the most powerful in California's history (MAGNITUDE of only 6.3), but it caused extensive damage because it occurred in a heavily settled area. The EPICENTER was located slightly offshore from Huntington Beach, but the shock was felt from San Diego to Santa Barbara and as much as 50 miles (80 km) inland. More than 120 people were killed, and hundreds more were injured. Damage was estimated at approximately $50 million. Most damage occurred in an area characterized by water-rich soil and structurally unsound buildings. The most strongly affected area extended from southern Los Angeles to Manhattan Beach, Anaheim, and Laguna Beach. Destruction was especially notable in such places as Compton, Huntington Park, and Long Beach. Severe damage to up to 75% of school buildings occurred, some to the point of complete collapse. If school had been in session, it is estimated that there would have been several thousand deaths. No fault displacement was noted.

El Centro, 1940

One of the most famous and powerful earthquakes in California history was the Imperial Valley earthquake of May 18, 1940. The earthquake was estimated at magnitude 7.1, generated effects of Mercalli intensity X, and had its epicenter southeast of El Centro. Horizontal displacement up to 15 feet (5 m) and vertical displacement of as much as four feet (1.3 m) were reported. Damage was widespread in the Imperial Valley and was estimated at $6 million on the UNITED STATES side of the border alone, not counting indirect effects such as those on agricultural irrigation systems. Surface ruptures occurred over some 40 miles (70 km). About four-fifths of the buildings in Imperial were damaged, and the city's water tanks were ruined. Almost half the buildings in nearby Brawley were damaged, and a city water tank collapsed at Holtville. The earthquake also was blamed for a fire at a hotel in Mexicali, Mexico.

One of California's largest earthquakes was in KERN COUNTY, 1952. It and the SAN FERNANDO earthquake of February 9, 1971, are described in detail elsewhere in this volume, as is the 1989 LOMA PRIETA earthquake in the San Francisco area and the 1994 NORTH RIDGE earthquake near Los Angeles.

Tsunamis in California

California, having a long coastline, is subject to occasional TSUNAMIS, or seismic sea waves. These waves may be generated by earthquakes along the California coast or elsewhere; the 1964 GOOD FRIDAY EARTHQUAKE in ALASKA, for example, produced a tsunami that caused extensive damage in CRESCENT CITY, California, near the OREGON border. One of the most remarkable tsunamis in California history occurred on December 17, 1896, when a wave that may have been seismic in origin struck Santa Barbara and carried away a large portion of a boulevard that had been reinforced especially to resist wave action. A large hill of sand between the boulevard and the usual high-tide mark was also carried away by the wave. There is a possibility, however, that this wave was produced by a storm rather than by seismic activity because there appears to have been no strong earthquake in the vicinity on this date.

Because of the earthquake potential of southern California, the "tsunamigenic," or tsunami-making, potential of faults there has received considerable attention. Generally speaking, faults along the southern California coast do not appear to pose a great tsunamigenic threat, but faults thought to be capable of producing tsunamis are found in some locations off the coast of southern California, notably in the vicinity of Point Arguello. In and near the Santa Lucia Bank, an elevated area on the continental slope near the California shore, faults are suspected of being able to produce powerful earthquakes and the tsunamis associated with them. Another offshore area with suspected tsunamigenic potential is located in the area of Santa Barbara Channel and the nearby Transverse Ranges. Powerful, destructive earthquakes have occurred in this area (the December 21, 1812, earthquake that caused such heavy damage to missions onshore, for example, is thought to have involved fault rupture under the Santa Barbara Channel), and future earthquakes of high magnitude might result in tsunamis of considerable destructive power. Tsunamis produced by submarine slides are another possibility along the southern California coast, although this particular hazard is thought to be small. The California coast remains vulnerable, however, to tsunamis involved with seismic disturbances elsewhere around the Pacific Basin, particularly in Alaska, where certain areas contain the potential for very powerful earthquakes that might produce large tsunamis.

California also has a history of recent volcanic activity, some of it during the 20th century. LASSEN PEAK, a volcano in northern California, erupted with great violence during World War I. Volcanic formations may be seen also at LONG VALLEY caldera. GEOTHERMAL ENERGY plays a small but interesting part in providing electric power for California.

Cameroon, Mount *Cameroon* A STRATOVOLCANO located 180 miles west of the capital city of Yaoundé. It is also known as Mount Faka and "Chariot of the Gods." It generally produces FLANK ERUPTIONS with LAVA FLOWS from small cinder cones. One eruption in the fifth century B.C. and observed by a Carthaginian ship is one of the earliest recorded volcanic eruptions in the world. Mount Cameroon is very active, having produced six eruptions in the 20th century alone. Its most recent activity lasted from March until June 1999. Thousands of people were evacuated, and hundreds of homes were destroyed during this eruption.

Campania *earthquake, Italy* On November 23, 1980, an earthquake of MAGNITUDE 6.8 struck southern ITALY. It killed as many as 4,575 people and injured 7,750. The severe damage to buildings, transportation, and services devastated 9,000 square miles (25,000 km²) and left more than 250,000 homeless.

Campi Phlegraei *See* PHLEGRAEAN FIELDS.

Campo Bianco *volcano, Lipari Islands* Campo Bianco (White Field) is a VOLCANIC CONE, also known as Monte Pelato, on the north end of the island of Lipari, just north of the island of Vulcano. The cone was named for the remarkable whiteness of the PUMICE deposits there. Prior to World War II, these deposits were an important source of pumice for industrial and other applications, but the unreliability of ocean transport during the war forced users to develop other sources for pumice, and the economic importance of Campo Bianco diminished. The CRATER of Campo Bianco has steep sides several hundred feet high and has been breached on the east, from which side a broad stream of lava extends down to the ocean. The LAVA FLOW is called Rocche Rosse ("Red Rocks") because oxidation has given it a red coating of iron oxide. A submarine cable running southeast from the community of Lipari, on the eastern side of the island, was broken on several occasions by eruptions associated with Vulcano between September 1889 and December 1892.

Canada Though not commonly associated in the public mind with earthquakes and volcanoes, Canada has both. The volcanic CASCADE MOUNTAINS extend northward from the UNITED STATES into Canada and include Mount GARIBALDI, and both eastern and western Canada have zones of pronounced seismic activity. BRITISH COLUMBIA has its share of earthquakes as a result of NORTH AMERICA's collision with the JUAN DE FUCA CRUSTAL PLATE along the CASCADIA SUBDUCTION ZONE off Canada's Pacific shore. Eastern Canada also is known for earthquakes, especially along the SAINT LAWRENCE VALLEY, which separates Canada from the northeast United States. Seismic activity is so commonplace in this region that it has become a leading natural "laboratory" for studying earthquakes and techniques for predicting them.

Cañadas, Las *caldera, Canary Islands* The Las Cañadas CALDERA is located on the island of Tenerife and is the site of the Pico Viejo and Pico de Tiede volcanoes. An eruption is believed to have occurred in 1492, and there is evidence of earlier eruptions around 1341, the 1390s, and 1444. Eruptive activity also occurred between 1704 and 1706. More than a year of seismic activity preceded an eruption in 1909. It lasted 10 days and produced LAVA FLOWS.

Canary Islands The volcanic Canary Islands are located on the CONTINENTAL SHELF of western AFRICA. They originate from a HOT SPOT that is being overridden by the ATLANTIC OCEAN part of the African crustal plate. Volcanoes active in recent times in this group include Chahgorra, on Tenerife, which erupted in 1798 for more than three months and emitted ashes, lava, and rocks. Some of the rocks ejected in this eruption are thought to have reached an altitude of more than a half mile before falling back to Earth. Other eruptions

on Tenerife in 1704 and 1706 are said to have destroyed harbor facilities on the island. On the island of La Palma is a great CALDERA approximately a mile (1.6 km) deep. It has had seven historic eruptions from several vents including Teneguia, Tahuya, and San Martín. Teneguia was the most recently active with an eruption in 1971. Tahuya is a cone that was formed during a three-month interval in 1585 similar to San Martín, which formed in two and one-half months in 1646.

The island of Lanzarote is a large SHIELD VOLCANO that was the site of a great series of eruptions that began on September 1, 1730. They produced a large hill of volcanic material overnight and proceeded to emit lava that overran several villages. On September 7, a very large, solid rock was thrust up with a loud noise from the bottom of a stream of lava. The rock changed the course of the lava so that the flow ran through the town of St. Catalina and several villages. When the lava reached the ocean four days later, fish were killed in large numbers. Later, three new CRATERS developed on the site of St. Catalina itself and cast out large volumes of ASH, sand, and stones. All the cattle on the island were killed on October 28 by vapors from the volcano that reportedly condensed from the air and fell as a lethal rain. A fierce storm also beset the island at this time. On January 10, 1731, an eruption produced a large hill of volcanic material that reportedly collapsed on that same day into the crater from which it arose and that was followed by several streams of lava, which then flowed to the sea. Between January and March, several new cones appeared and emitted lava. Altogether, approximately 30 new cones grew on the island. Another fish kill occurred in June, and there was apparently a SUBMARINE ERUPTION because smoke and fire were reported rising from the waters near the shore. The eruptions in this series continued through 1736 and required the evacuation of many residents to safer islands. A three-month eruption on the island of Lanzarote in 1824 was preceded by very strong earthquakes and appears to have resembled the eruption that produced Monte Nuovo near Naples.

Hierro is a large shield volcano that has had one historic eruption in 1793 from Volcán de Lomo Negro. The eruption originated from three FAULT zones and lasted about one month.

capable fault As defined for use in selecting sites for nuclear power facilities in the UNITED STATES, *capable fault* means a fault that is capable of movement in the "near" future. This, in turn, signifies a fault along which movement has occurred within a few thousand years. It is basically an ACTIVE FAULT but not necessarily in historic times.

Cape Ann *Massachusetts, United States* A major earthquake that damaged buildings over a wide area of eastern Massachusetts in 1755 is believed to have been centered off Cape Ann. Ground motions were reportedly so strong that they knocked standing people off their feet.

Cape Verde Islands The volcanic Cape Verde Islands are located west of AFRICA and were discovered by the Portuguese around the end of the 15th century. These volcanic islands were formed by a HOT SPOT under the east ATLANTIC OCEAN. Islands include São Nicolau, Boa Vista, Santiago, and Maio and the active volcanoes of Santo Antão, São Vicente, Brava, and FOGO. The volcano Fogo was in constant eruption from its discovery until 1760, making it a lighthouse for Atlantic sailors. It is noted for an eruption in 1847 in which no fewer than seven openings developed and sent out streams of lava. Its most recent eruption was in 1995, and it was quite intense.

Cap-Haitien *earthquake, Haiti* There was a terrible earthquake that struck the entire island of Hispaniola at 9–10 P.M. on May 7, 1842. The source of the earthquake is suspected to be the TRANSFORM margin that extends from PUERTO RICO and crosses Hispaniola as the Septentriovial Fault. This fault crosses the southern peninsula just to the south of Port-au-Prince. In addition to the shaking, a TSUNAMI was produced and struck especially hard at Port-de-Paix, where 200 people were killed by the wave. The ocean receded some 200 feet (60 m) and then roared back into the city with waves up to 16.5 feet (5 m). AFTERSHOCKS pounded the area for several days after the disaster. In all, at least 5,000 people lost their lives, but some estimates put the DEATH TOLL as high as 10,000.

capillarity A set of conditions in which a liquid, water being the most familiar example, is drawn up by surface tension into small tubes or interstices in rock. The drawing-up process is known as capillary action. The capillary fringe in SOIL represents a boundary zone between the water table and the dryer "zone of aeration," or "vadose zone," above it in which air as well as water fills spaces between soil particles. Water in the capillary fringe is known as capillary water. Capillary migration (also known as capillary flow and capillary movement) refers to water movement through capillaries. This concept can be applied to the movement of MAGMA through fractured rocks and also in upward movement of groundwater during LIQUEFACTION.

Caracas *earthquake, Venezuela* The March 26, 1812, earthquake in Caracas destroyed about 90% of the city and is thought to have killed some 10,000 people in the city itself, plus another 5,000 or so in the surrounding area. The number of casualties rose to approximately 20,000 by the middle of April, partly because of disease spread by contaminated water. The earthquake apparently was most destructive in the northern portion of Caracas, where two churches collapsed on worshipers (the earthquake occurred on Holy Thursday).

Caribbean Sea The Caribbean islands have been a hotbed of volcanic activity, notably the highly destructive eruptions of Mount PELÉE and SOUFRIÈRE in 1902. Volcanic islands make up the Lesser Antilles, a chain of islands that stretches some 500 miles (840 km) across the Caribbean Sea. The Lesser Antilles have two separate branches at their northern end: the Volcanic Caribbees, including active volcanoes, and just to their east, the Limestone Caribbees, made up of older volcanic lands crowned with limestone reefs.

These islands form an ISLAND ARC above a SUBDUCTION ZONE in which the ATLANTIC OCEAN crust of the NORTH AMERICAN CRUSTAL PLATE is being driven beneath the Caribbean Plate. To the north, the boundary between the two plates is a TRANSFORM margin that extends onto land as the MOTAGUA FAULT in GUATEMALA. In addition to Pelée and Soufrière (of which there are two), there are many other volcanoes that can become active at any time. Such is the case with SOUFRIÈRE HILLS, which destroyed the island resort of Montserrat in 1995–97. Other volcanoes include Saba, Quill, Liamuiga, Nevis Peak, Diable, Diablotin Morne, Micotrin, Patates, Qualibou, Kick-'em Jenny, and Saint Catherine. Earthquake activity has also been prominent in the Caribbean area. Moderate to deep earthquakes are generated along the BENIOFF ZONE in the subducting plate. Shallower earthquakes along both the transform margin to the north and within the islands are also common.

See also PORT ROYAL.

Cartago *earthquake, Costa Rica* The nation of COSTA RICA is extremely vulnerable to earthquakes because of its position in seismically active CENTRAL AMERICA. One of the most destructive earthquakes in the history of Costa Rica occurred on August 27, 1841, killing some 4,000 people out of a population of fewer than 20,000. Another earthquake on May 4, 1910, killed more than 200 people and caused extensive damage.

Cascade Mountains *United States and Canada* The Cascades are the location of some of the most beautiful volcanic mountains in the world, as well as the recently active volcano Mount SAINT HELENS. Volcanoes in the Cascades include Mount GARIBALDI, THREE SISTERS, Mount RAINIER, CRATER LAKE, MEDICINE LAKE volcano, Mount BAKER, Mount SHASTA, Mount HOOD, and LASSEN PEAK among others. These volcanoes were produced in some spectacular eruptions. With the rapid population growth in the area during the past century or so, these volcanoes become more dangerous by the day. About 100 to 150 miles (161 to 241 km) east of the PACIFIC OCEAN, the Cascades form a mountain chain from BRITISH COLUMBIA through the states of OREGON and WASHINGTON and into northern CALIFORNIA. Three rivers flow from east to west through the Cascades: the Fraser, Klamath, and Columbia.

Cascadia *earthquake and tsunami, Canada* On January 26, 1700, between 9 and 10 P.M., a huge earthquake of approximate RICHTER magnitude 9.0 resulted from massive movement on a MEGATHRUST in the CASCADIA SUBDUCTION ZONE off the western coast of CANADA (similar to the BANDA ACEH earthquake of 2004.) The reason geophysicists know the precise timing of this event is that the earthquake produced a TSUNAMI that crossed the PACIFIC OCEAN and struck JAPAN, where scientists made records. The tsunami struck the eastern coast of Japan on January 27 at 10 A.M. at a height of 6.5 to 10 feet (2 to 3 m) and continued pummeling the coast into the next day. This tsunami produced minor to moderate damage in several Cascadian villages. Oral traditions of the native peoples of Vancouver contained mention of a tsunami that destroyed a village at Pachena Bay and the destruction of houses by shaking in the Cowichan Lake area. Other evidence can be found in tree rings from the Pacific northwest, in tsunami scars and sand deposits in coastal WASHINGTON, and in TURBIDITE deposits of the appropriate age recovered from drill cores. It has been 300 years since the last magnitude 9 earthquake from Cascadia. The recurrence interval for such events is approximately 500 years. Odds are that the area is safe for the time being, but, at some point, there could be a disaster of catastrophic proportions along the northwest coast.

Cascadia Subduction Zone *Pacific Ocean* This structure is located off the Pacific coast of the northwest UNITED STATES and BRITISH COLUMBIA, CANADA, where the NORTH AMERICAN CRUSTAL PLATE overrides the JUAN DE FUCA

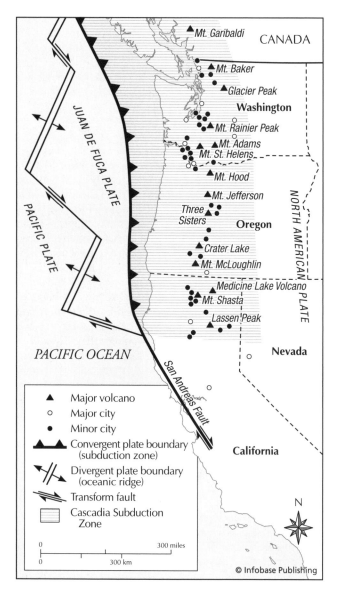

The plate tectonic relations in the Cascadia Subduction Zone and the volcanoes of the Cascade Range in the northwestern United States

CRUSTAL PLATE immediately to its west. The Cascadia subduction zone forms a MAGMATIC ARC in which magmatism and volcanism that produced the volcanoes of the CASCADE MOUNTAINS, including Mount HOOD, Mount BAKER, Mount RAINIER, and Mount SAINT HELENS among others is generated. The subduction zone is as active as ever, which means that all of these volcanoes have the potential to and are expected to spring back to life as Mount Saint Helens did in 1980. There is also a BENIOFF ZONE in the subduction zone defined by earthquake foci. Therefore there is a seismic risk in this area.

cataclasite Shallow fault-generated rock. At depths of four to about nine miles (6–15 km), faults produce earthquakes but leave no open spaces in the fault rocks. At shallower levels blocks of rock can be broken and left in the fault with large voids around them. The deeper FAULT rock is made by grinding up the rocks around the fault into microscopic particles and cementing them together. It is like grinding coffee. The resulting rock is very fine-grained, fragmental, and typically green from the cementing minerals.

Catania *lava flow, Sicily, Italy* This community is known for its early attempt to divert a flow of LAVA. During an eruption of Mount ETNA in 1669, a party of men covered themselves with wet hides for protection against the heat and made an attempt to divert the LAVA FLOW that was approaching Catania. At first, this tactic appeared successful. The new flow of lava moved toward the town of Paterno, several hundred of whose residents chased the men of Catania away from their lava-diversion project. The breach in the wall of the lava flow was short-lived; solidified lava filled the opening, and the lava flow continued toward Catania as before. The lava overran and destroyed a large portion of Catania, and large quantities of the same lava flow are still visible in the city.

catena A chain of CRATERS, of suspected volcanic origin, found on Mars.

cauldron A very large CALDERA relative to the other calderas in an area or relative to the type of volcano is called a cauldron. The processes within a cauldron are the same as those in calderas. However, in addition to their size, they tend to be active more frequently than the other calderas in an area. If they occur within a CALDERA COMPLEX, they tend to be the dominant caldera.

Cecchi seismoscope The Cecchi seismoscope, invented in Italy in 1875, is considered by many to be the first true SEISMOGRAPH. The instrument utilized two common pendulums to record horizontal movements—one pendulum vibrating east-west and the other north-south. Vertical movements were recorded by a mass suspended on a spiral spring, and rotary motions were recorded by a free-spinning dumbbell device. All motions were mechanically amplified three times to better represent the vibrations. The device had a clock that started in motion at the first vibration of an earthquake. A recording surface would slide under the pen at a rate of 0.4 inches (1 cm) per second for 20 seconds and then stop. The seismoscope was therefore of limited use and was obsolete by the mid-1880s.

Central America

Located along the "RING OF FIRE," the belt of intense earthquake and volcanic activity encircling the PACIFIC OCEAN basin, Central America is a narrow land bridge connecting North and SOUTH AMERICA and has numerous volcanoes as well as a history of highly destructive earthquakes. Central America is a MAGMATIC ARC in which a SUBDUCTION ZONE is formed by the COCOS PLATE driving beneath the Caribbean plate and part of the NORTH AMERICAN CRUSTAL PLATE. For shear power and destructiveness of the volcanoes and earthquakes, the Central American Arc is second only to the Sunda Arc of INDONESIA.

In GUATEMALA, for example, the earthquake of June 7, 1773, destroyed the city of Santiago and killed almost 60,000 people. Another in 1976 on the MOTAGUA FAULT killed 23,000 people. Volcanic events in Guatemala's history include the April 13, 1902, eruption of Tacona, which buried the city of Retalbulen and is thought to have caused approximately 1,000 deaths. Other famous volcanoes of intense activity include ATITLÁN, FUEGO, PACAYA, and SANTA MARÍA which had one of the largest eruptions of the 20th century.

San Salvador, capital of EL SALVADOR, has been struck repeatedly by major earthquakes, notably on May 3, 1965, when an earthquake of RICHTER magnitude 7.5 killed 100 people, injured several hundred, and left thousands homeless. An eruption of the volcano San Miguel in 1844 killed several hundred. The volcanoes of El Salvador include COATEPEQUE, ILOPANGO, IZALCO ("Lighthouse of the Pacific"), and SANTA ANA.

The city of Cartago in COSTA RICA was destroyed by an earthquake on May 4, 1910, with the loss of more than 200 lives. Volcanic activity in Costa Rica in recent years includes the July 30, 1968, eruption of ARENAL, with damage estimated at more than $40 million. A 1623 eruption of the volcano IRAZÚ killed hundreds and destroyed several

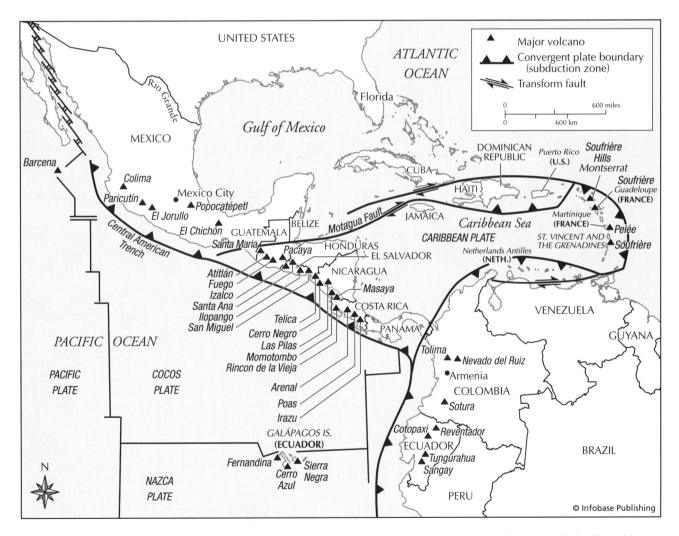

Map of the Central American and Caribbean arcs, including the lithospheric plates and plate boundaries. Lines are divergent boundaries, lines with triangles are subduction zones, and lines with arrows are transform margins. Also shown are many of the major volcanoes and a few cities that recently had major earthquakes.

villages. Other active volcanoes include POÁS and RINCÓN DE LA VIEJA.

NICARAGUA's history of earthquakes and volcanic eruptions is especially violent and tragic. The December 23, 1972, earthquake in MANAGUA measured 6.25 on the Richter scale of earthquake MAGNITUDE and killed some 11,000 people, leaving thousands more homeless. The January 1835 eruption of the volcano COSIGÜINA lasted three days and killed hundreds. An example of 20th-century volcanism in Nicaragua is the January 12, 1947, eruption of CERRO NEGRO, which destroyed almost 300 square miles (500 km²) of territory and caused dozens of deaths, as well as extensive damage to León City. The volcano MOMOTOMBO also destroyed León in another eruption. The seismic and volcanic history of Nicaragua was one of the factors that influenced the decision to construct an interocean canal in Panama, where geologic hazards are less extreme, rather than in adjacent Nicaragua. Other volcanoes of repute include CONCEPCIÓN, MASAYA, and SAN CRISTOBAL.

Although MEXICO is not part of Central America, it is part of the Central American Arc. The earthquakes and volcanoes of southern Mexico are the result of the subduction of the Cocos plate under the North American plate. The MEXICO CITY earthquake of 1985 killed more than 8,000 people. Volcanoes of the magmatic arc produced some famous eruptions including El CHICHÓN, which affected the world's climate in 1982. Other volcanoes include COLIMA, JORULLO, and POPOCATÉPETL.

Cerisy Peak   *See* LONG ISLAND.

Cerrito (**Cerrilito**)   *See* APOYO.

Cerro Azul *caldera, Galápagos Islands, Ecuador*   A small SHIELD VOLCANO on Isabela (Albemarle) Island, Cerro Azul has erupted in 1932, 1940, 1943, 1948, 1949 (uncertain), 1951, 1959, and 1979.

Cerro Negro *cinder cone, Nicaragua*   This basalt cinder cone, whose name means "Black Hill," is the most recent of its kind in the Western Hemisphere. It arose in 1850 and has erupted at least 20 times since then. The longest of these eruptions was in 1960 and lasted for three months. It has produced some spectacular STROMBOLIAN ERUPTIONS. It had a major eruption in 1995, which included LAVA FLOWS and the building of a LAVA DOME. As many as 12,000 people were evacuated. Cerro Negro again became active in 1998 just one week after a mudslide from nearby Casitas volcano killed 2,000 people along its flanks. Thousands of people were evacuated and hundreds of houses were damaged. FIRE FOUNTAINS 1,000 feet (305 m) high were reported. Cerro Negro has been intermittently active since then, including significant seismic activity.

Cerro Rico *volcano, Bolivia*   The remains of an ancient volcano, as are all volcanoes in BOLIVIA, laden with rich ores of silver and tin, Cerro Rico (Hill of Silver) yielded vast quantities of silver for the Spanish during their occupation of the former Inca Empire.

Aerial view of erupting Cerro Negro volcano, Nicaragua *(Courtesy of the USGS)*

Cha *caldera, Cape Verde Islands*   The Cha CALDERA is located atop the Fogo Island volcano. Although there is disagreement on the dates of eruptions in the historical record, it is said to have erupted almost continuously between 1500 and 1760. Sailors used it as a lighthouse because of its constant eruption. It erupted at least 10 times between 1500 and 1995. Most of the eruptions are from flank VENTS and of moderate size (VEI = 2). Only the 1847 eruption caused fatalities. An earthquake killed individuals in 1847, and powerful earthquakes and volcanic tremors occurred before an eruption in June 1951. Cha's most recent eruption was in 1995. It erupted explosively for nearly two months, emitting both LAVA and ASH, destroying 1.6 square miles (4 km²) of productive farmland, and causing the evacuation of 5,000 people.

Chang Heng seismoscope   The earliest-known seismoscope was invented by the Chinese philosopher Chang Heng (also known as Choko and Tokyo) in A.D. 132. This artifact is so famous that it appears in many college textbooks. The instrument had a six-foot (2-m)-wide wine jar at its center, with eight dragon heads at the upper part, each facing a principal direction of the compass. Beneath each dragon head was a toad on the floor facing upward with its mouth open. Each of the dragon mouths contained a ball that would drop into the mouth of the toad if vibrated in that direction. Some speculate that there was a pendulum suspended in the jar that accentuated the vibration. The Chang Heng seismoscope was said to have been able to detect earthquakes from as far away as 400 miles (640 km).

Chaos Craigs *volcanic domes, California, United States*   A collection of volcanic domes immediately to the north of Lassen Peak, Chaos Craigs, which includes four or more rhyodacite domes, formed approximately 1,000 to 1,200 years ago, according to one estimate based on radiocarbon dating of charcoal produced by PYROCLASTIC FLOWS. Each dome is about 1 mile (1.6 km) wide and some 1,800 feet (549 m) high. Large accumulations of debris surround the domes. Tremendous ROCKFALLS at one of the domes produced a pile of debris known as Chaos Jumbles. An explosion some 300 years ago is thought to have given rise to Chaos Jumbles.

Charleston *earthquake, South Carolina, United States*   The earthquake that struck Charleston on August 31, 1886, is believed to be the most destructive earthquake that has occurred on the Atlantic coast of the UNITED STATES in historical times. The shock was felt over an area of almost 3 million square miles (7.5 million km²), although damage was confined largely to the immediate vicinity of Charleston. Approximately 100 buildings were destroyed, and most brick structures were damaged. The earthquake apparently was most intense 12 miles (19 km) west of Charleston, where large amounts of sand and water spewed from FISSURES in the ground. One such outburst of water reportedly left behind a CRATER more than 20 feet (6 m) wide. The poorly consolidated SOIL under Charleston was responsible for much of the damage from the earthquake. Relatively minor earthquakes occurred earlier in the summer of 1886 but do not appear to have caused any great concern in Charleston. More notable shocks occurred on August 27 and 28, and several minutes before 10 P.M. on August 31, the major earthquake began, accompanied by a noise that later was compared to the sound of steam escaping from a boiler or of fast-moving street traffic at close range. The initial shock is believed to have lasted about 35 to 40 seconds, but recollections of the event do not agree on whether this shock occurred singly or consisted of several separate movements. The earthquake killed more than 20 people directly, and a greater number reportedly died later from exposure or from injuries received in the earthquake. Although some structures came through the earthquake with less damage than others and few buildings were shaken completely to the ground, cracks appeared in many stone structures, and safety required them to be demolished. Damage to railway tracks provided evidence of the ground motions: The rails showed both lateral and vertical displacement, and tracks were displaced toward the southeast. A study conducted soon after the earthquake concluded that the disturbance appeared to have two centers separated by approximately 13 miles (21 km). The main FOCUS of the

Collapsed buildings litter the street in Charleston, South Carolina, after the 1886 earthquake. It was the strongest historical earthquake on the east coast of the United States. *(Courtesy of the USGS)*

earthquake was estimated at about 10 to 14 miles (16 to 23 km) deep, and the secondary focus was figured to be about eight miles (13 km) deep. Reports indicate that some interesting events in distant parts of the country coincided with the earthquake, including a reduction in the yield of natural gas wells in Pennsylvania and the reactivation of a GEYSER in Wyoming's Yellowstone valley after four years of inactivity.

Chichón, El *volcano, Mexico*   The highly explosive 1982 eruption of El Chichón generated a controversy over the possible effects of that eruption on global climate and weather. Laser measurements of the cloud from El Chichón indicated that it was much more effective at blocking light than the cloud from Mount SAINT HELENS had been two years earlier. The eruption immediately preceded a disturbance in circulation patterns in the atmosphere and in the PACIFIC OCEAN. The change accompanied heavy rainfall in CALIFORNIA and drought in AUSTRALIA. Scientists suspected that the cloud from El Chichón might have played a role in the extraordinary weather of 1982–83, although such disturbances have been observed on other occasions without connection to volcanic events. Apparently the cloud from El Chichón did not cool global climate appreciably, but the cloud lingered in the stratosphere until 1985.

Chiginagak *stratovolcano, Alaska*   A symmetric andesitic-dacitic STRATOVOLCANO that is located 110 miles (177 km) south of King Salmon. There is an active FUMAROLE that has been venting steam since the early 1970s. Besides a documented 1971 eruption, there are a few reports of historic eruptions. There are, however, young PYROCLASTIC deposits on the flanks.

Chihli *earthquake, China*   A deadly earthquake struck the area of Chihli, CHINA, on September 27, A.D. 1290. This area is now known as Hebei Province, and it lies in the North China Plain in Inner Mongolia. It has always been one of the most populated areas of China. The earthquake was estimated to have had an approximate RICHTER magnitude of between 6.0 and 7.0. It is also estimated to have had a DEATH TOLL of approximately 100,000 people, making it one of the deadliest in history, but little was chronicled about the event, so the details of the event cannot be verified.

Chijiwa Bay *Japan*   Located near UNZEN volcano, Chijiwa Bay connects to the volcano by a GRABEN several miles wide, in an east to west trend. Volcanic and seismic activity have occurred in the Chijiwa Bay area frequently during the last several centuries. Earthquakes just before the 1792 eruption of Unzen began in late 1791 and were most noticeable along the west side of the Shimabara Peninsula, which extends south of the CALDERA. LANDSLIDES and sounds of detonations, compared to the noise of artillery, were reported. An especially powerful earthquake occurred on April 21, 1792, and was likened to the sound of naval gunfire. Several hundred earthquakes occurred on April 21 and 22. Thereafter, earthquake activity diminished temporarily, although a powerful earthquake took place on April 29. More than 60 houses and more than 280 stables were destroyed in a landslide in April near the

village of Nakakoba. This incident also reportedly damaged Shimabara castle. Strong earthquakes began again in late May and were accompanied by a great AVALANCHE that set off a destructive TSUNAMI. The earthquakes, avalanche, and tsunami together are thought to have killed some 14,000 people.

In 1922, a swarm of earthquakes killed more than 20 people. Other EARTHQUAKE SWARMS occurred between 1968 and 1974, and in 1984 a swarm of more than 6,000 events was recorded. Active FAULTS in the Chijiwa area include the Chijiwa Fault, to the northeast of the Chijiwa caldera, and the Fugendake, Futsu, and Kanahama Faults, all to the south of the Chijiwa Fault and east of the caldera.

Chile   Occupying a narrow band along the western shore of SOUTH AMERICA, Chile has experienced numerous powerful earthquakes and TSUNAMIS. Seismic activity in Chile is so frequent that the nation is the site of approximately 20% of the earthquakes that are recorded annually. Chile also contains numerous features associated with volcanic activity, such as copper mines and hot springs. Destruction from earthquakes in Chile has not been restricted to the country itself. In 1960, for example, an earthquake in Chile generated tsunamis that reportedly killed 60 people in HAWAII and 438 in JAPAN and the PHILIPPINE ISLANDS; this tsunami reached more than 1,500 feet (457 m) inland along the Chilean coast.

One of the most destructive earthquakes in Chilean history occurred on January 24, 1939. The earthquake lasted three minutes and shook a section of Chile more than 400 miles (644 km) long, killing perhaps 50,000 people, most of them children. Damage appears to have been most severe in the cities of CHILLÁN, Coihueco, and CONCEPCIÓN. One account says that only three buildings were left standing in Chillán out of more than 140 city blocks, and more than 300 persons died in Chillán at one location, a theater that collapsed on the audience. In Concepción, 70% of the buildings were destroyed, and coal mines collapsed on miners within them. More than a dozen cathedrals were demolished.

A series of earthquakes in central Chile in May 1960, which generated the aforementioned tsunami, affected more than 90,000 square miles (233,099 km²)and altered topography greatly in some locations. The ground surface of the earth in one place reportedly sank 1,000 feet (305 m) over an area of approximately 25 square miles (65 km²). Lakes vanished, new lakes formed, two small mountains were apparently leveled, and volcanic eruptions were reported in several places. An especially violent eruption occurred at Lake Ranco, where a LAVA FLOW entering the lake caused the lake to leave its banks. This earthquake occurred only three months after the catastrophic earthquake at AGADIR, Morocco, and the Chileans were able to benefit from the international relief effort already organized in response to that earthquake. Central Chile experienced still more destructive earthquakes on March 28, 1965. More than 400 people were killed, although damage was restricted to a small area. Much of the loss of life resulted from the collapse of a dam near the village of El Cobre. The resulting flood buried the village in mud to a depth of seven feet (2 m).

Chile also contains some active volcanoes. VILLARRICA is a powerful volcano that has produced four eruptions that resulted in fatalities mostly as the result of LAHARS. An earth-quake killed some 350 people in the village of Villarrica in 1575. Guallatiri has erupted four times, most recently in 1985. Ojos del Salado is the world's highest active volcano at 22,560 feet (6,900 m). Other volcanoes include Sollipulli, Tocorpuri, and La Torta.

Chilean 1960 earthquake On May 22, 1960, the largest recorded earthquake occurred off the coast of south-central CHILE in the Arauco Peninsula. It had a surface wave magnitude of 9.6 and a focal depth of 21 miles (33 km). The FOCUS was so deep that there were no areas of extreme intensity. Nonetheless, there were as many as 2,290 deaths and more than $500 million in damage as a result. One of every three houses in the nearby city of Valdivia was destroyed, and CONCEPCIÓN was also heavily damaged. There were some significant FORESHOCKS up to a day before the great earthquake. The MAIN SHOCK lasted for more than three minutes. AFTERSHOCKS continued to knock down damaged structures for weeks afterward and were still felt up to a year later. The maximum intensity of the earthquakes was felt along a north-south trend along the coast and also along the Reloncavi fault in the Chilean lakes region. FAULT PLANE SOLUTIONS indicate that the FAULTS largely underwent STRIKE-SLIP movement.

Damage to the landscape was impressive. Literally, thousands of sizeable LANDSLIDES littered the Chilean lakes region. Subsidence submerged thousands of acres of farmland and uplift elevated other areas up to seven feet (2 m). In the town of Valdivia, LIQUEFACTION resulted in the extrusion of such vast amounts of liquefied soil that the sheer weight depressed the land surface and caused houses to collapse. In another town the liquefied SOIL flowed into a harbor and engulfed an anchored ship. In some of the lakes, SEICHES with amplitudes up to three feet (9 m) were observed. The nearby once DORMANT Puyehue volcano sprung back to life a mere 48 hours after the quake. The ASH-steam cloud was shot 20,000 feet (6,000 m) into the air and similar eruptions continued for several weeks. Other volcanic activity was reported but they turned out to be MUD VOLCANOes with very high sand/mud blows. They just looked like real volcanoes.

The earthquake generated huge TSUNAMIS all over the PACIFIC OCEAN. The maximum wave heights are estimated at 80 feet (25 m), and debris was found inland as far as two miles (3 km). Coastal populations were evacuated early because they knew that earthquakes produced tsunamis. Unfortunately, the fourth waves were the highest and much higher than expected. People returning to their homes because they thought it was safe were killed, but so were many people evacuated to lower elevations. The wave crossed the Pacific Ocean with great energy at 400 miles per hour (690 km/hr). When it struck Hilo, HAWAII, 14 hours after the earthquake, the waves were still 34 ft (10.7 m). They killed 61 people, injured 282, and did $24 million in damage there. Again, it was the third wave that did the damage. Eight hours later, the waves reached JAPAN. In converging harbors, the wave maximum was 12 feet (4 m) and carried large fishing boats as much as 150 feet (46 m) inland. It killed 180 people there, did $450 million in property damage, and destroyed the homes and/or livelihoods of some 150,000 Japanese citizens.

Some grisly stories emerged from this Chilean earthquake and the tsunami associated with it, including a report of the murder of a six-year-old boy whose heart was torn from his body and offered as a sacrifice to the imagined gods of the sea by the Mapuche Indians. More than 100,000 Chileans were left homeless following this earthquake, and in some areas, the army had to disperse crowds rioting for food.

Chillán *earthquake, Chile* On January 25, 1939, at 3:30 A.M., a massive earthquake struck the region around the city of Chillán, CHILE, with devastating effects. The magnitude of the quake registered 8.3 on the RICHTER scale, with a duration of nearly three minutes. It produced damage of X on the modified MERCALLI scale. The area of damage was over 40,000 square miles (103,500 km²). Chile is above the Andean SUBDUCTION ZONE but is in an area of extensive uplift relative to its surroundings. As a result, it is likely the most seismically active area in SOUTH AMERICA.

The DEATH TOLL from this event is debatable. It was at least 28,000, but most sources claim that it was more like 50,000 people—with the majority children. In any event, it was one of the worst in South American history. Another 60,000 people are said to have been injured, and 700,000 were rendered homeless. In the 140 city blocks of Chillán, only three buildings were left standing. In nearby CONCEP-CIÓN, 70% of the buildings were destroyed. The cost of the tragedy was about $100 million, and Chile was forced to borrow money to rebuild. Chileans rebuilt using much better construction standards, and a middle class was established as a result. Chile actually improved its earthquake preparedness as the result of this tragedy and prevented subsequent earthquakes from causing as much damage and loss of life.

China China's history of destructive earthquakes is almost as long as its history. A partial list of seismic disasters in China includes a 1556 earthquake said to have taken more than 800,000 lives, another quake in 1920 that is believed to have killed almost 200,000, and in 1976, the great TANGSHAN earthquake that is supposed to have killed some 655,000. China's susceptibility to earthquakes has two sources. One is the northward movement of the Indian crustal plate, which collides with China in the vicinity of the Himalaya Mountains in the south. The other is the existence of the "RING OF FIRE," the circum-Pacific belt of intense earthquake and volcanic activity, just to the east of China. SEISMOLOGY is believed to have originated in China, and the Chinese have expended considerable time and effort studying the problem of quake prediction. (*See* KANSU.)

China also has several volcanoes. BAITOUSHAN lies along the Korean border and in A.D. 1060 produced one of the largest volcanic eruptions in the last 10,000 years. The Wudalian-chi volcanic field in eastern China has also produced historic eruptions.

Chios *earthquake, Greece* A major earthquake struck the island of Chios, GREECE, in the Aegean Sea off the coast of TURKEY at 11:40 A.M. on April 3, 1881. The modified MER-CALLI intensity of the event was IX through most of the island, and it had an estimated RICHTER magnitude of 6.5.

The event was foreshadowed by minor FORESHOCKS beginning in 1879 and occurring up to 10 times per day. There was a damaging foreshock just before the MAIN SHOCK. AFTERSHOCKS continued until 1884, 34 of which were particularly destructive. (One of these, produced a minor TSUNAMI.)

The earthquake was catastrophic. More than 7,000 people lost their lives and over 20,000 were injured out of a population of 80,000. Damage was estimated at more than $25 million in 2004 dollars. There were LANDSLIDES, FISSURES, water spouts of over three feet (1 m) in height, and local uplift that caused the sea level to drop up to 32 inches (80 cm). The most intense damage was felt in the southeastern part of the island. Forty thousand troops came to Chios from Turkey to bury the dead but had difficulty entering the harbor because there were so many floating bodies in the way.

Chirpoi *caldera, Kuril Islands, Russia* Chirpoi has a record of historical activity dating back to the 18th century. The island is believed to have emerged inside a pair of overlapping CALDERAS. Another volcanic island, Brat Chiropev, stands immediately to the south of Chirpoi and is believed to be a piece of a volcano that existed at the site before caldera formation. Two volcanoes occupy Chirpoi: Chernyi volcano, in the middle of the island, and Snow volcano, a PARASITIC CONE on the south side of Chernyi. Chernyi is named for one Captain Chernyi, a Cossack who visited the island in 1770. Snow volcano is thought to have arisen only a few years after his visit, prior to a visit by a Captain Golovin in 1811. Snow may have originated around the time of a very strong tectonic earthquake in January 1780. An eruption on Chirpoi was recorded in 1811, and another eruption may have occurred in 1854. Explosive eruptions are recorded in 1857, 1879, and 1960, and a LAVA FLOW was associated with the 1879 eruption.

Cinder Cone *California, United States* About one-half mile (1 km) wide and 600 feet (183 m) tall, Cinder Cone is a volcanic cone several miles northeast of LASSEN PEAK. LAVA flowed from Cinder Cone as recently as 1851. The LAVA FLOW from this eruption extends toward Butte Lake and is approximately three miles (5 km) long. Cinder Cone's lavas consist of a rare quartz basalt. Ordinarily, QUARTZ is seldom found in BASALT. The volcano appears to have formed during a series of eruptive episodes with many years separating them. *Cinder cone* is also a geological term for a volcanic cone made of cindery PYROCLASTIC material.

Geologist studying pyroclastic layers in the Aniakchak stratovolcano caldera. Each layer represents an eruption. By studying each layer, the history of the volcano is revealed. *(Courtesy of the USGS)*

Cinque Dente *caldera, Italy* Also known as Monastero, Cinque Dente is located on the island of Pantelleria, in the Strait of Sicily between Sicily and Tunisia. The island has two CALDERAS, Cinque Dente and La Vecchia. Cinque Dente experienced strong earthquakes in 1890. Fumarolic activity was observed, and uplift occurred along the northeastern shore of the island. Uplift continued into 1891. A fracture about 600 feet (183 m) long opened near the uplifted section of coast. More uplift and strong earthquakes occurred in mid-October 1891. On October 17, an explosive SUBMARINE ERUPTION began on the northwest side of the island. A second underwater eruption may have taken place south of the island in December.

Clarion Fracture Zone *Pacific Ocean* The Clarion Fracture Zone extends more than 2,000 miles (3,200 km) under the PACIFIC OCEAN in an east-west direction and appears to extend into the Mexican mainland as the FRACTURE ZONE along which the volcanoes COLIMA and POPOCATÉPETL are located.

clastic rock Clastic rock is sedimentary rock made up of fragments of other rocks. These fragments must have been derived through the weathering and erosion of existing rock; transported by water, wind, or ice; and deposited in some sedimentary environment. Individual fragments are known as clasts. These fragments or clasts are then lithified by a combination of compaction and cementation.

climate, volcanoes and Volcanic eruptions can have a dramatic short-term effect on the climate of Earth and possibly a long-term effect as well. A familiar example of the climatic impact of volcanic eruptions is the global cooling that followed the eruption of KRAKATOA in 1883. The explosion of the volcano cast large amounts of finely divided solid material into the upper atmosphere, where the dust intercepted incoming sunlight and thus reduced surface temperatures. A similar phenomenon was observed following the eruption of the volcano TAMBORA in 1815. Much remains to be learned about the physics and chemistry of the interaction between climate change and volcanic eruptions, but it has been suggested that a period of prolonged and voluminous eruptions of volcanoes might suffice to start an ice age, or glacial period.

An even more intriguing eruption has been suggested to have occurred in A.D. 535. Historical records report that there was an 18-month period when there was a worldwide haze. There were no sunny days. The worldwide temperature dropped by two degrees, and droughts and famines ensued. This encouraged many invasions including hordes sweeping from Asia into Europe, and the Turks invasion of Byzantium. It may have even heralded the plague because of increased rodent activity. Ice cores indicate a huge spike in worldwide SULFUR emissions at the time. It is postulated with good evidence that there was a tremendous volcanic eruption that caused the climate change. ASH thickness from that period indicate that the volcano was likely in INDONESIA, near Krakatoa.

Some of the largest continental volcanic emissions on Earth occurred during the breakup of the supercontinent PANGAEA. The huge Deccan, Karoo, and Pirana volcanic traps are massive FLOOD BASALT flows. The formation of these deposits coincides with a radical climate change and the demise of the dinosaurs. There may be large-scale and long-lasting climate effects as the result of excessive volcanic activity.

See also "YEAR WITHOUT A SUMMER."

clinker Rough-textured PYROCLASTIC rock or BRECCIA such as a piece of AA. It resembles a piece of slag from a furnace.

clinopyroxene A dark-colored mineral that is high in iron and magnesium. It is part of the SILICATE family with a single chain structure. It is very common in MAFIC igneous rocks. It is one of the main components in BASALT as well as GABBRO and PERIDOTITE. However, it is also common in DIORITE and ANDESITE and even occurs in some GRANITES. Orthopyroxene has a similar occurrence, but it has a square shape and is brownish. Clinopyroxene is very dark green to greenish black.

Coast Mountains *British Columbia (Canada) and Alaska (United States)* The Coast Mountains extend approximately 1,000 miles (1,600 km) along the Pacific shore of CANADA and ALASKA and reach altitudes of more than 10,000 feet (3,000 m) in some places. The mountains represent the boundary between the westward-moving crustal plate bearing NORTH AMERICA and the OCEANIC CRUST beneath the Pacific. Seismic and volcanic events occur frequently along this coast, especially in Alaska, where earthquakes and eruptions have caused great damage in the 20th century. The potential for generating TSUNAMIS is also great along the Coast Mountains; the 1964 GOOD FRIDAY EARTHQUAKE in Alaska, for example, was accompanied by a powerful tsunami that caused extensive damage along the Alaskan shore and was responsible for considerable destruction as far south as CRESCENT CITY, California. Another earthquake, the Yakutat Bay event of 1958, also produced large waves, in this case as a result of a slide.

See also LANDSLIDE.

Coatepeque *caldera, El Salvador* Located in a chain of volcanoes, Coatepeque CALDERA is situated just east of the SANTA ANA volcano. Two other volcanoes in the area are IZALCO and San Marcelino.

The caldera contains several LAVA DOMES. Santa Ana was very active in the 16th century. Explosive eruptions are thought to have occurred in 1520, 1524, 1570 (this date is not certain), and 1576. The volcano was relatively quiet until the late 19th century when eruptions were recorded in 1874, 1878, 1879, 1880, 1882 (this date also is not certain), and 1884. Other eruptions occurred in 1904 and 1920. Izalco has erupted every few years from the late 18th to the late 20th century. San Marcelino has been relatively quiet, with two eruptions on record, one in the 17th century and the other in 1722. The 1902 eruption of Izalco coincided, within three days, with the eruptions of Mount PELÉE on Martinique and of SOUFRIÈRE 2 on the island of Saint Vincent.

Cocos crustal plate The Cocos crustal plate is a small completely oceanic plate that lies beneath the PACIFIC OCEAN off the western shore of CENTRAL AMERICA. The subduction and destruction of the Cocos plate is the primary reason for the well-displayed volcanism and earthquake activity in Central America.

See also PLATE TECTONICS.

coda On an earthquake SEISMOGRAM, there are clearly recognizable P-WAVES, S-WAVES, and SURFACE WAVES, but there are also a whole train of arrivals that follows them. This train of waves is the coda. The waves may comprise REFLECTED OR REFRACTED waves or even waves that traveled through slower media, arriving later as a result of these additional travel processes.

cohesion The force that binds particles together in SOIL or sediment. In plain language, *cohesion* means how well soil or sediment "holds together" in an earthquake. Cohesion may be the product of surface tension of water between the grains in rock and soil or even a cementing agent.

Colima *volcano, Mexico* Colima is the most active volcano in MEXICO. It is located 75 miles (125 km) south of Guadalajara. Nearly 300,000 people live within 25 miles (40 km) of the volcano making it also the most dangerous volcano in Mexico. The complex is composed of two STRATO-VOLCANOES, Nevado de Colima and Volcán de Fuego (also known as Volcán de Colima). The last large explosive eruption was January 20–24, 1913. The volcano was DORMANT for 44 years before it awoke in 1961. It was active again in 1975, 1987, 1991, and 1994. In late 1998, explosive activity forced the evacuation of 250 people. It was then explosively active on and off for the next year and remains active today.

collapse of ancient societies by earthquakes Large earthquakes and associated TSUNAMIS have likely been responsible for several of the great and enigmatic historical catastrophes. This is especially true for the eastern MEDITER-RANEAN region, where plate boundary seismicity tends to be episodic. Large earthquakes tend to occur there over a 50- to 100-year interval, followed by a period of inactivity that may last hundreds to even thousands of years. The active periods are termed *seismic crises*, such as that in the fourth century A.D. Ancient cities were commonly protected by walls that could be destroyed during an earthquake; in turn leaving the city vulnerable to outside attacks. Such attacks included external enemies during active conflicts, such as Joshua at Jericho and the Arab attack on Jerusalem in 31 B.C., neighbors in long-term conflict, such as Mycenae's fall in 1200 B.C. and Saul's battle around 1020 B.C., and uprisings of enslaved people, such as with SPARTA and the Helots in 464 B.C.; Hattusas around 1200 B.C.; and Teotihuacán around A.D. 700. If a local earthquake occurred during a modest conflict, damage would have taken a few decades to repair, with minor if any shift of power. If a major earthquake occurred during a major military conflict, it would have taken hundreds of years to rebuild, with the city possibly plunged into a dark age, with efforts on pure survival rather than art and technology. It was during these incidents that the outcomes of conflicts could be decided by earthquakes and radical shifts of power possible.

Colombia One of the most destructive earthquakes in the history of the Americas occurred in Colombia on May 15, 1875. More than 16,000 people are believed to have been killed in this earthquake, which smashed Santiago and several other communities. A volcano is said to have erupted during this earthquake and caused extensive damage in CÚCUTA.

Colorado *United States* Although Colorado is characterized by dramatic orogenesis (mountain formation), it is not nearly as active from the seismic standpoint as some other states, such as CALIFORNIA. Colorado contains numerous signs of past volcanic activity. The San Juan Mountains are products of volcanoes. Eruptions there deposited hundreds of cubic miles of ASH, which contributed to the formation of a great plateau of BASALT and TUFF that became, after extensive erosion, the San Juan Mountains. The main volcanic unit is the Fisher Canyon Tuff. It was erupted 27.8 million years ago with a PYROCLASTIC output that dwarfed even the huge YELLOWSTONE deposits. It produced 7,500 cubic miles (3,000 km^3) of ash, which even exceeds the TOBA output of 7,000 cubic miles (2,800 km^3). It is the largest recognized output from a volcano. This volcanic output buried the Needle Mountains, which were later uncovered by erosion. Thermal baths are located at Ouray. The rich ore deposits of Colorado have played an important part in its history and that of the UNITED STATES as a whole.

Although FAULT movement and earthquakes played a major role in the development of the geology of Colorado, they are relatively infrequent and small at present. There was, however, an outbreak of regular earthquakes in Denver in the 1960s. It was determined that these earthquakes were actually caused by the pumping of waste into deep wells under high pressure. The earthquakes were caused by HYDROFRAC-TURING and ceased when the pumping stopped.

Colorado Plateau *United States* The Colorado Plateau occupies 130,000 square miles (336,698 km^2) between the Rocky Mountains and the BASIN AND RANGE PROVINCE in the UNITED STATES and exhibits abundant evidence of seismic and volcanic activity. The plateau covers portions of ARIZONA, NEW MEXICO, COLORADO, and UTAH and has numerous structures of volcanic origin, including VOLCANIC NECKS, cinder cones, mesas topped by LAVA FLOWS, and "dome mountains" formed by intrusive activity.

Volcanic activity on the Colorado Plateau is thought to have occurred as recently as the 11th century. The Datil section of the Colorado Plateau in Arizona and New Mexico contains thick flows of lava. Two spectacular volcanic formations are visible near Mount Taylor in this section. At Cabezon Peak, a volcanic neck 2,000 feet (610 m) wide has been exposed by erosion and towers above benches of sedimentary rock. Several miles away, a plug 500 feet (152 m) wide in a vent of an ancient volcano has been partly exposed by erosion. Immediately north of the Datil section is the Navajo section, characterized by large numbers of volcanic necks such as those at SHIP ROCK, New Mexico, and Monument Valley. The Canyon Lands Section is located north of the Navajo section and includes such signs of volcanism as the La Sal Mountains and Henry Mountains, which are believed to have been formed as MAGMA moving underground forced overlying rocks into domes. Some of the igneous rock that produced this effect may be seen at Mount Ellsworth and Mount Hilliers. The High Plateaus section on the northwestern border of the Colorado Plateau has many lava-capped plateaus separated by GRABENS. On the southwestern edge of the Colorado Plateau, the Grand Canyon section contains lava flows from the San Francisco Mountains and from some other volcanoes. FAULTS divide the western portion of this section into large blocks. Seismic activity along the boundaries of the Colorado Plateau indicate the plateau still may be rising with respect to the lands around it. The EPICENTERS of major earthquakes in the Colorado Plateau are concentrated along its southeastern and western edges. Along the southwestern, northern, and eastern boundaries of the plateau, earthquake activity is less pronounced. Earthquakes beneath the Colorado Plateau itself tend to be weak and infrequent.

Columbia Plateau *northwestern United States* A broad mass of volcanic rock on which Mount SAINT HELENS, Mount HOOD, and other famous volcanoes of the CASCADE MOUNTAINS rest, the Columbia Plateau is made up of numerous individual LAVA FLOWS and extends from northern WASHINGTON southward into northern CALIFORNIA and NEVADA. The plateau covers more than 200,000 square miles (500,000 km²). According to one hypothesis, the Columbia Plateau is the terrestrial equivalent of a lunar sea, or MARE, formed in the aftermath of a large meteorite impact in prehistoric times that sent molten rock flowing outward over the surface from the point of impact. In this scenario, the meteorite struck close to what is now the border between OREGON and IDAHO, blasting out large amounts of rock from the crust at the site. Removal of this crustal rock allowed molten material to rise from below and emerge on the surface as large flows of BASALTic lava, some of which extended to the PACIFIC OCEAN. Flow after flow of LAVA occurred and gradually built up the Columbia Plateau. One lasting result of the postulated meteorite impact was a HOT SPOT, an enduring area in which heat flows freely to the surface from Earth's interior. As the NORTH AMERICAN CRUSTAL PLATE moved westward over the hot spot, volcanism continued at that location, resulting in occasional, tremendous releases of ASH caused by buildup of water-rich MAGMA. Eruptions over the hot spot, if this scenario is correct, produced a series of CALDERAS across the SNAKE RIVER PLAIN in Idaho. The hot spot is now thought to underlie YELLOWSTONE NATIONAL PARK.

On the other hand, hot spots are not uncommon on Earth without meteorite impacts. The majority of geologists believe that NORTH AMERICA simply overrode a hot spot and that it is now under Yellowstone National Park. In addition, the BASIN AND RANGE PROVINCE is a classic example of the early stages of a DIVERGENT BOUNDARY. FLOOD BASALTS, which also resemble lunar mare and which are common throughout geologic history, are expected in these situations. The Columbia Plateau and Snake River Plain are classic examples of flood basalts. Therefore, impact theories, however intriguing, are not necessary to explain the Columbia Plateau. In addition, no volcanoes ever been shown to have

emerged from a meteorite impact. Until more definitive evidence for the impact theory appears, most geologists will stick with the conservative interpretation. This terrestrial versus extraterrestrial origin is an example of a typical argument for geological occurrences in the distant past. When supported with enough data, the more outlandish extraterrestrial arguments can explain drastic changes on Earth and warn of what is possible. There is now good evidence that an impact accompanied the extinction of the dinosaurs. There is far too little evidence for an impact origin of the Columbia Plateau and therefore the PLATE TECTONIC argument is still the one that is generally accepted.

Studies of one lava flow, the Roza Flow, indicate something of the conditions under which the Columbia Plateau is believed to have formed. The Roza Flow, which apparently occurred in two great releases of lava separated by several hundred years, is believed to have originated from a zone of FISSURES several miles wide and more than 100 miles (160 km) long in what is now the eastern half of Oregon and Washington. Lava flowed westward from the fissure zone at a temperature estimated at perhaps 2,000°F. Evidence of the lava's tremendous heat comes from deposits of volcanic glass more than 100 miles (161 km) from the origin of the flow. It appears that lava coming into contact with lake waters here formed glass so pure that very little cooling can have occurred during the lava's movement overland. This flow covered the land with about 100 feet (30 m) of basalt across a front perhaps 60 miles (100 km) wide. Another flow, the Pomona Flow, extended from its origin in Idaho to the Pacific Ocean more than 300 miles (483 km) distant.

columnar joints Cooling joints in LAVA FLOWS that look like columns. When a lava flow spreads across the land, it welds itself to the underlying SOIL and rock. As the lava crystallizes and begins to cool, it attempts to shrink. Because it is adhered to the ground, it cannot shrink, so stresses build up until the rock cracks to alleviate it. The cracks are generally vertical in profile but the tops make five- to six-sided polygons with sharp edges. In some areas where there is a quick succession of flows, the cracks may not be vertical but instead make a plumose (feather) pattern. Columnar joints may occur in any lava flow [not in TUFF (ASH)] but are most common in BASALT. Underwater flows cannot form columnar joints.

Comoro Islands *Africa* The Comoro Islands are a string of oceanic volcanoes between Madagascar and AFRICA. They were formed during the splitting away of Madagascar, likely as the result of a HOT SPOT. The youngest, Grande Comore, contains two basalt SHIELD VOLCANOes, Karthala and Massif de la Grille, both of which have been active in historical times. Karthala produced a PHREATIC ERUPTION in 1991. Little information is available on the eruptions.

composite volcanoes *See* VOLCANO.

compressional stress Compressional stress is force over an area where plates and/or rocks are being squeezed together. CONVERGENT margins are marked by the highest level of compressional stress on the Earth's surface. SUBDUCTION ZONES

Columnar jointing in a basalt flow in Devil's Postpile, California. As the lava cooled, it shrunk. The shrinking caused the rocks to break along vertical intersecting cracks that look like columns. *(Courtesy of NOAA)*

and continental collisions are areas of high compressional stress. They produce enormous amounts of earthquakes as the result of movement on the many REVERSE and THRUST FAULTS that result from compressional stress.

compressional wave   *See* P-WAVE.

Concepción (1) *earthquake, Chile*   The city of Concepción has suffered greatly from earthquakes on numerous occasions. An earthquake in the summer of 1757 reportedly killed 5,000 people, injured perhaps 10,000 more, and appears to have been accompanied by a TSUNAMI that submerged much of the community, but recorded details of the earthquake and associated events are few.

The earthquake of February 20, 1835, has gone down in history in part because one of the most famous naturalists of the 19th century, Charles Darwin, happened to be visiting CHILE at the time as a passenger on HMS *Beagle,* which was under the command of Captain FitzRoy. In his memoir *The Voyage of the Beagle,* Darwin described the earthquake

and its results (quoted in Appendix B). Darwin's speculations about the origins of the mountains in SOUTH AMERICA brought him very close to the formulation that German meteorologist, and geologist Alfred WEGENER would devise, early in the 20th century, to account for the global distribution of earthquakes and volcanic mountain ranges. This theory in turn would be modified to provide the modern theory of PLATE TECTONICS, which serves as the basis of modern geology.

Concepción (2) *volcano, Nicaragua*   A STRATOVOLCANO that has erupted at least 24 times since 1883. Eruptions include moderate-size explosions, the most recent in 1986. FUMAROLES are currently active from the area of the summit crater.

conduit   A pipelike channel in COUNTRY ROCK through which MAGMA rises from an underground reservoir (MAGMA CHAMBER) to the surface. The eruption of the molten rock produces a volcano.

cone   *See* VOLCANIC CONE.

cone-in-cone structure   Resurgent cone or cones that grow within an existing cone structure.

Connecticut *United States*   Connecticut has experienced numerous earthquakes in the 18th, 19th, and 20th centuries. Hartford experienced an earthquake on April 12, 1837, that caused bells to ring and startled residents into rushing outdoors. Much of Connecticut was shaken by an earthquake on August 9, 1840. The EPICENTER is thought to have been a few miles north of New Haven. The Hartford earthquake of November 14, 1925, caused widespread alarm and some damage to buildings and knocked objects off shelves at Windham; noises were reported at East Haddam. An earthquake caused considerable fright at Stamford on March 27, 1953. On November 3, 1968, an earthquake in southern Connecticut affected an area 30 miles (48 km) in length along the Connecticut River between Glastonbury and Lyme; some damage was reported at Madison, and effects of MERCALLI intensity V were noted at Chester, Deep River, and Essex.

Constantinople   *See* ISTANBUL; TURKEY.

contact metamorphism   MAGMA and LAVA range in temperature from about 800 to 1,400°C. The preexisting rocks that they come into contact with as they pass through or flow onto are typically less than 200° or at the surface, 25°C. When the molten rock comes into contact with the preexisting or COUNTRY ROCK, it is baked. This baking results in new mineral growth in the country rock. For volcanic flows, the SOIL underneath is welded by the heat. If a piece of country rock is pulled into a large accumulation of magma, as in a MAGMA CHAMBER, it can undergo an intense contact metamorphism called pyrometamorphism (*pyro* means "fire").

continent   A major mass of land that is composed of CONTINENTAL CRUST. The continents of Earth are NORTH AMERICA,

South America, Europe, Asia, Africa, Antarctica, and Australia. Continental crust is very light and floats high on the planet like a foam board on a swimming pool. OCEANIC CRUST sinks down because it is dense. ISOSTASY describes the balance between the two. Continents occupy about one-third of Earth's surface. Boundaries between continents may be the sites of intense seismic and volcanic activity, and crustal rock under the continents is generally much thicker than that beneath the oceans.

See also PLATE TECTONICS.

continental crust Light crust that underlies the continents. Continental crust is composed mainly of QUARTZ and FELDSPAR. It has a density of 2.65 grams per milliliter and is therefore much lighter than OCEANIC CRUST. Continental crust is much thicker than oceanic crust, typically ranging from 22 to 50 miles (35 to 80 km). It is much thicker under mountains than under the lowlands. The thickest crust is under the Himalaya Mountains. Continental crust is also much older than oceanic crust. It is not recycled in SUBDUCTION ZONES but instead is preserved. The oldest continental crust is preserved in the interiors, known as the cratons. Rocks there can be as old as 3.4 billion years, whereas ocean crust is no older than 200 million years.

continental drift *See* PLATE TECTONICS.

continental shelf The portion of a continent that is covered by ocean. A continental shelf exhibits a very gradual slope (only a fraction of a degree) and may be the site of considerable mineral wealth and other valuable natural resources. The continental shelf gives way to the continental slope in which the slope angle increases to several degrees. Strong earthquakes can occur on the continental shelf. The powerful 1927 Grand Banks earthquake, for example, generated a huge, fast-moving (60 miles [97 km] per hour) TURBIDITY CURRENT (an undersea LANDSLIDE) that destroyed several telegraph cables beneath the North Atlantic Ocean. The width of a continental shelf may vary greatly. Off the eastern shore of the United States, for example, the continental shelf is much broader than off the Pacific coast, where the ongoing collision between the North American crustal plate and the crust underlying the Pacific Ocean has resulted in a relatively narrow shelf. The shelf off California is especially vulnerable to undersea earthquakes because FAULTS involved in seismic activity extend offshore into the seafloor.

continuous deformation When earth materials are deformed by flow—that is, gradual and steady rather than by sudden failure. The deformation is said to be continuous as distinct from discontinuous deformation, which is required to produce earthquakes. Continuous deformation folds rock layers or produces MYLONITES, which are formed deeply within FAULT zones.

contour A line on a chart or map that links points of equal value, such as elevation or depth. Contour lines for elevation on a map of Earth's surface mean that it is a topographic map. Underwater contours for elevation on a map mean that it is a bathymetric map. A history of geologically recent vol-

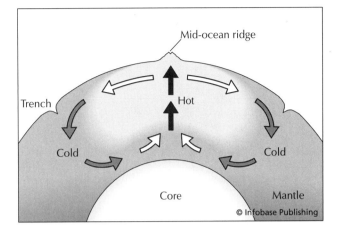

Hot material rises vertically to the top of the mantle, where it spreads out and cools. The cool material sinks back through the mantle, where it is heated and starts the cycle again. The driving force is density. Hot mantle material is light, and cool material is dense.

canic or FAULT activity in a given location can be interpreted using the contours on a topographic map.

convection current In general, a convection current is a circulation pattern illustrated by heating water in a pot on a stove. The water above a heat source heats up and becomes less dense. As a result, it rises to the top of the pot. There it spreads out on the surface of the water and cools. By cooling down, it becomes denser and sinks back to the bottom again along the sides of the pot. From there it moves back to the heat source where the cycle begins again. Convection plays a big role in geology from circulation in PLUTONS to PLATE TECTONICS. In plate tectonics, the ASTHENOSPHERE, which is gummy mantle, forms a convection cell. It rises up at the MID-OCEAN RIDGE above the heat source and sinks back down at the SUBDUCTION ZONE. The MANTLE convection drives the plates.

See also EARTH, INTERNAL STRUCTURE OF; PLATE TECTONICS.

convergent plate boundary A plate tectonic boundary where the two plates are driven together. If OCEANIC CRUST is being driven beneath the other plate, it is called a SUBDUCTION ZONE. The down-going ocean crust is being consumed in these zones. There are two types of subduction zone boundaries. In one, the ocean crust is driven beneath CONTINENTAL CRUST. This situation forms a MAGMATIC ARC, an example

(next page) Block diagrams showing the three different types of convergent boundaries: (A) ocean-ocean, (B) ocean-continent, and (C) continent-continent. An ocean-ocean convergent margin produces an island arc similar to the Aleutian Islands of Alaska. An ocean-continent convergent margin produces a magmatic arc similar to the Andes Mountains of South America. A continent-continent convergent margin produces an orogeny similar to the Himalayan Mountains of southern Asia.

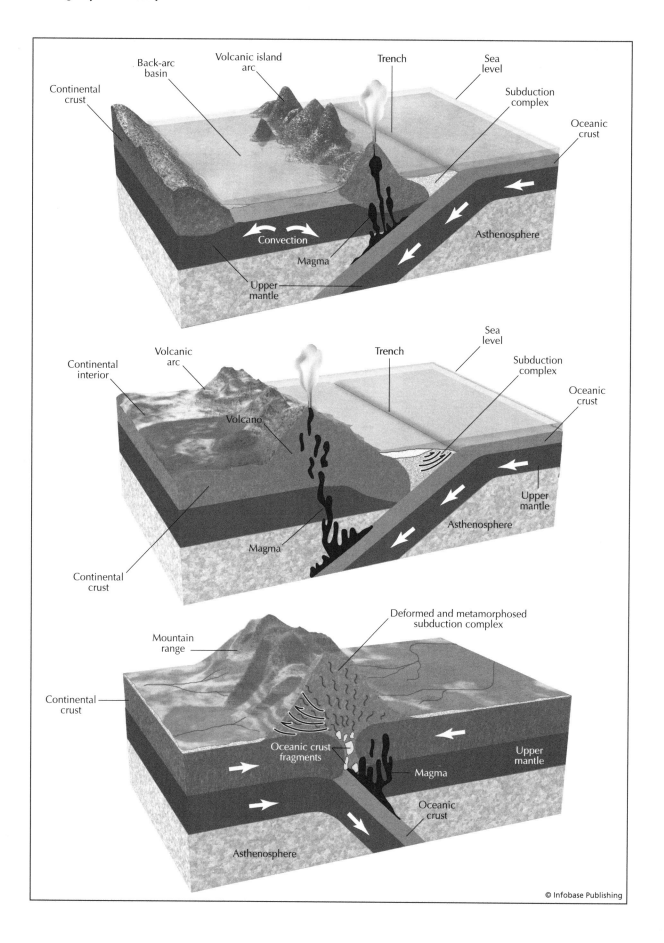

© Infobase Publishing

of which is the ANDES MOUNTAINS. If ocean crust is driven beneath ocean crust, an ISLAND ARC is formed. An example of this is the ALEUTIAN ISLANDS. The final type of convergent boundary is a continent-continent collision. Because continental crust is so light, it cannot be driven into the ASTHENOSPHERE as can oceanic crust. It is too buoyant. Instead, the crust is greatly thickened and pushed back onto itself in huge THRUST FAULTS. It winds up looking like shingles on a roof. The best example of a continent-continent CONVERGENT BOUNDARY is the Himalaya Mountains.

converted wave When seismic waves hit a boundary between two rock types of strongly varying seismic character, a whole new set of waves is generated. These waves are far weaker than the original wave and do not cause damage. However, they are recorded by local SEISMOGRAPHS at much lower AMPLITUDE.

Copahué *caldera, Chile* Copahué is located on the border between ARGENTINA and CHILE. The CALDERA includes a central cone with more than one CRATER. The level of the lake in Los Copahués crater reportedly fell more than 100 feet (30 m) between 1940 and 1945, and temperature and acidity increased.

Corbetti *See* ASAWA.

Cordillera Nevada *caldera, Chile* The Cordillera Nevada CALDERA has experienced eruptions in 1921–22, 1929, 1934, and 1960. The caldera is similar in some ways to the LONG VALLEY caldera in CALIFORNIA.

core *See* EARTH, INTERNAL STRUCTURE OF.

Corinth *earthquakes, Greece* The ancient city of Corinth, GREECE, and its modern successor have experienced continued seismic activity. The INTENSITY of the early earthquakes is debatable, but the city has an interesting earthquake history nonetheless. The earliest record of an earthquake in Corinth is from the summer of 420 B.C. The earthquake had an estimated RICHTER magnitude of less than 6.0, but there are no other details available. The next major earthquake, in 373 B.C., devastated the whole Gulf of Corinth. The earthquake had an estimated MAGNITUDE of 6.8 on the Richter scale and an X on the modified MERCALLI scale at the city of Helike. It generated a huge TSUNAMI that engulfed the city of Helike so that only the treetops were visible above the sea. Helike was later discovered by archaeologists, completely submerged in mud. Corinth was damaged by both the earthquake and tsunami, but there are no details available. The next major earthquake occurred during the night of June 20, A.D. 74. It had an estimated magnitude of 6.3; three temples were destroyed, but there are no other accounts.

There are reports of earthquakes from Corinth in A.D. 521, A.D. 543, and A.D. 580, with estimated Richter magnitudes of 6.3, 6.2, and 6.3, respectively. The real dilemma is with A.D. 856. Some sources claim that a huge earthquake destroyed Corinth that year, killing an estimated 45,000 people. This would be one of the greatest disasters in the area.

Other reliable sources do not even mention an earthquake in the region for that year. It remains unresolved.

The next important earthquake in Corinth occurred at 9 A.M. on February 21, 1858. Its estimated Richter magnitude was 6.5, and it caused destruction described as X on the modified Mercalli scale. The shock lasted more than 10 seconds, and the maximum damage area was about 21 miles (35 km) across. FISSURES opened in the ground, but the main damage came from a ROCKFALL into the city. It was reported that only 21 people died and 65 were injured, but the damage was so great and the AFTERSHOCKS so persistent (they lasted through May) that the city of Corinth, which was established by Sisyfos in 1438 B.C., was abandoned. The survivors of the earthquake established the city of New Corinth approximately eight miles (13 km) away. Unfortunately, that location was no safer. On April 22, 1928, at 8:13 P.M., an earthquake of Richter magnitude 6.3 caused damage of IX on the modified Mercalli scale. It destroyed 3,000 houses and made 15,000 homeless. It is considered to be the 33rd destruction of Corinth since it was established.

Cos *earthquake, Greece* After nearly two years of FORESHOCKS, a very destructive earthquake struck the small Greek island of Cos, off the Turkish coast, on October 3, 1493. The quake had an estimated RICHTER magnitude of 6.8 and an INTENSITY of X on the modified MERCALLI scale in the town of Kephalos. Many houses were destroyed in Antimacheia, Kardamaena, as well as Kephalos and two historic towers. The shock was felt in EGYPT and Israel and was damaging in TURKEY. The DEATH TOLL for this event was approximately 5,000 people.

Cosigüina *volcano, Nicaragua* Located on the Gulf of Fonseca, Cosigüina has a history of violent eruptions, notably that of 1835, which reportedly was heard from COLOMBIA to British Honduras, a distance greater than 1,000 miles (1,600 km). This eruption lasted more than four days. The caldera rim of Cosigüina rises to an altitude of more than 2,800 feet (853 m), and the inner walls of the CALDERA extend almost 3,000 feet (914 m) in places above a crater lake.

The 1835 eruption of Cosigüina occurred after a long period of apparent quiescence, and it was widely believed before this eruption that the volcano was EXTINCT. Much of the information on this eruption comes from the account of the commandant of the port of La Union in EL SALVADOR, approximately 30 miles (48 km) from the volcano. A large white cloud emerged from the summit of Cosigüina on the morning of January 20, 1835. The cloud then changed color, becoming first gray, then yellow, and then red. Evidently, no seismic activity preceded the eruption. Shortly before noon, the volcano's emissions had produced such darkness over the area that lamps had to be lighted in La Union. Fine PUMICE began to fall in great quantities that afternoon, and darkness in some places became virtually total. Several inches of ASH accumulated by late afternoon at one community in Honduras, some 40 miles (64 km) north of Cosigüina. Ash fell in El Salvador, more than 100 miles (161 km) northeast of the volcano, by nightfall. Darkness continued the following day, and residents of the area around the volcano felt strong earthquakes and heard noises from underground. Darkness covered the entire nation of Honduras, and chunks of pumice more than an inch in diameter fell some 20 miles (32 km) from the volcano. The winds changed direction on January 22 so that areas previously free of heavy ASHFALLS now were covered by ash. On this day, a tremendous noise, almost nonstop in some areas, emanated from the volcano.

Listeners compared it to the sound of artillery fire, and in Belize, soldiers prepared to cope with what sounded like naval gunfire offshore. There was a similar response in Guatemala City, where soldiers concluded an attack was imminent and braced for an assault. The noises were loud enough to cause distress at locations as far away as JAMAICA. Ash fell in MEXICO and along the border of COSTA RICA. The eruption started subsiding on January 23. Ashfalls stopped by January 27, but noises continued to be heard until the end of the month. FUMAROLIC activity may have continued in the CRATER for several decades after this eruption, which cast out tremendous quantities of solid material. Because much of the material was lightweight pumice, it floated on the waters and covered the sea for about 150 miles (241 km), according to one account.

Costa Rica   The CENTRAL AMERICAN nation of Costa Rica has a long history of volcanic activity, notably the 1963–65 eruption of IRAZU, which cost the Costa Rican economy an estimated $150 million. ARENAL is another famous volcano of Costa Rica. It was DORMANT from 1500 to 1968 when it awoke to become a killer volcano. It has been in near constant eruption ever since. POÁS is similarly in near continuous activity, having erupted 39 times since 1828. Both Poás and Arenal have become tourist attractions because of their constant low level activity. RINCÓN DE LA VIEJA is another active volcano. Costa Rica also is extremely vulnerable to earthquakes because of its position in seismically active Central America. One of the most destructive earthquakes in the history of Costa Rica occurred on August 27, 1841, killing about 4,000 people out of a population of fewer than 20,000. Another earthquake on May 4, 1910, killed more than 200 people and caused extensive damage.

Cotopaxi *volcano, Ecuador*   Located near QUITO, capital of ECUADOR, Cotopaxi stands about 19,300 feet (6,000 m) tall and has erupted at least 50 times since 1738. It is famed for its 1877 eruption in which an outpouring of LAVA from the CRATER is said to have melted ice on the summit and have sent meltwater rushing down the volcano's flanks in such quantities that the resulting flood affected lowlands more than 200 miles (322 km) distant. LAHARS alone traveled more than 60 miles (97 km). The most recent eruption of Cotopaxi was in 1904 although there is an unconfirmed eruption in 1942.

coulee   This expression refers to several phenomena in geology. One is a flow of viscous LAVA with a steep front. Another is a long, dry gorge that was carved by meltwater from a sheet of ice. GRAND COULEE in WASHINGTON in the UNITED STATES is a famous example of a coulee. Grand Coulee cuts through LAVA FLOWS and is the site of a dam. Coulees can also form from volcanoes melting glacial ice to form such a gorge. They are common around ice-covered volcanoes.

country rock   The preexisting rocks in an area into which a PLUTON intrudes is the country rock. The preexisting rock into which mineral deposit or vein is emplaced is also referred to as country rock.

cow-pie bomb   A volcanic bomb that is ejected while it is still very molten. It becomes nearly spherical in flight because of the physics of a liquid. It then lands splat on the ground forming a flattened circular-shaped bomb that resembles a cow pie.

crater   A roughly circular depression at a volcano's summit or on its flanks, from which LAVA and TEPHRA emerge during an eruption. They typically have a VENT or several vents in their centers.

Crater Lake *Oregon, United States*   One of the most spectacular CALDERAS in the world, Crater Lake occupies the BASAL WRECK of Mount Mazama, a volcano that evidently collapsed following an eruption many centuries ago, leaving a basin that filled eventually with water. The lake is about 35 miles (56 km) in circumference and some 1,900 feet (580 m) deep. A comparatively small volcano, Wizard Island, rises more than 700 feet (213 m) above the lake's surface and stands as evidence that Mazama's fall did not mean the end of volcanic activity at that site. Credit for "discovering" Crater Lake goes to John Wesley Hillman, who rode his mule up the outer slope of Crater Lake one day in 1853 and suddenly found himself at the rim of the caldera. The lake apparently was known to Native Americans, however, long before settlers of European descent came to OREGON.

See also CASCADE MOUNTAINS.

Craters of the Moon Monument *Idaho, United States*   Generated by the same northwestern volcanism that produced the COLUMBIA PLATEAU, Craters of the Moon Monument exhibits many different kinds of volcanic formations. Based on radiocarbon analysis, the LAVA FLOWS at the monument occurred 2,000 years ago. Samples for this analysis were collected in a novel fashion. Students dug tunnels under the lava flow in hopes of finding burnt plant material that had been both produced and then preserved undisturbed by the advancing LAVA. Some material was indeed preserved under the lava, which had advanced as a smoothly flowing mass and covered the plants in its path without displacing their charred remains. Craters of the Moon Monument includes an area of "tree molds" formed when lava surrounded the trunks of trees. The trees burned away, but the impressions of their trunks remained. Here again, charcoal derived from roots of the original tree was preserved under the lava, and radiocarbon dating yielded a date of approximately 2,200 years ago for the lava flow that overwhelmed the trees and produced the charcoal.

craton   A craton is a stable, generally interior portion of a CONTINENT. Typically, cratons are composed of very old (> 1.5 billion years old) CRUST that was highly metamorphosed and deformed during ancient orogenies but then remained tectonically quiet thereafter. Cratons are contrasted

The picturesque Crater Lake, in Oregon, fills the large collapsed caldera of Mount Mazama. Wizard Island in the middle of the lake is a small resurgent cone. *(Courtesy of the USGS)*

with mobile belts that occur along the margins of continents where the volcanic and seismic activity is concentrated. The mobile belts are much younger than the craton interiors. All of the continents have old interior cratons. Much of CANADA's interior is an example of a craton as compared with the Appalachians or Rockies, which are mobile belts.

Creede *caldera, Colorado, United States* Located in the San Juan Mountains, Creede CALDERA is almost 15 miles (24 km) in diameter and has a central mountain (Snowshoe Mountain) rising from the caldera floor.

creep The gradual DEFORMATION of rock or SOIL subjected to small but steady STRESS over a prolonged period. Creep leads to slow permanent deformation of a material. Creep can be observed in the tilting and toppling of retaining walls on slopes. Sometimes creep can force a material into gravitational instability and ultimately lead to a catastrophic event such as an AVALANCHE or LANDSLIDE.

Crescent City *tsunami, California, United States* Crescent City, in northern CALIFORNIA near the OREGON border, experienced heavy damage from the TSUNAMI that followed the 1964 GOOD FRIDAY EARTHQUAKE in ALASKA. Eleven of the 13 persons killed by that tsunami died in Crescent City, and damage was estimated at about $11 million. Current velocities were estimated at some 30 miles (48 km) per hour. More than 50 homes in Crescent City were destroyed, and 37 others were reported damaged. Some 40 businesses were destroyed completely, and more than 100 others experienced major or minor damage. The tsunami lifted houses from their foundations and deposited the houses in nearby streets. The harbor area experienced serious damage, including the capsizing of 15 boats. Buoyancy associated with the tsunami did considerable damage to a dock, which also was damaged by a loaded barge driven against it by the tsunami.

crust The rigid outer layer of Earth in which earthquake activity takes place. There are two types of crust, oceanic and continental. The crust varies greatly in thickness and composition. Under major mountain ranges, the crust may be up to 50 miles (80 km) in thickness (the Himalayas). Under the oceans, by contrast, the crust tends to be relatively thin (three to six miles [5–10 km] thick). The crust is the upper part of the LITHOSPHERE, which defines a plate. The lithosphere is divided into several major plates and many smaller plates, interactions among which produce earthquakes, volcanoes, and many other geological phenomena. On its surface, the crust exhibits a wide variety of materials and landforms, ranging from broad plains of sediment to rugged ranges of volcanic mountains and from carbonate rock such as limestone to dark igneous rocks like BASALT. The crust is widely broken by FAULTs, generated and expanded by a variety of processes related to PLATE TECTONICS. Motion of crustal rock along faults regularly generates earthquakes. Oceanic crustal rock is produced by solidification of MAGMA and lava along MID-OCEAN RIDGES, from which the newly formed crust moves outward on either side in much the same manner as goods on a conveyer belt. The rate of production of fresh OCEANIC CRUST at these sites varies from one location to another but is usually only a matter of inches per year. Oceanic crust is then destroyed in SUBDUCTION ZONES as fast as it is produced at the mid-ocean ridge. Therefore, it is typically less than 200 million years old, which by Earth standards is young. CONTINENTAL CRUST, on the other hand is light, old, and thick and is produced much more slowly by a variety of processes. The boundary between both kinds of crust and underlying mantle is known as the MOHOROVICIC DISCONTINUITY, or "Moho." It was found that velocities of seismic waves traveling through Earth's interior increase sharply at the depth of the Moho.

The Moho dips deeply under thick regions of crust, such as the Alpine Orogen, and rises relatively near the surface under thinner portions of crust, such as the mid-ocean ridge. The behavior of the crust during earthquakes depends on many different factors, including the MAGNITUDE of the earthquake, depth of the earthquake FOCUS, degree of faulting and folding, depth of overlying sediment, and the distribution of groundwater in that sediment. The extreme variability of the composition and structure of the crust means that two closely situated localities may respond to the disturbance from a single earthquake in greatly different ways. An area underlain by heavily faulted and folded rock may undergo a powerful earthquake but experience less damage than another locality, which undergoes more serious and widespread damage from a comparatively weak earthquake. In the latter case, the underlying rock provides a more effective medium for transmitting vibrations over long distances.

See also EARTH, INTERNAL STRUCTURE OF; PLATE TECTONICS; SEISMOLOGY.

Crystal Ice Cave *Idaho, United States*   The ice formations of Crystal Ice Cave are located in a volcanic FISSURE in the southern portion of IDAHO's Great Rift. LAVA flowed from this fissure at one time and covered an area of more than one square mile (2.6 km²) at an average of more than 20 feet (6 m) thick. One spectacular feature of Crystal Ice Cave is

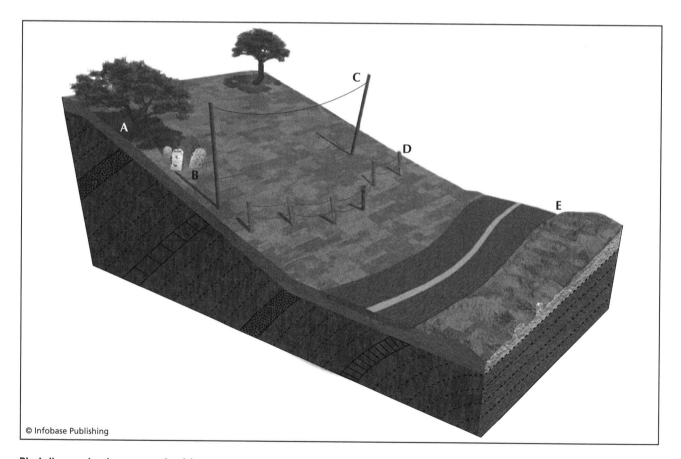

Block diagram showing some results of the mass movement creep, a slow downhill migration of soil and, in some cases, rock. Movement increases closer to the surface.

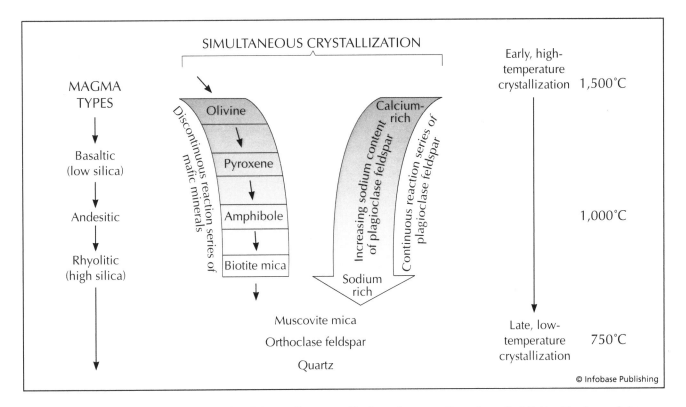

Diagram showing the order of crystallization of minerals in a cooling magma. The discontinuous reaction series on the left side of the diagram is for ferromagnesian or mafic minerals. At the highest temperature, olivine crystallizes, followed by pyroxene and then amphibole and biotite at progressively lower temperatures. This means that a mineral will crystallize from the magma only to a certain temperature, and then it will stop and a completely different mineral will start to crystallize. In reality, there can be slight overlaps. The right side of the diagram is the continuous reaction series. It occurs at the same time as the discontinuous side, but only one mineral is formed—plagioclase feldspar. The plagioclase, however, has a different composition, depending on the temperature that varies in the amount of sodium and calcium. At the lowest temperatures, quartz and k-feldspar crystallize last. The left column of the diagram shows the magma types associated with each group of minerals. It shows that mafic rocks form at high temperatures, whereas felsic rocks form at relatively low temperatures.

King's Bowl, a large depression thought to have been formed by a phreatic blast produced when groundwater came into contact with extremely hot rock.

See also PHREATIC ERUPTION.

crystalline A textural term for coarse-grained igneous and metamorphic rocks. It means that the minerals in the rock are fully formed (no glass) and large enough to see. All plutonic IGNEOUS ROCKS are crystalline. Some volcanic rocks are crystalline, but most are only part crystalline or not crystalline at all.

crystallization The process by which molten rock cools and solidifies, forming crystals of the igneous minerals that make up the rock. Crystallization involves the bonding of specific atoms into specific mineral structures, such as QUARTZ, FELDSPAR, PYROXENE, and OLIVINE. Contrast this process with simple hardening of liquid into a volcanic glass that has no crystalline structure.

See also IGNEOUS ROCK.

Cucuta *earthquake, Colombia* Although there are conflicting reports on the details, geophysicists agree that a massive earthquake struck the city of Cucuta, COLOMBIA, and the city of San Cristobal, Venezuela, in mid-May of 1875. The most consistent date was May 18, 1875, and the reported times were 10:10 A.M. and 4.15 P.M. The estimated RICHTER magnitude was 7.3 and the reported duration 45 seconds. Damage on the modified MERCALLI scale was estimated at X. The FOCUS was estimated to be at 12 miles (20 km) of depth on either the Eastern Frontal Fault or the Tarra Fault to the southeast or northwest of the city, respectively. (Both are seismically active.) It was reported that volcanic activity accompanied the earthquake. The DEATH TOLL for this event is also conflicting. There appear to have been at least 10,000 people killed and up to 16,000, making the earthquake one of the worst in the history of the region.

Cumbre Vieja *volcano, Canary Islands* After the BANDA ACEH earthquake and TSUNAMIS, interest in coastal safety against tsunamis increased dramatically. The Cumbre Vieja volcano is located across the Atlantic Ocean, quite a distance from the east coast of the United States, making it an unlikely threat. But since the teletsunami that generated in Banda Aceh killed people across the Indian Ocean, anything is possible. During Cumbre Vieja's eruption in 1949, the western flank of the volcano began a slow collapse. It slid farther during the 1971 eruption, to a precarious position. Researchers speculate that this flank may collapse during a subsequent eruption, causing a catastrophic landslide into the Atlantic Ocean. This landslide could act like the surface-to-water landslide that occurred in Lityua Bay in Alaska, which generated a huge seiche that climbed 1,700 feet (510 m) up the facing shore. They project that a wall of water some 300 feet (90 m) high could race across the Atlantic Ocean and devastate the east coast, the Caribbean Sea, and even West Africa. Certainly, if such a scenario were to occur, the death and destruction would make Banda Aceh look small. On the other hand, the Canary Islands have been in existence for many million of years, and there is no evidence of the Atlantic seaboard ever having been hit by any wave of such size.

D

dacite An intermediate volcanic rock, dacite is more QUARTZ-rich than ANDESITE. It is also less potassium-rich than RHYOLITE. They are typically fine-grained, but phenocrysts (large grains) include feldspars, BIOTITE, and HORNBLENDE. Ironically, dacite is the most common rock in eruptions along the continental side of SUBDUCTION ZONES between oceanic and continental crustal plates. These boundaries are also called Andean Margins. Even though andesite is named for the ANDES MOUNTAINS, most of the volcanic rocks there are dacites.

Daisetsu-Tokachi *graben, Japan* Now mostly filled, the Daisetsu-Tokachi GRABEN is a large depression on the island of Hokkaidō. Several VOLCANIC DOMES (Ushiorasahi-dake, Kuma-dake, Hokuchin-dake, Ryon-dake, Keigetsu-dake, Kuro-dake, Eboshi-dake, and Hakuun-dake) are found in the vicinity and are thought to occupy points along the rim of a buried Daisetsu CALDERA. The volcano Tokachi-dake occupies the south end of the graben. It has produced 17 clustered historic eruptions in 1670, 1857, 1887–89, 1925–31, 1952–62, and 1985–89. Tokachi-dake was active in the 1920s, starting with increased emissions of steam and other gases in 1923. Also in 1923, a SOLFATARA near the rim of the CRATER was replaced by a pool of either boiling water or melted SULFUR. A hot spring near the crater also exhibited a dramatic rise in temperature.

Activity at the volcano, including minor explosions, continued through late 1925 and early 1926, and a fairly strong explosion took place on April 5. By early May, minor earthquakes were occurring. An explosion and mudflow were observed on May 24, followed later that day by a more powerful explosion, another mudflow and a LANDSLIDE. The mudflows or LAHARS moved down the Hurano valley 12 miles (20 km) in 26 minutes. They destroyed 5,080 homes and killed at least 146 people. A strong earthquake on September 5 preceded another eruption by three days. Activity then diminished at the volcano, and by 1928 the only signs of continued volcanism were changes from time to time in the output of fumaroles in the crater of Tokachi-dake.

FUMAROLES became notably more active in August 1952 when a new crater, called Showa, formed and continued puffing for the next seven years. A strong earthquake off the coast of Hokkaido in 1962 preceded a large increase in fumarolic activity at the crater. Temperatures of fumaroles rose for about two months afterward. Earthquakes became more frequent in the vicinity of the volcano, and ROCKFALLS occurred along the east wall of the crater in late April and in May. Five people were killed by the rockfalls. Steaming and rumbling increased sharply on May 30, and numerous earthquakes occurred on May 31. Muddy hot water emerged from fumaroles on the northwest side of the central cone on June 29. Seismic and fumarolic activity then diminished until about midnight, when a large explosion took place. By late August, earthquake activity had dropped again to its normal level.

A very strong TECTONIC earthquake on May 16, 1968, almost 200 miles (322 km) from Tokachi, appears to have touched off renewed activity at the volcano. Just after the tectonic earthquake, a swarm of earthquakes occurred at Tokachi-dake. This activity diminished for a few months but intensified again in December 1968 and reached maximum activity in January and March 1969. The swarm of activity included several thousand recorded volcanic earthquakes. No eruption, however, occurred at this time. Small eruptions occurred in June 1985.

Dakataua *caldera, Papua New Guinea* The Dakataua CALDERA is situated on the Talasea Peninsula on the northern shore of New Britain Island. The caldera is thought to have formed only about 1,150 years ago. Inside the caldera is a central volcano, Mount Makalia, that contains a crater lake more than 300 feet (91 m) deep. Eruptions have occurred from ring FRACTURES in the caldera. LAVA flowed from the summit in an eruption about the year 1890. Warm springs exist at the base of Mount Makalia, and a cinder cone in the summit CRATER emits steam.

Damghan *earthquake, Iran* On December 22 A.D. 856, one of the greatest natural disasters of all time occurred. A massive earthquake struck IRAN in the province of Fars and in the region of Qumes, especially in the city of Damghan. The quake was felt as far away as Syria and Yemen. Apparently, this was a very seismically active period in this region because there are reports of numerous earthquakes for several years before this. The number and INTENSITY increased toward the end of 856, when strong earthquakes were reported from December 3 to the real MAIN SHOCK on December 22. There are reports of great FISSURES in which people were consumed, AVALANCHES in the mountains, surface ruptures, and heavily flowing springs that switched to bone-dry and back on a daily basis. In all, some 70 towns were destroyed. According to the U.S. Geological Survey, some 200,000 people lost their lives in this event, making it among the most devastating. Other sources list the December 22 event as having killed 48,690 people in Khoresan. It is unclear whether 200,000 were killed in the month or by the single event.

This was not the first report of a massive earthquake destroying Damghan. There was a massive earthquake in A.D. 662 that laid waste to the city. It reportedly killed some 40,000 people, but details are few. Damghan is an area of historical seismic activity that periodically produces a record disaster.

damping Damping is the loss of energy (and AMPLITUDE) in a moving wave as it passes through a medium by frictional forces. The friction converts the seismic energy into heat energy. The different rock and SOIL types cause differing amounts of damping of the seismic waves as they pass through.

dams Artificial and natural dams can present potential hazards during earthquakes and volcanic eruptions. Waters impounded behind a dam can be released with devastating effect if the dam should fail, or LANDSLIDES or other phenomena could raise the water level above the dam, sending water rushing along the valley downstream.

In the 1971 SAN FERNANDO earthquake in southern CALIFORNIA, for example, the Upper and Lower Van Norman Dams sustained severe damage. The Lower Van Norman Dam appears to have come close to failing. Most of the damage to the Lower Van Norman Dam consisted of a slope failure that dislodged a huge segment of the earth-fill embankment and sent it to the floor of the reservoir. The slide removed the dam's upstream concrete lining and crest, and one of two intake towers was destroyed. Discharge facilities at the Lower Van Norman Dam were opened to lower the water level behind the dam, and the area below the dam was evacuated for several days after the earthquake while water was drained from the reservoir. The Upper Van Norman Dam's crest sagged considerably, and the dam itself shifted about 6 feet (1.8 km) downstream.

Volcanic eruptions also contain a potential for catastrophes involving dams when an eruption occurs nearby.

The 1980 eruption of Mount SAINT HELENS, for example, occurred only several miles from Swift Reservoir. Had the tremendous AVALANCHE of debris from the eruption moved south toward the reservoir, instead of north and east, it might have overwhelmed the reservoir and resulted in the flooding of the lower valley of the Lewis River, a tributary of the Columbia River. Following the eruption, two debris dams in the Toutle and Cowlitz river systems were studied for their flood hazards. One of the dams failed the day after the theoretical failure analysis was carried out, and the hypothetical results agreed closely with the actual results.

Dasht-e Bayaz *earthquake, Iran* A massive earthquake struck Dasht-e Bayaz in the northeastern province of KHORASSAN, IRAN, on August 31, 1968. The MAGNITUDE of the MAIN SHOCK was 7.2 on the RICHTER scale, with a relatively shallow FOCUS of less than nine miles (15 km) in depth. The maximum modified MERCALLI intensity was X, occurring over a 30-mile (50-km) length. It was felt over an area of about 39,000 square miles (100,000 km^2). The main AFTERSHOCK occurred on September 1 at 6.4 magnitude, but during the following six weeks, there were nine aftershocks with magnitudes between 5.0 and 5.6 among the hundreds of smaller events. The fault that generated this series of events experienced LEFT-LATERAL strike-slip displacement of up to 15 feet (4.5 m) along 48 miles (80 km). Offset ridges, streams, and other features show this displacement.

Between 7,000 and 12,000 people were killed by the earthquake. Poor records and insufficient accounting are the reason that this number is so inexact. At least 12,000 houses were destroyed, leaving over 70,000 people homeless. The reason for the massive destruction and loss of life was primarily the shoddy construction of buildings and narrow streets. In order to prevent disease, bodies were quickly buried in a mass grave. In addition to the aid provided by Iran, Great Britain and TURKEY were instrumental in relief efforts.

dead fault Basically, an inactive FAULT. A dead fault is typically declared to have no chance of activity, whereas an inactive fault may become reactivated. However, declaring a fault incapable of producing earthquakes is a good way to risk having to eat one's words in a serious situation.

death toll There are horrifying death tolls reported for most of the earthquakes and volcanoes in this book. When the numbers are low, the death toll can be considered relatively accurate. Large events may not be so accurate because multiple sources must be tallied, and there is a lot of room for error. In the distant past, no one had time to count all of the bodies during a disaster, so the numbers tended to be very imprecise and could be in error by as much as an order of magnitude. More recent geologic events have had more accurate death tolls, but still have room for improvement. With inland earthquakes, where rescue and relief efforts are done properly, the numbers are usually accurate because bodies remain despite widespread destruction. With coastal earthquakes that spawn TSUNAMIS, there are often errors due to lack of evidence (bodies, buildings, and records swept out to sea). Volcanic eruptions can be equally inaccurate due to the severe destruction caused by the eruptions and LAVA FLOWS. Translations from other languages may also lead to errors that can then wind up throughout the literature. For example, one of the worst earthquake disasters on record occurred in China in the 1920s. Records have a death toll of 200,000. A new, more reliable source states that 200,000 livestock perished in this event but only 35,000 people. Death tolls may also be exaggerated by officials for political reasons. The great TANGSHAN earthquake in China had an official death toll of 249,000 people, but many scientists believe that it may have been closer to 655,000. The communist government at the time did not want to appear weak to its enemies and so reported a deflated death toll. On the other hand, when there is a lot of international aid available, some governments will inflate the numbers to get more money.

Death Valley *California, United States*   Death Valley—site of Death Valley National Monument—is famous for having the lowest point in NORTH AMERICA, the Badwater Depression, 282 feet (86 m) below sea level. Death Valley is part of the BASIN AND RANGE PROVINCE of the western UNITED STATES and shows abundant evidence of volcanic activity. Death Valley represents a block of crustal rock that is sinking so rapidly that erosion from surrounding areas cannot fill it fast enough to keep up with its descent. Some of the most recent volcanic activity in Death Valley is thought to have occurred several thousand years ago in the vicinity of Ubehebe Crater. About a half-mile wide, Ubehebe Crater appears to have originated in a PHREATIC ERUPTION in which a giant blast of steam followed contact between groundwater and MAGMA below the surface. Near the Black Mountains, the action of heated water on deposits of RHYOLITIC ash has created a spectacular display of colors.

Debal *earthquake, Pakistan*   There are historical accounts of a devastating earthquake that struck the Indus delta in present-day Sindh, Pakistan, near the city of Debal. Records are incomplete, but the event was recorded as having taken place between March 13, A.D. 893, and December 14, 894. Based upon the density of sand blows identified for this event compared to that from the nearby GUJERAT (Bhuj) event of 2001, the RICHTER magnitude of this event was estimated at 7.5. Historical documents state that the earthquake occurred at night, causing massive damage to the town of Debal. Only 100 buildings were left standing. Archaeological excavations of the area revealed human remains crushed in doorways and crouched in corners. There was even a case of a brick embedded in a human skull. All of the evidence pointed toward a huge disaster, and the DEATH TOLL was estimated as high as 150,000 people, making it one of the worst ever. Other historical documents pointed toward a great shift in power in the area and the abandonment of several cities near Debal. Geological studies indicated that the Indus and Nara Rivers abruptly shifted course to the west about this time, which might be consistent with a very large earthquake. The shift caused flooding and the abandonment of several major seaports on the river. It is interesting that this earthquake was approximately the same time as the ARDABIL, IRAN, earthquake, which was also reported to have killed 100,000 to 180,000 people.

debris avalanche   Avalanches are rapidly moving diffuse bodies of material on steep mountain slopes. They can be predominantly composed of snow (snow avalanche) or a mixture of materials such as SOIL, rocks, trees, ice, and other items all constituting debris. Debris avalanches are a type of MASS MOVEMENT characterized by velocities between 33.5 and 335 feet (10 and 100 m) per second. They originate on mountain slopes that collapse by any of a variety of mechanisms (earthquakes and volcanoes are the most common). The movement history can include a period of free fall, during which time the mass of debris can attain high velocities and travel long distances. The movement of the debris can generate strong wind by pushing air out of the way. These winds can be of hurricane force. Debris avalanches are devastating to all life that stands in their path. The scars left by debris avalanches are shaped like long narrow funnels. The trough-like chutes they leave look smooth from a distance but are usually pockmarked from the bouncing boulders.

debris flow   A moving mass of rock, soil, and organic material where more than half of the particles are larger than sand. Debris flows are included as LANDSLIDES and can move as slowly as feet per year to as quickly as hundreds of miles per hour. Earthquakes and volcanoes can trigger debris flows, but they are typically of the avalanche variety.

debris slide   Debris slides are masses of unconsolidated surficial material that break loose and slide down a sloped, underlying BEDROCK surface. They are a type of MASS WASTING. Both debris slides and ROCKSLIDES slip on an existing slide plane, such as a bedding plane or a FAULT plane. Debris slides are most common where thin layers of unconsolidated material mantle slope and becomes water saturated. They can

A debris avalanche is a very rapid mass movement down a steep slope. This debris avalanche flowed down a 1,500-foot (457-m) peak on the peninsula between Ugak and Kiliuda Bays, Alaska. Notice small overflow channels along the side of the slope. *(Courtesy of the USGS)*

then slip on the rock surface. Earthquakes can generate debris slides, and debris slides are common on the slopes of volcanoes where layers of ash can break free and slide.

decade volcanoes The major contribution of the International Association of Volcanology and Chemistry of the Earth's Interior (IAVCEI) to the International Decade for Natural Disaster Reduction (1990s) was the designation of decade volcanoes. The goal of IAVCEI's decade volcano project was to draw attention to a small group of active volcanoes and encourage a range of research and public awareness activities to enhance literacy on volcanoes and the hazards they pose. The following 16 volcanoes have been officially designated as decade volcanoes:

1. Avachinsky-Koryaksky, Kamchatka
2. Colima volcano, Mexico
3. Mount Etna, Italy
4. Galeras volcano, Columbia
5. Mauna Loa, Hawaii
6. Merapi volcano, Indonesia
7. Niragongo volcano, Republic of Congo
8. Mount Ranier, Washington
9. Sakurajima volcano, Japan
10. Santa Maria/Santiaguito volcano, Guatemala
11. Santorini volcano, Greece
12. Taal volcano, Philippines
13. Teide volcano, Canary Islands
14. Ulawun volcano, Papua New Guinea
15. Unzen volcano, Japan
16. Mount Vesuvius, Italy

Deccan Plateau *India* The Deccan Plateau is made up of FLOOD BASALTS and is comparable in form in many ways to the volcanic COLUMBIA PLATEAU and SNAKE RIVER PLAIN of the northwestern United States, but it is much larger. The Deccan Plateau was formed during the breakup of PANGAEA during the Cretaceous period. It is part of the early stages of the development of a DIVERGENT BOUNDARY. One theory is that this extensive volcanism is responsible for the mass extinction that included the demise of the dinosaurs. With the other contemporaneous flood basalts (Pirana and Karoo), this time was one of the most volcanically active periods in the last 500 million years. One theory suspects the Deccan Plateau and related volcanism of being an impact structure, created when a planetoid struck Earth's surface and generated a HOT SPOT that produced large flows of EXTRUSIVE ROCK.

Deception Island *South Shetland Islands, United Kingdom* A spectacular CALDERA in the middle of Deception Island (located near the Antarctic Peninsula) is thought to have been formed by tremendous eruptions of ANDESITE. Many VENTS are active along RING DIKES (FRACTURES) near the harbor. Records of eruptions at Deception Island before very recent times are probably incomplete because the island has no permanent settlements. When HMS *Chanticleer* stopped at Deception Island for several weeks in 1829, the captain reported hearing noises like that of "mountain torrents" emanating often from underground. The captain added

that these noises were so loud on one occasion that there was concern that they might damage instruments being used to conduct experiments at the island. Large numbers of FUMAROLES were visible in the caldera.

A LAVA FLOW may have been responsible for an observation in 1842 that the south side of the island appeared to be on fire. PYROCLASTIC deposits indicate that eruptions occurred in the first few years of the 20th century. Near a whaling station, the shore of the island subsided in 1923, and strange phenomena were observed in the bay, where water appeared to boil and paint was stripped from the hulls of ships. Several years later, in 1930, the floor of the harbor subsided some 15 feet (5 m) during an earthquake. Earthquakes preceded an eruption in 1967 that produced a new island, Yelcho Island, at the northern end of the harbor. Several weeks of seismic activity preceded an eruption in 1969, from vents along the east ring fracture in the caldera. Dramatic uplift was observed in 1970 along the northern part of the bay. The shoreline was moved more than 1,500 feet (457 m) in some places, and Yelcho Island was joined to the mainland. Satellite photos showed a plume from the island in 1987.

deep-focus earthquake An earthquake with a depth of focus from 186–435 miles (300–700 km). Normally, the depth of earthquake FOCI range from zero to approximately nine miles (15 km). Below that, faults undergo DUCTILE deformation which is aseismic. The main reason that the faults are ductile and aseismic is the increase in temperature with depth. In SUBDUCTION ZONES, however, the OCEANIC CRUST subducts so fast into the ASTHENOSPHERE that it doesn't have time to heat up. Therefore, it can still generate earthquakes even at such great depth. Deep-focus earthquakes originate only in subduction zones.

deep gas hypothesis *See* METHANE.

Deep Ocean Assessment and Reporting of Tsunamis program (DART) DART is an international early warning system for TSUNAMIS that are still distant from coasts. The equipment used in this system consists of seafloor seismic sensors that record changes in the weight of water above them to detect a tsunami passing over. Information is then sent to a buoy on the surface of the water. The buoy sends the data to the Geostationary Operational Environmental Satellite, which can broadcast information to land-based receivers. Coupled with the global seismic network, DART can prove effective in warning coastal communities about impending tsunamis, permitting full evacuations before their arrival.

deformation Deformation, or strain, is the change in shape or state of a material in response to an applied STRESS. Deformation may be BRITTLE, where a material breaks and loses cohesion, like broken glass; DUCTILE, where a material bends and stretches but does not lose cohesion, like clay or putty; or ELASTIC, where an object changes shape only when stress is applied but then returns to its original shape when stress is removed, like a rubber band. Brittle deformation generates earthquakes, and the seismic waves cause elastic deforma-

A tsunami receiver-transformer buoy being deployed from a research vessel in the Pacific Ocean. *(Courtesy of NOAA)*

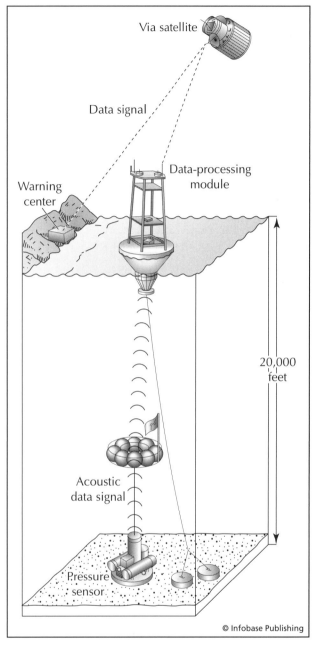

Diagram showing the components of DART. The ocean floor pressure sensor sends information to the fixed position buoy that processes the data and beams it to a satellite that in turn beams it to a warning center. Although complex, this is a real-time monitoring system.

tion of the materials through which they pass. Deformation may also be homogeneous, where it affects an entire body of material in the same way, or heterogeneous, where it is concentrated in a zone like a FAULT.

Delaware *United States*   The small state of Delaware is located in a region of minor seismic risk. Only a few notable earthquakes have occurred in Delaware. An earthquake on October 9, 1871, reportedly caused damage to windows and chimneys in Wilmington and was felt also in Oxford and New Castle, as well as in PENNSYLVANIA and NEW JERSEY. On March 25, 1879, an earthquake along the Delaware River was felt over an area of about 600 square miles (1,554 km²), in Dover and below Philadelphia, Pennsylvania. The affected area extended between Chester, Pennsylvania, and Salem, New Jersey. The earthquake of May 8, 1906, near Seaford, Delaware, shook buildings and was felt elsewhere in Delaware and in MARYLAND. A significant earthquake (over 4 on the RICHTER scale) took place in Wilmington in 1971.

Denali Fault *Alaska*   A powerful earthquake of MAGNITUDE 7.9 occurred on the Denali Fault, ALASKA, just after noon on November 3, 2002. Because the area was so sparsely populated, no one was killed, and there was little damage. The Denali Fault is a major DEXTRAL strike-slip fault with regular and intense seismic activity. There was a major FORESHOCK of magnitude 6.7 10 days prior to the MAIN SHOCK. When the quake occurred, it propagated eastward at 7,000 miles per hour (11,200 km/hr) to produce a SURFACE RUP-TURE some 209 miles (335 km) long and with up to 29 feet (8.7 m) of offset. Shaking lasted approximately 90 seconds, producing a 16–24-mile (9.6–14.5-km)-wide band of extensive LANDSLIDES along the fault and FISSURES that could have swallowed a bus. LIQUEFACTION caused SAND BOILS and other surface flow, and SEICHES were strong in Seattle, WASHINGTON, but identified as far away as LOUISIANA. YEL-LOWSTONE experienced an intense EARTHQUAKE SWARM as a result of the shock. Remarkably, the Trans-Alaska Oil Pipeline that crosses the Denali Fault, carrying 17% of the U.S. domestic oil supply, did not break even after the earthquake shifted the ground some 14 feet beneath the pipeline. If it had broken, it would have resulted in a major environmental and economic disaster for the United States. The pipeline was so well engineered that there was not even a disruption in supply. This is earthquake prevention at its best.

Denizli *earthquake, Turkey*   At 8:30 A.M. on February 25, 1702, a devastating earthquake shook southwestern TURKEY. The EPICENTER was at the kaza of Denizli, but the shock was violent in Smyrne and felt strongly in CHIOS. The estimated RICHTER magnitude was 7.0, and damage was estimated at X on the modified MERCALLI scale. The shock was so strong that it diverted the course of the Gumus Cay (River) near Eskihisar. There are reports of tent cities having been set up because there were so many buildings destroyed. The DEATH TOLL for this event was 12,000 people in Denizli alone.

Denizli was reportedly struck by yet another strong earthquake on November 19, 1717. It had an estimated MAGNITUDE of 6.6 and an intensity of IX on the modified Mercalli scale. It reportedly flattened 10 villages within a six-mile (10-km) radius of Denizli and killed 6,000 people within Denizli alone.

Deriba *caldera, Sudan*   The Deriba CALDERA occupies the summit of Jebel Marra Volcano in the western central portion of the Sudan. Although the geologic history of the volcano is uncertain, it has been suggested that OLIVINE basalt built up during early eruptions at the site and then was eroded away after activity ceased. When eruptions resumed in a second phase of activity, the LAVAS were more FELSIC. According to this scenario, the third and most recent phase of activity has involved the formation of many secondary cones. Deriba caldera is thought to have originated during this third phase. A central cone and a lake have formed in the caldera. Although it is not known exactly when the caldera formed, it may have originated in historical times. There have been no recorded eruptions at the caldera, but hot springs and FUMAROLES were reported in the CRATER in the mid-20th century.

Jebel Marra is located within the great Sahara. However, its higher elevation causes it to get much more rain than the surrounding area, a phenomenon called orographic precipitation. Not only is there a lake in the caldera but streams flow down the slopes but disappear into the desert.

Devils Tower *Wyoming, United States*   Devils Tower is a famous example of a VOLCANIC NECK. The volcano that formed Devils Tower was active some 40 million years ago

The famous landmark Devils Tower, in Wyoming, is a volcanic neck. There used to be a conical volcano around it, but weathering has left only the old conduit for lava. Because the cone was mostly ash and pyroclastics, it was more susceptible to weathering. Columnar joints give it the tree stump appearance. *(Courtesy of the USGS)*

and is an unusual composition called a phonolite. The neck cooled relatively quickly and at a very shallow level. The vertical COLUMNAR JOINTS all around it, just as in volcanic flows, are a result of this cooling history.

dextral fault   Also referred to as a right lateral STRIKE-SLIP FAULT. It is a strike-slip fault where the block on the opposite side of the fault from the observer moves to the right. There is only lateral and no vertical movement on these faults.

Dhamar *earthquake, Yemen*   At 9:13 P.M. on December 13, 1982, a strong earthquake struck northern Yemen. The EPICENTER was near the city of Dhamar, and the FOCUS was shallow, at about three miles (5 km) in depth. The RICHTER magnitude of the earthquake was 6.0, and the damage was VIII on the modified MERCALLI scale. The DEATH TOLL from this event was 2,800 people as the result of intense shaking of structures that were not designed to be earthquake-resistant. In all, the earthquake destroyed over 300 villages and left 700,000 people homeless. Total damage was estimated at about $2 billion.

diabase   The MAFIC plutonic rock, GABBRO, but with a particular texture. In these rocks, PYROXENE must occur in well-formed crystals, whereas PLAGIOCLASE fills the interstices. They commonly occur in dike feeders to BASALT volcanoes.

Damage to buildings as a result of the Dhamar, Yemen, earthquake in 1982 *(Courtesy of the USGS)*

Diamante *caldera, Chile* The Diamante CALDERA is located in south-central CHILE and is thought to have been formed by great eruptions of PYROCLASTIC FLOW deposits that covered several thousand square miles. The STRATOVOLCANO Maipo occupies part of the western portion of the caldera. Several eruptions were reported in the 19th and 20th centuries, but the accuracy of these reports is questionable because the "eruptions" may have been nothing but enduring plumes of steam. Its remote location makes information sparse and unreliable.

diamond Diamond is the hardest naturally occurring mineral and one of the most precious gemstones. Hardness means that it can scratch any other substance but that it cannot be scratched. Diamonds are made of pure carbon just like the mineral graphite, but the atoms are packed more tightly together under MANTLE pressures. Diamond is associated with plutonic-volcanic formations, particularly the famous KIMBERLITES, or "diamond pipes," of South Africa. Diamonds are believed to have been found first in Borneo and INDIA. Brazil was a major producer of diamonds in the 18th and 19th centuries. Diamonds were discovered in South Africa in 1867 on the shore of the Orange River near Hope Town, and several years later, primary deposits of diamonds were found on the plateau between the Modder and Vaal Rivers. These primary deposits were located in volcanic pipes containing a variety of PERIDOTITE (kimberlite). These pipes intersected the surface in circular or elliptical areas. Some diamonds were recovered on the surface from a weathered material called yellow ground, but others had to be extracted from the harder "blue ground" below. Important South African diamond mines include Bultfontein, De Beers, Du Toitspan, Kimberly, and Wesselton. For a time most of the world's annual output of natural diamonds came from the pipe mines, but, later, alluvial deposits in AFRICA surpassed the pipe mines in production.

Diamond production is not confined to Africa; diamonds have been recovered in parts of the world as widely separated as MEXICO, Siberia, and CANADA. Diamonds not used as gemstones have widespread industrial uses. Industrial-quality diamonds occur naturally in forms including ballas (spherical aggregates of diamond crystals), bort (crystals with irregular shape, bad flaws, and numerous inclusions; used in drilling and abrasives), and carbonado (a dark variety of diamond). Although a colorless diamond is pure carbon (as demonstrated by the fact that burning such a diamond in an oxygen atmosphere yields only carbon dioxide), other elements may occur in diamonds and affect the color of the stones; blue, green, and yellow gem-quality diamonds have been found.

Dieng Plateau *Java, Indonesia* The Dieng Plateau is an area of frequent volcanic activity and is thought to occupy an ancient CALDERA. The word *Dieng* means "Abode of the Gods" and as such, there are many temples located there. There have been at least 18 eruptions since 1375. Small explosive and PHREATIC ERUPTIONS (VEI = 1–2) characterize the area. An explosive eruption apparently occurred around the year A.D. 1300. An eruption in 1786 was preceded by several months of earthquakes and involved emissions of sulfurous vapor. That eruption killed 38 people and destroyed a village. An eruption in October 1826 was accompanied by sounds like cannon fire. Several days later, explosions occurred, and earthquakes were felt at a considerable distance. The source of this activity is uncertain. Except for a minor eruption in 1847, the Dieng Plateau was quiet until 1884, when SOLFATARAS and FUMAROLES became more active and mud eruptions occurred. Powerful earthquakes shook the area in 1924 and 1928. Eruptive activity started just after an especially damaging series of earthquakes in May 1928. Some 40 people were killed during this eruption. Gas emissions continued for the next nine years. In 1939, earthquakes preceded minor phreatic eruptions that killed 10 people. LAHARS were produced during this eruption. Late in 1944, several phreatic explosions occurred. One of these killed a large number of people; one estimate puts the DEATH TOLL at more than 117 with 250 injuries. A small ASHFALL took place in 1953, and a minor eruption occurred in 1954. An eruption in 1964 killed some 114 people. More than 149 people were killed in a large outburst of carbon dioxide and hydrogen sulfide gas in 1979. The people were asphyxiated as they fled. Of those who managed to escape, 1,000 were injured, 100 of whom required hospitalization. More than 15,000 people were evacuated during this event, and many livestock were killed. A minor EARTHQUAKE SWARM was noted in 1981, and another such swarm in 1984 caused some damage to property. Another earthquake was felt over a wide area around Dieng in 1986, but thermal activity appears to have been unaffected. A HYDROTHERMAL area in the Dieng Plateau has been investigated for GEOTHERMAL ENERGY.

dike A near vertical, tabular body of intrusive rock that cuts across the local COUNTRY ROCK strata at a high angle. As opposed to forcing their way into country rock, dikes passively fill cracks that are tectonically produced. Dikes may radiate outward in a starlike pattern from a central volcanic neck or column of rock left standing after erosion has removed the outer layers of the volcano. They can also form rings around volcanic centers. It just depends upon the STRESS field in the area.

dilatancy Defined as the increase in volume resulting from tiny (or any other size) cracks developing in a rock, dilatancy has been an important concept in attempts to predict earthquakes. There has been disagreement about the process or processes involved in the postulated relationship between dilatancy and seismic activity. According to one hypothesis, water moves into the area affected by the small cracks just before an earthquake, weakening the rock; in another hypothesis, most of the cracks close in the affected region just before an earthquake occurs.

dilational wave *See* P-WAVE.

Dinar *earthquake, Turkey* On October 1, 1995, an earthquake of RICHTER magnitude 6 occurred. There were 101 deaths and thousands injured. Some 50,000 people were left homeless as the result of the destruction of more than 4,500 houses and building.

diorite A PLUTON of INTERMEDIATE igneous rock. It is the plutonic equivalent of the volcanic rock ANDESITE or andesite to DACITE. Common minerals include PLAGIOCLASE, PYROXENE, and HORNBLENDE. These rocks are common in MAGMATIC ARCS, at least in the early stages. They form the lower part of the huge BATHOLITHS found there. They are commonly succeeded by GRANODIORITE and GRANITE. Examples of diorite plutons can be found in SIERRA NEVADA.

dip-slip fault A fault in which movement is in a dip sense. This means that all movement is in a vertical rather than in a lateral sense. Dip-slip faults include normal faults, REVERSE FAULTS, and a subset of reverse faults called THRUST FAULTS.

directivity Directivity is an effect of FAULT rupture where the ground motion from the earthquake is more intense in the direction of the rupture propagation than it is in any of the other directions from the earthquake source.

discordant Refers to tabular PLUTONS that are intruded across the existing strata. DIKES, for example, are discordant by definition. In contrast, SILLS are concordant because they are intruded along the layering. For plutons emplaced at shallow levels, like those that feed volcanoes, the distinction is relatively plain. For plutons emplaced at deep levels, however, the surrounding rocks are soft and flow with and around the pluton. The distinction can be complex.

dispersion (of waves) Dispersion is the spreading out of an earthquake wave train as a result of each WAVELENGTH in the train traveling with a different velocity.

displacement Displacement is the change in position of a reference point over time. In fault terms, it is the amount of movement that takes place on an individual fault through a single or multiple events. The displacement on a fault for a single earthquake might be centimeters to meters, either subsurface (BLIND) or shown by the offset of a surface feature. DUCTILE faults may undergo semi-continuous movement so do not have specific events but still experience displacement.

divergent boundary A boundary between lithospheric plates in which the plates move away from each other. The boundary can occur in one of two geometries, symmetric (pure shear) where both sides appear the same and asymmetric (simple shear model) where they do not. Divergent boundaries occur in two types, those on CONTINENTAL CRUST and those on OCEANIC CRUST. On continental crust, they are RIFT

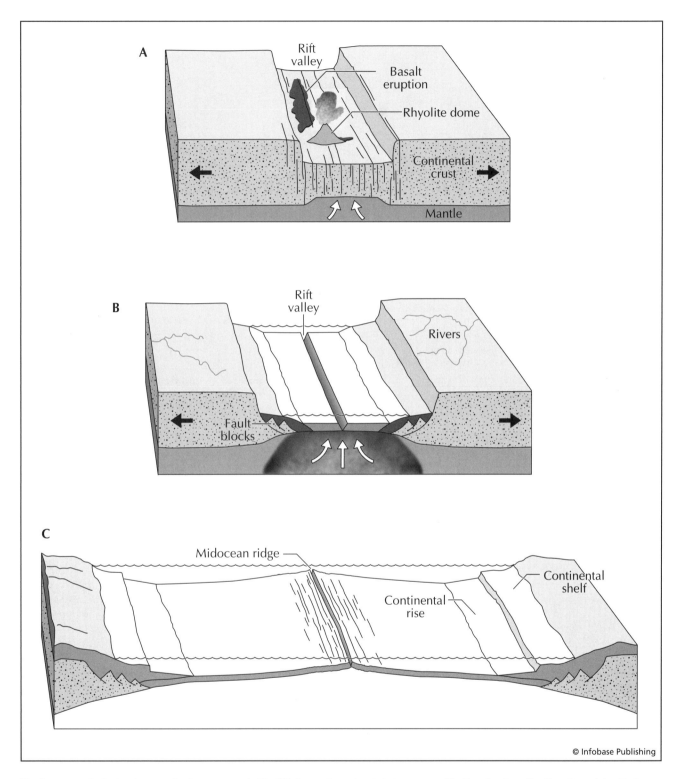

The three stages in the development of a divergent margin. First (A), the crust can elevate, but soon normal faults and grabens, flood basalts, and rhyolite domes form. The Basin and Range Province is in this stage. Next (B), a large depression or rift valley forms that eventually contains ocean crust and a narrow sea. The Gulf of California and the Red Sea are similar to this phase. Finally (C), a full ocean basin and a mid-ocean ridge such as that in the Atlantic Ocean is formed.

zones with four distinct stages. The stages begin with thermal bulging and uplift continue with subsidence and basin development. Finally, there is the development of oceanic crust and a MID-OCEAN RIDGE. Therefore, one type develops into the other with time. The early two stages include normal faulting and extensive earthquake activity. Later, in the second stage, there is FLOOD BASALT volcanism from FISSURE ERUPTIONS. These LAVA FLOWS form the most voluminous volcanic deposits on the continental crust. Examples include the COLUMBIA PLATEAU of WASHINGTON, the DECCAN traps of INDIA, the Piranha basin of SOUTH AMERICA, the WATCHUNG basalts of NEW JERSEY, and the Karoo deposits of AFRICA. Some areas also produce RHYOLITE volcanoes. Unlike the BASALTS that simply spill out of cracks in the ground, the rhyolites form well-developed cones. These volcanoes are highly explosive. The BASIN AND RANGE PROVINCE of the western UNITED STATES, the GREAT RIFT VALLEY system of Africa and the RIO GRANDE RIFT of TEXAS form good examples of these stages of rifting. The next two stages involve the development of an ocean basin. First is a restricted linear basin that contains a mid-ocean ridge and is very active. Later, the basin becomes wide and forms a full ocean basin. Volcanism is purely basaltic of the THOLEIITIC variety in these stages. Normal faults continue to be active in both stages, but TRANSFORM faults dominate as the earthquake producers in the full ocean basin. New oceanic crust is created in the mid-ocean ridge in both stages. The youngest part of the ocean basin is therefore at the mid-ocean ridge and becomes progressively older away from and toward the continents. The new crust is largely layers of PILLOW LAVAS cross cut by feeder dikes to the underwater volcanoes. The Red Sea is a good example of the early stage of basin development, and the ATLANTIC OCEAN is an example of the later stages.

Dominica *Lesser Antilles* Dominica is situated in the northern portion of the Lesser Antilles ISLAND ARC and has an unnamed CALDERA with a history of unrest dating back to the late 17th century. The caldera is estimated to be about six miles (10 km) wide. Several VENTS are present, including Grande Soufrière Hills, Morne Macaque (Micotrin), Morne Trois Pitons, the Valley of Desolation, and Watt Mountain. There is disagreement as to whether these vents belong to a single volcano. Other vents on the island include Foundland and Morne Anglais. Only one eruption has occurred here within historical times, a minor steam explosion in 1880.

There has been considerable earthquake activity at Dominica over the past several centuries, however, starting with the report of a moderately strong earthquake in 1673. Emissions of gas and numerous strong earthquakes were reported in 1765, and more strong quakes occurred in 1816 and 1838. In 1843 a very strong TECTONIC earthquake took place near Dominica and was felt both there and on nearby islands. Lesser earthquakes occurred at intervals of several years through 1849. A moderately strong earthquake in 1879 was accompanied by a minor PHREATIC ERUPTION and a black cloud rising from the area of SOLFATARAS. It is uncertain whether the earthquake was linked with the eruptive activity. Seismic swarms have been recorded at Dominica in 1967, 1971, 1974, and 1976. A moderately strong earthquake occurred approximately 100 miles (160 km) north of Dominica in 1976 and was viewed as the result of MAGMA movements.

dormant volcano Literally, a sleeping volcano. This volcano was active in the past and may well be active in the future but under the current conditions is inactive. Volcanoes may stay inactive for thousands of years and suddenly spring back to life. Dormancy is really not measured in geological terms but in historical terms. If the volcano has been active in historical times but is currently inactive, it is termed *dormant*. If the volcano has not been active in historical times, it is deemed *inactive*. The problem, however, is that there really is no distinction between active but not currently erupting and dormant.

ductile Deformation of rock that does not produce earthquakes. Ductile deformation of rock means that instead of breaking, it bends or flows. Ductile deformation depends upon composition and other factors. Clay is ductile at surface conditions. GRANITE, on the other hand, cracks and FRACTURES under surface conditions. At depth, under high temperature conditions, however, even granite is ductile. Because it can't rupture, no earthquakes can occur. This transition from brittle rupturing rock to ductile deformation occurs at about six to nine miles (10 to 15 km) depth for common CONTINENTAL CRUST. It is somewhat deeper for OCEANIC CRUST. This transition means that there are few earthquakes deeper than six to nine miles (10 to 15 km).

Duguna *See* ASAWA.

Dukono *volcanic complex, Halmahera, Indonesia* The Dukono volcanic complex is one of the most active volcanoes in INDONESIA. Its first recorded eruption was in 1550, and it was later active in 1719, 1868, and 1901. Dukono has been in nearly continuous eruption since 1933. The eruptions are moderately explosive (VEI = 3) and produced LAVA FLOWS and LAHARS. Other volcanoes in the complex include Ibu (one eruption in 1911), Todoko-ranu, and Jailolo. The only other active volcano is Gamkonora. It erupted 12 times between 1564 and 1987. The 1673 eruption was very strong (VEI = 5) and produced a large TSUNAMI that devastated several nearby coastal villages.

dust *See* VOLCANIC DUST.

E

Earth, internal structure of Earthquakes and volcanoes are manifestations of processes at work deep within Earth. These processes in turn are evidence of Earth's internal structure, models of which have undergone considerable evolution in the 20th century. On a large scale, the internal features of Earth may be categorized as follows:

1. *Core.* This is the innermost layer of Earth, a dense, approximately spherical mass of very hot rock that accounts for roughly one-third of the planet's mass and one-sixth of its volume. The core is divided into a solid inner core and a liquid outer layer.
2. *Mantle.* The mantle, the thick layer of rock surrounding the core, contains about two-thirds of Earth's mass and more than four-fifths of its volume. An outer layer of the mantle is called the ASTHENOSPHERE and is involved closely with driving PLATE TECTONICS. There is also a division within the upper and lower mantle based on mineral structure.
3. *Crust.* The crust, or surface layer, of Earth is very thin compared to the mantle and core (five to 80 miles [8 to 129 km] thick) and accounts for only a tiny fraction of Earth's total mass and volume. The crust floats on the asthenosphere by a process called ISOSTASY, in which the relatively lightweight rocks of the crust are supported by the denser rocks below. The crust is divided into a number of individual, rigid plates that interact with one another through various processes, generating earthquakes and volcanic activity. The crust contains many FAULTS that produce numerous earthquakes.

earthflow An earthflow is a type of MASS MOVEMENT. It is a viscous flow of saturated, fine-grained materials that moves downslope at speeds ranging from barely perceptible up to about 10 miles per hour (about 0.1 m/sec). Materials susceptible to earthflow include clay, fine sand and silt, and fine-grained PYROCLASTIC material (primarily ASH). The velocity and distance of the earthflow is controlled by water content, with faster and longer movements through higher water con-

tents. Some earthflows may continue to move for years. Earthflows normally begin when water content increases either through rain, melting (normal heating or VOLCANIC ERUPTION), or LIQUEFACTION during an earthquake. Shaking from an earthquake may also initiate flow in a saturated SOIL. The flows tend to bulge in the middle as they move, which in turn channels more fluid to the middle while the edges dry out. Earthflows will stop moving when the water content drops.

earthquake *See* SEISMOLOGY.

earthquake hazard An earthquake hazard is considered to be any of the damaging effects and processes of an earthquake that may affect the normal activities of people. Examples of earthquake hazards include surface faulting, ground shaking, LANDSLIDES, liquefaction, SLUMPing, fissures, AVALANCHES, TSUNAMIS, and SEICHES, among others.

earthquake light This phenomenon consists of a peculiar glow that sometimes is reportedly seen in the sky during earthquakes. What causes earthquake light is uncertain, but it has been suggested that methane escaping from underground during earthquakes is ignited somehow and burns near the surface, giving off light.

earthquake risk Earthquake risk is the probable damage to buildings, roads, services, and other infrastructure and the number of people that are expected to be killed or injured during an earthquake of a particular size in a particular location. Earthquake risk is a probabilistic model for a specific seismic event. It varies considerably with each area depending upon population and emergency readiness.

earthquake swarm Many earthquakes preceding a volcanic eruption. As MAGMA move upward in the CRUST, it pushes rock out of the way. It forces open cracks in the rock. Each

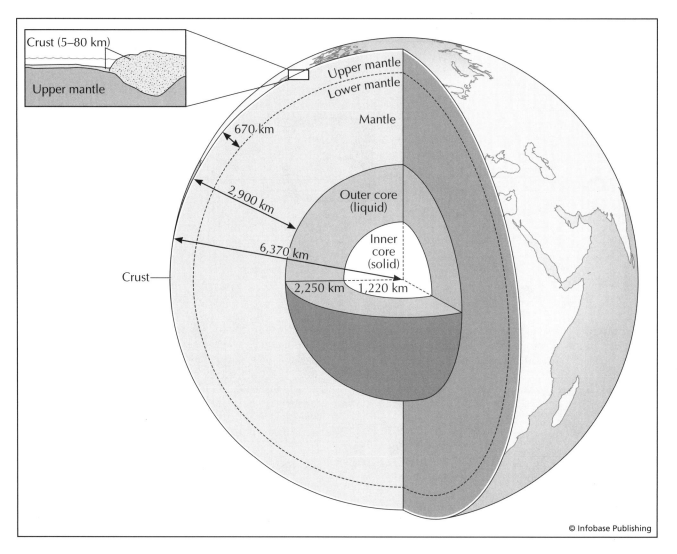

Diagram of Earth with a cutaway to show its internal structure. The center of Earth has an inner core (solid) surrounded by an outer core (liquid). The mantle encases the core and is divided into a lower and an upper mantle based on the physical character of the rocks and minerals. The composition is the same—peridotite. The crust is very thin in comparison with the rest of the planet so that the blowup inset is required to illustrate the difference between ocean crust and continental crust.

break results in a small earthquake. These earthquakes are typically of MAGNITUDES of 4 or less, but some larger earthquakes are possible. There can be hundreds to thousands of earthquakes in such swarms. VOLCANOLOGISTS use this kind of seismic activity to predict eruptions. The correlation is not foolproof. Many earthquake swarms do not foretell a coming eruption but rather just movement of magma.

East Pacific Rise   A MID-OCEAN RIDGE 6,000 miles (9,700 km) long, the East Pacific Rise is the site of remarkable geological formations and ecological communities on the seabed. An expedition to the East Pacific Rise in 1980 discovered that superheated (more than 700°F), mineral-laden water flowing from springs, or "hydrothermal VENTS," on the seabed had generated peculiar "chimneys" formed of minerals deposited when the hot water met the cold ambient waters of the deep

ocean. Around the vents lived giant worms and large clams and crabs in an ecosystem supported by bacteria feeding on hydrogen sulfide, oxygen and carbon dioxide in the water from the vents.

Eboshi-dake   *See* DAISETSU-TOKACHI.

economic effects of earthquakes and volcanoes   The economic effects of earthquakes and volcanic eruptions can be tremendous. After witnessing the effects of a powerful earthquake in CHILE, the 19th-century British naturalist Charles Darwin wrote:

> Earthquakes alone are sufficient to destroy the prosperity of any country. If beneath England the now inert subterranean forces should exert those powers,

which most assuredly in former geological ages they have exerted, how completely would the entire condition of the country be changed! What would become of the lofty houses, thickly packed cities, great manufactories, the beautiful public and private edifices? If the new period of disturbance were first to commence by some great earthquake in the dead of the night, how terrific would be the carnage! England would at once be bankrupt; all papers, records, and accounts would from that moment be lost. Government being unable to collect the taxes, and failing to maintain its authority, the hand of violence and rapine would remain uncontrolled. In every large town famine would go forth, pestilence and death following in its train.

A single earthquake or volcanic eruption can cause property damage in the hundreds of millions of dollars, and a very powerful shock or eruption occurring in or near a heavily populated area might cause damage into the billions of dollars. Of particular concern in our time is the anticipated earthquake in the TOKYO, JAPAN, metropolitan area. Tokyo has undergone major earthquakes at intervals of approximately 75 years during the last few centuries, and the most recent major earthquake there occurred in 1928. The cost of rebuilding Tokyo after the next "superquake" has been estimated at $1 trillion. In these days of an increasingly integrated global economy, a highly destructive earthquake in Tokyo might have drastic economic effects in many other countries as well, including the UNITED STATES.

Ecuador   Ecuador is located on the western shore of SOUTH AMERICA, where the westward-moving continental crustal plate encounters OCEANIC CRUST and generates earthquakes where the two plates collide. Ecuador has a history of highly destructive seismic activity, notably the great earthquake of August 5, 1949. This earthquake, of MAGNITUDE 7.5 on the RICHTER scale, was centered 25 miles (40 km) deep and affected an area of 1,500 square miles (3,885 km^2) along the eastern side of the ANDES, particularly Ecuador's highland plateau. The earthquake killed more than 6,000, injured some 20,000, and left approximately 100,000 people homeless. More than 50 cities and towns experienced severe damage, and total damage was estimated at more than $60 million.

Ecuador has some very active volcanoes as well including Altar, Antisana, Cayambe, Chimborazo, COTOPAXI, Cusin, GUAGUA PICHINCHA, Ilinzia, Imbabura, Mojanda, Ninahuilca, Pasochoa, REVENTADOR, Ruminahui, SANGAY, Sumaco, and Tungurahua. Chimborazo is the largest of the volcanoes, standing 20,700 feet (6,309 m). Sangay is probably the most active of the volcanoes, erupting nearly continuously since 1934. Guagua Pichincha is probably the most dangerous volcano because it overlooks QUITO, the capital city of Ecuador. The city has been devastated several times by the larger of its 25 historical eruptions. The worst eruption was in 1660 when a foot of ash blanketed the city.

Edo *earthquake and tsunamis, Japan*   A submarine earthquake struck an area south of the Boso Peninsula just east of TOKYO Bay on December 30, 1703. The estimated RICHTER magnitude for this event was 8.0. The quake destroyed the city of Edo, one of the more prosperous settlements in the area. Bathymetric studies along with drill coring in this area indicated that the seafloor was uplifted about 17 feet (5.1 m) in several places. As a result, some previously outlying reefs became dry land. Uplift also occurred along the coast at various locations, producing TSUNAMIS that struck the entire coast. There was extensive damage in nine provinces. The official reported DEATH TOLL for this disaster was 5,233 people, but there are some data that indicate a death toll as high as 37,000.

Egon *volcano, Indonesia*   Egon is among a group of volcanoes in the Flores Islands. The other volcanoes include Lewotobi, Ilimuda, and Lereboleng. Egon erupted in 1907 and likely in 1888–89. Lebwotobi had at least 19 eruptions between 1675 and 1991. Most eruptions are moderate in size with a VEI of 2–3. The largest eruptions were in 1869 and 1907.

Egypt   Egypt's position on the northern and northeastern edge of the African plate make it subject to earthquakes, though not directly associated with volcanoes. The Sinai border is along the Red Sea, which is an active DIVERGENT margin capable of sustaining active faults. It also faces the Eastern Mediterranean Sea, which is highly active by virtue of the north-dipping Hellenic Arc SUBDUCTION ZONE and impingement of TURKEY, which is being forced westward through EXTRUSION TECTONICS. These complex interactions yield complex faulting and earthquake activity.

Most of the historic activity is within the Nile valley and Nile Delta, where soft sediments amplify the waves, causing great destruction. Notable historic earthquakes include the A.D. 365 ALEXANDRIA event that destroyed the historic lighthouse and caused tens of thousands of deaths; the 1303 Alexandria earthquake that claimed 10,000 lives, and the 1754 CAIRO earthquake that claimed 40,000 lives. In all, there were 58 historical earthquakes between 2200 B.C. and A.D. 1900, with modified MERCALLI intensities between V and IX, 22 of which are well documented. Instrumental seismicity (recorded on a SEISMOGRAPH) was from 1900 to the present. Four distinct zones of activity were apparent from these data: Northern Red Sea–Gulf of Suez–Alexandria trending northwest-southeast, Gulf of Aqaba–Levant Fault trending north northeast–south southwest, Eastern Mediterranean–CAIRO–Faiyum trending northeast-southwest, and Egypt–Mediterranean Coast tending east-west. Most of the FAULTS in these zones experienced primarily STRIKE-SLIP motion. The most recent powerful earthquake was in Cairo in 1995 with a 7.0 MAGNITUDE.

ejecta   Solid volcanic fragmental material is ejecta regardless of composition or internal textural relations. The classification is based purely on the size of the fragments. Fine ash or volcanic dust is less than 0.2 inch (0.5 cm); coarse ash is between 0–.08 inch (0.01–0.2 cm). LAPILLI are 0.8–2.5 inches (0.2–6.4 cm) and along with ash is lithified into a TUFF. BOMBS are 2.5–10 inches (6.4–25 cm), and BLOCKS are larger than 10 inches (25 cm). Rocks of these ejecta are called AGGLOMERATE.

elastic rebound theory   This is the basic accepted theory for how earthquakes are generated. The idea is that FAULTS will remain locked with no slippage while TECTONIC forces slowly build up stress within the rocks. When the STRESS exceeds the strength of the rocks or locks on the fault, the fault slips, releasing or converting all of the stored stress energy in the form of heat and seismic waves in an earthquake.

elastic strain   Elastic STRAIN means that a material will strain as a direct function of the STRESS applied and then return to its original shape after the stress is released. This deformation is referred to as recoverable because it is not permanent. Springs and rubber bands exhibit elastic behavior: They stretch out as a function of how hard they are pulled and then return to their original shape when they are let go. Seismic waves cause elastic strain in rocks. The rocks deform as the wave passes through but return to their original shape afterward. The amount that rocks deform in this process is minuscule, especially in comparison with springs and rubber bands. Even if you could slow the wave down and watch it pass through a small piece of rock, you still wouldn't be able to see the rock change its shape. Both the change in and the return to shape are essentially instantaneous. For some substances, the return to original shape is not immediate. Cellophane behaves in this manner. It crushes immediately, but it will slowly uncrush after it is let go. This kind of slow recovery is referred to as anelastic.

Eldfell   *See* HEIMAEY.

Eldgja   *See* KATLA.

embryonic volcano   A VOLCANIC PIPE that is filled with volcanic BRECCIA but for which there is no surface expression. These features are considered to have formed through phreatic explosions. They are precursors to full volcanoes if activity continues in the area.

Emmons Lake *caldera, Aleutian Islands, Alaska, United States*   The Emmons Lake CALDERA is located near PAVLOF volcano in an area of frequent seismic and volcanic activity.

Emperor Seamounts *northern Pacific Ocean*   A chain of submerged volcanic mountains in the northern PACIFIC OCEAN, the Emperor Seamounts are part of a string of volcanic SEAMOUNTS and islands reaching almost from Hawaii to Russia's KAMCHATKA PENINSULA. The Emperor Seamounts and the HAWAIIAN ISLANDS are both part of this same chain, which formed as the PACIFIC crustal plate moved over a stationary HOT SPOT of volcanic activity. A sharp bend in the chain marks the division between the Emperor Seamounts and the Hawaiian Islands and indicates that the Pacific crustal plate changed its direction of motion abruptly, from north to northwest.
    See also PLATE TECTONICS.

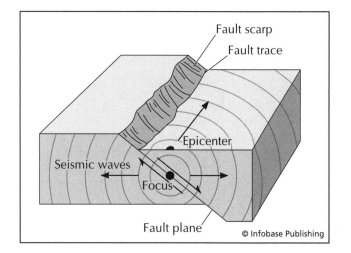

Block diagram showing a fault on which an earthquake is taking place. The hemispheres centered on the focus are seismic wave fronts. This is a normal fault.

emplacement   The mechanical process by which MAGMA is intruded into rocks of an area to form a PLUTON. There are several ways this can happen. The magma can eat into the rock above it by breaking off fragments of rock and pulling them into the magma. This pulling down of rock is called stoping, and the fragments are called XENOLITHS. Alternatively, magma can forcibly move the surrounding rock strata out of the way as it pushes upward. The magma can also filter up along fractures as it intrudes into the Earth.

epicenter   The location on Earth's surface directly above the HYPOCENTER, or FOCUS, of an earthquake. Typically, the area where the potential for the greatest damage occurs.
    See also SEISMOLOGY.

Erebus, Mount *Antarctica*   Located on Ross Island, Mount Erebus stands 12,448 feet (3,794 km) high and is the most active volcano in Antarctica. The cone on the volcano's summit occupies a CALDERA. The main CRATER is approximately 450 feet (137 m) deep and has a diameter of 1,500 to 1,800 feet (457 to 549 m). An inner crater is about 600 feet (183 m) wide and 300 feet (91 m) deep. The volcano was discovered by the British explorer James Ross on an expedition in 1841. Ross named it Mount Erebus after his flagship, and he named the nearby volcano Mount Terror after another ship on his expedition. Mount Erebus was evidently in eruption when Ross saw it. Flashes of red light were visible from the mountain by night, and by day black vapor surrounded the summit. On another expedition in 1908, a British team climbed successfully to the summit of Mount Erebus. The volcano is characterized by eruptions of the STROMBOLIAN variety, from pools of LAVA in the crater.

erosion caldera   A caldera that forms from erosional processes rather than a volcanic eruption. They are large depressions with steep sides. Typically, they result from the erosion of a preexisting volcanic caldera or a VOLCANIC CONE.

Penguins with Mount Erebus in the background. Mount Erebus is the most active volcano in Antarctica. *(Courtesy of the USGS)*

eruption Generally speaking, any volcanic activity that releases LAVA, TEPHRA, or gases. Eruptions may take many forms, from explosive, highly destructive eruptions like that of Mount SAINT HELENS in 1980 to comparatively harmless eruptions like those characteristic of volcanic activity in the HAWAIIAN ISLANDS. Volcanic eruptions fall into several categories:

- Hawaiian eruptions are gentle by volcanic standards and generate gently sloping volcanoes known as SHIELD VOLCANOES. Volcanoes of this kind erupt frequently and may form a lake of lava in the CRATER, from which fountains of lava arise.
- STROMBOLIAN ERUPTIONS, named after the STROMBOLI volcano and volcanic island off the Italian coast, are more energetic than Hawaiian eruptions but do not cause widespread damage. The volcano casts out some lava but little ASH.
- Icelandic eruptions do not involve the classic pattern of ash and lava emerging from a mountain peak. Instead, great quantities of highly fluid LAVA FLOW from FISSURES in Earth. These voluminous lava flows may fill valleys and form dams in river valleys, blocking the river's flow and causing flooding upstream. Icelandic eruptions are much like Hawaiian eruptions, except that in Icelandic eruptions, the layers of lava spread out horizontally and form great sheets rather than pile up into domelike formations.
- Peléan eruptions, named for the devastating eruption of Mount PELÉE in 1902, involve such thick, viscous lava that it plugs the throat of the volcano, so to speak, and forms a dome of solid lava in the crater. Meanwhile, pressure builds up inside the volcano until it explodes, sending a *NUÉE ARDENTE* sweeping down the flanks of the mountain.
- Vulcanian eruptions are named for VULCANO, a mountain near Stromboli, and resemble the classic picture of such events, with a large cloud of ash, shot through with bolts of lightning, arising from the volcano's summit. Eruptions of this kind are infrequent but strong.
- Vesuvian eruptions are much like the Vulcanian kind but also involve release of lava. Exceptionally violent Vesuvian eruptions are known as Plinian eruptions because they resemble the famous eruption of VESUVIUS in A.D. 79, described by the Roman naturalist Pliny the Younger. In a Plinian eruption, the volcano ejects vast quantities of ash and may send tephra to an altitude of 30 miles (50 km) or higher.

See also ICELAND.

eruption cycle The sequence of events that happens during an eruption episode. It depends on the volcano as to what the cycle will include and in what order. Typically, an eruption cycle begins with an EARTHQUAKE SWARM as the MAGMA moves into place. FUMAROLE activity commonly follows next as the groundwater is heated to the boiling point. The first eruption commonly includes gas emissions and a fair amount of ASH. Explosions begin and bombs are hurled from the volcano. LAVA FLOWS from the main vent in the middle stages of the eruption cycle. Later, LAVA may flow from other VENTS. Some volcanoes pour lava out continuously through the eruption. In this case, the volume tends to start out slow and increase in the middle of the eruption and then decrease again at the end. In others, eruptions are catastrophic with a huge amount of lava and EJECTA being emitted at once separated at once, punctuating otherwise relatively quiet periods. In subduction-related STRATOVOLCANOES and RHYOLITE volcanoes, the middle of the cycle commonly involves an explosion and loss of a significant portion of the volcanic cone.

The eruption column above the 1980 eruption of Mount Saint Helens *(Courtesy of the USGS)*

Later, it may be rebuilt in a resurgent cone and significant ash and ejecta emissions. In shield volcanoes as in HOT SPOTS or DIVERGENT BOUNDARIES, there are no explosions and loss of the cones. They produce far less ash than the subduction related volcanoes.

Erzincan *earthquake, Turkey* On December 26, 1939, the RIGHT-LATERAL strike-slip North Anatolian Fault produced another of its catastrophic earthquakes—this one in the city of Erzincan. The MAGNITUDE of the earthquake was 7.9 on the RICHTER scale. It produced surface ruptures, LANDSLIDES, ROCKFALLS, and LIQUEFACTION. It also produced significant TSUNAMIS on the Black Sea near Fasta Bay. There are reports that these waves achieved heights of 50–60 feet (17–20 m) locally. Strong AFTERSHOCKS continued for several day after the MAIN SHOCK, and minor aftershocks continued for months.

Some 33,000 people lost their lives in this disaster, and many tens of thousands were injured. Over 110,000 homes were heavily damaged. A large winter storm exacerbated this situation dropping snow on the ruins and bringing plummeting temperatures and strong winds. Many people died from exposure, and rescue efforts were hampered.

The Erzincan earthquake, although not the biggest, may have been the most significant in the history of TURKEY. As a result, the first seismic design codes for buildings were established in 1940, and the first earthquake zonation (hazard) map for Turkey was produce in 1942.

escarpment Also known simply as a SCARP, an escarpment is a steep cliff located at the edge of a flat or nearly flat area. An escarpment typically forms where there is vertical motion along a DIP-SLIP FAULT.

Etna, Mount *Sicily, Italy* One of the most famous volcanoes, Etna stands some 10,750 feet (3,500 m) tall and has a circumference greater than 90 miles (150 km). The volcano has a gentle slope and resembles a SHIELD VOLCANO, such as those of the HAWAIIAN ISLANDS, more than a steeply sloping volcano, such as FUJI or VESUVIUS. Numerous LAVA FLOWS have occurred from the flanks of the volcano. Charles Morris, in his 1902 survey of volcanoes, *The Volcano's Deadly Work*, described the activity of Etna:

> There is a great similarity in the character of the eruptions of Etna. Earthquakes presage the outburst, loud explosions follow, rifts and bocche del fuoco ["mouths of fire," or small lava vents] open in the sides of the mountain; smoke, sand, ashes and scoriae are discharged, the action localizes itself in one or more craters, cinders are thrown up and accumulate around the crater and cone, ultimately lava rises and frequently breaks down one side of the cone where resistance is least; then the eruption is at an end.

Etna has erupted at least 190 times from 1226 B.C. to the present. Most of these eruptions are moderate (VEI = 1–2). Virgil, Pindar, Seneca, Pythagoras, and Thucydides all mentioned the volcano in their writings. Perhaps the most famous eruption of Etna occurred in 1669, when powerful explosions destroyed part of the volcano's summit and LAVA from a FISSURE on the volcano's flank flowed some 10 miles to the sea and damaged the town of Catania. The 1669 eruption is notable for an attempt to divert a lava flow. About 50 townspeople from Catania reportedly dug a passage that drained away lava and apparently reduced the immediate threat to Catania, but when the flow changed direction and menaced the village of Paterno, inhabitants of that community drove away the Catanians and ended their attempt to divert the lava. This effort, though ultimately unsuccessful, showed that engineering could alter the direction of lava flows.

An eruption in 1755 was accompanied not only by emissions of lava but also by great flows of water, which resulted when lava came in contact with snow and ice on the summit. The water flowed down the flanks of the mountain, carrying along with it great quantities of unconsolidated volcanic material. The volume of water involved in this eruption was estimated at approximately 16 million cubic feet (453,070 m^3). It formed a channel two miles (3 km) wide and 34 feet (10 m) deep in places and flowed at the rate of approximately 40 miles (64 km) per hour. Another great eruption occurred in 1819. Although few deaths are attributed to Etna historically, in 1843, 36 people were killed in a phreatic explosion. An extremely violent eruption that lasted more than nine months occurred in 1852. This eruption is thought to have produced flows of lava more than 2 billion cubic feet (56,633,694 m^3) in volume over an area of some three square miles (7.8 km^2) More than three centuries after the attempt to divert a lava flow during the eruption of 1669, volcanologists in ITALY asked the UNITED STATES military to help with another such experiment during the eruption of 1992. The attempt was called Operation Volcano Buster and involved using explosives to blast a hole in a lava tunnel 6,000 feet (1,830 m) up the flank of Mount Etna. Helicopters then dropped large blocks of concrete into the hole in the hope of stopping the lava flow. The experiment, however, was considered only a partial success at best.

Etna is still quite active. The 1991–93 eruption was the largest in the last 300 years. In the 1995–96 eruption, explosions and FIRE FOUNTAINS were common. That eruption never really ended; it has just increased and decreased in intensity ever since. As of April 2000, air traffic routes are still being diverted around the erupting volcano.

Eurasian crustal plate One of the major crustal plates that are believed to occupy the surface of the earth, the Eurasian plate generally underlies the continents of Europe and Asia, although portions of Asia also are found in the ARABIAN CRUSTAL PLATE and INDO-AUSTRALIAN CRUSTAL PLATE. The borders of the Eurasian plate are noted for intense earthquake and volcanic activity. On its western edge, the MID-OCEAN RIDGE in the middle of the North ATLANTIC OCEAN is the site of volcanism, notably in areas such as ICELAND. The eastern edge of the Eurasian plate follows part of the trace of the "RING OF FIRE," the belt of seismic activity and volcanism that surrounds much of the PACIFIC OCEAN basin. Volcanic activity is especially pronounced along this border of the Eur-

asian plate in regions such as JAPAN and the KAMCHATKA PENINSULA of Russia. The southeast edge of the Eurasian plate displays abundant evidence of volcanic activity, both ancient and recent. This is the region of the highly destructive volcanoes of the Indonesian archipelago, such as KRAKATOA, the 1883 eruption of which has become virtually a synonym for a natural catastrophe. Along the southern border of the Eurasian plate, earthquakes are frequent and powerful where the Eurasian plate adjoins the Indo-Australian and Arabian plates. The ongoing collision between the Indian and Eurasian plates is believed to have generated the Himalaya Mountains, among other prominent features of southern Eurasia.

See also INDONESIA; IRAN; PHILIPPINE ISLANDS; PLATE TECTONICS.

explosive eruption An eruption that produces an energetic ejection of PYROCLASTIC material. ANDESITIC, DACITIC, and RHYOLITIC volcanoes produce the most pyroclastic material and the most explosive eruptions. There are two reasons for the explosivity. These LAVAS and MAGMAS are highly viscous and sticky. They tend to clog up the VOLCANIC PIPE. Clogged-up VENTS cause the pressure to increase in the volcano. The other factor is that these types of magmas carry a lot of water and other gasses. These components are referred to as VOLATILES. Under pressure, these volatiles are held in the magma as liquid. When the pressure is released, they can come out of the magma. Because it is so hot, they immediately turn to gas. Gas occupies 22.4 times as much space as the same amount of liquid at room temperature. In other words, boil one gallon of water and it will produce 22.4 gallons of steam. At volcanic temperatures, the increase in volume is even greater. This radical increase in volume is instantaneous and causes the volcano to explode. As an analogy, shake a bottle of soda. Nothing happens until you remove the cap and then it foams over or explodes. When you remove the cap, you release the pressure and the carbon dioxide, which was formerly held in the liquid, instantaneously turns to gas, producing the foam.

extinct volcano A volcano that is inactive and for which there is no chance of renewed activity. The transition from DORMANT to inactive and extinct is just a matter of time. If the volcano has not erupted in historic times and is not expected to erupt in the future, then it is considered EXTINCT. The distinction between inactive and extinct is unclear.

extrusion tectonics Otherwise known as escape tectonics, extrusion tectonics involve the lateral movement of landmasses out of the way of an ongoing continental collision. If a person were to place a piece of clay or putty in his or her hand and then smash the other fist into it. The clay would squirt out of the sides as the fist made impact. CONTINENTS do the same thing. As a continent moves toward another through SUBDUCTION, the OCEANIC CRUST eventually becomes exhausted, and the two continents impact. The subducting continent, here termed the *rigid indenter,* will attempt to drive beneath the other, and the margins of each continent will crumple with massive amounts of DEFORMATION, both faulting and folding. As the collision proceeds, the overriding continent develops a series of crossing STRIKE-SLIP

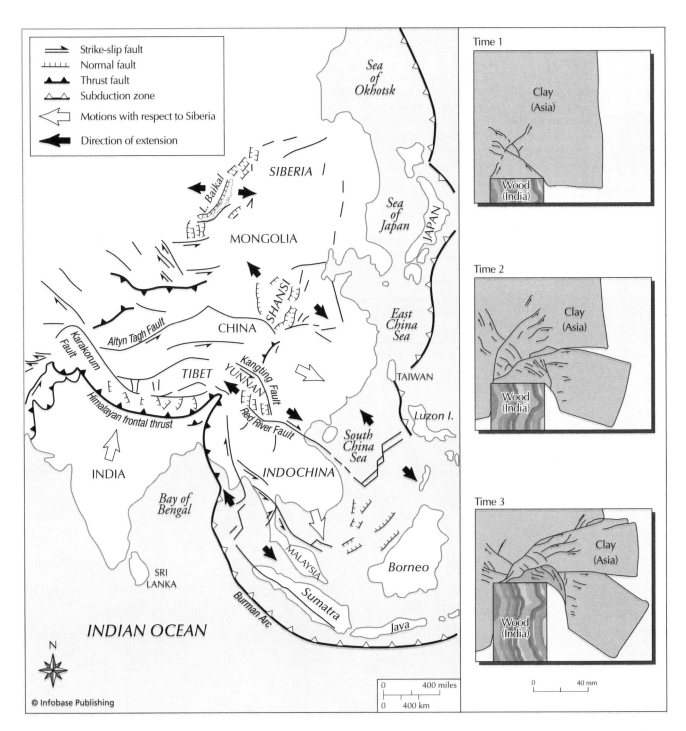

Map showing the classic area for escape tectonics, the Himalayan orogeny. The Indochina block was forced out of the way of the impinging Indian plate along the Red River Fault for about 25 million years and accommodating over 600 miles (1,000 km) of left-lateral strike-slip movement. When the Red River Fault became inactive 20 million years ago, left-lateral movement shifted to the Altyn Tagh Fault which also underwent many hundreds of kilometers of offset. A new fault connecting to Lake Baikal has initiated more recently. The three block diagrams show an analog model of escape tectonics. A wooden block (rigid indenter) models India by being driven into a block of clay that models Asia. The left side of the clay is fixed, but the right side (free face) is able to undergo tectonic escape. The sequence models the Himalayan collision.

FAULTS, one direction DEXTRAL and the other direction SINIS-
TRAL, behind the collision zone. This zone of crossing strike-
slip faults is known as a syntaxis. If one side of the overriding
plate faces onto an ocean basin, this is called a "free face."

the strike-slip faults that accommodate movement of land in
the direction of the free face will become very active, facili-
tating large-scale bulk movement of continent. The strike-slip
faults in the other direction will be less active. Once the first

piece of continent escapes the collision, the old strike-slip fault becomes inactive, and a new strike-slip fault begins to extrude a second, deeper piece out of the way.

The best example of extrusion tectonics can be found in the Himalayan orogeny. INDIA is the rigid indenter, Asia the overriding plate, and the PACIFIC OCEAN the free face. Upon impact of India with Asia, the huge Red River Fault began sinistral strike-slip movement that continued for about 25 million years. It moved Indochina to the east and completely out of the way of the collision and rotated it from an east-west orientation to a north-south orientation where it presently lies. The Red River Fault became inactive, and the Altyn Tagh Fault to the north then began sinistral strike-slip movement and pushed central China, the second piece, eastward and out of the way. At present, the Altyn Tagh Fault is becoming less active, and a new fault system that connects to Lake Baikal and Kamchatka, far to the north is initiating.

Another example of escape tectonics is in the Zagros orogeny. The ARABIAN CRUSTAL PLATE (rigid indenter) separated from AFRICA and drove northward until it collided with the margin of southern Asia. This collision produced the Zagros Mountains of IRAN and Iraq. As the two continents pushed together, TURKEY extruded westward and into the eastern MEDITERRANEAN SEA (free face), into which it now juts. Extrusion of Turkey westward is accommodated on the North Anatolian Fault, which undergoes dextral strike-slip movement and regularly produces some of the world's deadliest earthquakes.

extrusive rock IGNEOUS ROCK that flows, or has flowed, from a fissure or vent on Earth's surface.

See also VOLCANIC ROCK.

F

Falcon Island *Tonga* Also known as Fonua Foo, Falcon Island has been emerging and disappearing beneath the ocean surface for decades. HMS *Falcon* charted the island in 1865 and described it as a shoal. Eruptions in 1885 produced a cone approximately 300 feet (91 m) tall, but erosion destroyed much of the cone within months after eruptive activity ceased, and nine years later the cone had been reduced to a shoal again. Another eruption in 1896 raised a small cone about 100 feet (30 m) above the sea, but this too was destroyed in a matter of several years. Eruptions in 1927 and 1928 produced an island more than two miles (3 km) long, but a quarter-century later, the island had vanished.

Fantale *caldera, Ethiopia* The Fantale CALDERA is located in the main Ethiopian Rift near the point where the GREAT RIFT VALLEY joins the Afar Depression. Although Fantale does not appear to have been an especially destructive volcano in terms of fatalities and property damage, it apparently has a notorious reputation in the surrounding area, where parents are said to warn ill-behaved children that they will melt like the volcano. An eruption in the 13th century reportedly destroyed a village and a church. Several LAVA FLOWS occurred in the vicinity in 1820. Minor FUMAROLES were observed at numerous places in the caldera in 1930, and highly energetic fumarolic activity occurred in 1960, but these observations are not thought to represent a great departure from normal activity.

Farallon crustal plate A lithospheric (TECTONIC) plate of Earth that was believed to have occupied the space between the PACIFIC CRUSTAL PLATE and the NORTH AMERICAN CRUSTAL PLATE. It was located off what is now the coast of CALIFORNIA. The Farallon plate apparently was subducted and consumed as North America advanced westward. The subduction and destruction of the Farallon plate is thought to have been responsible for widespread volcanic activity in prehistoric California.

See also PLATE TECTONICS.

fault A break in Earth's crust along which displacement of rock occurs. Perhaps the world's most famous fault is the SAN ANDREAS FAULT, which lies along CALIFORNIA's Pacific coast for several hundred miles and is responsible for the earthquake that destroyed much of SAN FRANCISCO and neighboring communities in 1906. FAULTS are abundant in mountain belts and deep basins, although they also may occur in some flatlands. Some faults extend for hundreds of miles across Earth's surface and manifest themselves in dramatic ways. Other faults may be comparatively small.

Faults are described as having strike (the compass direction relative to north) and dip (the angle of inclination or slope relative to horizontal). Motion, or slip, can occur in several directions along an ACTIVE FAULT, and the various kinds of fault are named accordingly. A STRIKE-SLIP FAULT exhibits horizontal or lateral movement (along strike direction), whereas a DIP-SLIP FAULT displays vertical movement (along dip direction). An oblique-slip fault shows characteristics of both dip-slip and strike-slip faults. A normal fault is a type of dip-slip fault characterized by the rocks moving normally with respect to GRAVITY. If the fault is viewed as a hill, the rock on it falls down the hill just as gravity requires. In a REVERSE FAULT, rocks above the fault plane move up the hill which is the reverse of gravity.

There are two types of strike-slip faults as well. The faults are lines on a map or on the ground. The determination is made by standing on one side of the fault and looking across to the other side. If the opposite side of the fault moves to the left, it's a LEFT-LATERAL FAULT. If it moves to the right, it's a RIGHT-LATERAL FAULT. Oblique-slip faults can be named by the components of movement.

Some faults are clearly visible on Earth's surface. Evidence of them may take such forms as displaced hills, offsets in a shoreline across a fault, offsets in rivers, and elongated lakes along faults (for example, Lake San Andreas near San Francisco). Other faults are less evident, however, and in many parts of the world—notably the eastern United States, where bedrock lies buried under deep layers of sediment—

active faults may go unsuspected until movement along them generates an earthquake. This was apparently the case with the CHARLESTON, SOUTH CAROLINA, earthquake of 1883, which destroyed a large portion of the city, and the extremely powerful earthquakes of 1811–12 in the NEW MADRID FAULT ZONE in the Mississippi Valley.

There are various ways to identify an active fault and estimate its potential for surface rupture in future earthquakes. One approach is to study the history of faulting and movement. Another is to extreme ongoing earthquake activity and recurrence intervals (how often) for earthquakes. Measurements of strain by geodetic surveys are also use-

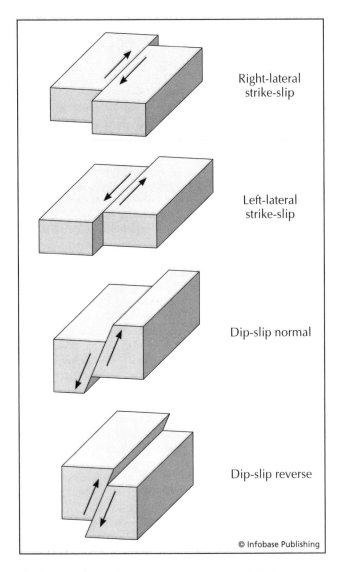

Block models of the various types of faults. Strike-slip faults have only lateral movement. Whether they are left- or right-lateral can be determined by standing on one side of the fault and looking across to the other. If the opposite side moves to the left, it is a left-lateral fault; if it moves to the right, it is a right-lateral fault. Dip-slip faults only move vertically and are classified based on gravity. It is normal for objects to slide downhill and reverse for objects to slide uphill (the hill being the fault plane).

ful. It is not always easy to differentiate between an active and an inactive fault because a single fault may vary greatly in its behavior over time. Whereas some faults are clearly active because they display measurable CREEP or generate frequent small earthquakes, not all faults are so consistently active. A fault that has shown no activity for many centuries may reactivate suddenly and slip 30 feet (9 m) or more in a single event. The determination of a fault as active or inactive may depend also, in part, on how much harm the fault may cause if it should move significantly again. A fault located near a nuclear power station, for example, may be evaluated on a different basis than a fault that poses less of a potential hazard to human activities. The evaluation of faults as active or inactive is made harder by the fact that not all active faults manifest themselves on the surface.

It is important to determine the slip rate along a fault in projecting future activity. The slip rate is a measure of how fast movement is occurring along a fault. It is also important to know how often earthquakes occur along a fault and how powerful the earthquakes are, as well how much slip occurs per earthquake. The slip rate can be derived from a study of recently formed features along a fault. The slip rate, however, represents only an average over time of movement along the fault and is not necessarily steady. Along a given fault, the slip rate may represent steady activity over a certain period or a few episodes of comparatively great and abrupt movement. In other words, a fault may be very (even catastrophically) active at some times and quiescent at others. As a rule, faults along the boundaries of major plates of Earth's crust exhibit greater slip rates than faults elsewhere. In southern California, slip rates tend to be greatest along the San Andreas and San Jacinto faults.

A fault's potential for generating destructive earthquakes may be judged partly on the basis of earthquake activity recorded by instruments. If such activity is clearly associated with a fault on an ongoing seismological record, then this information can help in associating the HYPOCENTERS of earthquakes with a particular fault at depths of a few miles or less below the surface. When numerous earthquakes can be linked to a given fault in this way, with their hypocenters on the fault at depth, then it can be concluded that the fault is capable of generating more earthquakes in the future. This approach cannot be used reliably to estimate a fault's potential for destructive earthquakes in all cases, however, because a fault can be quiescent but remain seismogenic, or capable of earthquake activity. Moreover, even an active fault is not necessarily dangerous to lives and property. A fault that generates many small earthquakes, for example, may pose little or no danger to residents of the area. Even when a fault exhibits surface rupture, its activity may be so gentle as to pose no particular threat to the surrounding community.

Sometimes risk assessment for a seismogenic fault requires estimating the maximum earthquake that the fault could produce. Here, the historical record of seismic activity is important because the magnitudes of earlier earthquakes may provide guidance in estimating the possible strength of other seismic events to come. (The strength of the FORT TEJON earthquake, for example, has become a benchmark to estimate the magnitude of future major earthquakes in southern California.) Geologists also may find it helpful to compare the history of seismic activity along a particular fault with the history of another fault comparable to it. The behavior of another fault under similar conditions may provide clues to the behavior of the specific fault concerned. The dimensions of a fault may provide some clue to the magnitude of future major earthquakes because big earthquakes are often associated with major faults. In several studies, the length of a fault and especially its surface rupture has been correlated with magnitude of earthquakes. A fault's dimensions are not a completely reliable guide to its seismogenic potential, however, because a fault may not be traceable for its entire length at the surface and because it is often difficult to foresee how far surface rupture may extend along a fault in an earthquake. In that event, the historical record may provide a better estimate of how powerful an earthquake a given fault might deliver. (In the Los Angeles area, a study of the history of earthquakes and the geology of the region indicates that an earthquake comparable in magnitude to the Fort Tejon earthquake of 1857 was about the most powerful seismic event that Los Angeles may expect.) Because the maximum possible earthquake along a fault may occur only rarely, at intervals of a century or longer on average, estimates of magnitude and frequency may concentrate instead on the most powerful earthquake that may be expected on a particular fault during a shorter interval. Typical intervals are 50 to 60 years because that is about the longest period many buildings are expected to last.

One problem with using the historical record to estimate the magnitude and frequency of potential earthquakes is that the historical record does not always extend far enough to make such estimates reliable. This is the case in portions of western North America where records of earthquake activity cover only a few decades. The historical record may be supplemented by studies of the geologic record, which can reveal evidence of prehistoric earthquakes in the form of surface faulting and disturbance to sediments. Such evidence may yield estimates of average slip rate and the average interval between major earthquakes along a specific fault. Studies of the San Andreas Fault in southern California have indicated that major earthquakes there have occurred at an average interval of approximately 50 years.

See also PLATE TECTONICS.

fault creep Motion that occurs gradually along a fault without producing an earthquake.

See also SEISMOLOGY.

fault plane solution When an earthquake occurs, it is recorded on SEISMOGRAPHS worldwide. SEISMOLOGISTS study the seismograph records (seismogram). By comparing the delay between the arrivals of the waves, they can determine the location of the FOCUS and EPICENTER. By studying whether the needle on the seismograph first went up or went down for each seismogram, they can determine what kind of fault it was (STRIKE-SLIP, DIP-SLIP, etc.). By combining these two pieces of information, seismologists can determine the fault plane solution or focal mechanism for the earthquake, that is, where and how the fault moved.

fault rocks There is a full classification system for fault rocks, just as if they were sedimentary or IGNEOUS ROCKS. Under shallow conditions 0–1.2 miles (0–2 km), the rocks in a fault are broken into large to small angular fragments with large gaps between them. This type of deposit is called fault gouge. If the gouge is cemented together with minerals precipitated from the groundwater, it becomes fault BRECCIA, a rock. Deeper in the ground 1.2–9.3 miles (2–15 km), the pressure is too high to leave gaps. Instead, the rock is pulverized and ground together like coffee grounds. It is fine-grained with no gaps and called CATACLASITE. Depending on the amount of leftover uncrushed rock within the cataclasite, the rock can be a protocataclasite (mostly uncrushed rock), cataclasite (about half and half), and ultracataclasite (all crushed rock). Below about 9.3 miles (15 km), rocks behave in a DUCTILE manner. They flow like silly putty in the fault zone rather than crushing like glass as with cataclasites. These rocks are called MYLONITES. They have the same classification as cataclasites. Fault rocks with mostly original rock and very little stretched-out grains is called a protomylonite. Half and half is a mylonite, and all stretched-out grains is an ultramylonite.

fault system In contrast to a single fault or several parallel faults, a series of interlinking faults is called a fault system. Many of the major faults in the world are really fault systems. The SAN ANDREAS FAULT is not a single fault but is part of a large system of faults that form a braided pattern. The CALLAVERAS FAULT and the HAYWARD FAULT among many others are really part of the San Andreas fault system. All of the faults have similar orientations, they are all related, and all are active at the same time (geologically speaking). Any one of them can produce an earthquake at any time. During the next 25 years or so they all will.

fault trace The surface expression of a fault. A fault is a 3D planar to curvi-planar structure. The fault trace is the intersection of the fault with the ground surface. In other words, if you look at a fault on a map or from an airplane, you are looking at the fault trace. For very steep faults, like the San Andreas, the trace is a linear feature. For shallow faults, especially in areas of rugged topography, the fault trace can be as winding as a mountain road.

Federal Emergency Management Agency (FEMA) FEMA is the official U.S. government agency that handles all natural disasters, including earthquakes and volcanic eruptions. When a disaster occurs, FEMA moves in quickly to coordinate rescue and relief operations where possible and to distribute federal aid to the affected communities all the way through the repair and rebuilding processes.

feldspar The most common mineral group in Earth's crust. All feldspars are SILICATES in which the atoms are arranged in a framework structure. There are two basic type of feldspar, PLAGIOCLASE and potassium (K)-feldspar (*see* K-FELDSPAR).

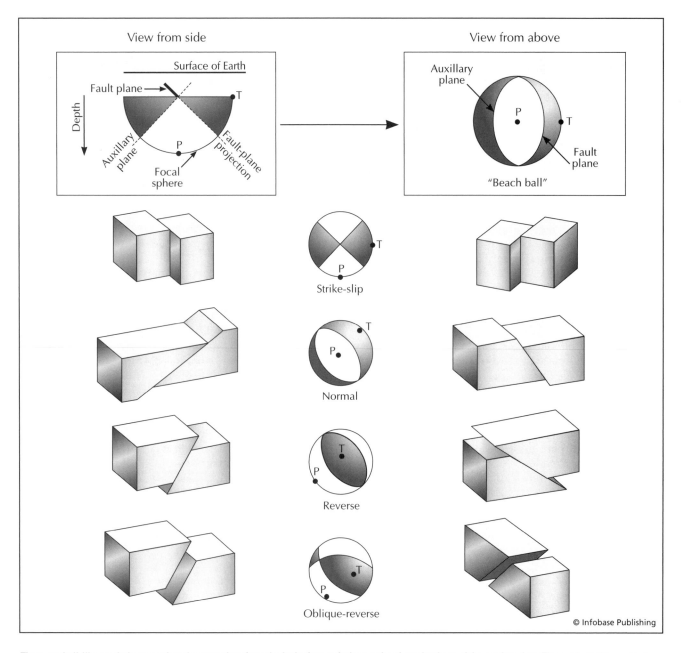

There are ball-like symbols on earthquake maps that show the fault plane solutions or focal mechanisms of the earthquakes. The symbols show which way the fault moved during the earthquake as shown by the diagram. The white stripes show compression, and the gray stripes show extension.

The main ions in plagioclase are calcium and sodium. Plagioclase occurs in probably 99 percent of all igneous rocks. Potassium feldspar primarily occurs in GRANITE and RHYOLITE.

felsic   An IGNEOUS ROCK that is high in FELDSPAR and SILICA (*fel* from *feldspar* and *si* from *silica*). These are light-colored rocks that form at the lowest temperatures possible for igneous rocks (750–800°C). The most common rock types include RHYOLITE (volcanic) and GRANITE (plutonic). These rocks must contain QUARTZ and K-FELDSPAR but commonly contain plagioclase as well. These three minerals are commonly referred to as the felsic minerals. Felsic rocks may also contain MUSCOVITE (white mica), BIOTITE (black mica), HORNBLENDE, and many other minerals in small quantity.

Fernandina *caldera, Galápagos Islands, Ecuador* The SHIELD VOLCANO Fernandina has a summit CALDERA whose floor collapsed following eruptions in 1968. Two days after a moderately strong earthquake about 200 miles (322 km) north of the island, Fernandina began to erupt on May 20, 1968. This eruption lasted several days. Further earthquakes on June 11 immediately preceded an explosive eruption that morning, followed by another explosion later in the day. The eruption continued for about one more day, and then the caldera floor started to collapse. The greatest subsidence (more than 1,000 feet [305 m]) occurred at the southeast end of the caldera. Several hundred moderately powerful earthquakes occurred during the collapse, which apparently took place steadily over a period of several days rather than all at once. The volume of the collapse was as much as 10 times greater than the small volume of MAGMA that erupted. Fernandina erupted again in 1978, an hour following an earthquake. Another eruption occurred between late 1980 and early 1982 but was so small that it went unnoticed until later. Activity resumed briefly in 1984, between March 30 and April 11. The most recent eruption of Fernandina was in 1995. LAVA flowed 2.5 to three miles (4–5 km) to the ocean where it killed many birds and fish by scalding them to death.

ferromagnesian   Minerals and rocks that are high in iron and magnesium. The term refers to SILICATE minerals. For igneous rocks, these minerals include PYROXENE, OLIVINE, HORNBLENDE, other amphiboles, and BIOTITE. They are all related in structure and chemistry. They are also referred to as MAFIC minerals because they are abundant in MAFIC igneous rocks. In ULTRAMAFIC rocks, there are essentially no other minerals. There are also a different group of metamorphic ferromagnesian rocks and minerals.

fiamme   Dark-colored, glassy lens-like structures in pyroclastic deposits that have a flame shape. They are thought to be formed by the collapse of PUMICE fragments during the formation of a welded TUFF by overlapping PYROCLASTIC FLOWS. (*Fiamme* is Italian for "flames.")

fire fountain   A prolonged and high spray of LAVA above a volcano's VENT. Also known as lava fountains, fire fountains can be as high as 1,000 feet (305 m) but are more commonly 100 feet (30 m). They can fill lava lakes or feed flows.
    See also HAWAIIAN ISLANDS.

first arrival   The first part of the waves to reach SEISMOGRAPH from a seismic source. These data are the main components in determining FAULT PLANE SOLUTIONS.

Fissures in a paved road at the head of a slump block generated by the 1970 Nevados Huascarán earthquake in Peru *(Courtesy of the USGS)*

first motion   The first motion is seen on a SEISMOGRAM during the first sign (record of ground motion—up or down) of the P-WAVE arrival of a particular event. The direction of the needle movement, up or down, gives information on the FAULT motion that caused the earthquake. Conventionally, upward motion indicates compression from the quadrant of fault slip (rock is moving toward the SEISMOGRAPH), and downward motion indicates dilation or rarefaction (rock is moving away from the seismograph). First motion studies are used to construct FAULT PLANE SOLUTIONS to determine the source mechanism of earthquakes.

Fisher *caldera, Aleutian Islands, Alaska, United States*   Fisher CALDERA is located on Unimak Island. A tremendous eruption believed to have occurred here several thousand years ago resulted in the formation of a caldera approximately 14 miles (8.5 km) long and eight miles (13 km) wide. In historical times, the record of Fisher's activity is uncertain. There is a poorly documented eruption in 1826. An exceptionally noisy eruption in 1850 may have involved Fisher. There is an active field of FUMAROLES in the caldera.

fissure   In general terms, a crack in Earth's crust. Fissures commonly form during strong earthquakes. When LAVA emerges in large quantities from a fissure, the phenomenon is knows as a FISSURE ERUPTION. The resulting flow of lava is called a fissure flow.

fissure eruption   A volcanic eruption that emanates from a fissure or set of fissures in Earth's crust. They are typically not explosive but can pour out vast quantities of LAVA at a high rate. The lava is very fluid and tends to flood the surrounding area. The composition is exclusively BASALTIC. Fissure eruptions can occur in SHIELD VOLCANOes as well.

flank eruption   In contrast to a SUMMIT ERUPTION where the explosion emits from the top of the volcano, a flank

Fire fountains approximately 30 feet (9 m) high along a fissure at Kilauea volcano, Hawaii, in 1979 *(Courtesy of the USGS)*

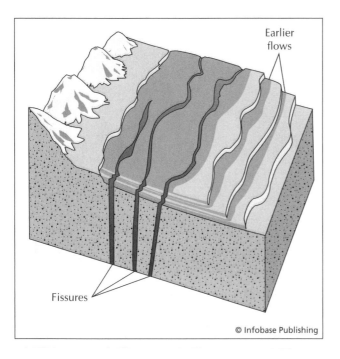

Diagram showing how fissure eruptions form. The fissures are cracks in the ground from which basalt lava pours and spreads over the landscape, forming flood basalts. Iceland has some good examples of fissure eruptions.

eruption comes out of the side of the volcano. In volcanoes with LAVA of high viscosity (sticky lava), the summit orifice or feeder pipe to the summit can become clogged. If MAGMA continues to move into the MAGMA CHAMBER beneath the volcano, the pressure will continue to increase. The pressure may exceed the strength of the wall of the volcano, the inflating of the volcano may cause a LANDSLIDE that reduces the strength of the wall, or there may be a FISSURE in the side of the cone that can't contain excessive pressure. In any of these cases, the side of the volcano will blow out in an explosion that typically exceeds that of a summit eruption. Summit eruptions can shoot ash and gases 10 miles (16 km) upward into the atmosphere. Now an even greater explosion shoots down the side of the volcano. The blast will knock everything down in its path resulting in "blow-down" of trees and structures. The advancing cloud is called a *NUÉE ARDENTE* or a PYROCLASTIC FLOW. It is composed of ASH, volcanic fragments, and toxic gases. It is at least 800°C and travels at 60 miles (97 km) per

A lava curtain composed of joined lava fountains from Mauna Loa, Hawaii, on March 24, 1984. The curtain height averages about 80 feet (25 m). Curtains are common in fissure eruptions, where they erupt in a long, thin series of cracks. *(Courtesy of the USGS)*

hour or more. Nothing survives it, and everything is immediately buried in ash and fragments.

flood basalt A wide, relatively thin sheet of BASALTic lava that flowed outward from a FISSURE in much the same manner as a flood of water. Flood basalts cover much of the Pacific Northwest of the United States as the COLUMBIA PLATEAU basalts. The WATCHUNG MOUNTAINS of New Jersey is also a flood basalt, as are the DECCAN Traps, the Karoo, and the Piranha traps. An individual flow may have a volume of more than a hundred cubic miles.

Florida *United States* The state of Florida is generally free from strong earthquakes, but a region of minor seismic risk is found in the northern portion of the state along the border with GEORGIA. One notable earthquake, of MERCALLI intensity VI, occurred in the vicinity of Saint Augustine on January 12, 1879, and affected an area of about 25,000 square miles (64,750 km². The earthquake rattled windows and doors in Daytona Beach and caused minor damage in Saint Augustine. On October 31, 1900, Jacksonville experienced an earthquake of Mercalli intensity V that included eight separate shocks.

focal mechanism *See* FAULT PLANE SOLUTION.

focus The underground point on a fault plane where an earthquake originates. Also known as the HYPOCENTER. Where the fault moves is where the energy is released. Earthquake waves are sent out in all directions from the focus. The EPICENTER sits on the ground surface directly above the focus.
  See also FAULT; SEISMOLOGY.

Fogo *See* CHA.

footwall In a DIP-SLIP FAULT, the FAULT plane is a sloping surface. The block (rock) beneath the fault is called the footwall. The block (rock) above the fault is the HANGING WALL. The relative motion of the hanging wall compared to the footwall determines what kind of fault it is, normal or REVERSE.

forearc All areas between the TRENCH and an ISLAND ARC are considered to be part of the forearc. Primarily, there is a wedge-shaped pile of sediment that is scraped off the subducting OCEANIC CRUST that is called a forearc prism. The movement on the SUBDUCTION ZONE in the forearc forms a MEGATHRUST that can produce TELETSUNAMIS, such as in the case of BANDA ACEH.

foreshock A small earthquake that foretells of a larger earthquake. The terms *foreshock* and *main shock* can only be distinguished in retrospect. When an earthquake occurs, it could be a foreshock if another larger earthquake follows it. Otherwise, if nothing follows it, then it was the main shock. In areas of high earthquake activity, small earthquakes are carefully monitored. This is because foreshocks commonly foretell of major earthquakes.

Fort Tejon *earthquake, California, United States* The Fort Tejon earthquake of January 9, 1857, is thought to be the most powerful earthquake to have struck the Los Ange-

les area since Spanish exploration began. Although no instruments were available to measure the earthquake's magnitude, comparisons with the more accurately measured 1906 SAN FRANCISCO earthquake indicate the Fort Tejon earthquake was considerably more powerful. The EPICENTER appears to have been located on the SAN ANDREAS FAULT near Cholame, and there is evidence of surface faulting for more than 216 miles (360 km) associated with this earthquake. The earthquake was felt for almost the entire width and length of CALIFORNIA from San Diego to the vicinity of Sacramento and was perceived as a strong shock in Yuma, ARIZONA. The earthquake is named for Fort Tejon, a U.S. Army installation built at Tejon Pass in the Tehachapi Mountains. Fort Tejon, a collection of buildings constructed of adobe, was destroyed by the earthquake. Although the earthquake caused no fatalities at the fort, the soldiers stationed there were forced to spend the rest of the winter in their tents because rebuilding was impossible during the cold weather. One fatality was reported at lower altitudes: An elderly woman was killed in a collapsing building. The earthquake is said to have had dramatic effects on local rivers, throwing the Mokelumne River out of its bed and making the Kern River reverse its flow and spill over its banks.

fractional fusion Partial melting of rock as it is heated. Imagine a mixture of marbles, nails, and wax. If heated to the melting temperature of wax, the wax will become liquid and flow out of the mix, leaving the solid marbles and nails behind. The same thing happens with rocks. A rock is a mixture of different kinds of minerals. As a solid rock is heated, the minerals with the lowest melting temperature will melt first. The liquid will flow away forming MAGMA and leaving the higher temperature minerals behind. The lowest temperature minerals are those that form FELSIC rocks, GRANITE, and RHYOLITE. The main minerals in these rocks are QUARTZ, K-FELDSPAR, and Na-PLAGIOCLASE (FELDSPAR).
  See also FRACTIONATION.

fractionation The minerals in MAGMA and LAVA crystallize as they cool. Different minerals have different temperatures of crystallization. OLIVINE crystallizes at a higher temperature than PYROXENE, which in turn is higher than HORNBLENDE and BIOTITE respectively. This sequence of mineral crystallization is called BOWENS REACTION SERIES. If the higher-temperature minerals crystallize and are removed, the composition of the liquid changes. MAFIC magma becomes more and more FELSIC with this process because the felsic minerals are formed at lower temperature. This process of driving the composition of the remaining liquid toward more felsic compositions is called fractionation.

fracture A break in rock. Under shallow conditions, rocks behave in an elastic manner. They deform as a function of the amount of STRESS that is applied to them. When the stress is released, they recover to their original size and shape. They behave like a rubber band or a spring. Just as with a rubber band, if they are stressed past their strength, they break. By breaking or fracturing, they release all of the stress that was built up. When you pull on rubber band, it requires your muscle power. That is the stress. When it breaks, the muscle is no longer required because the stress is released. In a rock, the energy that was involved in stressing the rock is abruptly released when the rock breaks (fractures), just like the rubber band. That is an earthquake. The released energy from stress is converted to seismic waves which travel away in all directions.

fracture zone A type of TRANSFORM FAULT on the ocean floor that is characterized by densely spaced fractures. Because the fractures result from earthquakes, these zones are seismically active.

free oscillations An earthquake not only sends out waves that damage the surface and arrive at SEISMOGRAPHS, they also cause the whole Earth to "ring" like a bell. The waves constructively interfere at the natural frequency of Earth. These oscillations cannot be heard and are barely detectable on seismographs.

frequency Frequency is the number of times something happens over a given period of time. In terms of an earthquake, it is the number of vibrations from seismic waves that pass through an area with time. Earthquake frequency is measured in cycles per second or hertz.

Friuli *earthquake, Italy* The Friuli region in northern ITALY was the site of highly destructive earthquakes beginning May 6, 1976. The first earthquake struck Friuli in the evening and was estimated at 6.0 on the RICHTER scale. More than 900 people were killed, and some 50,000 were left homeless. Another strong earthquake in September shook down reconstructed buildings.

Fuego *volcano, Guatemala* Fuego is the historically most active volcano in CENTRAL AMERICA. It has erupted at least 60 times since 1524. The eruptions are typically violent and of vulcanian type with VEI of 2–3. All are SUMMIT ERUPTIONS. Eruptions typically occur in clusters lasting 20–70 years and spaced by 80–170 years. The most recent major eruption was in 1974. It was active for 10 days and produced numerous glowing AVALANCHES that moved down the slopes at 35 miles (56 km) per hour. In 1977–79, Fuego underwent persistent low-level activity.

Fuji, Mount *Japan* Mount Fuji, known to the Japanese as Fuji-san (not Fujiyama), is an andesitic volcano that is a product of the SUBDUCTION ZONE formed by the convergence of the PACIFIC and EURASIAN CRUSTAL PLATES. Mount Fuji is revered as the most beautiful volcano in the world and even appears on Japanese currency. It is perfectly symmetrical. Mount Fuji has erupted at least 16 times since A.D. 781. Although Mount Fuji has been less active in recent centuries than some other volcanoes in JAPAN, some of the eruptions were impressive. In 1707, the heat from the eruption resulted in fractured and shattered rocks, and fallout from the eruption covered the present location of TOKYO to a depth of several inches. The two largest eruptions were in 1050 B.C. and 930 B.C. with VEI = 5. Eruptions of Mount Fuji created a naturally partitioned set of lakes at the foot of the volcano.

Once the lakes were one single large lake, but LAVA FLOWS divided the lake into five smaller lakes, named Kawaguchi, Motosu, Sai, Shooji, and Yamanaka.

See also ANDESITE.

Fukui *earthquakes, Japan* Just as the devastating effects of World War II were being addressed in the city of Fukui, JAPAN, a strong earthquake rocked the area. The quake occurred on June 28, 1948, at 5:13 P.M. It registered a 7.3 on the RICHTER scale. Large FISSURES opened to 1.5 feet (0.5 m) in width in the center of town, and LANDSLIDES dropped rubble on the roads outside of town. The postwar makeshift communication lines and water lines were severed by the quake, leaving the city completely cut off. Probably the major hazards from this event were the fires from overturned charcoal stoves that stormed through town. In the end, some 5,390 people lost their lives, and 21,750 were injured. More than 39,000 homes were destroyed at a cost of over $1 billion.

fumarole A hole in Earth's surface through which hot water, steam, and various hot gases escape from underground.

Fumaroles at Spirit Lake on Mount Saint Helens, Washington, after the eruption in 1980. Steam rises almost continuously from these vents. *(Courtesy of the USGS)*

Volcanologists collecting data from active fumaroles in Alaska *(Courtesy of the USGS)*

The driving force for the steam is hot underground MAGMA. Groundwater seeps into contact or near-contact with the MAGMA and is heated into the steam. GEYSERS form in much the same way, but the geometry of the water reservoir is such that it allows the water to flash boil.

    See also HYDROTHERMAL ACTIVITY; VALLEY OF TEN THOUSAND SMOKES.

Fuppushi *See* AKAN.

Furebetsu *See* AKAN.

Furnas *caldera, Azores* The Furnas CALDERA is located on San Miguel Island in the AZORES and is noted for a violent eruption in 1630 (VEI = 4). Earthquakes on September 2–3 of that year were followed by the eruption of two fiery clouds. Some 200 people are thought to have died in this eruption, most of them killed by MUDFLOWS. Several other eruptions appear to have occurred within the last 1,000 years.

G

gabbro A MAFIC rock that is characterized by coarse grains. Gabbro is composed of PLAGIOCLASE and PYROXENE and less commonly OLIVINE and HORNBLENDE. Gabbro is the plutonic equivalent of BASALT. DIKES and SILLS of gabbro are common in areas of basalt volcanism. They make up a large proportion of OCEANIC CRUST. There are also some large gabbro bodies on CONTINENTAL CRUST as well. The Palisades Sill in New York–New Jersey is composed of diabase, which is a textural variety of gabbro. STOCKS and BATHOLITHS (circular PLUTONS) of gabbro are rare but can yield significant quantities of platinum and chromium.

Gadamsa *See* ASAWA.

Gademota *See* ASAWA.

Galápagos Islands *Ecuador* The Galápagos Islands are located off the west coast of SOUTH AMERICA and are the summits of volcanoes along a HOT SPOT. The islands are perhaps best known for their giant tortoises and for remarkable variations in the bodily forms of animal populations from one island to another. The most active of the volcanoes in the Galápagos Islands is FERNANDINA, a major eruption of

Flooded caldera rim forming an arcuate island in the Galápagos, Ecuador *(Courtesy of NOAA)*

Small resurgent dome volcanoes within a larger caldera scatter across the landscape of the Galápagos Islands, Ecuador, in the Pacific Ocean. *(Courtesy of NOAA)*

which took place in 1968 and again in 1995. Other volcanoes include Alcedo, CERRO AZUL, Darwin, Ecuador, Sierra Negra, and Wolf, all on Isabela Island.

Galung Gung *volcano, Java* Not quite 7,000 feet (2,134 m) high, Galung Gung is adjacent to the Plain of Ten Thousand Hills, which actually number several thousand less. They are thought to have been formed by a LAHAR that accompanied one of Galung Gung's eruptions. Galung Gung erupted twice in 1822 (VEI = 5) and destroyed no fewer than 114 villages and killed more than 4,000 people. Since then it has erupted four times, most recently in 1984. The 1982 eruption (VEI = 4) killed about 68 people and caused $15 million in property damage.

Gamalama *volcano, Halmahera, Indonesia* Also known as Peak of Ternate, Gamalama STRATOVOLCANO forms a seven-mile (11-kilometer)-wide island. It has had at least 70 eruptions since 1538. Eruptions from 1771 to 1775 caused fatalities including a NUÉE ARDENTE that killed more than 1,300 people in 1775. Other eruptions that resulted in fatalities occurred in 1838, 1871, and 1962. The most recent eruption was in 1993. Most of the eruptions are explosive with VEIs of 2–3.

Gansu *earthquake, China* An earthquake estimated at MAGNITUDE 8.6 on the RICHTER scale struck the Chinese province of Gansu (formerly Kansu) on December 16, 1920, killing approximately 180,000 people. An additional 20,000 deaths during the following months were attributed to lack of shelter during bitter winter weather. Natural dams formed by LANDSLIDEs had to be destroyed to prevent flooding.

Garibaldi, Mount *British Columbia, Canada* A volcanic mountain at the northern end of the CASCADE RANGE.

Gaua Island *Vanuatu* Located in the northern portion of the Vanuatu archipelago, Gaua Island is a STRATOVOLCANO with a CALDERA occupied by a lake, Steaming Hill Lake, and by a postcaldera cone, Mount Gharat. From 1963 to 1982, Mount Gharat underwent at least 13 eruptions. Most eruptions were small (VEI = 2) and lasted only one to two days. The 1963 and 1973–74 eruptions lasted several months.

Geger Halang *caldera, Java, Indonesia* The Geger Halang CALDERA is located in central Java near Juningan and Telaga and has a history of eruptive activity dating back to the late 17th century. How the caldera originated is uncertain. Possibly the caldera collapsed, but LANDSLIDE activity has also

Static electricity generated during an eruption produces lightning. *(Courtesy of the USGS)*

been put forward to account for the calderas formation. Before the collapse, a STRATOVOLCANO is thought to have occupied the site of the caldera. The stratovolcano Cereme is situated on the caldera's northern rim. It has erupted six times between 1698 and 1951. An explosive eruption in 1698 reportedly caused many deaths. Other eruptions are recorded in 1772, 1775, and 1805. Earthquakes and subsidence in 1876 at a site several kilometers away from the caldera may have been unrelated to volcanism at Geger Halang. Emissions of SULFUR gas increased in 1917 as did FUMAROLIC activity in 1924. A series of eruptions in 1937 and 1938 was accompanied by a large number of strong earthquakes. An eruption in 1951 consisted of a single detonation followed by an emission of thick smoke. TECTONIC earthquakes occurred in 1973 but evidently were unrelated to volcanism at the caldera.

Gemini Seamount *volcano, eastern New Hebrides* A SUBMARINE VOLCANO that has been erupting regularly since 1996. It was reported to have produced explosions every three to nine minutes. These eruptions produce steam and ASH.

geodesy Geodesy is the science of determining the size and shape of the Earth, both on the large and local scale, and the precise locating of points on the surface. Geodedic surveys are conducted at all scales on a regular basis, now with the aid of satellite telemetry. These surveys can help to identify areas that are bulging or subsiding, which may be precursors to earthquakes and volcanic eruptions. They can also determine uplift and subsidence as the result of seismic and volcanic events.

geology Literally translated, geology means "science of the Earth," but it is more than that. It is a composite science that encompasses all other sciences—biology, physics, and chemistry as well as many aspects of engineering—and even has subdisciplines of each with composite names (geochemistry, GEOPHYSICS, etc.). It is the study of the chemical and engineering properties of the materials of the Earth—rocks, minerals, soils, fluids (water, oil, etc.), and gases. It is the study of the internal and external processes that affect the planet, including PLATE TECTONICS, deformation, surface process (GEOMORPHOLOGY), MAGMA generation and movement, and even environmental processes. It is also the study of the history of the planet, the internal and external changes that have taken place through time, and the history of life through fossils (paleontology).   In the 19th century, geology was by far the most influential of the sciences, and most scientists were geologists. In the 20th (and 21st) century, humans relied on the expertise of geologists to provide virtually all energy needs. It is geologists who find all of the oil, natural gas, and nuclear energy. Now geologists have added environmental analysis and cleanup to their repertoire. Considering that when a particle, gas, or fluid settles to the ground or penetrates the surface, it becomes the realm of the geologist, virtually all of the Earth is covered by this science.

geomorphology   The study of surface landforms and their evolution. These forms are grouped by the type of environment in which they occur. Deserts form dunes and ripples, mesas and buttes, and related features. Beaches form dunes, spits, and strands of sand. Glacial environments include outwash plains, gravel piles and other features. Rivers have gravel bars and sand. The landforms and surface deposits are in the realm of geomorphology.

geophysical monitoring   Monitoring of volcanoes using geophysical techniques to predict eruptions. Swarms of MICROEARTHQUAKES or small earthquakes record the movement of MAGMA upward in the MAGMA CHAMBERS prior to eruption. By monitoring seismic activity, geologists can predict eruptions and recommend evacuation.

geophysicist   Scientist who studies the physical properties of Earth. It is the geophysicist who monitors seismic activity on FAULTS. They determine the source and MAGNITUDE of earthquakes as well as the resulting damage. Their work is used to construct seismic risk maps. Eventually, their studies may lead to reliable earthquake prediction. Geophysicists also do the geophysical and topographic monitoring of volcanoes to help predict eruptions.

geophysics   Science of the physical properties of Earth including SEISMOLOGY, the study of earthquakes.

Georgia *United States*   Although the state of Georgia is not itself highly susceptible to earthquakes, it has been affected strongly on occasion by powerful earthquakes in neighboring states, such as the NEW MADRID earthquake and the CHARLESTON, South Carolina, earthquake. An earthquake in northern Georgia on November 1, 1875, affected an area of about 150 by 200 miles (241 by 322 km). Portions of the state, especially along its coast, are vulnerable to damage from lique-faction in the event of future strong earthquakes.

geothermal energy   Geothermal energy may be defined as energy that is naturally given off by Earth. Sources of enhanced geothermal energy may include GEYSERS, hot springs, and VOLCANOES. Although facilities for exploiting geothermal energy have been built at numerous locations in several countries. The potential for utilizing this source of energy appears to be unlimited. Considering that at just a few feet underground, the temperature remains a constant 55°F. There is great potential for supplemental cooling in the sum-

mer and heating in the winter. Obviously, other systems must be utilized as well, but it is much easier to warm a house from 55°F than from 25°F. On the other hand, in some areas, the geothermal energy can provide all or nearly all of the energy needs of a community. In these areas, heat from within Earth is concentrated in formations at or near the surface where the heat is stored and can be tapped conveniently.

The geology of geothermal energy resources is complex, but commercially exploitable geothermal resources tend to be concentrated along certain areas. Of particular interest are areas of crustal spreading, where new CRUST solidifies from molten rock rising to the surface, and along plate boundaries, where converging plates give rise to conditions favoring increased HEAT FLOW to the surface. In either case, MAGMA rising from below brings large amounts of heat toward the surface. Heat from the magma converts underground water, whether natural or introduced, into geothermal systems, which may be dominated either by liquid or by vapor. The former tend to be high-pressured, whereas the latter are comparatively low-pressured. Geothermal systems have been used for centuries on a small scale for heating, but application of geothermal energy to other uses began only in the late 18th and early 19th centuries, starting in ITALY.

Development of geothermal energy resources in NEW ZEALAND began shortly after World War II, and the first such facility there was finished in 1958, at the Wairakei fields. Wairakei derives its energy from a huge volume of hot rock believed to be supplied with water by rainwater seeping down from the surface. Development is thought to have touched only part of New Zealand's geothermal energy resources. Geothermal heating facilities are widespread in ICELAND, which lies along the MID-ATLANTIC RIDGE and is the site of intense and frequent volcanic activity. In this arrangement, hot water from underground is distributed from central locations to users and achieves good results with less pollution that other heating systems based on combustion of fossil fuels. Geothermal heating in Iceland also is used in greenhouses and in various industrial processes, such as drying seaweed and washing wool. The preeminent geothermal power facility in the United States is located at the geysers in northern CALIFORNIA near SAN FRANCISCO and is operated by Pacific Gas and Electric Company. Other sources of geothermal energy exist at numerous locations in the United States, and some are used for heating, but other factors, such as distance from major population centers, restrict the availability of these heat sources for generation of power or other large-scale commercial applications. JAPAN and MEXICO also have, or have plans to develop, geothermal power facilities.

Generating electricity using geothermal energy causes various problems such as mineral-rich water corroding metals and building up on turbines. In some locations, earthquakes are associated with high heat flow from Earth's interior. Sometimes the earthquakes are only minor, but some may be much more powerful. Earthquakes in geothermal areas are thought to be connected to motion on FAULTS along which geothermal fluids flow. Volcanic activity also is associated with areas suitable for geothermal power production because volcanoes themselves constitute areas of high heat flow.

See also LARDARELLO; PLATE TECTONICS.

geyser A jet of water that erupts on an occasional basis from a small opening in Earth's surface. (The hole itself also may be called a geyser.) In a typical geyser, an underground heat source (typically a MAGMA body) sits beneath rock strata with large interconnected void spaces—in other words, something like a cave system. Groundwater seeps into these voids until they are full of water. The water is quickly heated above the boiling point, but it cannot boil because the pressure from all of the water prevents it. High pressure can prevent liquids from boiling. The water heats up until even the water near the surface begins to boil. When it starts to turn to steam, the pressure is released, and the water in the voids flash boils all at once. The quick conversion from water to steam causes an explosion. It creates a fountain of very hot water and steam that may reach height of several hundred feet that empties the void space of water. The cycle of vapor generation and eruption then starts over as groundwater begins to seep back into the void space. Among the most famous geysers is Old Faithful at YELLOWSTONE NATIONAL PARK in WYOMING, United States. Some geysers derive the vapor for their operation from a different source-release of dissolved carbon dioxide.

Gharat, Mount *See* GAUA ISLAND.

Gilan *earthquake, Iran* A massive earthquake struck northwest IRAN in the area along the Caspian Sea on June 21, 1990, at 12:30 A.M. It was one of the worst disasters in the modern history of the region. The MAIN SHOCK registered a 7.7 on the RICHTER scale but strong AFTERSHOCKS with MAGNITUDES up to 6.5 continued to rock the area for the following four days and continued to cause loss of life into July. The main shock was felt throughout northwestern Iran and even in Azerbaijan, USSR. The source of stress for this event was the continuing collision of the ARABIAN PLATE with EURASIA, which forms the Zagros orogeny. The FOCUS of the earthquake was 11.5 miles (19 km) in depth.

The casualties from this event were enormous. Upward of 50,000 people lost their lives, and 200,000 people were injured. Over one-half million people were left homeless, and the total damage was $8 billion. SURFACE WAVES shook loose AVALANCHES and ROCKFALLS from the mountains. Large LANDSLIDES struck the Rascht-Qazvin-Zanjan area, causing great destruction and rupturing a dam that caused flooding and casualties in the valley below. More than 100 towns were destroyed or heavily damaged in this event. Bad weather immediately following the event hampered rescue operations, undoubtedly increasing the DEATH TOLL.

glaciation The formation of glaciers, great fields of ice resulting from the accumulation and compression of snowfall over long periods, has helped shape the peaks of numerous volcanoes. A case in point is OREGON's Mount HOOD, where glaciers have carved valleys in the flanks of what apparently once was a conical mountain. The water contained in glacial ice can melt and generate destructive LAHARS during eruptions. MUDFLOWS pose a particular threat to settled areas near the volcanoes of the CASCADE MOUNTAINS in the Pacific Northwest of the UNITED STATES. AVALANCHES also may occur when eruptive activity melts glacial ice on a volcanoes

summit and weakens the structure of the ice sheets. At three locations in the United States, Mount WRANGELL in ALASKA and Mount BAKER and Mount RAINIER in WASHINGTON, heat and steam from the volcanoes have created a network of caves and passageways in the glacial ice at the summits of the volcanoes. Volcanoes under continental ice sheets can create huge volumes of meltwater that can cause major floods. In ICELAND, the volcano GRIMSVÖTN has accumulated up to three cubic kilometers of meltwater. When the water was released in an ice-dam break (JÖKULHLAUP), floods devastated the country. Floodwaters from a glacial lake in what is now western MONTANA are thought to have played an important part in shaping the landscape of the COLUMBIA PLATEAU in Washington State, stripping away SOIL and scouring the land down to bedrock. In some places, the floodwaters apparently removed whole layers of BASALT and thus produced clearly visible MESAS and terraces along the Columbia River.

global positioning system (GPS) GPS uses satellites to find exact locations and elevations on the surface of the Earth. The device locates satellite signals and triangulates its position when it has located enough of them. The position is commonly read in terms of latitude and longitude, but some units also have the capability to plot the location on a map and store each position as a way station. Previously, satellite signals were scrambled for security reasons, and a location was only accurate to about 165 feet (50 m). The scrambling was discontinued several years ago, and GPS data are now much more accurate. There is a difference between research-grade GPS and those mounted in cars or sold in sporting goods stores. The highly accurate GPS units determine locations to within one centimeter and are the type that are used for earthquake and volcano studies.

gold One of the most economically valuable minerals, gold is highly DUCTILE and malleable, a good conductor of electricity, and resistant even to powerful solvents. Gold is found in native (or pure) form as ores in areas of volcanic activity, where chemical FRACTIONATION and HYDROTHERMAL activity combine to concentrate the metal. Gold often occurs as minerals called tellurides and also may be found in native form in veins of QUARTZ. Deposits of gold along the ANDES MOUNTAINS played an important part in the conquest and colonization of SOUTH AMERICA's Inca civilization by the Spanish empire in the early 16th century. Drawn to the Andes by reports of abundant gold, Spanish troops under the leadership of Francisco Pizarro conquered the Inca by simply taking their emperor Atahuallpa prisoner. The Spaniards demanded a heavy ransom in gold for their imperial captive and, when the ransom was delivered, had Atahuallpa killed. The Inca gold reportedly was mined largely from placer deposits, which occur when bits of gold are eroded away from their original location (the mother lode) and laid down in sediments from which they may be extracted more easily than from solid rock.

See also PLATE TECTONICS.

Good Friday earthquake *Alaska, United States* One of the most powerful (M = 9.2) and destructive earth-

A boat washed far inland and beached by the surge of the tsunami generated by the Good Friday earthquake, Alaska, 1964 *(Courtesy of NOAA)*

A large slump broke an elementary school in half during the Good Friday earthquake in Alaska, March 27, 1964. The earthquake had a magnitude of 9.2 and caused over $538 million in property damage. *(Courtesy of NOAA)*

quakes of the 20th century, the Good Friday earthquake struck the south coast of ALASKA along Prince William Sound on March 27, 1964, and lasted between three and five minutes. The earthquake killed more than 100 people and involved displacements of as much as 50 feet (15 m) along various faults. Estimates of property damage from the earthquake range in the hundreds of millions of dollars. The Good Friday earthquake was notable for the TSUNAMI that accompanied it. The wave destroyed the waterfront at Seward and, at Kodiak, wiped out much of

Chaotic jumble of houses, trees, and landscape resulting from slumping and surface failure during the 1964 Good Friday earthquake in Alaska *(Courtesy of the USGS)*

the downtown area and destroyed almost half of the local fishing fleet.

The tsunami caused extensive damage as far south as CRESCENT CITY, California. More than 2,000 LANDSLIDES and AVALANCHES were attributed to this earthquake, and at one location at Shattered Peak in the Chugach Mountains, an avalanche was found to have traveled several miles atop a cushion of air across a glacier. Later examination showed that the air cushion had preserved structures on the glaciers surface as the avalanche passed. The earthquake had dramatic effects on lakes in Alaska; movement of lake waters cast chunks of ice onto the shore and reportedly caused damage to trees as high as 30 feet (9 m). Saltwater invaded certain freshwater lakes along the shore. The Good Friday earthquake coincided with unusual observations in other portions of the UNITED STATES; the water level at one well in SOUTH DAKOTA, for example, is said to have fluctuated more than 20 feet (6 m). Similar, though less dramatic, fluctuations were reported from PUERTO RICO and AUSTRALIA at the time of the Good Friday earthquake.

Gorely Khrebet *caldera, Kamchatka, Russia* The collapse that created the Gorely Khrebet CALDERA is thought to have had a volume of perhaps five cubic miles (21 km³). The caldera is located on the edge of a large, negative-gravity anomaly that is evidence of the existence of a huge, buried caldera. The Gorely Khrebet volcano occupies the center of the caldera. Eruptions have been recorded in 1828, 1832, 1855, 1869 (uncertain), and 1929–31. SOLFATARIC activity was reported in 1947. Temperatures of FUMAROLES increased in 1960–61, and small eruptions of ASH followed. Fumarolic activity diminished for several years and then increased again in the late 1970s as new fumaroles appeared and temperatures in fumaroles rose. Gas plumes rose to heights of several hundred meters in 1979. In June 1980, PHREATIC ERUPTIONS began, and others took place over the following months. Another eruption, similar to that of 1980–81, occurred in 1984–85.

graben A valley formed by the down-dropping of a FAULT block along normal faults. A half-graben means only one normal fault, but a full graben requires two normal faults that slant toward each other. Grabens typically form in association with DIVERGENT BOUNDARY. There are many good examples of grabens worldwide. Every basin in the BASIN AND RANGE PROVINCE is a graben. The Newark Basin in NEW JERSEY, like the Hartford Basin in CONNECTICUT and MASSACHUSETTS, and Gettysburg Basin in PENNSYLVANIA are all examples of half-grabens. The Rhine River valley and Viking graben in the North Sea are famous examples of grabens. During the 1964 GOOD FRIDAY EARTHQUAKE in ALASKA, grabens formed and caused extensive damage to property. In one incident, a building toppled off the edge of a HORST, or elevated block, and landed upside-down inside a graben.

Grand Banks *earthquake, Newfoundland, Canada* This very powerful earthquake occurred on November 18, 1929, and was centered under the Grand Banks of Newfoundland, a rich fishing area off the east coast of CANADA, east of Nova Scotia. The earthquake measured 7.2 on the RICHTER scale. Felt all throughout the New England region of the UNITED STATES and parts of Canada south of the SAINT LAWRENCE RIVER and the Strait of Belle Isle, the earthquake stopped clocks and shook objects from shelves on land and subjected ships at sea to a powerful shaking. Because the EPICENTER was some 250 miles (402 km) from the nearest land, very little damage was done to inhabited areas. A TSUNAMI associated with this earthquake struck three hours later and reportedly was responsible for extensive damage and the loss of 27 lives at Placentia Bay, Newfoundland. Minor waves were reported as far south as the shores of SOUTH CAROLINA. The earthquake broke submarine telegraph cables that were laid across the area of the earthquakes epicenter in 28 places. The breaks were caused by sediments that were shaken loose and that slid down the continental slope like an undersea LANDSLIDE but called a TURBIDITY CURRENT. By timing the point at which each of the lines went dead and plotting their position under water, it was determined that these turbidity flows achieved speeds of up to 50 knots. For scientific purposes, this earthquake provided an excellent laboratory experiment.

Grand Coulee *canyon, Washington, United States* A deep canyon cut into lavas by ancient floodwaters, Grand Coulee is about 25 miles (40 km) long and 800 feet (244 m) deep in places. The Grand Coulee lies approximately northeast to southwest between the Columbia River and the Quincy Basin and is the site of the Grand Coulee Dam. The waters that carved the Grand Coulee in the lavas are thought to have originated from the melting of a glacier immediately to the north.

See also COULEE.

granite A crystalline intrusive igneous rock, granite is one of the most common PLUTONIC rocks. It is commonly used as a building material because of its availability and resistance to weathering and wear. Granite displays an interlocking pattern of crystals of such minerals as QUARTZ, potassium (K-)feldspar (*see* K-FELDSPAR), some PLAGIOCLASE, and possible BIOTITE, HORNBLENDE, and/or MUSCOVITE. Granite comprises an entire family of rocks. Coarse-grained granite cooled slowly from the molten to the solid state and thus gave large crystals time to form. Fine-grained granite, on the other hand, cooled more quickly so that crystals in the rock had less opportunity to grow. Granitic rocks often are found in large circular to elliptical PLUTONS called BATHOLITHS.

See also INTRUSION.

granitization Conversion of metamorphic rocks to GRANITE by excessive heat and pressure and aided by water moving upward from deep inside Earth. Granitization is believed to be responsible for forming PLUTONS that show no clear boundary with surrounding COUNTRY ROCK.

granodiorite The plutonic equivalent of DACITE-ANDESITE (intermediate composition). These rocks commonly form large BATHOLITHS in the roots of MAGMATIC ARCS. The SIERRA NEVADA batholith is largely composed of granodiorite, as are undoubtedly the roots of the ANDES MOUNTAINS.

gravity   Gravity is one of the fundamental properties of physics and the main contributor to the disaster in earthquakes and volcanoes. Gravity is the ultimate cause of all MASS WASTING, including LANDSLIDES and AVALANCHES. It is the reason that LAVA, LAHARS, and NUÉE ARDENTEs come down the slopes of volcanoes and wreak havoc. When seafloor faulting takes place, gravity is behind the ocean water readjusting and thus producing a TSUNAMI. It is also the reason that buildings fall when damaged.

GEOPHYSICISTS study detailed variations in gravity to determine subsurface features and processes. It is especially useful for locating faults and PLUTONs and in PLATE TECTONIC applications. Microgravity studies can even be used to track movements of magma.

Great Basin   *United States*   The Great Basin is an area in CALIFORNIA, NEVADA, and western UTAH where streams find no outlet but drain instead into lower portions of the basin itself. DEATH VALLEY occupies part of the Great Basin. It is part of the BASIN AND RANGE PROVINCE.

Great Rift   *United States*   An area of parallel fractures in present-day IDAHO from which LAVA FLOWS emerged and covered an area of some 600 square miles (1,554 km²). The most recent volcanic activity along the Great Rift is believed to have occurred approximately 2,000 years ago.

Great Rift Valley   *Africa*   A region of East AFRICA characterized by RIFTing and associated volcanic activity. The Great Rift Valley is the inactive arm of the TRIPLE JUNCTION that includes the Red Sea and the Gulf of Aden as the active arms with fully developed MID-OCEAN RIDGES. The Great Rift Valley crosses the countries of Eritrea, Ethiopia, Kenya, Uganda, Tanzania, Rwanda, Burundi, Malawi, and Zambia. Some of the worlds most famous volcanoes and CALDERAS are located along the Great Rift Valley. Most of the volcanoes are BASALTic, but RHYOLITIC volcanoes are also present. The active rifting also produces earthquakes on a regular basis.

See also ASAWA; FANTALE.

Greece   With the exception of ITALY, Greece is the most active country in Europe in terms of earthquakes and volcanoes and one of the more active in the world. Greece owes its activity to its PLATE TECTONIC position in the eastern MEDITERRANEAN basin. The Hellenic Arc marks a SUBDUCTION ZONE where African OCEANIC CRUST is being consumed beneath Eurasian crust along a small ISLAND ARC. TURKEY is moving westward into Greece primarily along the North Anatolian Fault through the processes of EXTRUSION TECTONICS. The western margin of Greece along the Aegean Sea is being affected by the compressional activity that affects Italy. Basically, Greece is being squeezed from all sides.

There are several dormant and inactive volcanoes along the Hellenic Arc. The only large eruption was from THIRA (Santorini) in 1,500 B.C., which may have been one of the greatest in historical times. Earthquakes, on the other hand, occur in Greece frequently. Deep FOCUS earthquakes are exclusively associated with the Hellenic Arc and and define a BENIOFF ZONE beneath the Mediterranean Sea. Shallow focus

earthquakes occur all over. The RECURRENCE INTERVAL studies for Greece show that there is on average, one MAGNITUDE 6.3 earthquake per year. Earthquakes with magnitudes of 8.0 or greater have a recurrence interval of only 49 years. Earthquakes in SPARTA, CORINTH, CHIOS, and Helike are examples of truly devastating events. Because Greece is surrounded by water, there is also great potential for TSUNAMIs. Waves with RUNUP heights of over 100 feet (30 m) have been recorded in the 20th century. The Gulf of Corinth is especially susceptible. It is a great credit to the ingenuity of the Greeks that these powerful events have not caused greater devastation in terms of loss of life and property.

Grimsvötn   *volcano, Iceland*   Grimsvötn has erupted 45 times since records were kept. The 1934 eruption of Grimsvötn, located underneath the Vatnajökull glacier, is a good example of a JÖKULHLAUP, or subglacial volcanic eruption. In late March, an unseasonable increase in the volume of flow in the Skeidara River, which carried away meltwater from the glacier, was among the initial signs of the eruption. The river was muddy and smelled of SULFUR—further evidence of an eruption. The flow of water from beneath the glacier increased during the next several days, and portions of the glacier broke away and were carried off with the water. So great was the flow that is spurted under high pressure from openings in the glacier. The eruption is believed to have created a reservoir of meltwater under the glacier. As this pool of water drained away, the glacial ice overlying the volcano subsided until, soon after the jökulhlaup ended, eruptive activity broke through the ice, and the volcano expelled gas and ASH into the air, depositing a light ASHFALL over several thousand square miles. Every time Grimsvötn erupts, flooding from these meltwaters occurs. In the 1996 eruption, a glacial outburst released an enormous amount of dammed-up meltwater at a rate of 1.5 million cubic feet (45,000 m³) of water per second. Damage from this massive flood was estimated at $12 million. The most recent eruption began in December 1998.

ground failure   Any failure of the surface of the Earth as the result of shaking from the passing of seismic waves (primarily SURFACE WAVES). Ground failure includes LANDSLIDES, LIQUEFACTION, SLUMPing, FISSURES, lateral spreading, and any other resultant MASS WASTING. It typically does not include ROCKFALLS and ROCKSLIDES or any other steep slope process. Ground failure is commonly responsible for the majority of the damage to human-made structures such as buildings, bridges and roads.

ground motion   In general terms, any shaking of Earth's surface resulting from a seismic disturbance. Most ground motion that is felt is caused by SURFACE WAVES (LOVE WAVES and RAYLEIGH WAVES). However, BODY WAVES may also cause ground motion to a lesser degree.

groundwater   In general, any water moving beneath Earth's surface. Groundwater plays an important role in LIQUEFACTION, the process responsible for much of the damage caused by earthquakes in localities where structures are built on unconsolidated soil with groundwater close to the surface. Groundwater is derived largely from rainwater that has percolated downward through the soil and is confined to within about 3,000 feet (914 m) of Earth's surface. Groundwater may travel through porous underground layers or conduits called aquifers and emerge at the surface in the form of natural springs. Groundwater also provides the fuel for steam explosions in GEYSERS and in FUMAROLES.

Guagua Pichincha *volcano, Ecuador* This volcano, near the capital city of QUITO, is therefore the most dangerous volcano in ECUADOR. Guagua has had 25 historical eruptions several of which devastated Quito. In 1660, more than 10 inches (25 cm) of ASH and volcanic fragments blanketed the city. The last eruption was in 1993. It killed a VOLCANOLOGIST who was studying the volcano at the time.

Guatemala Guatemala's location in CENTRAL AMERICA places it in one of the regions most susceptible to earthquakes and volcanic eruptions. Notable earthquakes in Guatemala's history include that of the night of April 8, 1902. This earthquake lasted perhaps 30 to 40 seconds and caused some 2,000 deaths in Quetzaltenango. The earthquake coincided with a torrential rainstorm, and the city lost electrical power and lighting. The resulting darkness reportedly led to numerous deaths when townspeople, fleeing buildings and running into the street, were unable to see where they were going and perished when walls fell on them. The heavy rainfall also resulted in deaths from drowning. Shocks and rainfall continued for three days, making relief work difficult or impossible. This earthquake occurred at approximately the same time as the great eruptions of Mount PELÉE and SOUFRIÈRE in the CARIBBEAN.

The other major earthquake was on the MOTAGUA FAULT in 1976. It had a magnitude of 7.5 on the RICHTER scale and killed more than 23,000 people. It was felt over 38,610 square miles (100,000 km²) and caused $1.1 billion in property damage.

Guatemala has several volcanoes including Acatenango, Agua, ATITLÁN, FUEGO, PACAYA, SANTA MARÍA and Santiaguito, and TOLIMAN. Most of the volcanoes are inactive or infrequently active. However, Fuego has erupted more than 60 times since 1524, making it the most active volcano in Central America. The 1974 eruption was the most voluminous in recent history. It has shown activity as recently as 1999. Pacaya is also active, having erupted 23 times since 1565. It has been erupting nearly continuously since 1965. Santa María produced the second largest eruption of the 20th century in 1902. It produced 1.3 cubic miles (5.5. km³) of DACITIC EJECTA and had a VEI of 6.

Gunung Baru *See* SEGARA ANAK.

Gunung Rinjani *See* SEGARA ANAK.

Gutenberg, Beno (1889–1960) U.S. SEISMOLOGIST Gutenberg, who relocated from Germany to the United States in the years just before World War II, was a colleague of Charles RICHTER and used earthquake data to estimate the diameter of Earth's CORE. The boundary between the core and MANTLE is called the Gutenberg discontinuity.

Gutenberg discontinuity The zone that is marked by a radical shift in the velocity of seismic waves deep within the Earth that coincides with the boundary between the MANTLE and CORE. Thus discontinuity was named after the famous seismologist Beno GUTENBERG.

guyot An undersea, flat-topped volcanic mountain that does not extend to the ocean surface. Geologists believe that guyots once reached the surface and formed islands but wave action reduced them to subsurface levels. Guyots are often found in chains with older, lower mountains at one end and younger, taller guyots at the other. Guyots can have circular reefs on their tops outlining the edges of the island.

Hakkoda *volcano, Japan* The Hakkoda volcano is composed of stratocones and LAVA DOMES the highest of which is Ootake. It is located 12.5 miles (20 km) southeast of Aomori City in northeastern JAPAN. It has not been active in historic times.

Hakone *caldera, Japan* The Hakone CALDERA is located near Sagami Bay on the central island of Honshū near TOKYO. The caldera is thought to have formed through several explosive eruptions. A large eruption, possibly phreatic, is believed to have occurred perhaps 3,000 years ago, along with a PYROCLASTIC FLOW or LANDSLIDE that created a natural dam and thus formed Lake Ashinoko. A lava plug formed afterward. Apparently, no eruptions of LAVA have happened here since approximately 950 B.C. Although no extremely powerful eruptions have occurred at Hakone within historical times, this caldera is interesting for its solfataric and seismic activity. Minor earthquakes occurred in early 1917. Other earthquakes in the 1950s and 1960s helped establish correlations between seismic activity and very high hydrothermal temperatures, as well as between earthquakes and intensified activity of FUMAROLES. There are various explanations for these correlations. One is that hot water or steam rising from great depths causes earthquakes as it expands on the way to the surface. In some areas, it has been suggestion, very hot water underground weakens the rock so that strain is released as earthquakes in these thermal areas. A worker was killed Hakone in 1933 in an explosion at a SOLFATARA.

Hakuun-dake *See* DAISETSU-TOKACHI.

hanging wall The block of rock above the fault plane in a DIP-SLIP FAULT. *See also* FOOTWALL.

harmonic tremor (archaic) Also known as a volcanic tremor, a harmonic tremor is a small earthquake, observed in the vicinity of active volcanoes, that indicates molten rock is flowing beneath the surface. Such a tremor has a frequency of several cycles per second. Scientists are not certain how harmonic tremors are generated, but the vibrations have been linked tentatively to turbulence in MAGMA flowing underground and to the emergence of gas bubbles from the molten rock. Although eruptions do not necessarily follow the occurrence of harmonic tremors, the 1980 eruption of Mount SAINT HELENS in WASHINGTON State was preceded by harmonic tremors that told geologists that magma appeared to be forcing its way upward toward the surface.

See also SEISMOLOGY; VOLCANISM.

Haroharo *caldera, North Island, New Zealand* The Haroharo CALDERA and the Okataina volcanic center are located in the TAUPO VOLCANIC ZONE on NEW ZEALAND's North Island. The area is noted for abundant volcanic and geothermal activity. Haroharo and Okataina have undergone considerable seismic and volcanic activity in the past two centuries. An eruption of TARAWERA on the southern margin of Haroharo in 1886 involved PHREATIC ERUPTIONS that were generated when molten rock underground heated subterranean water. There were apparently some early signs of an approaching eruption, notably earthquakes, increased GEYSER activity and rising temperatures at hot springs. The eruption began an hour after a series of strong earthquakes on June 10, 1886. It produced an estimated 0.3 cubic miles (1.3 km³) of ASH in both summit and lateral explosions as well as more than 0.5 cubic miles (2 km³) of BASALTic lava. Two villages (Te Ariki and Te Wairoa) were buried, and more than 150 people were killed. A strong geyser, Waimangu, became active in 1900, and very large HYDROTHERMAL explosions occurred at the geyser in 1915 and 1917. This latter blast occurred unexpectedly and hurled out rocks and mud, damaging a government tourist department facility nearby. More earthquakes took place in 1962, and in 1973 a small hydrothermal explosion at Waimangu followed an increase in ground temperatures and in the activity of hot springs and FUMAROLES. A fairly strong earthquake occurred about 100 miles (161 km) south-southeast of Waimungu approximately a half-hour before the explosion, but the earthquake may have been unrelated to the explosion. Earthquake activity increased at Okataina in 1982 and 1983, and considerable seismic activity continued through the late 1980s.

Hawaii *See* HAWAIIAN ISLANDS.

Hawaiian Islands *United States* The Hawaiian Islands constitute an archipelago in the midst of the PACIFIC CRUSTAL PLATE and are the sites of some of the most intensively studied volcanic activity. The relatively gentle character of eruptions of Hawaiian volcanoes, compared to the more explosive and destructive activity of volcanoes such as BEZYMIANNY and Mount SAINT HELENS among many others, makes them easy to observe. Hawaiian volcanoes include MAUNA KEA, the tallest volcano on Earth (from its origin on the ocean floor); MAUNA LOA, and KILAUEA. However, all Hawaiian Islands are of volcanic origin. Unlike the MAGMA that supplies volcanoes with histories of explosive eruptions, Hawaiian magma tends to be low in dissolved gases. It is also low in SILICA and rich in iron and magnesium. LAVAS from Hawaiian volcanoes are generally BASALTic and extremely fluid. The Hawaiian chain includes the LOIHI SEAMOUNT, a volcano that represents the youngest peak among the islands but that has not yet reached the surface of the ocean.

Hawaii, the major island of the chain, is made up of several volcanic structures: Kohala Mountains, at the northern tip of the island; Mauna Kea, in the north central sector; Mauna Loa, immediately to the south of Mauna Kea; Kilauea, on the southeast shore; and Hualalai, on the western shore near Kealakekua Bay. Mauna Kea is the taller of the two principal volcanoes on the island (the other principal volcano is Mauna Loa) and is thought to be

inactive. Mauna Loa is active and is situated atop two RIFT zones trending northeast and southwest from the mountain. These rift zones have emitted LAVA FLOWS frequently in the last two centuries, and in some cases, the flows from the northeastern rift zone have come very close to the city of HILO, about 50 miles (80.5 km) northeast of the volcano's summit. The fast-flowing lava from Mauna Loa's eruptions may reach velocities of more than 30 miles (48 km) per hour. In some locations where lava from Mauna Loa has reached the sea, it cooled rapidly into dark volcanic glass that has been fragmented by wave action and formed striking beaches of black sand. A spectacular eruption of Mauna Loa in 1950 lasted two weeks and covered more than 30 square miles (78 km^2) of the island with lava. Fountains of molten rock shot several hundreds of feet into the air from several miles of FISSURES along the southwestern rift zone. The emissions of lava caused considerable property damage and flowed across the coastal highway. Kilauea, to the southeast of Mauna Loa, erupts more often than Mauna Loa but is perhaps the safest active volcano on Earth to approach because of its predictable and nonexplosive eruptions. On the floor of the Kilauea CALDERA is a huge lava lake inside Halemaumaua Crater. The level of the lava lake in the crater fluctuates. On several occasions in the 20th century, lava has spilled out of the CRATER. Most of the time, however, the lava remains at a safe level, and sightseers may approach the rim of the crater in safety. At night, visitors may see a dazzling display on the crater floor as solidified lava on the surface cracks and allows the glowing molten rock underneath to be seen, forming dramatic patterns of light. Two rift zones trending east and southwest from Kilauea have been responsible for most eruptive activity at the volcano in the recent past. A curious effect of the lava flows produced "lava fossil trees" along the eastern rift zone where lava flowed around live trees and sheathed their trunks. When the trees themselves died and were carried away, the sheaths, or "lava trees," remained. These columnar formations may be seen at Lava Tree State Monument. In 1790 along the southwestern rift zone, toxic gas from an eruption caused numerous deaths.

Oahu, site of the Japanese attack on the United States naval base at Pearl Harbor in 1941, was formed by eruptions of two large volcanoes. One had its caldera where the Waianae Range on the western side of the island stands today. Mount Kaala, a flat-topped mountain approximately 4,000 feet (1,219 m) high at the northwest corner of the island, comprises a portion of the original Waianae volcano. The second major volcano caldera is thought to have been located on the eastern side of the island at what is now Kaneohe Bay along the Koolau Range. This latter volcano is known for its great numbers of basalt DIKES that may be seen along roadsides in the eastern part of Oahu. The famous Diamond Head appears to be the product of lesser and later volcanism than that at Koo-Kohala Maunalau. Diamond Head is part of a chain of volcanic formations that also includes Sugarloaf and Round Top. Another chain of volcanic formations extends along the southeastern shore of Oahu

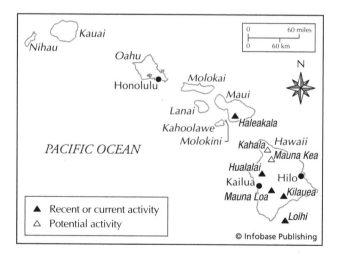

Map of the Hawaiian Islands showing the location of active and potentially active volcanoes as well as several major cities

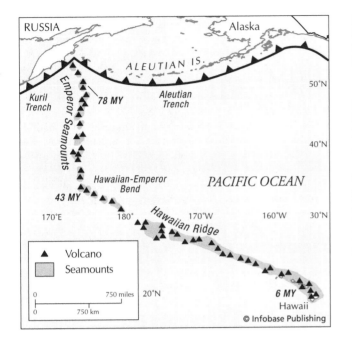

RUSSIA

Alaska

ALEUTIAN IS.

50°N

Kuril Trench

Emperor Seamounts

78 MY

Aleutian Trench

40°N

Hawaiian-Emperor Bend

43 MY

170°E   180°   170°W   160°W   30°N

Hawaiian Ridge

PACIFIC OCEAN

▲ Volcano

▢ Seamounts

0   750 miles   20°N
0   750 km

6 MY

Hawaii

© Infobase Publishing

Map of the Hawaiian ridge and Emperor Seamounts in the Pacific Ocean. As the Pacific plate moved over the Hawaiian hot spot, it left a chain of seamounts. To the southeast, the volcanoes are progressively younger.

from Koko Head crater near Hanauma Bay to Makapuu Head. Other craters on Oahu include Punchbowl (site of the National Memorial Cemetery of the Pacific), Makalapa, and Aliamanu.

The island of Kauai is the location of a single great volcano called Waiualaele, more than 5,000 feet (1,524 m) high and surmounted by a caldera some 10 miles (16 km) wide. Rainfall has eroded the deep Waimea Canyon in the flank of the volcano, exposing the layered evidences of past eruptions. Erosion also has produced spectacular formations along nearby Honopu Valley.

The island of Maui has two volcanoes. On the southeastern side of the island, Haleakala, the larger of the two volcanoes at more than 10,000 feet (3,000 m) high, has an exceptionally large crater and still emits steam from time to time. A lava flow that passed through a breach in the crater,

A lava cascade (lava waterfall) at night in 1969. The lava flow is from Mauna Ulu, Phase 2, on Kilauea volcano, in Hawaii. The source of the lava is the fountain in the background. *(Courtesy of the USGS)*

Lava from an eruption of Kilauea engulfed a tree and cooled around it, leaving a fossilized lava tree when the rest of the lava flowed away. *(Courtesy of NOAA)*

the Kaupo Valley gap, reached the sea. The other volcano on Maui is West Maui, on the island's northwestern side. The summit of West Maui, Puu Kukui, is almost 6,000 feet (1,830 m) high.

Molokai Island has two volcanoes and is remarkable for its tall cliffs eroded by wave action along its shores. A comparatively recent addition to the island is the Kalaupapa Peninsula, formed by eruption of a small volcano.

TSUNAMIS are an occasional hazard in the Hawaiian Islands, which stand exposed in mid-Pacific to seismic sea waves accompanying earthquakes along the "RING OF FIRE" surrounding the PACIFIC OCEAN basin. In 1946, a tsunami resulting from an earthquake in the ALEUTIAN ISLANDS caused extensive damage at Hilo; it attained great heights at various points in the Hawaiian chain: 45 feet (14 m) on the northern shore of Kauai, 55 feet (17 m) on the northern coast of Hawaii and 35 feet (11 m) on the northwest coast of Oahu. The islands were battered by another powerful tsunami in 1960, this one approaching from the direction of the CHILEAN 1960 EARTHQUAKE. Waves reached heights of 35 feet

(11 m) on Hawaii, 13 feet (4 m) on Oahu, and 14 feet (4.3 m) on Kauai. Both waves among others caused significant loss of life as well as property damage.

Most earthquake activity in Hawaii occurs in the vicinity of the island of Hawaii. The documented history of earthquakes in Hawaii extends back to the early 19th century. An earthquake on February 19, 1834, knocked down stone walls on the island of Hawaii, stopped clocks, and made it difficult for standing persons to remain upright. On December 12, 1838, another earthquake similar in its effects to the 1834 earthquake shook Hawaii. On April 2, 1868, one of the strongest and most destructive earthquakes in the history of Hawaii occurred near the southern coast of the island. The earthquake caused heavy damage to wooden homes and straw houses. Numerous walls were shaken down in Hilo, and LANDSLIDES occurred as well. Fissures formed in the ground, and mud appeared in brooks. Effects of the earthquake were especially remarkable at Kohala, where ground waves one to two feet (0.3–0.6 m) in amplitude were reported, and the shock stopped machinery at a sugar mill, including a big 75-horsepower engine that was operating under a full head of steam at the time of the earthquake. On Maui and Lanai, more than 100 miles (161 km) from the EPICENTER, rumbling noises were reported, and buildings shook. In Honolulu, the shock was strong enough to stop clocks. The earthquake was felt more than 300 miles (483 km) distant on Kauai and Oahu.

On Hawaii, a tsunami estimated at 60 feet (18.3 m) high or more came ashore along the south coast and reportedly swept over the tops of palm trees; there were several fatalities, and numerous houses were destroyed. This wave was 10 feet (3 m) high when it reached Hilo and eight feet (2.5 m) high at Kealakekua. A fissure more than two miles (3 km) long formed at Kohuku, and a volcanic eruption occurred at this fissure on April 7. The earthquake was preceded by FORESHOCKS, one of which, on March 28, was sufficiently powerful to knock down stone walls.

On April 26, 1973, an earthquake of MAGNITUDE 6.2, centered near the northeastern shore of the island of Hawaii, affected a large area and was felt as far away as Kauai. The earthquake generated effects of MERCALLI intensity VII and caused an estimated $5.6 million in property damage in the area of Hilo. There were 11 injuries, but no one was killed. Two people died in the November 29, 1975, earthquake (magnitude 7.2) that was felt all through the island of Hawaii and also on Oahu, Molokai, and Lanai. Damage was estimated at $4.1 million, including some $1.5 million in damage from a tsunami that came ashore as a wave some 18 feet (6 m) high near the epicenter of the earthquake. Numerous foreshocks and AFTERSHOCKS occurred, and a small eruption was noted at Kilauea volcano less than an hour after the earthquake.

See also PLATE TECTONICS.

Hayward Fault *California* The Hayward Fault extends through the East Bay suburbs near SAN FRANCISCO and was responsible for powerful earthquakes in 1836 and 1868 that destroyed buildings in San Francisco and Oakland. Although less famous than the nearby SAN ANDREAS FAULT, the Hayward Fault has the potential to cause tremendous destruction in a future earthquake because the East Bay is so densely populated. The fault underlies the University of California campus at Berkeley and lies directly under the Memorial Stadium. Also, many major highways in the San Francisco Bay area either cross the Hayward Fault or may be affected by damage from any strong earthquakes involving that fault in the future. Among medical facilities in Contra Costa and Alameda Counties, eight acute care hospitals, particularly important in the aftermath of any major earthquake, are at risk because they are located within a mile of the Hayward Fault. These hospitals represent about 30% of such hospitals available to the counties. One projection of the effects of an earthquake of RICHTER magnitude 7.5 along the Hayward Fault puts the possible number of fatalities as high as 7,000, with the potential for more than 13,000 injuries requiring hospitalization and lesser injuries affecting more than 130,000 individuals. This projection involves an earthquake in early afternoon. Another projection for the hours just after midnight puts fatalities lower, at perhaps 1,500, and injuries at about 50,000, with more than 4,000 of those requiring hospitalization. An earthquake of magnitude 7.5 on the Hayward Fault is expected to cause the greatest damage on the west side of the fault, within an area about five miles (8 km) from the fault. This area encompasses such heavily settled areas as San Leandro and San Lorenzo.

heat flow The flow of heat from Earth's interior to the surface varies greatly from one location on the surface to another. The total heat release at the surface has been expressed as the sum of the heat transmitted by conduction (transmission of heat directly through stationary masses of rock) and heat carried by mass transport (that is, the migration of MAGMA from lower levels to the surface). Some areas, such as active volcanoes, represent an unusually vigorous flow of heat to the surface, whereas other parts of Earth's surface are relatively cool. The rate of heat flow to Earth's surface is measured in units called HFUs (heat flow units). One HFU equals one-millionth (that is, 0.000001) calorie per square centimeter per second. Depending on how it is estimated, the mean global rate ranges between about 1.0 and 1.5 HFUs, although in some areas of active volcanism and HYDROTHERMAL activity, measurements well above 200 and even 700 HFUs have been made. Various factors are believed to influence the rate of heat flow. For example, radioactivity and the heat released by decay of RADIONUCLIDES may vary from one location to another. Spatial variations in temperature of the MANTLE also may account for some differences in heat flow. Heat Flow shows great variability in North America, with the western half of the nation generally showing higher and more variable HFU values than the east. The dividing line between east and west has been drawn approximately at the eastern boundary of the COLORADO PLATEAU. The higher heat flow in the west is reflected in the relative abundance of hot springs, GEYSERS, and other hydrothermal activity there, as well as the existence of numerous volcanoes, both active and extinct. Eurasia likewise shows great geographical variations in heat flow. HOT SPOTS are found, for example, in the Alps, the Carpathian Mountains, and the Caucasus Mountains and in Russia's KAMCHATKA PENINSULA, with its numerous volcanoes. High heat flow shows up in areas of strong hydrothermal activity, such as LARDERELLO in ITALY.

Some highly radioactive rock units also have high heat flow. Certain granites with high uranium content, such as the Conway Granite, NEW HAMPSHIRE, have high enough heat flow to be commercial.

Hebgen Lake *Montana, United States*   On August 17, 1959, an earthquake of MAGNITUDE 7.1 on the RICHTER scale and locally INTENSITY X on the MERCALLI scale struck MONTANA. It was the largest historical earthquake in Montana. Fortunately, it was in a very sparsely populated camping area and only 28 people were killed. It caused dramatic changes in the shoreline of the large recreational Hebgen Lake. The lake bed tilted and displaced the lake toward the north, submerging docks and other waterfront property along the northern shore and lifting the southern shore out of the water. The earthquake caused considerable disturbance in the waters of the lake creating a SEICHE with waves up to 20 feet (6 m) high and a period of 17 minutes. The seiche was still detectable 11 hours later. Near the lake, an ESCARPMENT some 10 feet (3 m) high formed along a FAULT more than 10 miles (16 km) in length. This same earthquake caused a huge ROCKSLIDE, 2,000 feet (610 m) by 1,300 feet (396 m) and containing almost 40 million cubic feet (1.17 million m³) of rock, along the south wall of the gorge of the Madison River in the Madison Range. This rockslide dammed the river and created a lake that was some 200 feet (61 m) deep, named Earthquake Lake. The lake grew to such proportions as to threaten to undermine the weakened dam of Hebgen Lake.

A landslide generated by an earthquake dammed the river to make a lake, Hebgen Lake, Madison County, Montana, in 1959. The light-colored area on the hillslope to the left is the landslide scar, and the light-colored area that dams the lake is the landslide deposit. *(Courtesy of the USGS)*

The Army Corps of Engineers cut a 250-foot (76-m) channel across the slide to release the water from Earthquake Lake. If the dam of Lake Hebgen had collapsed, damage would have been far greater.

More than 20 feet (6 m) of vertical displacement was seen near Red Canyon Creek. Part of State Highway 287 was lost into Hebgen Lake during the earthquake, and damage to roads and timber alone was estimated at more than $11 million. This same earthquake also shook YELLOWSTONE NATIONAL PARK severely, and fresh GEYSERS started erupting there whereas others became inactive. A ranger station at Yellowstone reported more than 150 AFTERSHOCKS during the first day following the MAIN SHOCK, and aftershocks kept occurring for months thereafter. Some of the aftershocks were themselves notable earthquakes.

Hector Mine *earthquake, California* On October 16, 1999, an earthquake of RICHTER magnitude 7.1 occurred. The earthquake was on the Lavic Lake Fault in the southern Mojave Desert where 2.8–4.7 feet (.85–1.4 m) of right lateral STRIKE-SLIP offset was observed. Damage to an Amtrak line was reported.

Heimaey *volcano, Iceland* The 1973 eruption of the volcano Heimaey near the port of Vestmannaeyjar in ICELAND's Vestmannaeyjar archipelago is one of the most famous eruptions of modern times and one of a series of more than a dozen eruptions in Iceland since midcentury. The island is about five miles (8 km) long by three miles (4.8 km) wide and is made of BASALT only several thousand years old.

The eruption of Heimaey began in the early morning of January 23, 1973, on the east side of the island along a FISSURE roughly north-south and extending more than a mile in length. FIRE FOUNTAINS emanated from the fissure at first, and volcanic activity occurred underwater at the north and south ends of the fissure for the first several days. Later, the eruption became concentrated in a relatively small portion of the fissure near the community of Helgafell. In about two days, a cinder cone more than 300 feet (91 m) high formed. This cone was named Eldfell (fire mountain) and expelled TEPHRA at a rate of more than 100 cubic yards (79 m³) per second. The tephra fell in large amounts on nearby Vestmannaeyjar. Less than a month after the eruption began, the volcano's output of tephra diminished, and LAVA began to flow from the volcano. A large LAVA FLOW encroached on Vestmannaeyjar, and it looked for a time as if the lava might fill the harbor on the north side of the island. SUBMARINE VOLCANISM continued during this period and broke a power cable that carried electricity from the mainland as well as a freshwater pipeline that served the island.

In late February, Eldfell stood more than 600 feet (183 m) tall, and lava was flowing gradually along a front, ranging from north to east of the volcano. Within several weeks, this flow stood more than 60 feet (18.3 m) high at some locations along its front. The average depth of the flow exceeded 100 feet (30 m), and at places the lava was more than 300 feet (91 m) deep. By April, the lava flow was about 1,000 yards (925 m) long and equally wide and moving at a speed of several yards daily. Another lava flow originated from the volcano in

An aa flow advances through a town at Heimaey, Iceland, in 1973. Notice the white rubble at the base of the black lava flow. It is a house that has just been crushed. *(Courtesy of the USGS)*

late March and began to move northwest. This flow destroyed numerous homes and the town power facility. Portions of the cone broke away and were carried away with the initial lava flow, including one especially large block known as *Flakkarinn* (the wanderer). Submarine volcanism continued through late May, and the eruption finally subsided in July.

The eruption produced an estimated 300 million cubic yards (237 million m³) of lava and more than 25 million cubic yards (20 million m³) of tephra. Among the other products of the eruption were concentrations of poisonous gases, mostly carbon dioxide, with some carbon monoxide and methane that accumulated in low portions of eastern Vestmannaeyjar and killed one person. The origin of the poisonous gases is uncertain, although it appears likely that the gases migrated from the conduit of the volcano through older volcanic rock and into Vestmannaeyjar. Bulldozers pushed a wall of tephra into place between the vent and the community to halt the gas, and a lengthy trench was dug to carry away steam, but neither of these measures worked with total effectiveness. A rapid evacuation from the island saved almost all of the more than 5,000 residents of the island from the harm. Property damage was extensive. Residences near the volcano were destroyed by tephra and lava. Flows of lava from the volcano also destroyed a fish-freezing facility, caused damage to two other such facilities and wrecked a large number of homes in the eastern portion of the community.

The 1973 eruption on Heimaey was notable for efforts to counteract and control the flows of lava. Experiments and observations performed early in the eruption of Heimaey, and also at the earlier eruption of the volcano SURTSEY, indicated that spraying seawater on the lava could cool and harden the molten rock and thus impede its flow. Because the northward- and eastward-moving lava posed the greatest threat to property and activities on the island, including the operation of the harbor, lava-control efforts concentrated on that front. One approach was to spray water on the advancing lava. Another was to build a barrier of lava on the northwest side of the lava flow to block its progress into Vestmannaeyjar. Early in February, about two weeks after the eruption started, a water-spraying operation began. Results were encouraging, and in March a ship capable of directing large amounts of water onto the lava flow was moved into the harbor. Additional pumping equipment was brought in from the United States. Water was pumped straight onto the lava near the harbor and was carried to various portions of the flow through a system of metal and plastic pipes almost 20 miles (32 km) long. The lava-control effort used a combination of water-spraying and earth-moving activities.

The cooling operation caused a dramatic change in the appearance of the lava flow. When left to cool naturally, the lava solidified into a reddish mass with a surface covered with volcanic BOMBS and varying about three feet in relief. After water cooling, the surface turned gray or black and displayed much greater relief, as much as 15 feet (5 m). Cooling the lava with water created peculiar difficulties, such as reduced visibility caused by the large quantities of steam produced by the water contacting the molten rock. Some 8 million cubic yards (6.3 million m³) of water were directed onto the lava flows at Heimaey and are through to have solidified some 5 million cubic yards (4 million m³) of lava. Water cooling appears to have accelerated the solidification of the lava by as much as 100 times. The cooling operation lasted until early July and cost less than $2 million.

One beneficial result of the 1973 eruption of Heimaey was a heating system for Vestmannaeyjar that utilized heat from the still-cooling lava flows. Initial investigations indicated that heat from lava and scoria deposits could be used to supply space heating for the community. Early experiments along these lines met with success, and houses began to be connected with a heating system that exploited the heat from the lava and tephra. Projects were under way by 1979 to exploit heat in several areas of the fresh lava flows. In each area, a set of steam wells extending down into the tephra was connected to a heat exchanger that circulated heated water through the town central heating system. This system was serving almost all the homes on Heimaey by the early 1980s.

The evacuated population of Heimaey returned gradually to the island after the 1973 eruption. Approximately 80% of the island residents returned by early 1975. Hundreds of new homes were built to replace those destroyed by the eruption, and tephra deposited by the volcano was used as landfill for the building of many of those homes. Tephra also was used to expand runway facilities on Heimaey's airport, and lava from the 1973 eruption now serves as a breakwater in the harbor.

Hekla *volcano, Iceland* Hekla, or Gateway to Hell, as the first Icelanders called it, is the most active volcano in ICELAND. Since the settling of Iceland in A.D. 1104 it has erupted 167 times, including 15 major eruptions, in 1104, 1158, 1206, 1222, 1300, 1341, 1389, 1440, 1510, 1554, 1597, 1636, 1693, 1725, 1766, 1845, 1878, 1913, 1947, 1970, 1980, 1981, 1991, and others. The 1300 eruption lasted a full year and is the second largest TEPHRA eruption in Iceland's history. The 1510 eruption is reported to have shot rocks 25 miles (40 km) away killing one person in the process. The 1693 eruption produced some 75,000 cubic yards (60,000

m³) of tephra per second. The eruption lasted for seven months. The 1766 eruption was the largest observed. BOMBS as long as 18 inches (46 cm) were shot nine to 12 miles (14.5 to 19 km) away. Lava poured out in all directions and killed much livestock and wildlife. The 1845 eruption of Hekla began on September 2 and was preceded by strong earthquakes. Tephra was extruded at a rate of 25,000 cubic yards (20,000 m³) per second at the initiation of the eruption. One flow of LAVA from this eruption was measured at 22 miles (35 km) long, one mile (1.61 km) wide at one point, and 40 to 50 feet (12 to 15 m) deep. No humans were reported killed in this eruption, although numerous cattle died.

One remarkably large mass of PUMICE, its weight estimated at almost a half-ton, was carried four to five miles by this eruption. Ice and snow that melted in the eruption flooded rivers. Perhaps the most destructive aspect of the eruption was its effect on pastureland, much of which was covered by volcanic ASH and thus made unusable by animals. Even where ash did not cover the land, herbage became toxic and killed cattle that ate it.

After 100 years of inactivity, Hekla erupted in 1947. It began with earthquakes and a 19-mile (31 km)-high eruption column. FISSURES opened and BASALTic lava poured out at a rate of 4,400 cubic yards (3,500 m³) per second. Several CRATERS formed during this eruption. The ash from the 1970 eruption had such a high fluorine content that some 7,000 sheep in the area died of poisoning.

The most recent eruption of Hekla was in the winter of 2000. Earthquakes preceded 45,000-foot (13,716 m) steam columns that preceded the fissure eruptions. Lava poured out of the fissures and flowed about one meter per minute and was three miles (5 km) long at the time of writing.

Helike *See* CORINTH.

Herculaneum *See* POMPEII AND HERCULANEUM.

Herdubreid *volcano, Iceland* A TABLE MOUNTAIN, Herdubreid is believed to have been formed by eruptions underneath a glacier, thus explaining its flattop.

Hibok-hibok *volcano, Philippines* This STRATOVOLCANO is also called Catarman and has had four historic eruptions. The eruptions were in 1827, 1862, 1871–75, and 1948–53. The 1862 eruptions caused 326 fatalities from *NUÉE ARDENTE*s. The 1871–75 eruption produced the LAVA DOME called Mount Vulcan. SULFUR was mined from one of the craters prior to the 1948–53 eruption.

Hilo *volcanoes and tsunamis, Hawaii, United States* The city of Hilo has a history of trouble from volcanic eruptions and TSUNAMIS. For example, a LAVA FLOW from MAUNA LOA in 1933 posed such imminent danger to Hilo that the U.S. Army Air Force attempted to halt or divert the flow by dropping aerial bombs on selected locations. Whether as a result of the bombardment or from other causes, or possibly both reasons, the lava flow stopped. Another lava flow in 1881 also threatened Hilo but stopped near the edge of the city. The tsunami that struck Hilo on April 1, 1946, did extensive damage in the downtown area and is believed to have killed more than 175 people. Another tsunami, this one from the CHILEAN 1960 EARTHQUAKE, killed more than 60 people in Hilo and came ashore as a wave estimated at 30 feet (9 m) high.

Hobicha   *See* ASAWA.

Hokuchin-dake   *See* DAISETSU-TOKACHI.

holohaline   The technical term for a glassy volcanic rock.

Hood, Mount *Oregon, United States*   One of the most beautiful volcanoes, Mount Hood stands more than 11,000 feet (3,353 m) tall and is located near Portland and other major urban areas in the Pacific Northwest of the UNITED STATES Mount Hood had four major eruptive periods in the past 15,000 years, and further activity is possible. During the last eruptive period in the late 18th and early 19th centuries (250 to 180 years ago). Numerous *NUÉE ARDENTE*s and LAHARS were generated along the southwest flank of the mountain. A small dome was formed during this eruption. Eruptions about the time of the Civil War cast out PUMICE. Because Mount Hood is located in such a populated area, it is a potentially very dangerous volcano.

Hooke's law   A spring law. There is a straight line relationship between STRESS and STRAIN. In other words, the amount that the spring is stretched is directly proportional how hard the spring is pulled. This type of DEFORMATION is referred to as elastic. Rock deforms in this manner as seismic wave pass through it. It stretches or compresses as the wave is passing through and then returns to its original shape after the wave is through.

hornblende   A common mineral of the amphibole group, this iron- and magnesium-rich SILICATE mineral with a double-chain structure is black and elongate and has a distinctive diamond or rhombic shape. It is common in such intermediate igneous rocks as DIORITE, ANDESITE, DACITE, GRANODIORITE, and some GRANITE.

hornito   A tall, thin SPATTER CONE, *hornito* translates as "small oven" from the Spanish. They typically resemble chimneys rather than the conical shape of most spatter cones.

horse   A block of rock surrounded on all sides by related FAULTS. The term *horse* was defined for THRUST-FAULT systems where slivers of rock are systematically stripped up along a fault and stacked like shingles on a roof. The slivers are the horses because they are bounded on all sides by thrust faults. The stack is called a duplex structure. Later, the horse and duplex terminology was applied to strike-slip systems as well. The SAN ANDREAS FAULT system has strike-slip horses and duplexes.

horst   A FAULT block left elevated while the land around it drops down as the result of faulting. Normal faults form the slopes on either sides of a ridge and drop the land surface down to form the surrounding valleys. The classic examples of horsts are the ranges in the BASIN AND RANGE PROVINCE. The basins there are GRABENS. These normal faults produce earthquakes, and an area characterized by horsts and grabens is commonly associated with BASALT and sometimes RHYOLITE volcanoes.

hot, dry rock   A plan to tap GEOTHERMAL ENERGY involves pumping water underground into areas of hot, dry rock, where temperature increases quickly with depth. The rocks are fractured by pumping fluid into them down a well (HYDROFRACTURING), and water is circulated through the fractured rock. The heated water then is returned to the surface through a separate conduit in the form of liquid water or steam and used to generate electricity.

hot spot   Otherwise known as a MANTLE plume, an area of intense, localized igneous activity. Hot spots may occur deep beneath the interior of crustal plates, far from the well-defined belts of volcanic activity that commonly mark plate boundaries. The HAWAIIAN ISLANDS have been generated by the movement of a crustal plate over an underlying hot spot. There is a chain of islands that are smaller to the northwest and connect to a chain of seamounts (both Hawaiian and EMPEROR SEAMOUNTS) that extend across the PACIFIC CRUSTAL PLATE. The islands and seamounts are progressively older to the northwest. The trajectory of the seamounts and age relations show that the Pacific plate is moving northwest at about 4 inches (10 cm) per year. Although the origins of hot spots are not entirely understood, it has been shown that they are located above plumes of rock rising through the mantle of Earth.

See also PLATE TECTONICS.

Huaraz *earthquake, Peru*   On May 31, 1970, a devastating earthquake of MAGNITUDE 7.8 struck Huaraz, Peru. It killed some 66,794 people and caused more than $250 million in property damage. Several towns near the EPICENTER were essentially destroyed. The earthquake triggered both LANDSLIDES and floods from backed-up rivers. This earthquake was one of the greatest disasters to strike the Southern Hemisphere.

Huascarán *avalanche, Peru*   Some say that it was a minor earthquake that began a series of cataclysmic events that devastated a coastal area of PERU, in 1962 but it may have simply been a melting glacier. On January 10, 1962, at 6:13 P.M., 3 million tons of ice from Glacier 511 broke loose at an elevation of 21,834 feet (6,550 m) and barreled down the steep slopes of the ANDES MOUNTAINS. The ice broke loose a large amount of rock and SOIL and then dropped into the gorge of Callejón de Huailas at speeds of hundreds of miles per hour, becoming a true STURTZSTROM. The wind generated by the mass reached hurricane-force and swept up trees, sheep, and anything else not tied down. The avalanche collected all varieties of debris, including houses from the town of Yanamachico, where 800 residents were killed. As it slowed to 60 miles per hour (100 km/hr) into a DEBRIS AVALANCHE, it demolished the town of Ranrahirca, with 2,700 residents, before coming to rest. The melting ice combined with soil and rock dust generated by the enormous impacts to form waist-deep mud. In all, more than 3,500 people lost their lives in this catastrophe.

Hungary Hungary is a landlocked country in eastern Europe in the southern part of the CONTINENT. It is generally flat with thinner than normal CRUST and pieces of both AFRICA and Europe within its geology. It has no current volcanic activity and is considered to be of moderate seismic risk. Seismicity is primarily in the northwestern part of the country, along a border with the Alps (Balkans). Hungary also has a lesser area of seismicity along its eastern margin in the Carpathian (Transylvanian) Mountains. The oldest-documented earthquake was in the Szombathely area on September 7, A.D. 456. It had an estimated modified MERCALLI intensity of IX and a RICHTER magnitude of 6.1. Many people were reportedly killed. The city that has most commonly been struck by earthquakes in Komaron (events occurred in 1599, 1763, 1783, 1806, and 1851). The strongest event was on June 28, 1763, with a modified Mercalli INTENSITY of X and an estimated Richter MAGNITUDE of 6.3. Some 63 people were killed in this earthquake, and 102 people were injured. The strongest modern quake was in Pishkolt in 1834, with an estimated Richter magnitude of 6.8. Most earthquakes in Hungary, however, and 5.5 or less, and recurrence of these larger earthquakes is typically one or two per century.

hydrofracturing Pumping water into the ground at high pressure can cause the rock to FRACTURE. If this fracturing is large enough, it produces earthquakes. An example of earthquakes resulting from this process was inadvertently tested in Denver, COLORADO. At the Rocky Mountain Arsenal, deep holes 11,800 feet (3,600 m) were drilled for waste disposal. The waste was pumped down (injected) the wells at high pressure. Soon thereafter, earthquakes began to occur regularly in Denver. These earthquakes (up to 4.3 on the RICHTER scale) were large enough to cause concern among the local residents. The relation was suspected and injection was suspended after almost three years of operation. When the injection stopped, so did the earthquakes. A similar situation occurred in Cleveland, OHIO. The relationship was not quite as clear.

hydrothermal activity In general terms, hydrothermal activity is any process involving extremely hot water underground. Hydrothermal activity is involved in the deposition of numerous minerals, including QUARTZ, GOLD, and galena, a widespread one of lead. Hydrothermal activity differs from geothermal activity in that the latter involves the movement of the waters and need not involve heating of water by contact with a still-hot body of IGNEOUS ROCK. Hot spring activity is also related to hydrothermal activity.

See also GEOTHERMAL ENERGY; GEYSER.

hypabyssal Refers to a shallow PLUTON into volcanic field. As MAGMA feeds volcanoes, some does not make it out of the volcano. Instead, it may intrude the surrounding volcanic strata as a pluton. To be hypabyssal, it must be very shallow.

hypocenter Also known as the FOCUS, the hypocenter is the underground point of origin of an earthquake and where the FAULT moved.

See also SEISMOLOGY.

I

ice cap Glacier atop a volcano. Many mountains have ice caps. At higher altitude, the air is colder and whereas precipitation may fall as rain on the slopes, it will be snow at the peak if the mountain is high enough. At higher latitudes, snow can occur at any elevation. When ice caps are atop volcanoes, they pose a different threat. When Mount SAINT HELENS erupted in 1980, tremendous amounts of ASH at about 800°C landed on the ice cap. The water went from frozen to boiling upon contact. Huge LAHARS of boiling mud swept down the mountain, destroying everything in their path. Although MUDFLOWS can result from several types of situations, lahars of boiling ASH can only result if there is an ice cap.

Iceland The island nation of Iceland is an exposed segment of the Mid-Atlantic Ridge and is the site of intense volcanic activity. The island is essentially all BASALT and is, in effect, being torn in two by the spreading action that occurs along the MID-OCEAN RIDGE. Iceland's LAKI Fissure, a RIFT VALLEY, is evidence of this process. Volcanic eruptions is Iceland generally occur as LAVA FLOWS from FISSURES in the ground rather than as outbursts from volcanic mountains. The most famous eruption in Iceland's history took place in June 1783 when strong earthquakes preceded an eruption that cast large quantities of ASH into the air, obscuring the sun and harming crops as far away as Norway and Scotland. Two days later, lava erupted from the ground along the Laki Fissure and filled the valley of the nearby Skafta River, forming a natural dam that caused extensive flooding upstream. Flowing southwest toward the sea some 15 miles (24 km) away, the lava from this eruption covered the land to an average depth of 100 feet (30 m). The lava flow was 10 miles (16 km) wide at maximum. Another eruption on August 8 sent lava flowing southeast and filled the Hverfisfljöt river valley, causing more flooding.

SURTSEY achieved worldwide prominence when a SUBMARINE ERUPTION built a new volcanic island between 1963 and 1967. It was accompanied by several impressive PHRE-ATIC ERUPTIONS of firework-quality incandescence. Scientists and the press flocked to the eruption at the time. An eruption of Eldfell on the Island of HEIMAEY also made the evening news. In 1973, a vigorous eruption of Eldfell encroached on the fishing village of Vestmannaeyjar. Even worse, it threatened to block the harbor, which would cut off the fleet. The advancing lava flows were sprayed with water until they hardened and stopped. It was very dramatic to see the struggle between the advancing lava and the Icelanders.

Another impressive eruption in Iceland occurred of GRIMSVÖTN in 1996. A subglacial eruption produced a tremendous amount of meltwater that was dammed up. When the dam burst (JÖKULHLAUP), 57,000 cubic yards (45,000 m^3) of water per second went coursing a valley, destroying everything in its path.

The most active volcano in Iceland is HEKLA. It has erupted an estimated 167 times since A.D. 1104. Its most recent eruption began on February 28, 2000, and continues today.

See also ASKJA; BARDARBUNGA; HERDUBREID; KEAFLA; KOTLUGJA; KVERKFJÖLL; ÖRAEFA-JÖKULL; SKAPTARJÖKULL; TORFAJÖKULL.

Idaho United States Idaho experienced severe earthquakes in 1916 and 1944, but only minor damage was reported. Volcanic landforms are abundant in Idaho, notably in the SNAKE RIVER PLAIN in the southern portion of the state.

See also CRATERS OF THE MOON MONUMENT.

Idaho Batholith A tremendous mass of igneous rock some 250 miles (402 km) long and as wide as about 100 miles, (160 km) the Idaho Batholith is associated with rich deposits of metal ores at the Coeur d'Alene mining area.

See also BATHOLITH.

igneous breccia A breccia that is made up of pieces of volcanic rock, broken up by explosions at the VENT and lithified. To be a breccia, a rock must be an aggregate of angular fragments that are greater than 0.08 inch (2 mm) but typically much larger, and up to boulder size.

igneous complex Intimately associated PLUTONS and/or volcanoes in both space and time. They are also related by chemistry. Any large volcanic field or plutonic complex qualifies as an igneous complex.

igneous rock A rock that solidifies directly from a molten or partially molten state is known as an igneous rock. (The word *igneous* is derived from *ignis,* the Latin word for "fire.") Igneous rocks are one of the three principal categories of rock, the others being METAMORPHIC ROCK and SEDIMENTARY ROCK. OBSIDIAN, PUMICE, and SCORIA are familiar examples of igneous rocks, many of which have considerable economic importance. Igneous rocks make up some of the most famous and spectacular landforms on Earth, including the CASCADE MOUNTAINS in the UNITED STATES and CANADA, the HAWAIIAN ISLANDS, the PALISADES near New York City, and the formations of Monument Valley.

Igneous rocks are compositionally classified as ULTRAMAFIC, MAFIC, INTERMEDIATE, and FELSIC. The term *felsic* derives from FELDSPAR, a dominant SILICATE mineral in such rocks, and high SILICA is the *si.* Besides K-FELDSPAR and PLAGIOCLASE, these rocks contain QUARTZ and can contain BIOTITE, MUSCOVITE, HORNBLENDE, and others. Mafic rocks have high concentrations of magnesium (*mg*) and iron (*fe*). Common minerals in these rocks are PYROXENE, plagioclase, and some OLIVINE. Felsic rocks are generally light in color, whereas mafic rocks and minerals tend to be darker. Intermediate rocks are intermediate in color and composition. They contain plagioclase and quartz and commonly contain hornblende, biotite, and/or pyroxene. Ultramafic rocks contain only pyroxene and/or olivine and are very dark in color.

Grain size differs greatly among igneous rocks. Coarse-grained rocks are called PHANERITIC (from the Greek *phanero-,* meaning "visible"), whereas fine-grained rocks are called APHANITIC (from the Greek *aphan-,* meaning "invisible"). GRANITE is a familiar example of phaneritic rock, and BASALT is a widespread aphanitic rock. Grain size and degree of crystallinity depend on how quickly the molten rock cools. The finer the grain size, the faster the cooling; large grain size, on the other hand, means slower cooling. In PEGMATITES, very coarse-grained rocks, grains can be several feet long. At the other extreme are volcanic glasses, such as obsidian, which have an amorphous, grainless structure because rapid

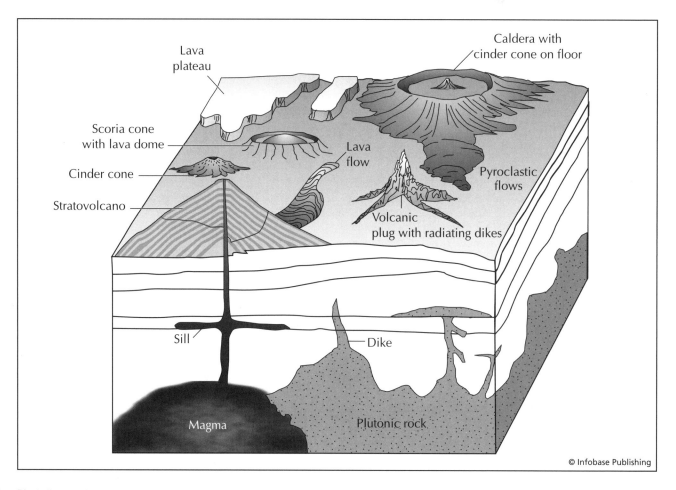

Block diagram showing several varieties of volcanic landforms and pluton types

cooling gave crystal structure no opportunity to form in the solidifying lava.

Igneous rocks can be classified as ACIDIC or BASIC but those terms are holdovers from the 19th century when silicic acid was believed to play a role in depositing silica (silicon dioxide, or SiO_2) in igneous rocks. A rock with high silica content was therefore thought to be acidic, as opposed to basic rock with less silica in it.

Extrusive igneous rocks are those that emerge from the earth and solidify at the surface. Extrusive igneous rocks may occur as LAVA FLOWS, the result of molten rock flowing in sheet form across the surface, or volcanic EJECTA, pieces of rock blasted from the throats of volcanoes and cast upward and outward through the air.

See also ASH; MAGMA; PLUTON; TEPHRA.

ignimbrite A deposit of PYROCLASTIC, PUMICE-rich, EJECTA-rich rock laid down by a PYROCLASTIC FLOW. A given ignimbrite may extend more than 60 miles (97 km) and be 30 feet (9 m) thick or deeper.

Ijen *caldera, Java, Indonesia* Located at the eastern end of Java, the giant Ijen CALDERA (about 12 miles [19 km] wide) is the site of several volcanoes. Kawah Ijen volcano (also known as Ijen Crater) stands inside the caldera, and Merapi and Raung volcanoes have arisen along a ring fracture at the edge of the caldera. A lake occupies the CRATER at Kawah Ijen. Raung tends to exhibit modest eruptions of ASH and occasional flows of LAVA, preceded by minor earthquakes. PHREATIC ERUPTIONS, many of them occurring through the crater lake, characterize Kawah Ijen. Eruptions of Kawah Ijen are associated with increases in seismic activity and in the temperature of the crater lake.

Ijen appears to be a noisy volcanic site; many reports of eruptions mention loud noises from the volcanoes. The historical record of volcanic activity at Ijen dates back to 1586 and includes dozens of eruptions. The 1593 eruption of Raung had a VEI = 5 and caused fatalities. In one eruption in 1638, a loud noise that appeared to come from Raung, was comparable to thousands of artillery pieces going off at once. Earthquakes were continuous and so powerful that a person could hardly stand upright. Rivers reportedly dried up for several days and then resumed flowing in a powerful flood. These floods and LAHARS killed more than 1,000 people. In 1817, the crater lake in Kawah Ijen collapsed, producing MUDFLOWS that engulfed three villages and killed an unknown number of people. When Raung underwent an ASH eruption in 1890, witnesses reported strong vibrations and rumbling noises. A 1902 eruption of Raung was accompanied by a very loud noise like cannon fire. Roaring sounds were heard at Raung during a period of increased activity in 1913–14, and two small earthquakes from Raung were reported in 1915. Water from the lake at Kawah Ijen flowed over a sluice after a large TECTONIC earthquake near Bali in February 1917; the following month, water in the lake was spouting some 30 feet (9 m) into the air in a muddy, noisy display. The temperature of the lake rose to scalding intensities in March but dropped back to lukewarm levels by late 1917. The lake heated up again in 1921. Raung erupted in

1921 and again in 1931. A 1952 eruption at Ijen followed immediately after several strong local earthquakes. An eruption cloud appeared, and "boiling" activity was observed in the lake. A minor earthquake preceded an eruption of Raung in 1982; in 1985, 1993, 1995, and 1997, Raung underwent ash explosions.

Ijen is especially interesting because of a relationship seen between lake level and eruptive activity. On several occasions when the lake level has been lowered by artificial means, fumarolic activity under the lake has increased, lake temperature has risen, and sometimes HYDROTHERMAL or phreatic eruptions have occurred. These phenomena are thought to result from reduced pressure resulting from lowering the lake level.

Ikeda *See* ATA.

Iliamna *volcano, Alaska, United States* Located in the eastern ALEUTIAN ISLANDS, Iliamna has erupted on several occasions in historical times including 1778, 1867, 1876, and 1953. It is suspected to also have erupted in 1768, 1786, 1793, 1843, 1933, 1947, and 1952. Iliamna is one of many Alaskan volcanoes characterized by explosive eruptions. An EARTHQUAKE SWARM began beneath Iliamna in 1996 but without eruption.

Illinois *United States* The state of Illinois has undergone numerous strong earthquakes since it was settled by Americans of European descent. One notable earthquake was the Fort Dearborn earthquake of August 24, 1804, felt at Fort Wayne, INDIANA, some 200 miles (322 km) away. A series of strong earthquakes occurred in southern Illinois in 1882–83. A severe earthquake at Cairo on August 2, 1887, stopped clocks and was felt over a large area of the Midwest. Illinois is affected by seismic instability in the Mississippi Valley, notably the NEW MADRID FAULT ZONE, where a future earthquake comparable to those of 1811–12 could cause tremendous damage in Illinois and other midwestern states.

Ilopango *caldera, El Salvador* The Ilopango CALDERA is about five miles (8 km) wide and seven miles (11.2 km) long and lies between two faults of a GRABEN that extends parallel to the string of volcanoes in EL SALVADOR. The caldera is thought to have formed during a huge eruption in A.D. 260. Apparently, there was a sophisticated Mayan culture occupying the highlands of the area at the time. The violent eruption of Ilopango destroyed everything in a 60-mile (97-km) radius around the volcano and appears to have killed thousands of people. Lake Ilopango occupies the caldera, and a LAVA DOME breaks the surface of the lake at one point to form the Islas Quemadas, or "burned islands." The Islas Quemadas formed during an eruption in January 1880. The lake level rose so dramatically at this time that the Jiboa river valley was flooded on January 9, destroying the town of Atuscatla and killing numerous cattle. This rise in the lake is thought to have been due to the growth of a lava dome underwater. The lake level subsided after this flood, but SULFUR gas began to emerge from the middle of the lake, and in late January an eruption of ASH and glowing rock occurred. Within three days, the Islas Quemadas formed. Strong earthquakes in

1879 preceded this eruption. Another powerful earthquake occurred in February, was felt all through El Salvador, and preceded an intense smell of sulfur in the Ilopango area. Ilopango appears to have been inactive since 1884.

impact structure   In recent years, geologists have recognized that Earth's surface bears the marks of numerous impacts. Major impacts, involving meteorites perhaps a hundred yards in diameter or larger, are by no means infrequent. The impact that created the famous Meteor Crater in ARIZONA (also known as the Barringer Crater after D. M. Barringer, a geologist who investigated its origins) is believed to have occurred only a few thousand years ago. Evidence for an impact origin of the Arizona crater began to accumulate in the late 19th century. A mineralogist named A. E. Foote gathered large numbers of iron meteorites, some of them containing minuscule diamonds, at the site of the crater. After Foote completed a report on his findings at the site, G. K. Gilbert of the U.S. Geological Survey examined the crater and found evidence to indicate it was the result of a large object falling from outer space and striking Earth. (The alternative view was that some kind of volcanic explosion had generated the crater.)

A search for commercially workable deposits of nickel and iron led around 1902 to a strong interest in mining the crater. Within the next decade, drilling and excavation of mine shafts at the crater established that there was no workable body of metal buried at the site. Pieces of meteoritic material were found in a BRECCIA extending several hundred feet beneath the surface, but no substantial body of metal appeared. These explorations revealed that the volcanic explanation for the origin of the crater was flawed because an unaffected layer of sandstone was discovered under the breccia. Later studies of the crater indicated that a meteorite some 140 feet (43 m) in diameter blasted out the formation on striking Earth at a velocity of approximately 45,000 miles (72,400 km) per hour. The energy released in the impact is thought to have been equivalent to the explosion of a 15-megaton thermonuclear weapon.

When a planetoid (in effect, a giant meteorite) strikes the surface of Earth, the impact tends to generate a characteristic structure often called an impact crater, although a more general term is *tistron* or *astrobleme*. A classic impact character has the following characteristics: round or roughly square shape; depressed central area, with perhaps a central peak where rebound has occurred following impact; a raised rim with overturned strata outside the rim; and clastic material generated by the impact distributed in and around the crater. Arizona's Meteor Crater exhibits these characteristics, except for the central peak. Other indicators of an impact origin include the presence of shatter cones, peculiar conical structures produced in rock by the tremendous forces released on impact, and minerals, including DIAMOND (found near Meteor Crater) and stishkovite and coesite, two highly dense forms of QUARTZ. A slaglike material associated with impact craters is known as impactite. PSEUDOTACHYLITE is a rock believed to form from melting at impact sites.

Many well-preserved impact structures are found on the Canadian Shield, including the Manicouagan formation in Quebec, now the site of a ring-shaped reservoir, and Ontario's Brent Crater, some two miles (3 km) wide and discovered in a study of aerial photos taken by the Royal Canadian Air Force (RCAF). One of the first impact craters identified in CANADA was Chubb Crater, named after Frederick Chubb, a prospector who in 1950 noticed a remarkable round lake in an RCAF photo of northern Canada and saw that the lake lay in a formation that looked much like a CALDERA. If a caldera did exist there, Chubb figured, it might be worth investigating for diamonds, which are known to occur in some volcanic formations. Chubb went to see V. B. Meen of the Royal Ontario Museum of Geology and Mineralogy in Toronto and asked Meen if he thought the lake occupied a volcanic formation. Meen thought the lake instead had formed inside an impact crater like Meteor Crater in Arizona. An expedition to the site was formed, funded by a newspaper publishing company, and Chubb and Meen left Toronto in an amphibious plane on July 17, 1950, to investigate the lake. Four other men (the pilot, an engineer, a reporter, and a photographer) accompanied Chubb and Meen. On reaching the lake, the men found its surface too icy to allow a safe landing, and the pilot landed the aircraft on another lake in the vicinity. Chubb and Meen made their way from the landing site to the alleged impact crater, a study of which revealed quickly that the formation was not volcanic in origin but rather was formed by meteorite impact.

It has been suggested that much larger features of the Canadian landscape are also impact structures, including the Gulf of Saint Lawrence and the Nastapoka island arc in Hudson Bay. In the UNITED STATES confirmed and possible impact structures (besides Meteor Crater) are found at Serpent Mount, OHIO; Wells Creek, TENNESSEE; Decaturville, MISSOURI; Odessa and Sierra Madera, TEXAS; Kentland, INDIANA; Manson, IOWA; Red Wing Creek, SOUTH DAKOTA; Panther Mountain, NEW YORK; and Chesapeake Bay, MARYLAND, among other sites. One recently identified impact structure in the United States is the Beaverhead astrobleme in MONTANA. Believed to have been about 45 miles (72 km) in diameter when formed, the Beaverhead astrobleme is not immediately apparent to observers because various processes have reworked the area extensively since the crater originated. Impressive shatter cones are present in strata extending some 15 miles (24 km) from north to south at the Beaverhead site, which also exhibits pseudotachylite, as well as suevite, a "microbreccia," characterized by very small fractures in the rock.

Other identified or suspected impact structures are distributed widely around the globe. Germany's Ries Kessel is widely recognized as an impact structure, as is the Vredevoort Ring in South Africa. The Vredevoort Ring, often identified as the largest demonstrated astrobleme on Earth, appears to be left over form an impact that created a crater roughly the size of the state of WEST VIRGINIA. An abundance of shatter cones played an important part in identifying the Vredevoort Ring as an impact structure. The Chixilub Crater in MEXICO to the Gulf of Mexico may be even larger. That impact has been proposed to have caused the massive extinction event that occurred at the end of the Mesozoic. That event included the extinction of the dinosaurs. The impact can be seen in sediments worldwide by high levels of iridium and shocked quarts. Shocked quartz has microscopic crossed fractures and can only be produced from the excessive and paid increases in pressure.

The physics of a planetoid impact is complex but may be summarized in general terms as follows. A planetoid several hundred feet wide or larger is not slowed greatly by passing through Earth's atmosphere and strikes the surface with much of its original velocity undiminished. Depending on the mass and velocity of the planetoid, the impact may release enough energy to vaporize much or all of the planetoid, producing the equivalent of a huge nuclear explosion. The impact may create a crater miles in diameter, as is thought to have happened in the cases of Manicouagan and Vredevoort. EJECTA from the point of impact may fall great distances away from the crater. Secondary meteorites, or sizable bodies of rock cast out by the primary impact, may bring about further destruction, albeit on a smaller scale, when they fall to Earth.

The physics of an impact on land are considerably different from those of an impact at sea. On land, a giant meteorite would cause widespread destruction through seismic effects as well as from heat and other radiation from the fireball produced on impact. A land impact also might alter climate by casting large amounts of particulates (finely pulverized rock and smoke from fires started by heat from the fireball) into the upper atmosphere where the material could intercept incoming sunlight and reduce temperatures at the surface. The damage from an ocean impact, however, is projected to be much greater because the sea acts in ways to amplify the destructive effects of the planetoid strike. Some of the energy released by the impact is transferred to the water as a TSUNAMI, that may travel around the globe many times before eventually dissipating.

Imperial Valley *earthquake, California*   On October 15, 1979, an earthquake of RICHTER magnitude 6.9 struck the Imperial Valley of central CALIFORNIA. It caused $30 million and injured 91 people. There was extensive surface faulting, LIQUEFACTION, and ground-DEFORMATION slumping. The extensive agricultural industry in the area suffered heavy losses from these surface effects. Canals, irrigation ditches, and subsurface drain tiles were all disrupted by these effects.

incandescent   The glowing of very hot PYROCLASTIC material as it shot from the mouth of a volcano. If there is much glowing debris, then the eruption is referred to as incandescent as are the pyroclastics. Incandescent eruptions are especially spectacular at night.

inclusion   A piece of older COUNTRY ROCK or a remnant of the melting or crystallization process contained inside igneous rock. The IGNEOUS ROCK may or may not be related otherwise to the inclusion.

India   The Indian subcontinent has been the site of numerous strong and destructive earthquakes. India's seismic potential is largely the result of an ongoing collision between the Indian subcontinent and the Asian landmass to its north. The Himalayas mark the boundary between the Indian and

Asian plates. Earthquakes in India tend to take large numbers of lives because of the country's extremely high population density. For example, an earthquake in Kitchen on June 16, 1819, wiped out 1,500, and one in ASSAM on June 12, 1897, killed about 1,000. On May 31, 1935, in QUETTA, an earthquake measuring 7.5 on the RICHTER scale killed some 60,000 people. An earthquake measuring 6.4 on the Richter scale struck a remote part of the state of Maharashta on September 30, 1993, at 3:56 A.M. Approximately 50 villages within a 50-mile (80-km) radius were damaged and 16 were completely leveled within a few seconds. Many of the people killed were asleep in badly constructed stone houses that collapsed on top of them. Although the exact death toll was not known in October 1993, it was thought to be in the tens of thousands. Although there are no volcanoes in India, the Deccan traps are one of the largest volcanic deposits in the world. The Bhuj earthquake in 2001 killed 20,000 people and injured 166,000.

Indiana *United States*   Strong earthquakes affect Indiana occasionally, although the state is not as susceptible to earthquakes as high-risk areas such as CALIFORNIA. Strong shocks were reported at Albany on August 6–7, 1827, and another powerful earthquake on September 27, 1909, shook Indianapolis and was felt over much of the Midwest.

Indo-Australian crustal plate   A major plate of Earth, the Indo-Australian plate is involved in a SUBDUCTION ZONE with the EURASIAN plate and several microplates in the East Indies. The plate includes AUSTRALIA, the Indian subcontinent, and part of the Indian Ocean. The subduction zone underlies INDONESIA, which includes some of the most destructive volcanoes in the world. This subduction zone includes a significant component of STRIKE-SLIP motion. Major earthquakes are commonplace and mainly associated with deformation on the Segmanko Fault. The Banda arc is another part of an active boundary. The boundary between INDIA and Asia is marked by the Himalaya Mountains—the classic continent-continent CONVERGENT MARGIN on Earth. Although there are no longer volcanoes along that margin, there is significant seismic activity. There are major REVERSE FAULTS that are seismogenic.

See also PLATE TECTONICS.

Indonesia   An archipelago of volcanic islands between AUSTRALIA and the main Asian landmass. Indonesia is an ISLAND ARC beneath which OCEANIC CRUST from the Australian plate is actively subducting into the Java trench. The island arc is called the Sunda arc and it is the longest continuous arc on Earth. This TECTONIC activity results in a very active CONVERGENT MARGIN with both earthquakes and volcanoes. The islands of Indonesia include Java, Sumatra, Halhamera, Sulawesi (Celebes), Timor, Irian Jaya (on New Guinea), Flores Islands, and part of Borneo among others. Indonesia has more than 200 active volcanoes by one estimate. Indonesia was the site of the 1883 eruption of KRAKATOA, one of the most famous and violent volcanic events in history. The eruption of TAMBORA in 1815 cast tremendous amounts of finely divided solid material into the atmosphere and was impli-

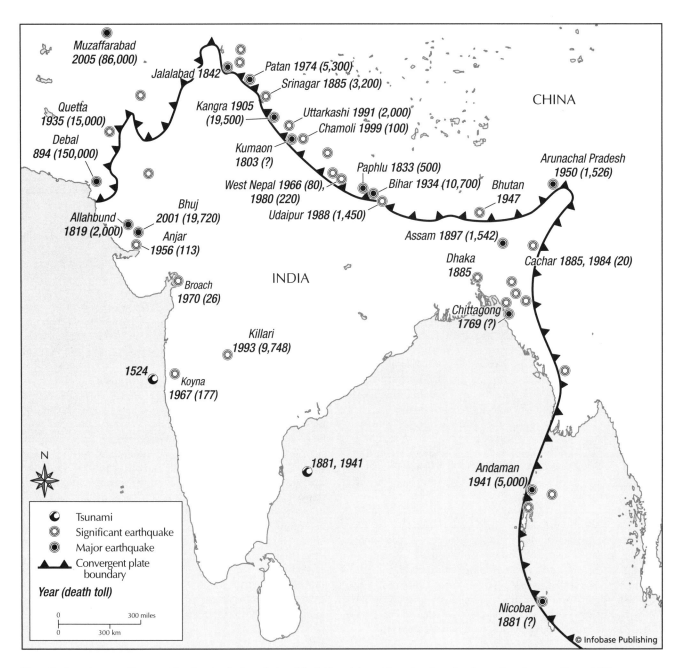

Map of India and surrounding countries showing the boundaries of the Indian plate and the epicenters of major earthquakes over the past 200 years, including the names of the epicentral cities and the approximate death toll in parentheses. Lines with triangles represent subduction zones, with triangles on the overriding plate.

cated in the unusually cold climatic conditions that affected much of the Northern Hemisphere over the following year. Other volcanoes include DIENG, Dukono, GALUNG GUNG, IJEN, MAKIAN, Raung, and Semeru among many others. Earthquakes and TSUNAMI activity also occur frequently in the Indonesian archipelago. The long STRIKE-SLIP Segmanko fault has similar activity to the SAN ANDREAS FAULT and adds additional seismic activity to that of the SUBDUCTION ZONE. There are probably five earthquakes of magnitude 7 per year in Indonesia. One of the world's worst disasters occurred at

BANDA ACEH on December 26, 2004. A huge TSUNAMI was generated by a massive MAGNITUDE 9.0 earthquake, killing an astounding 283,000 people. The largest AFTERSHOCK ever recorded was at magnitude 8.7 in March 2004.

See also BANDA API; SEGARA ANAK; SUKARIA; SUNDA; SUOH DEPRESSION; TENGGER; TOBA; TONDANO.

injection Forceable intrusion of MAGMA into COUNTRY ROCK. As magma is forced upward in Earth's CRUST, it forcibly squeezes into the preexisting rock. It may push the rock

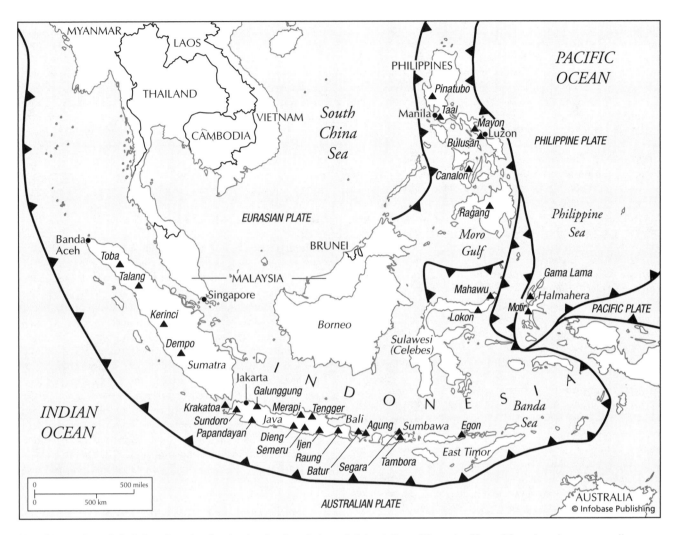

Map of the southeast Asia–Indonesia region showing the plate boundaries and plate relations of the region. Many of the major volcanoes as well as several of the major cities are also shown. Lines with triangles represent subduction zones, with triangles on the overriding plate.

out of the way. It may force open existing cracks or create its own cracks through which to flow. It may consume the rock into itself by melting it and ASSIMILATING it. The magma may intrude by any combination of these. All constitute injection. Waste and water can also be injected into rock, but the source of the pressure is from above rather than below as it is with magma.

injection well A deep well in which material (usually waste) is pumped at very high pressure. Wells are usually used for extracting fluids from the earth. Examples are water wells and oil wells. Injection wells are used to put fluids and gas into the earth. In the petroleum industry, they are used to FRACTURE the rock around the bottom of the well (HYDRO-FRACTURING) so oil and gas will flow more freely to the well. In waste disposal, they are used to put hazardous material out of the biosphere. However, these well may spur earthquakes through hydrofracturing if they are not placed properly.

inner core Solid Fe/Ni core in the center of Earth. It lies within the OUTER CORE. There is likely an interaction between the outer and inner core, based on different rates of spinning within the Earth. This interaction forms a self-exciting dynamo within the core, which produces Earth's magnetic field. It may also produce electrical impulses in the earth.

insolation In general terms, insolation means the amount of solar radiation reaching the Earth's surface. Volcanoes may affect insolation by injecting large quantities of ASH and aerosols into the upper atmosphere during eruptions and thus intercepting sunlight before it can reach the ground. The result is to diminish insolation and reduce temperatures at Earthy's surface. This cooling effect has been observed following several major volcanic eruptions over the past several centuries.

See also KRAKATOA; "YEAR WITHOUT A SUMMER."

intensity   In SEISMOLOGY, *intensity* refers to an earthquakes effects at a particular location. Intensity, which may be measured by factors such as the presence (or absence) of cracks in walls, and even of mass panic, is distinguished from MAGNITUDE, which refers to the amount of energy released by an earthquake. Magnitude is expressed by the strength of the vibrations of the ground. Several scales are used to measure earthquake intensity, including the famous MERCALLI scale.

interference   Interaction of the amplitude of waves in which there is addition (positive interference) increasing the effect of the wave or subtraction (negative interference) reducing or eliminating the effect. In terms of a SEISMOGRAPH, two seismic signals can arrive at a given point at the same time, and one can mask the other.

intermediate   A chemical range for IGNEOUS ROCKS. Intermediate compositions lie between FELSIC and MAFIC composition. PLUTONS of intermediate rocks include GRANODIORITE, DIORITE, and tonalite among others. Volcanic igneous rocks include ANDESITE, DACITE, and TRACHYTE. Intermediate rocks are the primary igneous rocks produced in CONVERGENT BOUNDARIES. ISLAND ARCS and MAGMATIC ARCS are full of intermediate igneous rocks.

interplate coupling   FAULTS between two tectonic plates accumulate STRESS and then release it in a series of earthquakes. Weak interplate coupling means that the stress will be released in a series of regular smaller earthquakes, whereas strong interplate coupling means that the stress will be allowed to accumulate. The accumulating stress is very dangerous because at some point it will exceed the strength of the rock and produce a massive earthquake. Small earthquakes can be tolerated by settlements of people with less destruction and injury. Large events cannot.

interstice   A void or pore space inside a soil or rock. These interstices may be filled with gas (typically air) or liquid (groundwater). The number and degree of interstices or pores in a rock constitute porosity. The interconnectivity of these interstices constitutes permeability. During earthquakes, the degree of these two quantities determine the extent of LIQUEFACTION in an area.

intraplate   As opposed to interplate, which means at a boundary between plates, intraplate means something located within a tectonic plate. Both earthquakes and volcanoes can occur at intraplate locations, but they are not as common as interplate examples. HAWAII, for example, is a volcano at an intraplate location. These volcanoes can be destructive, but they are not nearly as dangerous as interplate volcanoes. The strongest historical earthquake in the continental UNITED STATES was not on the SAN ANDREAS FAULT in CALIFORNIA but instead at NEW MADRID in MISSOURI, an intraplate earthquake. If an equivalent of the 1812 and 1814 event recurs, it is expected to cause the deaths of 250,000 people, making it one of the worst natural disasters ever. Intraplate earthquakes do not recur at regular intervals, but they can occur virtually anywhere and at any time without warning.

intrusion   A body of IGNEOUS ROCK that has entered as molten rock into surrounding solid COUNTRY ROCK. Intrusions form PLUTONS in rocks. Intrusion has a forcible rather than passive connotation. Intrusion also refers to the process by which such a body of igneous rock, known as intrusive rock, is formed. Intrusion is not restricted to molten rock but also may occur in salt deposits that pierce into conformable strata. Salt may become buoyant and rise through the CRUST by virtue of gravitational instability due to its low density and DUCTILE behavior.

intrusive rock   *See* INTRUSION.

Inyo Craters   *See* MONO LAKE.

Io *Jupiter's moon*   Io is remarkable for having active volcanoes. U.S. space probes have detected and photographed the plumes of volcanoes ejecting SULFUR high above the moon's surface. Images returned to Earth by *Voyager* space probes in 1979 and 1980 showed that the whole visible surface of Io appeared to be affected to one degree or another by volcanism. These images showed several plumes of volcanic material, almost 200 miles (322 km) high and about 600 miles (966 km) wide, rising from volcanoes. Approximately 200 CALDERAS also were identified on the face of Io. Some of these calderas are believed to contain lakes of LAVA. Apparent LAVA FLOWS on the surface of Io are colored black, red, and orange. These colors are believed to be the result of temperatures at which the molten material cooled and solidified. Because these colors are those of molten sulfur as it cools, the lava flows of Io are thought to be made at least partly of sulfur.

Other materials also may be included in the MAGMA and lava because sulfur alone apparently does not have the strength to form some of the steepsided features on Io's face. It has been suggested that BASALT may make up part of the surface rocks of Io. The moon appears to have a very high internal heat flow compared to that of Earth. Io's interior is

Io is the innermost moon of Jupiter and the most volcanically active place in the solar system. Several lava flows at over 90 miles (150 km) in length were erupted from the Ra Patera volcano. *(Courtesy of NASA)*

thought to be kept hot by gravitational influence from Jupiter and from its moon Europa. Evidence of extraterrestrial volcanic activity has been discovered on VENUS and MARS as well as on Io.

Iowa *United States* Iowa is not characterized by frequent seismic activity but has experienced strong earthquakes on occasion. A severe earthquake at Sioux City on October 9, 1872, was accompanied by a noise like thunder.

Iran Iran is located in an area of intense seismic activity and has been the site of numerous highly destructive earth-quakes. These are not always extremely powerful earthquakes; many are moderate but shallow. The northern part of Iran is the most seismically active. Iran's active FAULTS include, but are not limited to, the Ferdows, Kubbanan, and Nayband Faults as well as portions of the Shahrud Fault. The Tehran area reportedly has experienced some earthquakes that are of interest for being evidently unrelated to surface faults. Strong earthquakes have struck Iran repeatedly within historical times, notably in 1611, 1619, 1755, 1879, 1881, 1883, 1911, 1930, 1968, 1969, 1972, and 1977. The March 21, 1977, earthquake (magnitude 7.0), which occurred in southeast Iran, caused widespread damage and killed about

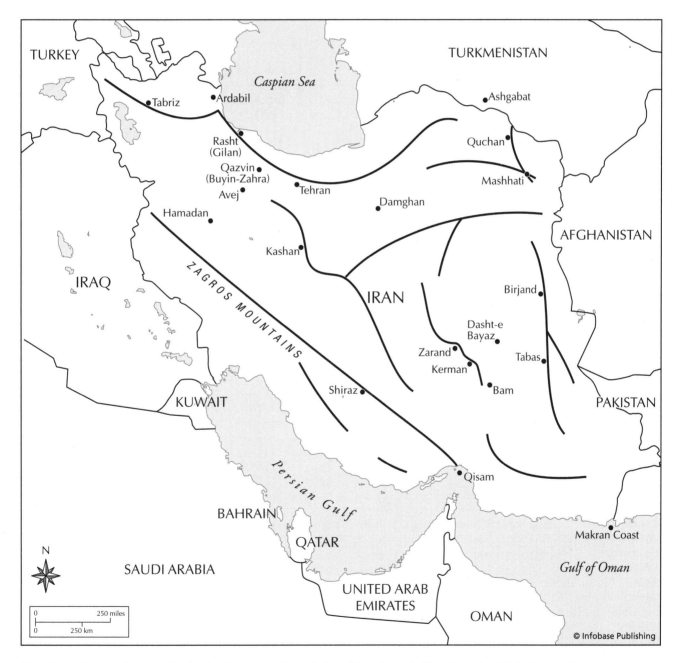

Map of Iran and surrounding countries showing the location of major faults and the epicentral cities of the major historical earthquakes

180 people. A strong AFTERSHOCK occurred approximately 90 minutes after the MAIN SHOCK, and aftershocks continued for weeks afterward. Another earthquake, in TABAS in September 1978, killed 25,000 and was suspected of having some link to the underground testing of a Soviet nuclear device some hours before near Semipalatinsk (now Semey) in Siberia, approximately 1,500 miles (2,414 km) away. Believed to have yielded 10 megatons, the Russian bomb test was unusually powerful and allegedly was set off at a shallow depth of only about one mile (1.61 km). The coincidence between the test and the earthquake was remarkable, as was the unusual set of characteristics observed in the Iranian earthquake. It was very shallow (centered only about 10 miles (16 km) underground) and reportedly had no aftershocks. Also, a peculiar report from Iran concerned access to the site of the earthquake. As a rule, SEISMOLOGISTS are admitted to be affected area following a large earthquake to assess the damage; but in this case, access was denied, leading some observers to wonder if there was something especially sensitive and deserving of secrecy that had occurred at Tabas. Other recent earthquakes in Iran include DASHT-E BAYAZ in 1968, in which some 12,000 were killed; GILAN in 1990, which had a DEATH TOLL of over 50,000; and the BAM earthquake of 2003, in which the loss of life was 43,200. Although the death toll from these recent earthquakes is far less than the hundreds of thousands reported of those from antiquity, they are still among the deadliest in modern times. With the frequency of these deadly quakes, it is no exaggeration to dub Iran as one of the seismically most dangerous countries on Earth.

Irazu *volcano, Costa Rica*   The STRATOVOLCANO Irazu has a double CRATER and has undergone explosive eruptions some 23 times since 1723. Eruptions between 1963 and 1965, the most recent activity generated many ASHFALLS that caused considerable harm to agriculture and affected the city of San Juan. Resulting LAHARS killed at least 40 people and destroyed 400 houses.

Irian Jaya *earthquake, Indonesia*   On February 17, 1996, an earthquake of MAGNITUDE 8.1 occurred. At least 108 people were killed and 423 were injured. There were 5,043 houses destroyed or damaged in the EPICENTRAL area. TSUNAMIS generated from the earthquake reached heights of 22 feet (7 m).

Ischia *volcano, Italy*   The volcanic island Ischia stands close to VESUVIUS, in the Bay of Naples, and has been active since earliest historical times. Resurgent uplift at Ischia, similar to that observed at IWO JIMA, had led some observers to speculate that Ischia has a CALDERA, although this interpretation is not certain. There has been considered long-term ground deformation at Ischia. One sign of this deformation is the site of a Roman metal foundry on the northeast side of the island, now about 15 to 20 feet (5–6.5 m) underwater. On the islands southern side, a beach has been uplifted almost 100 feet (30 m) in places above the present water level. At one point on the island, thermal baths built in the late 18th or early 19th century have risen up to approximately 20 feet (6.5 m), at a rate of more than an inch (2 cm) per year, since

their construction. In the 20th century, the south side of the island appears to have started subsiding, thus reversing the earlier trend of uplift. At the same time, Monte Epomeo, the highest point on the island, appears to be undergoing continuing uplift.

Although uplift at Ischia may be volcanic in origin, it also is possible that TECTONIC activity is involved. Subsidence at Ischia, however, is more difficult to explain in terms of volcanic activity. Major explosive eruptions in prehistoric times appear to have laid down the Tufo Verde, or "Green Tuff," more than 3,000 feet (914 m) thick. If a caldera does exist at Ischia, it may have formed during the eruption that deposited the Tufo Verde. An eruption around 470 B.C. drove away a Syracusan colony on the island. Residents of the island had to flee yet again in an eruption that occurred between approximately 400 B.C. and 350 B.C. This eruption is said to have followed earthquake activity. A TSUNAMI may have accompanied an eruption, possibly of Monte Epomeo, around 350 B.C. Another eruption may have taken place in 91 B.C., although there is some question whether this eruption involved Ischia, VULSINI, or ROCCAMONFINA. The eruptive history of Ischia over the next thousand years is sketchy, but eruptions appear to have occurred around A.D. 80, 180–222, and 284–305. Approximately 700 people were killed in a ROCKSLIDE in A.D. 1228. There was considerable earthquake activity in 1302 as well as an explosive eruption and a flow of lava. In either 1557 or 1559, an earthquake caused a church at Campagnano to collapse. Another church in the same community collapsed during an earthquake in 1762, and a church at Rotaro was demolished in an earthquake in 1767. In the community of Casamicciola (formerly Campagnano), several houses and seven of their occupants were destroyed in a 1796 earthquake. Strong earthquakes shook the island in 1827 and 1828. After one earthquake, residents of the island slept outdoors for more than two weeks. Earthquakes occurred again at Ischia in 1841, 1863, and 1867. Seismic events between 1881 and 1883 may have accompanied an increase in the temperature of wells on the island, but this is not certain. A moderate earthquake occurred in 1961.

island arc   This is an arcuate (in map view), or crescent-shaped, string of islands that forms at an ocean-ocean CONVERGENT PLATE BOUNDARY. Island arcs are SUBDUCTION ZONES where OCEANIC CRUST is going beneath other oceanic crust. The downgoing crust melts at a certain depth producing vast quantities of MAGMA. The magma rises through the overlying MANTLE and crust and forms volcanoes on the ocean floor behind the trench. The volcanoes grow into islands. Alaska's ALEUTIAN ISLAND chain is a familiar example of an ISLAND ARC among many others.

isoseismal   SEISMOLOGISTS evaluate the modified MERCALLI intensity of each earthquake for the area around each EPICENTER. The highest intensities usually occur near the epicenter and the lowest farther away. INTENSITY numbers, I to as high as XII, are determined for the locations evaluated. Seismologists then put contour lines on the map to separate the areas with similar intensity numbers. These lines represent equal intensities and are called isoseismal lines.

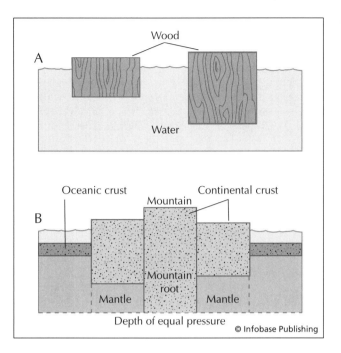

Illustrations of the principle of isostasy. (A) Shows two blocks of wood of different sizes illustrating the large amount of block required below the water surface to produce height above the water. (B) Shows the large amount of continent required in the mantle to support mountain ranges.

isostacy Lithospheric plates float on the soft, gumlike ASTHENOSPHERE. The depth at which the plates sink into the asthenosphere is dependent upon the density and thickness of the CRUST. These relations constitute isostasy. OCEANIC CRUST is dense and sinks deeply into the asthenosphere. CONTINENTAL CRUST is light and floats higher. That is why the oceans are on ocean crust—it is at a much lower elevation on the planet and water goes to the lowest point. To support mountain ranges, the continental crust must sink deeply into the asthenosphere as well. Typical continental crust at sea level is about 35 miles (56 km) thick. To support the 25,000–30,000-foot (7,600–9,100 m) Himalayan Mountains, the crust thickens to about 80 miles (129 km). Another example of isostasy is glacial rebound. During the last ice age, a mile (1.6-km)-thick sheet of ice extended from the Arctic to about New York City. It was so heavy that it pressed the crust deeper into the asthenosphere. After the ice melted and retreated the land very slowly rose back up. Castles that were built on the ocean in the Middle Ages are now one-quarter to a half-mile (0.5–1 km) inland. Lake Champlain in NEW YORK is purely freshwater, and yet there are whale bones in the sediments from several thousand years ago. The lake was open to the SAINT LAWRENCE seaway after the glacier retreated but the land rebounded and was elevated enough to cut off the lake from the ocean.

Istanbul *earthquakes, Turkey* The great Turkish city of Constantinople and its successor, Istanbul, were devastated on several occasions by massive earthquakes. The city seems to have been struck in blocks of earthquakes at various points in its history. The first well-documented event was on January 26, A.D. 447. It had an estimated RICHTER magnitude of 7.3. and a modified MERCALLI intensity of VIII. This earthquake was responsible for extensive subsidence and uplift to the point that ocean floor was exposed. There was another large quake on September 25, A.D. 447, of estimated MAGNITUDE 7.2 and INTENSITY IX that was followed by 90 days of strong AFTERSHOCKS. Magnitude 6.5–7.0 events also occurred in 542, 554, and A.D. 557. All caused great damage (VIII–IX on the modified Mercalli scale). The A.D. 740 event on October 26 was a bit stronger, with a Richter magnitude of 7.6 and modified Mercalli intensity of IX and strong aftershocks for 12 months. There were intermediate events on March 17, 780; May 23, 862; January 9, 869; October 26, 989; September 23, 1063; March 11, 1231; June 11, 1296; and October 18, 1343, for which there is little information.

The earthquake of September 10, 1509, on the other hand, was considerable. It was estimated to have had a magnitude of 7.4 on the Richter scale and an intensity of X on the modified Mercalli scale. It was said that not a single house in Istanbul was left undamaged. The city's fortifications were also severely damaged, with 49 towers destroyed. Some 109 mosques were ruined. Great FISSURES opened in the city, and SAND BOILS were common. Subsidence along the coast resulted in flooding, and reports of large waves suggest that TSUNAMIS may have formed. Within Istanbul, at least 5,000 people lost their lives and more than 10,000 were injured. The DEATH TOLL for the entire event was estimated at 13,000 people.

Istanbul was again struck on May 22, 1766, at 5:30 A.M. The earthquake had an estimated magnitude of 7.3 on the Richter scale and a maximum intensity of IX on the modified Mercalli scale. The quake was said to have produced a significant tsunami in the Sea of Marmara that ravaged the coast and especially the city of IZMIT. The death toll for this event was estimated at greater than 4,500 people.

Italy The Italian Peninsula and its nearby islands and waters are home to a number of famous volcanoes, notably VESUVIUS, which has erupted on numerous occasions within historical times and caused widespread destruction, including the ruin of the cities of POMPEII AND HERCULANEUM. The other major and famous volcano is Mount ETNA in Sicily. It, too, is very commonly active. Eruption types are named after two Italian volcanoes, STROMBOLI and VULCANO. Earthquake activity is also frequently destructive in Italy, which is located in the seismically active MEDITERRANEAN basin. Between two earthquakes in CALABRIA in 1783 and 1857, more than 110,000 people were killed. Because of the dense population and the large amount of ancient ruins that are susceptible to being broken and destroyed, earthquakes in Italy are typically well covered by the press. GEOTHERMAL ENERGY has been harnessed in Italy. The MESSINA earthquake of 1908 was perhaps the worst natural disaster in European history. Some 160,000 people were killed by the terrible quake and subsequent TSUNAMI, leaving Messina to be dubbed "city of the dead."

The AVEZZANO earthquake of 1915 was also devastating, with a DEATH TOLL of 30,000–35,000. These frequent and horrifying events make Italy, by far, the most dangerous country in Europe in terms of natural disasters.

See also ISCHIA; LARDERELLO; PHLEGRAEAN FIELDS; ROCCAMONFINA; VULSINI.

Iturup *earthquake, Kuril Islands, Russia* On October 4, 1994, an earthquake of MAGNITUDE 8.2 occurred. Only 10 people were killed, owing to the sparse population in the area. However, resulting TSUNAMIS affected coastlines throughout the PACIFIC OCEAN. Wave heights ranged from 12 feet (3.5 m) to eight inches (17 cm), depending upon location.

Iwo Jima *caldera, Japan* This small island, a famous World War II battleground, occupies a submerged CALDERA. The name *Iwo Jima* means "Sulfur Island." Much of the island is taken up by Motoyama, a volcanic cone with a high proportion of TUFF to LAVA. Mount Suribachi, at the southern tip of the island, is made up of tuff and lava. Iwo Jima has a history of PHREATIC ERUPTIONS reaching back to 1889, and the barrenness of certain PUMICE deposits seen on the island during World War II indicated the deposits were reasonably fresh. FUMAROLES are abundant on Iwo Jima and tend to be very hot; one test drilling had to be suspended because the intense heat ruined the drill bits. The island appears to have undergone dramatic uplift over the past several hundred years. Estimates put the rate of uplift at perhaps seven or eight inches (18–20 cm) per year. It is not clear whether the uplift at Iwo Jima may represent an early indication of another caldera-generating eruption in the future. Iwo Jima has had at least 10 historical eruptions. The most recent eruption was in 1982 and was small and phreatic.

Izalco *volcano, El Salvador* A young STRATOVOLCANO on the south side of the SANTA ANA volcano in the western part of the country. It has had semicontinuous eruptions since 1770. In total, there have been at least 51 eruptive periods. These eruptions consist of small explosions, ejection of cinder showers and BOMBS and LAVA FLOWS from lateral vents. As the result of the activity, it came to be known as the Lighthouse of the Pacific. However, right after a hotel was built to accommodate tourists coming to see this attraction, activity waned and it is quiet today. Its last eruption was a small FLANK ERUPTION in 1966.

See also COATEPEQUE.

Izmir *earthquake, Turkey* The ancient city of Smyrna, now called Izmir, was destroyed several times by earthquakes. One of the most devastating of these earthquakes occurred at 11:45 A.M. on July 10, 1688. The shock lasted for 20–30 seconds and had an estimated MAGNITUDE of 6.8. The INTENSITY was estimated at X on the modified MERCALLI scale. The shaking destroyed 75% of all buildings, leaving only three of 17 mosques standing, and fire ravaged the city. SUBSIDENCE east of the city was so great that the fort of Sacak Burnu, which had previously been situated on a peninsula, wound up on an islet separated from the mainland by a 100-foot (30-m) stretch of sea. On average, the sea encroached the land by 24 inches (60 cm). More than 5,000 people lost their lives in Smyrna, but the DEATH TOLL for the entire event may have reached 20,000.

Among the numerous other damaging earthquakes to have struck the city of Smyrna, the only other one with such a high death toll was in A.D. 688. It was said to have also killed 20,000 people, but there are few details available.

Izmit *earthquake, Turkey* On August 17, 1999, there was an earthquake of MAGNITUDE 7.4. At least 30,000 people were killed and there was more than $6.5 billion in property damage. However, the total economic impact ranged between $10 and $20 billion. More than 50,000 people were injured, and more than 600,000 were left homeless. It was the largest and most deadly earthquake in TURKEY in more than 60 years. A maximum displacement of 16 feet (5 m) of right-lateral STRIKE-SLIP movement occurred along a 75-mile (120-km) segment of the nearly vertical North Anatolian Fault. Izmit was one of 11 major earthquakes (larger than magnitude 6.7) along some 621 miles (1,000 km) of the North Anatolian Fault.

Izu-Ōshima *caldera, Japan* An island located near the point where the PACIFIC CRUSTAL PLATE meets the EURASIAN and Philippine crustal plates, Izu-Ōshima is often active, and the historical record of eruptions, mostly explosive but some with LAVA FLOWS as well, reaches back to the seventh century A.D. There have been at least 74 eruptions in historic times, the last of which was in 1990. Most of the eruptions are STROMBOLIAN in nature. The CALDERA is less than three miles (4.8 km) wide and contains a volcano, Miharayama, with a summit crater some 450 feet (137 m).

Izu-Ōshima is known for the dramatic and periodic rise and fall of the floor of Miharayama's summit CRATER. The crater floor has been known to rise and fall more than 1,200 feet (366 m) during 20 years. Periods of rising are correlated with eruptions. Very strong earthquakes may follow these episodes by several years. For example, uplift began in 1908, followed by eruptions in 1912–14 and 1919; a powerful earthquake occurred in 1923. Another episode of uplift began in 1933; eruptions followed in 1950–51 and another major earthquake in 1953. Yet another period of uplift started in 1963; Izu-Ōshima erupted in 1974 and again in 1976, and a major earthquake is anticipated. It has been suggested that TECTONIC forces are responsible for the dramatic up-and-down motion of the crater floor. Tectonic activity, according to this model, may squeeze an underground MAGMA reservoir (somewhat in the manner of a tube of toothpaste) and cause the crater floor to rise; when pressure is relieved, the crater floor then falls. Because Izu-Ōshima thus may serve as a natural indicator of tectonic stress, it is studied closely as a possible guide to future strong earthquakes in the Tokyo region. Marked changes in the magnetic field have been observed before and after eruptions at Izu-Ōshima. These changes may be due to demagnetization of a reservoir of magma about three miles (4.8 km) down.

See also PLATE TECTONICS.

Izu Peninsula *Japan* The Izu Peninsula is located a few miles south of TOKYO, at the point of intersection of the EUR-ASIAN, Philippine, and PACIFIC CRUSTAL PLATES and close to Izu-Ōshima. Known for its strong seismic and volcanic activity, the Izu Peninsula has numerous volcanic cones and domes. Some of the domes are thought to have appeared only about 3,000 years ago. The KANTO earthquake of 1923 occurred only a few miles north of the Izu Peninsula. Instability on the Izu Peninsula in recent times may be due to movement of MAGMA underground.

J

Japan The islands of the Japanese archipelago occupy one of the most concentrated areas of seismic and volcanic activity in the world. The result of a collision between the PACIFIC CRUSTAL PLATE to the east, the EURASIAN CRUSTAL PLATE to the west, and the Philippine plate to the south, the extraordinary history of earthquakes and volcanism has had a profound impact on the history of Japan. Time and again, cities in Japan have been damaged heavily or destroyed entirely by earthquakes and volcanic eruptions. Perhaps the most famous earthquakes in Japanese history is the one that struck TOKYO in 1923, killing more than 140,000 and destroying much of the city.

Japan's volcanoes are numerous and spectacular. One of them, Mount FUJI, or Fuji-san, is located within sight of Tokyo and has become a symbol of Japan. An interesting feature of Mount Fuji is a set of lakes, Kawaguchi, Yamanaka, Sai, Shooji, and Motosu at the foot of the mountain; originally these separate lakes were a single large lake, but eruptions produced lava flows that divided the lake into five portions. Other notable volcanoes and CALDERAS in Japan include ASAMAYAMA, ASO, BANDAI-SAN, BAYONNAISE, IZU-ŌSHIMA, SAKURAJIMA, TARUMAI, UNZEN, and USU.

Japan is an excellent example of a chain of volcanoes formed along SUBDUCTION ZONES. Three volcanic arcs intersect in Japan, the Kurile arc, the Izu-Bonin (Marianas) arc, and the Ryuku arc. The volcanoes of the archipelago are located 50–120 miles (80–195 km) west of the deep ocean trench that marks the actual boundaries between the crustal plates. The subduction of crustal rock and the forces generated and released by that ongoing process create the intense volcanism and seismicity of Japan. The tallest volcanoes in Japan, as a rule, are located along the eastern side of the belt of volcanism, close to the subduction zone. Hot springs and other geothermal phenomena are commonplace in Japan; indeed, hot springs provide popular attractions for vacationers in many years. Japan has several major volcanic zones:

1. *Kirishima zone.* This zone is located in the southern portion of the Japanese islands.
2. *Hakusan zone.* Just to the north of the Kirishima zone, the Hakusan zone lies along the southwestern coast of Japan and includes portions of the southern island of Kyūshū and the main island of Honshū.
3. *Norikura zone.* This small zone is located southwest of Tokyo and incorporates portions of two mountain ranges, known as the central Alps and northern Alps of Japan.
4. *Fuji zone.* Located in the vicinity of Tokyo, the Fuji zone includes much of the KANTO Mountains.
5. *Chokai zone.* Lying along the northwestern coast of Japan, the Chokai zone takes in the Dewa Mountains.
6. *Nasu zone.* A long, narrow strip of volcanically active land, the Nazu zone extends from the northern island of Hokkaidō down the northeast coast of Honshū, Japan's main island, and reaches well into the interior of Honshū, stopping just a few miles from Tokyo, The Oou and Mikuni mountain ranges are included in this zone.
7. *Chishima zone.* Located in Japan's far north, on the island of Hokkaidō, the Chishima zone extends from central Hokkaidō outward into the Pacific.

Altogether, these volcanic zones occupy much of the area of Japan and contain some 200 volcanoes.

The Japanese coast is vulnerable to TSUNAMIS which may be generated along the Japanese coast or elsewhere in the PACIFIC OCEAN basin and cause great damage and loss of life when the waves come ashore in Japan. The Japanese shoreline has a long history of devastating tsunamis. One wave at Sanriku in 1896 reportedly reached the shore as a 100-foot (30-m)-high wave that killed some 25,000 people and swept away approximately 10,000 buildings. Other devastating earthquakes include EDO in 1703, in which up to 37,000 people died; SINANO in 1847, in which 12,000 people were killed; MINO-OWARI in 1891, with a DEATH TOLL of nearly 7,500; and TANGO in 1927, which killed

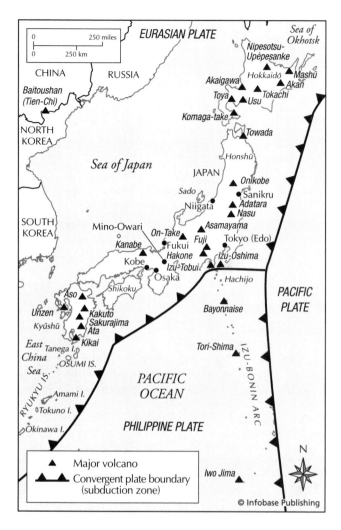

Map of the Japan region showing the plate geometry as well as the major volcanoes and several cities that have experienced major earthquakes

2,900 people. Even recent earthquakes, such as KOBE in 1995 (more than 5,500 dead), have had high death tolls. This continued death and destruction from earthquakes means that Japan is still vulnerable to natural disasters. Japan's rise to its current status as an economic superpower has generated cause for concern about the possible economic impact of the next powerful earthquake that is expected to strike the Tokyo area in the near future. Because the economy of Japan is tied now so intimately to the economic of other nations, especially the United States, it has been suggested that a major natural catastrophe such as an earthquake in Japan could have dire consequences for much of the rest of the world.

See also ECONOMIC EFFECTS OF EARTHQUAKES AND VOLCANOES.

Jawa *earthquake, Indonesia* On June 2, 1994, an earthquake of MAGNITUDE 7.8 occurred just south of Jawa. It caused more than 250 deaths and 423 injuries. About 1,500 houses were damaged or destroyed, as were 278 boats. The earthquake produced TSUNAMIS with up to 1,640 feet (500 m) of RUN-UP. Most of the damage was done by the tsunamis.

Jebel Marra *See* DERIBA.

jökulhlaup Translated from the ICELANDIC as "glacier burst," jökulhlaup refers to an outpouring of meltwater produced when a volcano erupts beneath a covering of glacial ice, as happens from time to time in Iceland. The jökulhlaup may start slowly as volcanic heat melts ice beneath the surface of the glacier and forms a buried pool of meltwater. Soon that water makes its way out from under the ice and flows toward the sea. As the waters drain away, the overlying ice settles. Sometimes the ice around the edge of the volcano can dam and contain the meltwater only to release it in a catastrophic flood.

See also GRIMSVÖTN.

Jorullo *volcano, Mexico* Jorullo is located only about 50 miles (80 km) from PARICUTÍN and resembles that volcano in many ways, although the birth of Jorullo was studied less intensively than that of Paricutín because the area was sparsely populated when Jorullo emerged in 1759. The eruption that gave rise to Jorullo was preceded by earthquakes that began in June, continued through late September, and were strong enough to cause widespread distress among the population and damage to a chapel. On September 29, ASH and steam erupted from a ravine. Condensation of steam produced a muddy rain that fell on a nearby hacienda, as the eruption cloud darkened the sky, and a strong odor of SULFUR spread over the countryside. The eruption, which soon required the evacuation of the hacienda, continued for two days before LAVA (or what is believed to have been lava, on the basis of a local plantation administrator's written reports) emerged from the volcano. The eruption destroyed forests in the vicinity and, several days after it began, was reportedly casting out large rocks, some of them the size of cattle. A new vent, one of three that eventually formed appeared on September 12, Accumulating ash and contaminated water caused great harm of flora and livestock during October and early November. No written records of the eruption appear to have been made from firsthand observation after late 1759, although eruptive activity reportedly continued through early 1760 and recurred intermittently until 1775. Lava extruded from Jorullo covered several square miles.

Juan de Fuca crustal plate A small plate of Earth's crust compared to the adjacent PACIFIC and NORTH AMERICAN CRUSTAL PLATES, and Juan de Fuca plate is nonetheless important to the UNITED STATES and CANADA because its ongoing collision with North America has created the CASCADIA SUBDUCTION ZONE. This structure is the source of seismic and volcanic activity affecting the Canadian province of BRITISH COLUMBIA and the states of WASHINGTON and OREGON in the northwest United States as well as parts of northern CALIFORNIA. Subduction and consequent melting of the descending crustal plate are thought to have generated

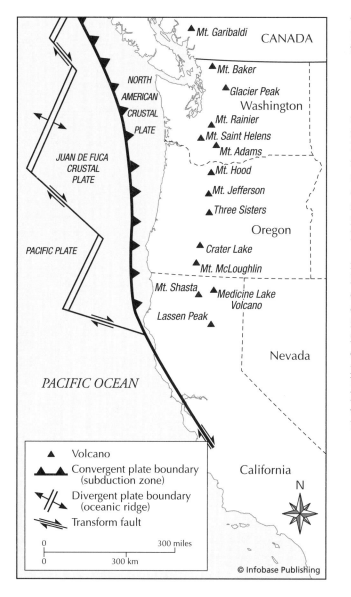

The Juan de Fuca crustal plate adjoins a subduction zone off the Pacific Northwest coast of the United States. Also shown are the volcanoes of the Cascade Range.

the MAGMA that formed the volcanoes of the Cascade Mountains, including Mount HOOD, Mount SAINT HELENS, Mount RAINIER, and MAZAMA. The magma involved in eruptions of these volcanoes has been rich in dissolved gases, as a result of water-rich sediments being subducted along with the Juan de Fuca plate. The water thus drawn into Earth's interior rises back to the surface dissolved in molten rock, generating highly explosive eruptions such as that of Mount Saint Helens in 1980.

Although there is abundant evidence of volcanism connected with the Juan de Fuca plate, there is no history of highly destructive earthquakes such as those recorded frequently in California. The reason for this relative absence of strong earthquakes along the Cascadia subduction zone is uncertain, but it has been suggested that the descending Juan de Fuca plate merely "slides" smoothly on its way downward, without such powerful disturbances as threaten portions of California. It is worth noting, however, that the United States government considers the coast of the Pacific Northwest an area of high earthquake potential and has warned that a powerful earthquake remains a district possibility for the future.

Kagoshima Bay *See* SAKURAJIMA.

Kakuto *caldera, Japan* Located close to the AIRA and KIKAI calderas, Kakuto CALDERA is about six by nine miles (9.6 by 14.4 km) wide and is situated northwest of KIRISHIMA volcano. GEOTHERMAL fields are situated nearby. Although there have been no eruptions in Kakuto caldera itself within historical times, Kirishima has a record of dozens of explosive and PHREATIC ERUPTIONS dating back to the eighth century A.D.

Activity at Kirishima and seismic activity at Kakuto appear to be interrelated. In 1913–14, several EARTHQUAKE SWARMS at Kakuto were followed by minor eruptions at Kirishima at about the same time as a major eruption of nearby SAKURAJIMA (although there may have been no connection between eruptions at the two sites). This activity coincided closely with a strong earthquake about 60 miles (97 km) to the east. Earthquake swarms were noted again in the caldera in 1915, but Kirishima did not erupt this time. Several minor volcanic events in 1923, 1934, and 1936 did not accompany earthquake activity in the caldera. Likewise, a phreatic eruption in 1959 was preceded by only slight earthquake activity or none at all. Earthquake swarms occurred after this eruption.

In 1961, a strong earthquake about 60 miles (97 km) east of Kirishima preceded earthquake swarms in the vicinity of the caldera. Earthquake activity, some of it moderately strong, occurred between 1966 and 1969. Data collected between 1967 and 1969 indicate perhaps a couple of feet of uplift to the west, or possibly a comparable amount of subsidence to the east. A minor phreatic eruption without any outstanding seismic precursors occurred at a hot spring at the base of Kirishima in 1971. An earthquake swarm took place in the area later that year. FUMAROLES showed increased activity in the early 1970s near Kirishima. Considerable earthquake activity continued from 1974 to 1979.

In the winter of 1978–79, a small flow of SULFUR, several inches wide and about 150 feet (45 m) long, is thought to have occurred. Between 1980 and 1983, fumaroles became more active, another small sulfur flow occurred, and a pit inside one crater at Kirishima cast out material resembling tar. The most recent eruption of Kirishima was in 1992.

Kamchatka Peninsula *Russia* The Kamchatka Peninsula, on the east shore of Russia, is a MAGMATIC ARC in the "RING OF FIRE" around the PACIFIC OCEAN basin. It connects the KURIL and ALEUTIAN ISLAND arcs, but it is on a peninsula of CONTINENTAL CRUST. The SUBDUCTION ZONE at Kamchatka involves the Pacific plate to the east being driven beneath the EURASIAN CRUSTAL PLATE to the west. The margin is relatively straight between the Kurils and Kamchatka, but takes a nearly right angle turn at the boundary with the Aleutian Islands. The surrounding landmasses may control the shape of this bend, but it is also at the point where the EMPEROR SEAMOUNTS are being subducted into the Kuril-Kamchatka-Aleutian trench. The eastern margin of Kamchatka is quite mountainous, but there is a FAULT-controlled valley down the middle of the island and parallel to the trench. Much of the seismic and volcanic activity is controlled by these faults.

Kamchatka is noted for its typical subduction zone volcanic activity, notably the eruption of BEZYMIANNY in 1955 that closely resembled Mount SAINT HELENS. KLIUCHEVSKOI is another very active volcano that has made itself an aviation hazard by spewing heavy eruption columns high into the atmosphere on a regular basis. Other important volcanoes of Kamchatka include GORELY, KARYMSKY, KORYAKSKY, KRASHENINNIKOV, KRONOTZKY, OPALA, SHEVELUCH, and TOBLACHIK among many others. Several of these volcanoes are quite active. Most of the volcanoes are so remote that there is very little information about their eruptive histories.

Kamchatka is also seismically active. The faults bounding the central valley produce regular moderate to strong earthquakes. The BENIOFF ZONE on the downgoing Pacific plate also produces regular deep earthquakes. Again,

An earthquake of magnitude 9.0 struck the Kamchatka Peninsula in 1952 and generated a tsunami that flooded this area of Hawaii. *(Courtesy of the USGS)*

because the area is so remote and contains a relatively small population, records are poor, and hazards to human life are minimal.

Even though the lowlands may be green for part of the year, the stratovolcanoes of Kamchatka are invariably snow- and ice-covered throughout the year. *(Courtesy of NOAA)*

Kangra *earthquake, India* The first major natural disaster of the 20th century was an earthquake that struck the northwest Indian Himalayas, with its EPICENTER near the city of Kangra. The MAIN SHOCK struck at 6:19 A.M. on April 4, 1905, and lasted for two to three minutes. The estimated RICHTER magnitude was 7.8, although some placed it as high as 8.0 The source of the earthquake appeared to have been one of the low-angle THRUST FAULTS that accommodates shortening of the crust through the continuing Himalayan orogeny. The FOCUS of the event was estimated at a lower crustal depth of 20 to 39 miles (34 to 64 km).

The DEATH TOLL from this earthquake was a staggering 20,000 people. Over 100,000 buildings were destroyed, and the estimated cost of destruction was 2.5 million, in 1905 rupees. Because the area is in the mountains, much of the destruction was caused by severe ROCKFALLS, DEBRIS AVALANCHES, and LANDSLIDES. One rockfall in the town of Bawar even created a new lake. SLUMPING was also a problem, creating large FISSURES on the tops of slopes. An unusual and detrimental aspect in this event was that virtually all government officials were killed. Lengthy delays therefore surrounded coordinated rescue and relief efforts, and many people died in the rubble who normally could have been saved.

Kansas *United States* Kansas does not have a history of frequent and destructive seismic activity but has undergone strong earthquakes from time to time. The Lawrenceville earthquake of April 24, 1867, was characterized by strong shocks and a wave some two feet (0.61 m) high reported on the Kansas River. The earthquake caused some damage to buildings and was felt widely through the Midwest.

Kanto *earthquake, Tokyo, Japan* The "great Kanto earthquake" on September 1, 1923, destroyed much of TOKYO and the neighboring city of Yokohama. The earthquake, estimated at between MAGNITUDES 7.9 and 8.3 on the RICHTER scale, struck one minute before noon and had its epicenter 57 miles (98 km) southeast of Tokyo in Sagami Bay. The earthquake and subsequent TSUNAMIS (as high as 30 feet [9 m]) and widespread fires are believed to have killed 140,000 people. The 1923 earthquake was one in a series of major earthquakes that have struck Tokyo at intervals of approximately 70 to 75 years in the past several centuries: in 1633, 1703, 1782, 1853, and 1923.

Edward Seidensticker, in his story of Tokyo, *Low City, High City,* reports that the initial shocks of the 1923 earthquake knocked out SEISMOGRAPHS at the central weather bureau office in Tokyo. The seismograph at Tokyo Imperial University made the only record of the series of more than 1,700 earthquakes that struck the Tokyo area during the following three days, some 237 to which were strong enough to be felt. He adds that the earthquake damaged or destroyed almost 75% of buildings in Tokyo, whereas the fire affected almost 67%. Firefighting equipment was destroyed by the

earthquake, and so fires burned out of control. The cities were built largely of wood, and oil stored in aboveground tanks added to the conflagration. The fire reportedly generated a curious phenomenon called fire tornadoes. These were blazing storms with winds strong enough to lift a person off the ground. More than 35,000 people reportedly were killed at a single location, a park on the east bank of the Sumida River, when a storm of fire descended upon refugees who had gathered there. In all, some 100,000 people were killed, 40,000 were injured, and 400,000 houses were totally destroyed.

Destruction was most widespread in the Low City, the eastern and generally lower-class portion of Tokyo, built on flat ground consisting largely of unconsolidated and LIQUEFACTION-prone sediment and landfill. The comparatively affluent High City built on hilly and more stable ground in the western half of Tokyo came through the earthquake and fire with less damage. The Low City had also suffered extensive damage in another earthquake in 1855, but because of the magnitude of this damage, the 1923 earthquake marked a sharp decline in the influence of Low City culture and hence a change in Tokyo's social history as well.

Persecution of the unpopular Korean minority in Japan is said to have accounted for several hundred additional deaths following the 1923 earthquake. The traditional Japanese prejudice against Koreans made them a convenient target for public anger. Koreans were accused of arson and of poisoning wells and were sentenced to death, but the government reportedly took a dim view of the persecution and killings because of possible effects on international opinion.

Only the sturdier buildings of downtown Tokyo, Japan, such as the Imperial Hotel (closest building, to right), survived the massive 1923 Kanto earthquake and resulting fires. *(Courtesy of the USGS)*

An atomic bomb blast could not have inflicted more damage to the cities of Tokyo and Yokohama, Japan, than the great Kanto earthquake of 1923. *(Courtesy of the USGS)*

Subsidence and uplift in Sagami Bay was described as dramatic. In the center of the bay, subsidence of 300 to 600 feet (91 to 183 m) was reported. North of the bay, uplift of some 750 feet (229 m) was reported. However, the preearthquake bathymetry to which the subsidence and uplift was compared may not have been accurate.

Yokohama is about 40 miles (64 km) from the epicenter of the earthquake. About 20% of its buildings collapsed during the earthquake. Some 208 fires broke out within minutes and spread throughout the city. Because the earthquake broke all of the water mains and wrecked the firefighting equipment, the fire spread unchecked. When the disaster was over, nearly 27,000 people were dead, 40,000 were injured, and some 70,000 houses were destroyed.

Karangetang, Mount *Indonesia* This STRATOVOLCANO is very active. Its last eruption was in 1997 and included LAVA FLOWS and *NUÉE ARDENTE*s. Not many people were killed (three), but the slopes had to be evacuated. During the 1974 eruption, the entire population of the island had to be evacuated. It also erupted in 1992.

Karkar *volcanic island, near Papua New Guinea* Located in the western portion of the Bismarck volcanic arc, the STRATOVOLCANO Karkar exhibits a double, nested CALDERA. It has erupted on approximately 10 occasions since the mid-17th century. An eruption involving an impressive "flame" from the volcano summit was reported in 1643. In 1885, smoke and ASH reportedly issued from Karkar. Numerous earthquakes on June 17, 1895, were followed by an eruption that continued on an intermittent basis through August. A moderately strong earthquake occurred under the northern coast of the island in 1930. An explosive eruption may have occurred in 1962, but this information is uncertain.

No further eruptions were reported until 1974 and 1975, when great amounts of LAVA emanated from the volcano. Earthquake activity started to intensify in 1976, and in 1978 a periodic variety of earthquake, called a banded tremor because of its "banded" appearance on seismograph records, began occurring at Karkar. Also in 1978, ground temperatures increased dramatically and reached the level of incandescence in some areas. FUMAROLES also became more active during this time. PHREATIC ERUPTIONS started in 1979 and included an eruption in March of that year that killed two VOLCANOLOGISTS. It has been suggested that banded tremors precede, and may provide early warning of, phreatic eruptions because the tremors may result from processes similar to GEYSERS underground.

Karymsky *volcano, Kamchatka, Russia* Karymsky is a STRATOVOLCANO that has erupted 29 times since 1771. Most eruptions are moderate (VEI = 2), but recent eruptions in 1960–64, 1965–67, and 1970–82 were larger (VEI = 3). One eruption shot ASH to an altitude of 33,000 feet (10,000 m).

Karymsky is located in a remote part of the KAMCHATKA PEN-INSULA, so no villages or towns are directly threatened by it.

Kashmir *earthquake, India* A tremendous earthquake was said to have struck the Kashmir region in September of 1555. Based upon the damage reported, the event was estimated to have had an INTENSITY of XII, the maximum number on the modified MERCALLI scale. A huge catastrophic LANDSLIDE was blamed for most of the death and destruction. Reports of huge FISSURES that swallowed houses were also reported, and the Earth's surface was said to have rippled like ocean waves. Lateral translation of areas was described in whole villages exchanging places and fields winding up on opposite sides of houses from their pre-quake position. If these accounts were accurate, there must have been a phenomenal amount of STRIKE-SLIP movement. The DEATH TOLL for this event was also a matter of debate. One account put the number at 600, which seems low for the event, whereas another account claimed 60,000 death, which appears high. If the 60,000 figure is accurate, this was one of the deadliest earthquakes in the region.

Katla *volcano, Iceland* A subglacial volcano, Katla has a history of eruptions dating back to the 10th century and has erupted on the average every 42 years since the late 16th century. A tremendous eruption from the Eldgja FISSURE was reported in A.D. 934. It lasted three to eight years and produced an estimated 4.8 cubic miles (19.6 km³) of BASALTic lava making it the largest FLOOD BASALT in historical times. The amount of sulfuric acid aerosol delivered into the atmosphere by this eruption produced a tremendous environmental impact. These fissures are adjacent to the LAKI fissures. Minor earthquakes preceded another eruption in 1625. Earthquakes occurred again near Katla before eruptions in 1660, 1721, 1755, 1823, 1860, and 1918. The 1918 eruption produced a water flow of about 12 million feet (3,660,000 m) per second for two days. Another eruption has been expected in recent years but has not occurred. It has been suggested that the eruptions of SURTSEY from 1963–67 and HEIMAEY in 1973 may have relieved pressure at great depth and thus delayed the anticipated eruption of Katla.

Katmai, Mount *Alaska, United States* The eruption of Katmai on June 6, 1912, is one of the most famous events in volcanology. The eruption was the largest in North America in the 20th century, with a VEI = 6. In the week before the eruption, earthquakes shook the area around the volcano. Early on the afternoon of June 6, the mountain emitted a

The flooded caldera of the great Katmai volcano, Alaska *(Courtesy of NOAA)*

thick cloud of ASH that cast the land around the volcano into darkness. Great explosions could be heard at a distance of as much as 800 miles (1,300 km) from Katmai. Ash began to fall and accumulated to such depths on the ocean that rafts of PUMICE were capable of bearing a man's weight. Visibility was reduced to less than 100 feet (30 m). On one ship caught in the ASHFALL, the air was so murky that the helmsman had difficulty seeing the compass in front of him. Ashfalls reached Juneau, 750 miles (1,200 km) southeast of the volcano, and the Yukon Valley, some 1,000 miles (1,600 km) north of Katmai.

The volcano emitted large quantities of SULFUR oxides, which, when dissolved in moisture in the air, produced a solution of sulfuric acid that burned eyes in the vicinity of the volcano and ruined clothes hanging on lines to dry in Vancouver, British Columbia, some 1,500 miles (2,400 km) southeast of Katmai. After the eruption, Katmai was found to have collapsed. The summit had lost its inner supports and fallen inward, creating a CALDERA about three miles (4.8 km) wide and a half-mile (0.8 km) deep. MAGMA had been diverted from Katmai through another volcano called Novarupta. Novarupta put out enough incandescent vapor and ash to cover more than 40 square miles (104 km²) of land to a depth of hundreds of feet. Huge PYROCLASTIC FLOWS from Katmai's eruption created the VALLEY OF TEN THOUSAND SMOKES, a plain of TEPHRA from which hot gases and water vapor arose for years afterwards. The 1912 eruption removed approximately 1.5 cubic miles (6.25 km³) of rock from Katmai. It produced more than 7.2. cubic miles (30 km³) of ash in only 60 hours. It covered more than 46,332 square miles (120,000 km²). Comparatively minor eruptions of Katmai occurred in the 1920s, the 1950s, and 1968.

Kawah Baru   *See* SUNDA.

Kawah Batu   *See* SUNDA.

Kawah Ijen   *See* IJEN.

Kawah Ratu   *See* SUNDA.

Kawah Upas   *See* SUNDA.

Kelud (or Kelut) *volcano, Java, Indonesia*   Also known as Gunung Kelud, this volcano has erupted on at least 15 occasions during the last two centuries, but Chinese residents recorded five eruptions in the 14th century, so the history of eruption is long. Kelud is known for its highly dangerous LAHARS (volcanic LANDSLIDES or MUDFLOWS). Kelud is an outstanding case study in using engineering to reduce hazards from volcanoes, lahars specifically. Because the volcano crater lake was involved in generating lahars, a plan was devised for draining the lake through a tunnel. This plan was carried out in the 1920s, and a tunnel drainage system was finished in 1928. This system worked so effectively that an eruption in 1951 generated no lahars. Other attempts to counteract lahars included a wall that was built in 1905 to direct mudflows away from settled areas, but the wall was destroyed in the 1919 eruption.

The most recent eruption was in 1990. It was explosive (VEI = 4) with a heavy TEPHRA fall. PYROCLASTIC FLOWS traveled some four to five miles (6.4–8 m) from the volcano, and small BOMBS were shot some 35 miles (56 km) away in the explosion. The weight of the tephra destroyed more than 500 houses and 50 schools. Some 32 people were killed, and several tens of thousands of residents were evacuated.

Kentucky *United States*   Kentucky is not a very seismically active state but does have a history of occasional strong earthquakes. One such earthquake occurred near Hillman on December 27, 1841, and was accompanied by a rumbling noise and a great disturbance on a river.

Kerinici *volcano, Sumatra, Indonesia*   The ANDESITIC stratovolcano has erupted at least 20 times between 1838 and 1970. The eruptions are small (VEI = 1–2), explosive, and typically PHREATIC. They commonly involve the CRATER lake on Kerinici.

Kern County *earthquake, California*   The Kern County earthquake of July 21, 1952, was one of the most powerful and destructive in the history of CALIFORNIA and the strongest in the United States since the great SAN FRANCISCO earthquake of 1906. Estimated at RICHTER magnitude 7.7, the earthquake generated effects of up to MERCALLI intensity XI and caused an estimated $50 million in property damage. The earthquake did its most serious damage to Arvin and Tehachapi and significant amounts to Bakersfield. It resulted in extensive damage along the Southern Pacific Railroad near Bealville where tunnels built of reinforced concrete with walls 1.5 feet (0.5 m) thick were twisted and cracked and even caved in. At one facility alone, the earthquake caused several million dollars in damage from fire. Twelve people died in this earthquake, nine of them in a single incident involving the fall of a brick wall at Tehachapi; 18 people required hospitalization, and injuries numbered in the hundreds. The shock was felt in NEVADA and ARIZONA as well as in California. In LOS ANGELES and Long Beach, up to 100 miles (161 km) away, tall buildings swayed into each other causing walls to collapse from the collision. The earthquake focus was on the White Wolf fault, which lies perpendicular to the SAN ANDREAS FAULT but which is produced from DEFORMATION related to movement on it. More than 100 AFTERSHOCKS of magnitude 4 or greater were recorded, 47 of those on July 21 alone. Extensive slumping of the land surface, LANDSLIDES, MUD VOLCANOes, and rapid and extreme fluctuations of water levels in wells were some of the observed geologic effects.

K-feldspar   A common mineral in FELSIC igneous rocks. K-feldspar is a SILICATE with a framework-type atomic structure. It is light-colored and commonly pink or white, but it can be green in some areas. It crystallizes at relatively low temperatures for an IGNEOUS ROCK. FELDSPAR is the most common mineral group on Earth.

Khorasan *earthquakes, Iran*   The area of Khorasan, IRAN, has experienced numerous devastating earthquakes throughout its history. One such earthquake struck on June 16, 1653.

It had an estimated RICHTER magnitude of 6.5 and an intensity of IX on the modified MERCALLI scale. AFTERSHOCKS continued at regular intervals and moderate-to-strong intensities through August of that year. The most damage was in the towns of Mashad and Nishabur, where more than 5,600 people were killed.

The most destructive earthquake reported for Khorasan occurred in 1101. It was estimated to have had a Richter MAGNITUDE of 6.5 and an INTENSITY of X on the modified Mercalli scale. It was reported to have killed 60,000 people, but few details exist on this event.

The most recent devastating earthquake in Khorasan (in the town of DASHT-E BAYAZ) occurred on August 31, 1968, at 10:47 A.M. The recorded magnitude was 7.3, and the intensity reached X in the most affected areas. The FOCUS of this event was estimated at eight to 15 miles (13 to 25 km) in depth. More than 100 villages were said to have reduced to rubble, causing upward of $35 million in damages. The DEATH TOLL for this event was estimated at 12,000 to 15,000, though some sources claimed 20,000. Many tens of thousands were injured in the event, and more than 100,000 people were left homeless.

Kii Peninsula *Japan* The Kii Peninsula is located southwest of the Izu Peninsula and has been the site of several powerful earthquakes in the late 19th and 20th centuries: Very strong earthquakes occurred here in 1899, 1948, and 1952. Other powerful earthquakes have occurred immediately off the coast of the Kii Peninsula in 1944 and 1946. Swarms of earthquakes were observed along the west side of the Kii Peninsula before the KANTO earthquake of 1923. Other EARTHQUAKE SWARMS occurred on the Kii Peninsula in the early 1950s, again just before a powerful earthquake, this one some 150 miles (240 km) southeast of TOKYO in 1953. Earthquakes on the peninsula are believed to be TECTONIC in origin.

Kikai *caldera, Japan* The largely submerged Kikai caldera lies at the south tip of the island of Kyūshū. The CALDERA occupies part of a large GRABEN that also includes the AIRA, ASO, ATA, and KAKUTO calderas. Parts of the rim of which forms three small islands, Tokara-Iwo-Jima, Take-Sime, and Syowa-Iwo-Jima. The caldera is thought to have formed in three cycles of eruptions, the most recent of which is considered one of the great eruptions in recent geologic history. This eruption was about 6,300 years ago and was one of the largest eruptions on Earth in the past 10,000 years. It had an estimated VEI of 7 and emitted more than 36 cubic miles (150 km³) of ASH. It sent PYROCLASTIC FLOWS 62 miles (100 km) across the sea. Kikai has been comparatively quiet in the 20th century, but an eruption that lasted for several months in late 1934 produced a small island. The eruption on September 20, 1934, was said to have been especially dramatic, occurring as it did during the passage of a typhoon. Pressure associated with a surge of water accompanying the tropical storm may have hastened this eruption, although the connection has not been proven. Eruptive activity continued through late 1936. Strong earthquakes accompanied the eruptions between 1934 and 1936. The latest eruption was in 1988, and it was minor.

Kilauea *volcano, Hawaii, United States* Located on the island of HAWAII, the largest island in the Hawaiian chain, Kilauea is perhaps the most intensively studied volcano in the world. Because eruptions here are gentle compared to the explosive eruptions of many other volcanoes, such as Mount SAINT HELENS, scientists have been able to study Kilauea at close range and at great length. MAGMA here is low in dissolved gases and flows freely, almost like water, on occasions when the LAVA spills out of the CRATER and engulfs the surrounding land.

Kilauea is one of five volcanoes on the island of Hawaii; the others are MAUNA LOA, MAUNA KEA, Kohala, and Hualalai. Kilauea has erupted on dozens of occasions within the past several centuries. In 1866 the crater of Kilauea surprised observers when the lava lake drained away and disappeared.

Kilimanjaro, Mount *Tanzania* The giant STRATOVOLCANO is the highest mountain in AFRICA reaching 19,335 feet (5,893 m). Kibo is the youngest and highest cone. Kilimanjaro is located at the south end of the GREAT RIFT VALLEY of Africa. Its foundation is a basaltic SHIELD VOLCANO, but the cone is BASALT and ANDESITE. It has not been active in historical times but steam and SULFUR are still frequently emitted.

Killari *earthquake, India* In the predawn of September 29, 1993, at 3:56 A.M., a devastating earthquake struck the area around Killari in the Maharashta region of INDIA. The MAGNITUDE of the quake was 6.3 and yielded a local INTENSITY that was IX but more commonly VIII. The duration of shaking was 30 to 40 seconds, and ground ACCELERATION was estimated at 0.2 g. The source of the tremor was a THRUST FAULT that crossed the area in a northwest-southeast line. The rupture zone was 3.3 miles (5.5 km) long and produced a small SCARP of 32 inches (0.8 m) in height. The FOCUS was a shallow 4.2 miles (7 km) in depth. This was the deadliest intraplate earthquake in India up to that point. Some 25 vil-

A line of spatter cones at Kilauea volcano, Hawaii, on February 25, 1983. The alignment of cones reflects that they are being fed by a planar fracture conduit from below the surface. This planar conduit will become a dike when the magma solidifies within. *(Courtesy of NOAA)*

lages around the EPICENTER were completely flattened, causing a staggering $1.3 billion in damage. The DEATH TOLL was 9,782 people, with more than 30,000 injured. The reason that the death toll was so high for a relatively moderate earthquake was that the area was considered essentially aseismic prior to this event. There were some 125 FORESHOCKS for years before the MAIN SHOCK, but not even historical records gave any indication of what was to come. Later, archaeologists found that a thriving early city at Killari had been destroyed by an earthquake at least 1,500 years earlier.

kimberlite Explosive PLUTONIC igneous rock that can contain DIAMONDS. Kimberlites originate in the MANTLE. They are gas- and water-charged. Besides FERROMAGNESIAN minerals, they can contain carbon in the form of diamonds. At lower pressures, the carbon would be graphite, but the extreme pressure packs the molecules tightly together making them diamond. A FRACTURE must open in the crust for this mass to be able to reach the surface. It travels up the crack and is emplaced at the surface in a funnel-shaped body of broken material called a diatreme. The reason the kimberlite is broken up is because the gas and water charge make it travel at speeds up to mach 2 (twice the speed of sound). These absurd speeds ensure that the diamonds will remain intact along the journey. Otherwise, if it went slowly, they would convert to graphite. Kimberlites are named for Kimberly, South Africa, where there are large diamond-bearing kimberlites.

kinematics Kinematics describes the general movement patterns and directions of FAULTS and plates at a given time or over a period of time. It is an interpretation or determination of sense or direction. For example, the kinematics of a given fault may be that it is DEXTRAL strike-slip.

Kirishima *Kyūshū, Japan* A SHIELD VOLCANO with at least 69 eruptions between A.D. 742 and 1992. It erupted 19 times since 1700, making it one of JAPAN's most active volcanoes.

The eruptions are purely PYROCLASTIC and emerge from 20 distinct eruptive centers. The 1992 eruption came from Shinmoe-dake, which is the most active vent. Five of the eruptions caused fatalities. Kirishima is part of the KAKUTO caldera. It is the site of Japan's first national park.

Kliuchevskoi *volcano, Kamchatka, Russia* One of the most active volcanoes on the KAMCHATKA PENINSULA and part a group of a dozen closely spaced volcanoes, this STRATOVOLCANO has had at least 80 eruptions since its discovery in 1697. Eruptive activity between 1937 and 1939 involved several locations on the volcano, including the summit and vents along the mountain flanks. BEZYMIANNY is included in this cluster of volcanoes. The previously DORMANT volcano erupted in 1955 in a Peléan-type event. The volcano has erupted 34 times since then. It erupted in 1994, 1997, and 1999–2000. Typical VEI for Kliuchevskoi eruptions is 4, which is moderate to strong.

In the 1994 eruption, an ASH column was shot nine to 12 miles (15–20 km) into the air and then swept 350 miles (565 km) to the southeast at 85 miles (140 km) per hour. Very few

people live even near Kliuchevskoi, so it is not a direct threat to people and documentation of the eruptions is sparse. However, by virtue of the enormous amount of ash it erupts, it is a hazard to aircraft. It is the culprit in several cases of clogged jet engines, and in general, they try to avoid it.

Kobe *earthquake, Japan* On January 16, 1995, a devastating earthquake of MAGNITUDE 6.8 occurred in the Kobe area. More than 5,502 people were killed, and 36,896 were injured. As the result of the destruction or damage of more than 200,000 buildings, about 310,000 people were evacuated to temporary shelters. There were numerous fires as well as disruption of water, power, and gas service. LIQUEFACTION and LANDSLIDES were also common. Surface faulting of up to five feet (1.5 m) right-lateral STRIKE-SLIP offset was observed along a 5.6 miles (9 km) FAULT length. Because Kobe is a large manufacturing area for computer chips, this earthquake caused a temporary shortage and prices doubled in places. This earthquake was well reported by the news media.

Komaga-take *volcano, Japan* Komaga-take is an ANDESITE stratovolcano 18 miles (29 km) north of Hakodate City. It was formed 30,000–40,000 years ago and has produced several large PUMICE-flow eruptions in historical times. These eruptions were in 1640, 1856, and 1929. As a result of the 1640 eruption, a DEBRIS AVALANCHE fell into the sea and formed a large TSUNAMI that killed 700 people. Because Hakodate City has a population of more than 320,000, Komaga-take is considered a dangerous volcano.

komatiite There are no ULTRAMAFIC lavas produced today. There were ultramafic LAVAS in the distant past. In 3-billion-year-old crust in AUSTRALIA, there are komatiites that are ultramafic volcanic deposits. These rocks have a distinctive "spinafex" texture, which is a type of grass. The OLIVINE forms long blades that looks like grass. The minerals in ultramafic rocks form at very high temperatures, higher than in any other type of rocks. The existence of ultramafic lava means that Earth was much hotter at the time of eruption of the komatiites. The overall temperature of Earth must have cooled during the past 3 billion years.

K'one *caldera, Ethiopia* Also known as Garibaldi, the K'one CALDERA COMPLEX is located east of Addis Ababa, in the Main Ethiopian Rift. LAVA FLOWS are thought to have occurred here around 1820 in a caldera collapse, and FUMAROLES and hot springs exist in the K'one area today.

Koryaksky *volcano, Kamchatka, Russia* The andesitic STRATOVOLCANO is one of the more active on the KAMCHATKA PENINSULA. It erupted in 1895, 1926, and 1956 with moderate explosions (VEI = 3) and PYROCLASTIC FLOWS. Its remote location results in very little information on these eruptions.

Kotlugja *volcano, Iceland* An 1860 eruption of Kotlugja was reportedly visible 180 miles (290 km) away at night and audible at a distance of 100 miles (161 km). During one eruption, lightning from the cloud is said to have killed two men and 11 horses.

Krafla *volcano, Iceland* Krafla underwent a series of eruptions in the 1720s but was less active thereafter until eruptions began again in 1975. In total, it has had 29 historic eruptions, the last one in 1985. In one eruption in 1977, LAVA emerged from a steam well at a GEOTHERMAL facility. Krafla is not a lofty, majestic volcano. Subsidence has been an important factor in its formation. According to one model of activity at Krafla, the flow of MAGMA into the area of the volcano is pulled apart by the DIVERGENT BOUNDARY so that large volcanoes like those of JAPAN or the CASCADE MOUNTAINS do not have an opportunity to form. The Krafla volcano is stretched out, so to speak, approximately as fast as it is built up by eruptions. Activity at Krafla has been interpreted as the result of magma flowing from great depths into a central reservoir or reservoirs and from there into rifts. It is one of 35 rifts in a chain.

The eruptive history of Krafla has been documented for almost three centuries. In May 1724, powerful earthquakes in the Krafla CALDERA were followed by a phreato-magmatic eruption that expelled PUMICE and SCORIA. Earthquakes also preceded volcanism at Krafla in January and April 1725. Strong earthquakes and changes in the level of nearby Lake Myvatn were reported in September 1875, and these signs are thought to have indicated an intrusion of magma to the south of Krafla. Lava erupted from the caldera in August 1727, and the flows extended to the north and south. Another eruption of lava, following hours of seismic activity, took place in April 1728, followed several hours later by yet another eruption a couple of miles to the south. A lava eruption in December 1728 was followed by a much more powerful eruption of lava in January 1729. More than one LAVA FLOW was reported during the next several months, but it is not known whether the eruption was intermittent or continuous. After the eruptive activity in 1729, Krafla appears to have remained quiet until 1746, when earthquakes and changes in lake level at Myvatn accompanied some kind of volcanism in the caldera.

Krakatoa (Krakatau) *volcano, Sunda Strait, Indonesia* The catastrophic 1883 eruption of Krakatoa is one of the most famous volcanic events in history. An island in the Sunda Strait between Java and Sumatra, Krakatoa began building up to an eruption in May 1883. An earthquake on May 20 was followed by a venting of steam and ASH from the volcano. As the eruption continued, it became clear that the activity did not center on Rakata, the tallest peak on the island at slightly more than 2,600 feet (792 m) but rather on comparatively tiny Perboewatan, a 300-foot (91-m) cone at the opposite side of the island (north end). An exploration party from nearby Batavia arrived at Krakatoa on May 27 and found the eruption had enlarged the CALDERA to a diameter of more than a half-mile (0.8 km); the floor was flat, some 500 feet (152 m) across, and covered by a crust of solidified LAVA. Steam emerged with a loud noise from a hole about 150 feet (46 m) wide in the middle of the floor. The eruption continued through July and August, accompanied by

This classic drawing shows the eruption of Krakatoa, in the Sunda Strait, Indonesia, in 1883. It was one of the largest volcanic explosions in history. The explosion was heard 3,000 miles (5,000 km) away. The shock waves were so strong that every barometer on Earth registered them. It generated 120-foot (37-m)-high tsunamis that killed more than 36,000 people. *(Courtesy of NOAA)*

loud explosions and mild earthquakes. PUMICE accumulated on the waters around Krakatoa and is mentioned in the logbooks of vessels that passed near the island during the eruption. The captain of the ship *Idomène* noted on August 11 passing through great fields of pumice, and several days later the captain of the bark *West Australia* mentioned in his log that the ship had encountered large amounts of "lava," by which he appears to have meant floating pumice. He added that some pieces of rock were several feet in diameter.

After the island destroyed itself, a postmortem on the eruption reached the conclusion that pressure had built up inside the volcano from gas escaping from MAGMA. The gases inside Krakatoa are believed to have had no adequate outlet because already viscous magma came in contact with seawater at the surface, cooled and solidified and thus blocked the vent. Gases from the interior could not escape and the pressure increased. The situation, Professor J. W. Judd of the British Royal Society's committee on the Krakatoa eruption suggested in a report on the event, was comparable to closing the safety valve on a steam boiler while the boiler continued to build up pressure inside. Eventually, the "boiler" in this case, the volcano had to give way and explode. The explosion began with a series of powerful blasts on August 26–27.

Loud explosions were heard at Batavia and Butzenbourg, about 100 miles (161 km) from Krakatoa, early on the afternoon of August 26. A few minutes later, the captain of the ship *Medea* about 75 miles (121 km) from the island, saw smoke and vapor rise from the volcano to an estimated altitude of more than 15 miles (24 km). At about 5 P.M., some four hours after the final series of explosions began, the noise of Krakatoa's destruction was audible all over the island of Java. Around sunset, the captain of the ship *Charles Bal*, some 10 miles (16 km) south of Krakatoa, was so intimidated by the eruption that he ordered the ship to retreat eastward. About this time, large chunks of pumice began landing on the decks of the *Charles Bal*. Meanwhile, the captain of the ship *Sir Robert Sale*, also in the vicinity of Krakatoa, witnessed spectacular displays of lightning in the eruption cloud and detected an odor of sulfur in the air. The SULFUR odor became intense during the night, as bright flashes of light from the volcano lit up the surrounding area in the manner of a giant strobe light. A thick cloud from Krakatoa covered Batavia soon after dawn on August 27, and a muddy rain began to fall, followed by a fall of small clumps of dust held together by moisture. This strange precipitation ended in midafternoon. Although atmospheric effects hid Krakatoa from view during this period, the sound of its final destruction was apparent. Four great explosions, starting at 5:30 A.M. on August 27 and ending shortly before 11:00 A.M., sent pieces of rock showering down on ships in the vicinity and were heard as far away as Rodrigues, an island in the Indian Ocean some 3,000 miles (4,800 km) away from Krakatoa. Heat and hot gases emitted from EJECTA allowed them to glide on a cushion of air across the surface of the water like a hovercraft for as many as 60 miles (97 km) where they came ashore on islands and started fires. The explosions destroyed the north half of Krakatoa and cut Rakata in two vertically from base to summit. The shock waves were registered by every barometer on Earth.

The explosive destruction of Krakatoa generated TSUNAMIS that caused extensive destruction and loss of life along shores in the region. In some places, the waves are estimated to have come ashore at heights of 120 feet (37 m). The men of the ship *Loudon* turned their ship's bow toward the approaching tsunami and rode out the wave's passage and then watched as the tsunami engulfed and destroyed a town. Similar scenes occurred along the coasts of Sumatra and Java near Krakatoa. In one location, a town some 10 miles (16 km) from the water was submerged by the wave. Some 3,000 residents of Karang Antoe were killed by the tsunami, as were more than 2,000 at Anjer and Batavia and 1,500 at Bantan. The total loss of life from the tsunami that followed Krakatoa's explosion is uncertain, but estimates range in the tens of thousands (most recently 36,000). The tsunami from Krakatoa was strong enough to twirl ships around their anchors at the harbor in Colombo, Ceylon (now Sri Lanka). Some deaths from this eruption were unrelated to the tsunami.

In 1990, more than a century after the eruption, an expedition to Krakatoa, including a team from the University of Rhode Island Graduate School of Oceanography, examined the role of PYROCLASTIC FLOWS in generating the tsunamis that originated from Krakatoa during the 1883 eruption. This expedition took samples of clastic material from the ocean floor around Krakatoa. Sediment cores taken in this way showed evidence of a large, poorly sorted main deposit, overlain by another deposit of volcanic gravel and sand. Pieces of pumice large enough to impede the action of the sediment-coring mechanism were reported, and the investigators noted, in addition to the chunks of pumice, pieces of rock more than 12 inches (31 cm) across in some cases. These large fragments of rock are contained in a matrix with a silty and sandy character, the same as deposits on land. The similarity between the submarine and terrestrial deposits indicated that the underwater deposits originated as pyroclastic flows. The expedition also determined that submarine mounds near the Steers and Calmeyer island platforms just north of Krakatoa were made up of pyroclastics from the eruption in 1883. On the Calmeyer island platform, the investigators noticed signs of TURBIDITY CURRENTS that may have resulted when pyroclastic material at high temperature came in contact with seawater, causing steam explosions that gave rise to ash-filled turbidity currents. At one point on the seafloor west of Krakatoa, the expedition found a pyroclastic flow deposit some 300 feet (91 m) thick laid down by the eruption, including fragments of pumice several feet in diameter.

The 1990 expedition found evidence that one or more pyroclastic surges preceded the tsunami from Krakatoa, and two or more additional pyroclastic surges came after the tsunami. Deposits included pieces of coral that appear to have been torn from the seafloor and deposited with the pyroclastic material by the tsunami. The investigators concluded that deposits on land from Krakatoa's 1883 eruption are made up largely of material from pyroclastic flows as are underwater deposits from that eruption. The pattern of these deposits shows they were laid down in all directions around the island, not principally to the north, as had been presumed before. Accumulations are greatest to the west of Krakatoa. Deposition of pyroclastic material on the west side of the volcano is thought to have generated powerful tsunamis.

That the meeting between ocean and pyroclastic flows was not gentle is indicated by the small amount of mixing between seawater and pyroclastics. The pyroclastic materials show poor sorting, evidence that there was little mixing between them and the water. Had mixing occurred on a significant scale, water action would have resulted in more effective sorting of material. This pattern of poor sorting persuaded the scientists on the 1990 expedition that masses of pyroclastic material from the volcano in 1883 displaced water in such amounts as to cause the destructive tsunamis associated with the eruption of Krakatoa. Apparently, five or more units of pyroclastic flow occurred in the eruption. The results of the 1990 expedition indicate that one particular tsunami, at approximately 10 A.M. on August 27, 1883, was linked with the formation of the caldera at Krakatoa during an especially violent portion of the eruption, possibly caused when large amounts of magma came into contact with seawater.

The 1883 eruption of Krakatoa, specifically the fine ash cast into the upper atmosphere by the eruption, has been implicated in the cooler-than-usual winter that followed the destruction. Mean annual global temperature dropped by one-half to one degree Fahrenheit in 1884, and this cool period is believed to have continued through the 1880s and part of the 1890s. The high-altitude ash cloud from Krakatoa is suspected of having intercepted enough incoming sunlight to lower temperatures on a worldwide scale.

Atmospheric phenomena attributed to dust from Krakatoa included vivid sunsets and a blue tinge to the Sun. Even a "green sun" was reported in India.

The 1883 eruption of Krakatoa is especially notable for the distances at which noises from the eruption were heard. The captain of a ship 500 miles (805 km) away from the eruption, at Surabaya, heard the explosions clearly. At Singapore, more than 500 miles (805 km) from Krakatoa, noises from the eruption were interpreted as the sound of a ship firing its guns as a call of distress, and two rescue ships were sent out to see what the problem was. Similar circumstances convinced officials at Macassar and Timor in Indonesia, approximately 1,000 and 1,350 miles (1,600 and 2,173 km) respectively from Krakatoa, to send out ships to rescue what they thought was a vessel in distress. Tribes-people in Borneo thought the noise from Krakatoa's eruption was the sound of a Dutch attack on them. On the island of Rodrigues, about 3,000 miles (4,830 km) from Krakatoa, the noise of the eruption was clearly audible and resembled the firing of heavy guns.

An eruption in 1927 generated a new island, called Anak Krakatoa, or "child of Krakatoa," in the midst of the three islands Rakata, Panjang, and Sertung that occupy what once was the site of the original island of Krakatoa. That volcano has had several periods of activity since 1927. Its latest activity began in 1994 and was still going in 1999. It is still a very dangerous volcano.

Krasheninnikov *caldera, Kamchatka, Russia* Several great flows of LAVA are believed to have emanated from this CALDERA in the past 3,000 years: the Vodopadny flow (about 3,000 years ago), the Ozerny flow (1,500 years ago), the Yuzhny flow (less than 700 years ago), and the Molodny flow (less than 500 years ago). The area is seismically active, and a FUMAROLE was observed in 1963.

Krozontzky *volcano, Kamchatka, Russia* An especially beautiful and symmetrical volcano, Krozontzky erupted in 1922 and 1923. Other than that, it was considered an EXTINCT volcano.

Kuma-dake *See* DAISETSU-TOKACHI.

Kurikoma *See* ONIKOBE.

Kuril Islands *Russia* The Kuril Island chain extends more than 700 miles (1,130 km), from northern JAPAN to the KAMCHATKA PENINSULA. The ISLAND ARC occupies the bound-

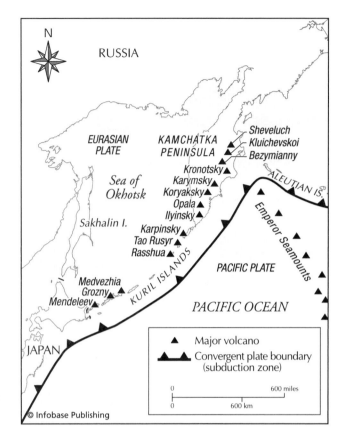

Map of the Kamchatka Peninsula and Kuril Islands showing the plate relations and several of the major volcanoes in the area

ary between the Sea of Okhotsk and the Pacific Ocean and consists of two chains of islands, the Greater Kurils and the Lesser Kurils. The island arc is formed by a SUBDUCTION ZONE in which the PACIFIC CRUSTAL PLATE is being subducted beneath the EURASIAN PLATE. There is significant volcanic and seismic activity as a result. Recently active CALDERAS in the Kurils include CHIRPOI, Golovnin, Ketoi, L'Vinaya Past, MEDVEZHII, MENDELEEV, Nemo Peak, RASSHUA, TAO-RUSYR, Uratman, and Zavaritsky.

Kuril Lake *caldera, Kamchatka, Russia* The Kuril Lake CALDERA occupies part of the larger Pauzhetka caldera. The STRATOVOLCANO Illinsky is located on the northeastern rim of the Kuril Lake caldera. Other, nearby volcanoes are Iavinsky, Kambalny, and Zheltovsky. GEOTHERMAL ENERGY has been exploited in this region.

Kuro-dake *See* DAISETSU-TOKACHI.

Kusatsu-Shirane *Honshū, Japan* A STRATOVOLCANO that has erupted at least 17 times from 1805 to 1989. The volcano has many summit craters that contain lakes. Nearly all of the eruptions were PHREATIC, many involving the crater lakes. Most of the eruptions were small to medium (VEI = 1–2) in size, at best, except for an eruption in 1932 that generated a LAHAR that caused two deaths.

The Kuril Islands contain some of the best-formed stratovolcanoes in the world. Here a resurgent volcano sits in the lake of a flooded collapsed caldera. *(Courtesy of NOAA)*

Kutcharo *caldera, Japan*   The Kutcharo CALDERA is located in northern JAPAN near the southern end of the volcanic KURIL ISLANDS. Kutcharo Lake now occupies much of the caldera. The island STRATOVOLCANO Nakajima stands in the western portion of the lake. To the southeast of the lake is Atosanupuri, a stratovolcano with an associated set of domes. The volcano Mashū, which has a crater lake similar to the one found at CRATER LAKE in OREGON, is located to the east of Atosanupuri, a stratovolcano on the rim of the Kutcharo caldera. Mashū is thought to have erupted most recently some 1,030 years ago. Although no eruptions have occurred in the area of Kutcharo caldera in recent decades, there has been considerable seismic activity, and the caldera has been studied as a possible indicator of major TECTONIC earthquakes.

Kverkfjöll *volcano and caldera, Iceland*   The Kverkfjöll volcano has a history of possible subglacial eruptions dating from the 17th century. Confirming these eruptions has been difficult, however, and the principal evidence for them appears to be floods of water. Two CALDERAS at Kverkfjöll are believed to lie under the ice.

Kwangtung *earthquake, China*   On February 13, 1918, at 6:07 A.M., a destructive earthquake occurred in Kwangtung, China, in the Kwangtung region about 180 miles (300 km) from Hong Kong. The MAGNITUDE of the event was estimated at 7.3 on the RICHTER scale.

Normally, the Kwangtung area is outside of the danger zone for the damaging earthquakes for which China is infamous. It is at an intracontinental location with minimal seismic activity. The city was relatively unprepared for the 1918 earthquake. It was for this reason that it was so deadly. The DEATH TOLL for this event was 10,000 people, and there were tens of thousands of injuries.

K-wave   A P-WAVE traveling in the OUTER CORE of Earth.

L

lahar In general terms, a volcanic MUDFLOW. A lahar is made up of water-saturated PYROCLASTIC material that moves down the side of a volcano and floods the countryside. A common occurrence of lahars is that the hot ASH of a volcano erupts onto an ICE CAP and melts it. The meltwater mixes with the hot ash to form a boiling hot slurry. It then flows down the slopes and into the valleys below and floods them. Lahars cause significant death and destruction during eruptions. The eruption of Mount SAINT HELENS produced very impressive lahars.

An interesting feature of lahars is exemplified by Mount PINATUBO in the PHILIPPINES. The eruption of Mount Pinatubo took place in 1991 and it was a very strong eruption (VEI = 5). The main cause of death was not the volcano but a typhoon that struck two days later and swept up the still-hot ash into boiling hot mudflows. In fact, the deeper ash is still hot enough nine years later that if another typhoon strikes, erosion will cut into it and again produce a boiling mudflow.

Mudflow/lahar from the Mount Saint Helens eruption, May 18, 1980. The mudflow came down a river valley and destroyed a bridge. It filled the old river with mud. **(Courtesy of the USGS)**

Because such occurrences are unrelated to volcanic activity, they should be considered mudflows rather than lahars.

Laki *fissure zone, Iceland* A massive eruption of the Laki FISSURE zone in 1783 was accompanied by the largest release of LAVA (calculated at 3.5 cubic miles [14.731 km^3]) in recorded history. It covered an area of 218 square miles (565 km^2). At peak eruption, lava poured from the fissures at a rate of 295,000 cubic feet (8,600 m^3) per second. The Amazon River is the fastest-flowing river in the world, and it flows 340,000 cubic feet (10,000 m^3) per second. The eruption also released gases that contaminated grass in ICELAND and caused the death of hundreds of thousands of livestock. Some 9,000 residents of Iceland, or about 25% of the population at the time, died in a famine that followed the eruption. Apparently the particulates released during this eruption remained in suspension over Iceland and portions of northern Europe for months after the eruption, resulting in a peculiar haze that was compared to fog.

Lamington, Mount *Papua New Guinea* The site of one of the most violent eruptions of the 20th century, Mount Lamington is located in the eastern portion of Papua New Guinea, near the communities of Buna and Popondetta, approximately 25 miles (40 km) from the coast. The mountain rises almost 6,000 feet (1,830 m) above sea level and has a deep CRATER from which a river flows. A large portion of the mountain appears to have been laid down by mudflows and *NUÉE ARDENTE*s. The powerful eruption (VEI = 5) of 1951 at Mount Lamington took many observers by surprise because the volcano had been DORMANT as it had no other historical activity to that point. Preliminary activity at the volcano either was not interpreted as eruptive or simply went unobserved. On January 15, 1951, LANDSLIDES were noted on the crater walls. Vapor emanated from the crater for the next two days, and EARTHQUAKE SWARMS affected the surrounding area. The volcano's gas output increased greatly on January 18, and ASH came from the mountain as well.

Underground noises intensified as the emissions increased. Earthquakes occurred with increasing frequency until they became virtually nonstop. Early on the morning of January 19, a dazzling display of electrical phenomena took place. After daybreak, the top of the cone was seen to be covered with ash. On January 20, a great ash cloud rose to an altitude of 5 miles (8 km) or higher. At 10:40 A.M. on January 21, a giant cloud rose from the volcano and reached an altitude of 20,000 feet (6,000 m) in less than a half-hour.

The base of the cloud spread out rapidly over the adjacent land. The *nuée ardente* from this stage of the eruption caused the greatest devastation to the north of the volcano and is thought to have moved at an average speed of about 60 miles (97 km) per hour, though probably much faster in certain areas. Phenomena similar to tornadoes are believed to have accompanied the *nuée ardente* and produced dramatic disparities in effects from one place to another. The *nuée ardente* laid waste some 68 square miles (176 km²) of land and killed almost 3,000 people in the vicinity on January 21. (Total number of deaths attributed to this eruption were 6,000.) Heated dust laden with steam probably accounted for most of the deaths. Many survivors suffered severe burns. The stiff condition of bodies found after the passage of the *nuée ardente* indicate the heat from the cloud was intense. Two zones of destruction were observed in the area affected by the *nuée ardente*. Close to the crater, destruction was virtually total because of the combination of intense heat and high velocity; in this zone, some trees and buildings were carried away completely. Close to the crater, the *nuée ardente* produced a powerful scouring action that effectively ground away everything in its path, down to the level of the abraded SOIL. A comparatively small outer zone was affected more by heat than by blast. Valleys around the volcano were filled with hot ash that retained its heat for months after the January 21 eruption. Wood buried under this hot ash would catch fire when exposed to the air weeks later.

At the time of the catastrophic eruption of January 21, Mount Lamington, whose noises had been intermittent up to this time, began to give off a steady roar that was audible on New Britain, 200 miles (322 km) to the north. A few minutes before 9 P.M., another strong eruption occurred. Activity subsided for the following three days and then began again on January 25. Further explosive events occurred in February and March. Less powerful eruptions followed on an intermittent basis. Dome formation began in the crater of Mount Lamington soon after the January 21 eruption. In about a month and a half, a dome more than 1,000 feet (305 m) high arose, sometimes at an average rate of more than four feet (1.2 m) per hour. An eruption on March 5 demolished this dome, but it grew back higher than before by the middle of May. Eventually, the dome reached a height of approximately 1,800 feet (549 m) above the floor of the crater. This eruptive period lasted until 1956. The devastated area recovered quickly after the 1951–56 eruptions, and by the mid-1960s, vegetation reportedly had regrown completely so that the area affected by the *nuée ardente* looked identical to adjacent land.

Land of the Giant Craters *Tanzania* Located on the edge of the GREAT RIFT VALLEY near the volcano NGORONGORO,

the Land of the Giant Craters is a plateau made up of EJECTA from nearby volcanoes.

landslide Many different kinds of earth movement are described by the comprehensive term *landslide,* which refers in general to a down-slope movement of unconsolidated material under the influence of gravity. Landslides may include MUDFLOWS, ROCKFALLS, AVALANCHES, and numerous other phenomena. Landslides are commonly associated with earthquakes, especially in mountainous regions, although landslides may occur in relatively flat country when conditions allow mass movement of unconsolidated material down a gentle slope. Landslides may also occur in volcanoes. The largest landslide ever observed began the 1980 eruption of Mount SAINT HELENS. Much of the death and destruction in earthquakes is caused by fast-moving landslides that drop from elevated areas.

In the 1964 GOOD FRIDAY EARTHQUAKE of ALASKA, for example, numerous buildings on a plain (flatland) near Anchorage were destroyed by landslides when the earthquake allowed material at and near the surface to slide along underlying wet clay. In some locations, the landslides produced by this earthquake were rotational, meaning that the ground surface tilted as blocks of Earth rotated while sliding downslope. Elsewhere, nonrotational movement was observed, as Earth broke into an up-and-down pattern of HORSTs, or relatively elevated blocks, and GRABENS, or depressed areas. In one widely publicized case, a school building toppled off the edge of a horst and landed upside-down in the adjacent graben. Another Alaskan earthquake, in 1958, generated a landslide that caused a spectacular and highly destructive wave to form in LITUYA BAY. This wave reached more than 1,700 feet (518 m) up the side of the valley where the slide occurred. (The Empire State Building in New York City, by comparison, is only about 1,000 feet (305 m) high.)

The Columbia River gorge between OREGON and WASHINGTON shows evidence of tremendous landslides within recent geologic time, such as the BONNEVILLE SLIDE, thought to have occurred about A.D. 1100 and to have involved

Scars in the mountains and lighter-colored rock avalanche deposits at the foot of the mountains show the general geometry of landslides in Alaska as the result of the 1964 Good Friday earthquake. *(Courtesy of the USGS)*

almost a half cubic mile (0.13 km³) of rock, damming the Columbia River temporarily. In the greater Los Angeles area, potential landslides pose a special problem because they stand to block important highways following a major earthquake there in the future, thereby isolating the region.

Landslides may occur either on dry land or on undersea slopes bearing unconsolidated material. Certain submarine landslides are known as TURBIDITY CURRENTS. One famous turbidity current was generated by the GRAND BANKS earthquake off Newfoundland in 1929 and severed a number of submarine telegraph cables on the bottom of the ATLANTIC OCEAN. Later, the recorded times of cable breaks allowed the maximum velocity of the turbidity current to be calculated at some 55 knots.

landslide hazard reduction   There are numerous areas that are plagued by LANDSLIDES. Geologists map these areas to determine their landslide potential on a fine scale. The areas with the highest landslide potential are recommended for landslide hazard reduction. There are numerous methods to reduce the landslide risk depending upon the severity of the hazard posed. There are simple methods such as planting vegetation to hold the slope together or digging drainage ditches to reduce the fluid pressure. More involved methods involve slope reduction and even the building of retaining walls or the cementing of a slope. In some instances, more than one method may be required to address the problem. Each of these methods is considered to be landslide hazard reduction, and some towns and cities have long-term programs for this practice.

lapilli   Pieces of PYROCLASTIC material, EJECTA, shot out of a volcano and deposited and ranging in diameter from about 0.08 inches (2 mm) to about 2.5 inches (64 mm).

Larderello *hydrothermal activity area, Italy*   Larderello, a HYDROTHERMAL activity area, is noted for its GEOTHERMAL power facility, which began to generate electricity for a chemical plant on the site in 1904 and, by World War II, was

Lapilli are moderate-size ejecta between ash- and bomb-size volcanic fragments. The dime is shown for scale. *(Courtesy of the USGS)*

producing almost 1,000 million kilowatt-hours of electricity each year. The power facilities at Larderello were wrecked by German forces in 1945 but were rebuilt. Larderello also has been the location of a large and profitable chemical industry that began by extracting borax from natural steam. Count Francesco Larderel is credited with the idea of using energy from the natural steam at the site to concentrate boric acid solutions. The hydrothermal activity at Larderello appears to stem from MAGMA underlying an area of several dozen square miles.

Lassen Peak *California, United States*   Located in northern CALIFORNIA just south of MEDICINE LAKE volcano and Mount SHASTA, Lassen Peak is the southernmost volcano in the CASCADE RANGE. Lassen Peak stands some 14,500 feet (4,460 m) high and was considered the most recently active volcano in the 48 contiguous United States until the eruption of Mount SAINT HELENS in 1980. Lassen Peak includes three active volcanoes and exhibits HYDROTHERMAL phenomena such as steam VENTS and hot springs. The eruption of Lassen Peak in 1914 began without warning and commenced with minor releases of ASH. Continuing eruptive activity over about a year cast blocks of rock from the CRATER. Explosive eruptions in the spring of 1915 were accompanied by emissions of LAVA, and one eruption on May 22 deposited TEPHRA in NEVADA, some 200 miles (322 km) east of the volcano. LAHARs descended along the flanks of the mountain, and on May 22 a *NUÉE ARDENTE* gave rise to a large MUDFLOW that required the prompt evacuation of a valley near the mountain, with no reported deaths. Further eruptions, though less intense, continued through 1917. Its last eruption was in 1921. Volcanic formations include CINDER CONE, which erupted in the 1850s and expelled both ash and lava; and the SULPHUR Works, where hydrogen sulfide escapes from hot springs and steam vents.

Lassen Peak differs from the other volcanoes of the Cascade Range in that it is not a STRATOVOLCANO but rather a very large dome surrounded on the south and west sides by smaller domes. A few miles south of Lassen Peak, one finds the remains of a huge stratovolcano, known as Brokeoff volcano, that has a diameter up to 15 miles (24 km) at its base and is believed to have stood some 11,000 feet (3,353 m) tall. The collapse of Brokeoff volcano created a CALDERA more than two miles (3 km) wide and left a dramatic fault scarp visible today at Brokeoff Mountain. Lava from Brokeoff volcano now mostly surrounds Lassen Peak, and these vast outpourings of lava from the ancient volcano may have been partly responsible for its subsequent collapse.

lateral blast   An effect sometimes seen in explosive eruptions of volcanoes, a lateral blast occurs when a volcano explodes sideways, directing the force of the explosion through a side VENT and down the volcano in a *NUÉE ARDENTE* rather than vertically in a SUMMIT ERUPTION. The *nuée ardente* spreads over the landscape and is probably the most destructive type of event. The reason that a lateral blast occurs is that the main vent that leads to the summit CALDERA clogs up. Very sticky MAGMA like RHYOLITE, DACITE, and some ANDESITE tend to cause clogs especially when there are many crystallized minerals in them. If it does, pressure builds up inside the volcano. Eventually a FRACTURE will open in a weak spot in the side of the volcano through which the pressure is released in a very strong explosion. This phenomenon was observed in the 1980 eruption of Mount SAINT HELENS, when the north flank of the volcano disintegrated and released pressure from inside the mountain, resulting in a blast that sent low-level clouds rushing over the adjacent terrain, destroying large amounts of timber. The archetypal eruption that produced a lateral blast was Mount PELÉE in 1902. A "Pelean eruption" is the most powerful type of volcanic eruption.

lateral spread and flow   In MASS WASTING, many forms of LANDSLIDES occur on gentle slopes, and the moving SOIL mass exhibits fluidlike behavior as it descends, spreading out laterally at the base of the hill. These terms, *lateral spread* and *flow*, refer to the movement of such masses.

latite   An INTERMEDIATE volcanic rock named for Latium, Italy. It is PORPHYRITIC (mixed large and small grain sizes) with large grains of PLAGIOCLASE and K-spar in a glassy matrix. It is relatively QUARTZ-poor. Compositionally, it fits between ANDESITE and TRACHYTE.

lava   Lava is molten rock that flows out onto Earth's surface. Molten rock at depth is called MAGMA. The difference between magma and lava is (1) that lava is at the surface and (2) that all of the VOLATILES including water, SULFUR compounds, and carbon dioxide are released in the eruption. Therefore, magma contains all of the volatiles, and lava does not. The word *lava* appears to be derived from the Italian word *lavare*, "to wash away." The Italian verb is not restricted to the action of lava but also may refer to a flood of water or even to a moving crowd of people. Over the centuries, "lava" has been used to refer to a variety of volcanic phenomena, including mudflows. Now, however, its use is generally confined to fluid, molten rock reaching Earth's surface and to the formations produced by such rock during volcanic eruptions.

Where lava solidifies after flowing for some distance, the resulting deposit is called a LAVA FLOW. Lava flows on dry land may form AA, characterized by a broken, sharp-edged surface, or PAHOEHOE, which has a peculiar ropy texture. Submarine lavas consist of rounded masses called PILLOW LAVA, formed as molten rock comes into contact with cool seawater. Lava flows show a wide range of grain sizes, depending on their rate of cooling and how long they spend in the MAGMA CHAMBER. The longer in the chamber and the slower the cooling, the larger the grain size. Open spaces or VESICLES may form in lava that has a high concentration of dissolved gases. An example of such vesicular volcanic rock is PUMICE, which is light enough to float on water. Another widespread, vesicular volcanic rock is SCORIA. Some lava flows include XENOLITHS, foreign material picked up and carried along by the lava. LAVA TUBES, hollow tunnels within a lava flow, form as surrounding lava cools and molten rock in the tube drains away. In some cases, lava tubes are large enough to accommodate humans. SPATTER CONES, small vol-

The foreground is a rope-textured pahoehoe lava, and the background is an encroaching slow-moving aa lava flow in Hawaii. *(Courtesy of the USGS)*

canic "mountains" several feet high, may form out of lava spewing from vents in the ground. One impressive phenomenon of some lava flows is COLUMNAR JOINTING, in which solidified lava forms vertical, prismatic columns several feet wide. Giant's Causeway in Ireland is a spectacular example of columnar jointing. Molds are commonplace features of lava flows and are formed when lava surrounds an organism such as a tree and hardens. When the tree decays and disappears, the mold remains in the lava flow. ASHFALLS also may produce molds by burying an animal or other organism, of which a mold remains after the organisms body decays. Excavations at POMPEII AND HERCULANEUM have revealed detailed molds of animals and humans who were killed in the eruption of VESUVIUS in A.D. 79. Plaster was poured into the molds, providing images of the eruption's victims.

Lavas vary in their chemical composition and are categorized according to their SILICA (silicon dioxide) concentration. Lavas containing 66% or more silica are called FELSIC or ACIDIC. MAFIC or BASIC lavas, at the other end of the spectrum, have a silica content of 52% or less. INTERMEDIATE lavas contain between 52% and 66% silica. The behavior of lava during an eruption depends largely on its silica content. As a rule, felsic lavas are "thicker" in character, flow with difficulty and are involved in explosive eruptions. (The explosive character of these eruptions stems from the difficulty dissolved gases have in escaping from the highly viscous lava. In some situations, gas coming out of solution from magma will accumulate and build up pressure until an explosive event occurs. Well-known examples of such eruptions are those of Mount SAINT HELENS in 1980 and BEZYMIANNY in 1952.) Slow-flowing, felsic lavas containing about 74% silica are called rhyolites and are often pink or gray in color. RHYOLITE is commonly found in LAVA DOMES. Mafic lavas flow more readily than felsic lavas and allow gases to escape easily, thus avoiding explosions.

Mafic rocks are also darker than felsic lavas. The volcanoes of the HAWAIIAN ISLANDS are familiar examples of volcanoes that emit mafic lava, BASALT. In between these two

Fountain and lava drain back into the volcanic vent at the end of the Mauna Ulu phase of the Kilauea eruption, Hawaii, October 20, 1969. The lava drains back into the vent after the eruption begins to subside. *(Courtesy of NOAA)*

extremes of chemical composition are the intermediate lavas, known as ANDESITES, named for the ANDES MOUNTAINS.

Besides silica, lava may contain many chemical components, including aluminum, calcium, iron, magnesium, potassium, and sodium, depending upon the rock. Felsic lavas typically are rich in alkalis (potassium and sodium) and poor in calcium, iron, and magnesium. These proportions are reversed in basic lavas. Lava emanating from a volcano may differ in composition from one time to another during the course of eruption. Also, a single volcano may emit lavas of different compositions from one eruption to another. These differences are thought to involve a process known as FRACTIONATION and ASSIMILATION. Fractionation refers to the dividing up of magma, or molten rock still underground, by removing components through CRYSTALLIZATION and settling. Initially, the composition of the magma is largely homogeneous. As heavier components of the magma are used to make mafic minerals during higher temperature crystallization, these heavier minerals settle out of solution. The crystallization sequence generally follows BOWENS REACTION SERIES. Therefore, the character of the remaining magma changes accordingly. The remaining liquid becomes more and more felsic. Assimilation involves the removal, melting, and mixing of COUNTRY ROCK from the walls of a magma chamber underground. The mixing of different kinds of magma that have intruded into a single magma chamber is another source of compositional change.

There is a distinct difference between lavas of the CONTINENTS and lavas of the ocean floor. Continental lavas tend to have higher concentrations of silicon and aluminum. When basaltic magma from the MANTLE rises toward the surface and passes through continental rocks, the molten rock takes on some characteristics of the continental crust through the process of assimilation. The result is a more intermediate lava or a "contaminated" lava. Earth's surface is covered in some places by huge and spectacular outpourings of basaltic lava. These deposits are known as FLOOD

BASALTS and may cover thousands of square miles. In the northwest United States, flood basalts have covered large areas of the states of IDAHO, OREGON, and WASHINGTON. Great flood basalts are also found in the Deccan plateau of India, Pihrana of SOUTH AMERICA, and Karoo of AFRICA. Flood basalts may originate as fissure flows, when cracks in Earth's crust disgorge huge quantities of highly fluid, fast-flowing lava. Such floods of molten rock may travel many miles in a single day.

Lava flows can cause great destruction when an eruption occurs in or near a heavily populated area. Such incidents are commonplace in southern Europe and along the Pacific rim, where large communities exist near active volcanoes. Especially dangerous in this respect are volcanoes such as VESUVIUS in ITALY and Mount ETNA in Sicily. There have been numerous attempts to divert lava flows before they reach populated areas, but these efforts have met with varying degrees of success.

Lava flows tend to cool gradually because rock is an inefficient conductor of heat. When a relatively cool layer of solidified lava forms on the surface of a lava flow, that layer may provide insulation that allows the lava below to remain in a fluid state for months or even years. Many factors may affect the rate of cooling of a lava flow, however, such as rainfall, cooling through cracks, and even flowing into the sea.

lava dome   This structure is made up of extremely viscous LAVA that moves upward through the VENT of a volcano and forms a domed, craggy surface. They are therefore common in RHYOLITE, DACITE, and some ANDESITE volcanoes. Lava-dome formation commonly follows an explosive eruption of a volcano, when gas-rich MAGMA has been expelled and thicker, more viscous lava flows out. A lava dome is commonly surrounded by debris dislodged from the surface of the dome during its growth.

The lava domes formed at Mount SAINT HELENS in WASHINGTON State have received extensive study. Out of five episodes of explosive activity at Mount Saint Helens in 1980 and 1981, a lava dome formed in three of these

Two lava dome fountains during the Mauna Ulu of Kilauea eruption on June 29, 1970 *(Courtesy of NOAA)*

episodes after explosive activity diminished. The first dome grew for at least a week after an explosive eruption on June 12, 1980, and eventually grew to be more than 1,000 feet (305 m) in diameter and some 150 feet (46 m) high. Part of this dome was destroyed during an explosive eruption in July. A new and smaller dome formed following another explosive eruption on August 7. An eruption in October was followed by the formation of a third dome. Analyses of samples from the dome indicated the magma was very low in SULFUR and oxygen, possibly because gases had escaped from the magma before the dome formed. Study of the dome also made possible an estimate of the thermal energy yield of the Mount Saint Helens eruptions of May through October 1980. This estimate puts these eruptions in roughly the same range of energy yield as the eruptions of HEKLA in 1970 and MAUNA LOA in 1950. Evidence of lava domes on the planet VENUS was returned by the U.S. *Magellan* space probe in 1990. Radar images of a portion of the surface of Venus showed a chain of seven domelike structures, each about 15 miles (24 km) wide and as high as about 2,200 feet (671 m).

lava field   An area that is covered by LAVA FLOWS, forming a field of them. Lava fields are common around active volcanoes.

lava flow   A single event of flowing lava. Always the result of an eruption, but an eruption does not guarantee a lava flow. A lava flow must include molten rock, so PYROCLASTIC eruptions or PHREATIC ERUPTIONS do not define a lava flow. In BASALT volcanoes, like those in HAWAII, nearly every eruption includes a lava flow. In INTERMEDIATE and FELSIC volcanoes, most eruptions do not include lava flows. In the Hawaiian eruptions, the lava flows quickly, and it is very hot (1,400°C). However, it is so predictable that they rarely inflict casualties. In felsic and intermediate volcanoes, if they are to the point of flowing lava, evacuation should have been finished long before.

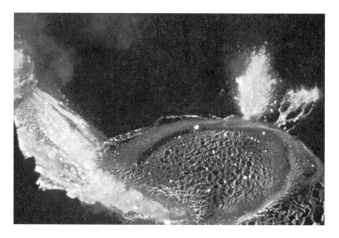

Lava cascades falling into a lava lake in Pauhi's west pit in the Mauna Ulu eruptions at Kilauea, Hawaii, November 11, 1973. The black crust cracks, revealing the glowing lava underneath. The dark ring represents a change in the filling of the lava lake. *(Courtesy of NOAA)*

lava tube   A tube or cavity within a LAVA FLOW. As a lava flow moves down the slope of the volcano, the top cools and hardens. There is still liquid lava underneath that can flow out. Therefore, there is a gap underneath the crust where the lava flowed out. The lava tube is the resulting cavelike feature. Lava tubes are common in BASALT volcanoes. The lava must have low enough viscosity to be able to flow out from under the crust. Sticky, high-viscosity lava will not flow out. There are abundant lava tubes in the Hawaiian volcanic fields. The roof of these cavelike structures can collapse leaving a hole or skylight. People can climb down into the lava tubes through the skylights.

left-lateral fault   A STRIKE-SLIP FAULT in which the block on the opposite side of the fault from where you are standing

A ribbonlike lava flow from the Kilauea east rift, Hawaii, on November 27, 1989. These rivers of lava can flow tens of miles per hour on steep slopes but are slower on flat areas. *(Courtesy of NASA)*

Skylight in a lava tube near Kalapana, Hawaii, on February 2, 1989. The lava tube formed when a crust developed on a flow. The liquid lava flowed out, leaving the cavelike lava tube. The skylight is formed when the ceiling above the lava tube collapses. *(Courtesy of NOAA)*

A lava tube is made because a solidified crust develops on top of the flow while the liquid lava drains from underneath. This photo shows the lava draining and pouring into the Pacific Ocean. The flow is from Kilauea's East Rift on November 27, 1989. *(Courtesy of NOAA)*

moves to your left. They are also known as sinistral strike-slip faults.

Lehmann discontinuity   There is a sharp change in seismic velocity deep within the Earth. This boundary was named the Lehmann discontinuity after the first woman GEOPHYSICIST, Dr. Inge Lehmann from Denmark. It is interpreted to represent the boundary between the INNER CORE and OUTER CORE of the Earth.

lherzolite   A type of PERIDOTITE that is dominantly composed of both orthopyroxene and clinopyroxene. They are ULTRAMAFIC, PLUTONIC rocks. Lherzolites have been found in fragments of mantle XENOLITHS. It is thought that much of the MANTLE is composed of lherzolite.

Libya   The country lies along the MEDITERRANEAN SEA and as such is at the northern margin of the African plate. Interactions between the African and European plates in this

area make northern Libya seismically active. Early records of earthquakes date back to Roman times, when two devastating earthquakes destroyed most of Cyrene. The earthquakes were dated at A.D. 262 and 365 and were said to have killed tens of thousands of people each, causing utter chaos in the Roman Empire. The Middle Ages also had reports of large earthquakes that caused great destruction. Notably, the earthquake of A.D. 704 in Sabha, southern Libya, which destroyed several towns and a village. The first well-documented disaster was the earthquake of A.D. 1183, which destroyed Tripoli and reportedly caused the deaths of some 20,000 people. Tripoli experienced several less intense tremors in 1803, 1811, and 1903. One of the major seismically active structures in Libya is called the Hun Graben. On April 19, 1935, an earthquake of RICHTER magnitude 7.1 struck the area and was followed by numerous aftershocks, including one of magnitude 6.5 and another 6.0. In 1941, it was hit by another earthquake of magnitude 5.6. The Gulf of Sirt is another active area, with an earthquake of magnitude 5.6 in 1939. A recent example of Libyan seismicity occurred on February 21, 1963, when a large earthquake struck the village of Al Maraj. The quake measured a 5.3 on the Richter scale and was followed by several aftershocks. The death toll was 300–500 people, and some 12,000 were left homeless.

Lipari Islands   *Tyrrhenian Sea, Italy*   Located near the tip of the Italian Peninsula, just north of the island of Sicily, the Lipari Islands (also known as the Aeolian Islands) are a cluster of volcanic islands, of which the most famous are Lipari, which contains CAMPO BIANCO, STROMBOLI, and VULCANO. Other islands in the group include Alicudi, Basiluzzo, Filicudi, Panaria, and Salina. The volcanoes are seen as part of a line of volcanic mountains extending along the west shore of the Italian Peninsula, parallel to the Apennines.

liquefaction   One of the principal causes of property damage from earthquakes, liquefaction occurs when earthquake vibrations make SOIL lose its COHESION and behave temporarily as a liquid, largely as the result of the addition of GROUNDWATER that is forced upward from depth. Structures built upon soil undergoing liquefaction may suffer severe damage. For this reason, the character of underlying material is, or should be, a major safety consideration when planning construction in areas known for frequent and/or energetic earthquake activity. Liquefaction commonly takes place during earthquakes in poorly consolidated soil where groundwater exists close to the surface. These conditions prevail in many areas subject to frequent, strong earthquake activity. Water between particles of soil is known as pore water, and the pressure of pore water rises with the passage of waves from an earthquake because groundwater is driven toward the ground surface and adds to this pore moisture. The rapid addition of large amounts of groundwater makes the ordinarily solid soil assume a liquid condition temporarily. It takes on the consistency of quicksand.

Damage may result in several ways. When liquefaction occurs on slopes as small as perhaps 3°, large masses of soil may flow downslope, either in liquefied form or as large blocks of still-cohesive soil sliding on an underlying, liquefied layer of

These apartment buildings in Niigata, Japan, were built to be earthquake-proof. During the June 16, 1964, earthquake of magnitude 7.4, the buildings withstood the shaking, but liquefaction compromised the foundations, and the buildings fell over still perfectly intact. *(Courtesy of NOAA)*

soil. In some cases, vast quantities of material are transported over distances of several miles. Lateral spread is another source of property damage from liquefaction. In this phenomenon, blocks of soil at the surface slide sideways because of liquefaction of an underlying layer. Lateral spread may not be as dramatic as flow failure and may result in displacements of only a few feet. Even such seemingly small displacements can be highly destructive, especially if they sever underground water lines required to fight postearthquake fires in urban areas. Ground oscillation occurs when liquefaction of subsurface layers of soil allows blocks of soil on the surface to oscillate and display a variety of earthquake-related effects, such as wavelike rippling and FISSURES. This kind of ground motion may cause severe damage to buildings, which as a rule are not designed and constructed to withstand strong ground oscillation.

Finally, soil simply may lose its bearing strength, or the ability to support overlying structures. This occurs when soil at the surface undergoes liquefaction; structures built on it respond in a variety of ways, such as sinking or rolling over. Multistory buildings have been known to capsize, much in the same manner as a ship at sea, during such episodes of liquefaction (see NIIGATA). Another effect associated with this loss of bearing strength is the rise of buoyant underground structures such as fuel tanks or septic tanks. In evaluating the

potential of a given locality to liquefaction, geologists must consider susceptibility to liquefaction (that is, the character of materials near the surface and their response to an earthquake) and the opportunity for liquefaction (how often the locality is struck by earthquakes and their severity). This determination of liquefaction potential may have to be made on a site-by-site basis rather than a regional basis because conditions may differ greatly from one location to another, in some cases within a distance of feet or yards.

In general, the soils most susceptible to liquefaction have the following characteristics: They were deposited relatively recently; they are made of loose and generally fine sediment (such as sand or silt) that lacks cohesion; and they contain water that rises close to the surface. Particle-size distribution is important because liquefaction potential tends to diminish as particle size increases. This relation means a gravelly soil is less likely to liquefy than a sandy or silty soil is, and deposits of boulders or of mixed cobbles and gravel are unlikely to liquefy at all. At the other extreme of particle size, clay can also be resistant to liquefaction if hardened. The depth to groundwater is also important because soils in which groundwater approaches to within several feet of the surface are more susceptible to liquefaction than are soils where groundwater remains several tens of feet below the surface. Many areas of the United States with histories of strong earthquake activity also show all three of the aforementioned soil characteristics that are associated with high liquefaction potential. The Mississippi River valley; the areas around CHARLESTON, SOUTH CAROLINA, and Boston, MASSACHUSETTS; and portions of CALIFORNIA all exhibit the combination of silty or sandy soil, deposited recently and containing groundwater near the surface.

Two other factors influencing the area extent and degree of liquefaction are, of course, the MAGNITUDE and duration of an earthquake. The more powerful an earthquake, the more intense its associated ground motion is likely to be and therefore the greater the danger from liquefaction of susceptible soils. Proximity to the origin of an earthquake is also a factor; a strong earthquake is likely to cause liquefaction at greater distances than a comparatively mild one.

Lisbon *earthquake, Portugal* One of the most destructive earthquakes in history, the great Lisbon earthquake may have killed some 70,000 people. The shocks began on the morning of November 1, 1755, All Saints' Day, and reportedly killed numerous worshipers in the churches of Lisbon as the buildings collapsed. Three shocks are believed to have struck Lisbon, shaking down most structures in the city. The destruction raised large quantities of dust, which blocked sunlight and cast the city briefly into darkness. The shocks reportedly generated a series of TSUNAMIS that devastated Lisbon's waterfront. Large numbers of refugees made their way to the waterfront, specifically to a large marble quay. Although the tsunami appears to have spared the quay itself from destruction, one report indicates that the quay subsided into the waters, leaving not a single survivor behind nor even a trace of the vessels moored to it. Fire following the earthquake added to destruction from the quake itself but appears to have reduced the threat of pestilence by consuming the bodies of numerous victims.

King Joseph was away from Lisbon but was in the vicinity, at Belem, when the earthquake struck. He and his court reportedly were so frightened by the shocks that they spent the night outdoors in their carriages. The next morning, the king returned to Lisbon to take stock of the damage and restore order to the shattered city, which had suffered from looting and murders in the aftermath of the earthquakes. The royal palace had survived with its kitchens intact, and his cooks began to prepare food for the hungry populace. The government purchased large amounts of grain from the area around Lisbon and distributed it at modest cost or free of charge to the public. Hundreds of AFTERSHOCKS during the following year are said to have kept the public's nerves on edge, but eventually the city recovered, and a decade after the earthquake, Lisbon had been rebuilt.

The Lisbon earthquake was felt over much of Europe; it appears to have generated LANDSLIDES in the mountains of Portugal and to have been felt as a powerful disturbance in ALGERIA and MOROCCO. In Morocco, a FISSURE is said to have opened that sent a village of more than 8,000 residents along the shore tumbling into the water. Tsunamis from the Lisbon earthquake allegedly came ashore as waves more than 20 feet (6 m) high in the Lesser Antilles, on the opposite side of the ATLANTIC OCEAN. Strong earthquakes also hit North Africa about the time of the Lisbon catastrophe. Water levels in lakes in Scotland and Switzerland rose and fell by several feet immediately after the Lisbon earthquake.

lithology In general terms, the description of rock, especially its texture (grain size and relation among the grains), mineral components (composition), and physical character (layering, fractures, etc.). A low-density, "foamy" volcanic rock such as PUMICE, for example, has a much different lithology from volcanic glass, or OBSIDIAN, though they may have the same composition.

lithosphere The outer, most "rigid" layer of Earth, including the CRUST and the uppermost layer of the MANTLE. The lithosphere is divided into a series of plates that are able to drift around Earth. These lithospheric plates float on the ASTHENOSPHERE, which is a soft layer in the upper mantle. The lithosphere sits beneath the atmosphere.

See also EARTH, INTERNAL STRUCTURE OF.

Little Sitkin *volcanic island, Alaska, United States* Little Sitkin is located in the ALEUTIAN ISLANDS and has two CALDERAS. Eruptions were reported in 1776 and between 1828 and 1830, but little is known about these events.

littoral cone When flowing LAVA hits water, it hardens almost instantly. Therefore, it becomes stiff and can pile up. In LAVA FLOWS that reach the ocean, a littoral cone forms where the lava hits the water. These cones are typically only five to 10 feet (1.5 to 3 m) high. Because they are in the surf zone, they are typically eroded away quickly.

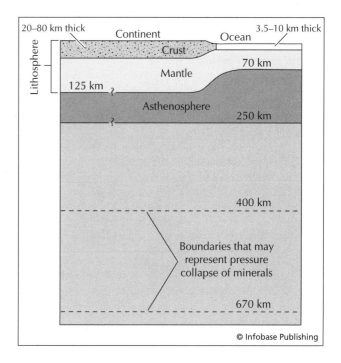

Diagram showing the layers of the shallow part of the Earth. The crust (continent and ocean) and uppermost mantle comprise the lithosphere that floats on the asthenosphere. The upper mantle is also shown.

The bare area on the right side of the photo was once as forested as the left side of the photo before the immense 1957 seiche stripped away the trees 1,700 feet (524 m) up the hillslope facing onto Lituya Bay, Alaska. *(Courtesy of the USGS)*

Lituya Bay *Alaska, United States*   Located on the southeastern coast of ALASKA, Lituya Bay is famous for the destruction caused by a wave that resulted from a LANDSLIDE associated with a powerful earthquake in 1958. A ROCKSLIDE along the shore of the bay set off a wave that destroyed timber as high as 1,720 feet (524 m). At various other locations along the shore of the bay, the wave swept away forest at elevations of 600 (183 m), 160 (49 m), 130 (40 m), and 90 feet (27.4 m). An island in the bay was denuded of timber up to an elevation of 80 feet (24.4 m). Three boats were in the bay at the time of the incident. One was destroyed, the wave carried the second across a spit at the mouth of the bay and into the ocean, and the third boat survived the wave in the shelter of the island mentioned earlier. Return waves in the bay formed a small SEICHE.

Liwa *earthquake, southern Sumatra, Indonesia*   On February 15, 1994, an earthquake of MAGNITUDE 6.6 occurred. At least 207 people were killed, and 2,000 were injured. Between direct damage, mudslides, and LANDSLIDES, more than 6,000 homes and buildings and 75,000 people were left homeless. Damage is estimated at $169 million.

local magnitude   The recorded MAGNITUDE of an earthquake less than 62 miles (100 km) from EPICENTER. Because seismic waves move away from the earthquake in a spherical front, they quickly lose their destructive power. It is like the rings that form on a lake when a rock is thrown in. The height of the wave is high when the ring is small near the source. As the ring grows larger, the height of the wave becomes progressively smaller very quickly. The reason is that all of the energy must be spread over a rapidly increasing circle. At any place, energy reflected in amplitude of the wave must be small. Seismic waves do the same thing. The local magnitude really monitors the area around an earthquake

where the greatest destruction occurs because that is where the waves have the greatest magnitude.

locked fault   A fault may be locked if the frictional strength exceeds the applied stress for a long period of time. Typically, something prevents an active fault from slipping in a segment, and there are no earthquakes for a much longer than usual period of time. An ASPERITY, an intrusion or other rock body, or intersecting faults are the most common culprits. In some cases, a bulge or elevated area may develop over the locked segment. This is the suspected case in the Palmdale bulge along the SAN ANDREAS FAULT in California. The problem is that when the stress finally overcomes the obstruction to movement, the resulting earthquake is much stronger than usual.

loess   Loess is windblown dust from deserts or other arid regions that accumulates on downwind slopes in non-arid areas. This fine dust is eroded by wind and carried in suspension until it encounters an area of lower wind velocity. Large hills and mountains can create enough of a wind shadow to allow deposition locally. Otherwise, the dust remains aloft and can circulate in the atmosphere for a long period of time. Where it accumulates as loess, typically on slopes, it can be quite thick but not particularly stable as the result of the fine size. CHINA has particularly thick loess deposits. It is deposits that have failed in strong earthquakes, destroying both the communities that inhabited them as well as those in the valleys below. Some of the largest death tolls from earthquakes occurred in the GANSU province from this process.

Loihi Seamount *Hawaiian island chain, United States*   The Loihi Seamount is located southeast of the island of Hawaii and is some 3,000 feet (914 m) below sea level. It is the youngest volcano in the Hawaiian chain. The seamount appears to have undergone recent volcanic activity and has TWO CRATERS on its summit. Earthquake activity under Loihi in the 1970s may have been linked with eruptions. It is estimated that Loihi will reach the surface of the PACIFIC OCEAN in a few tens of thousands of years.

Lokon *volcano, Sulawesi (Celebes), Indonesia*   An ANDESITE stratovolcano with at least 21 historic eruptions. The most recent eruption was in 1992. Most eruptions are small to moderate (VEI = 2) and involve pyroclastic emissions.

Loma Prieta *earthquake, California*   Almost equal in fame to the 1906 catastrophe was the less powerful but nonetheless highly destructive earthquake that struck the SAN FRANCISCO Bay area on October 17, 1989. This earthquake was the largest in northern CALIFORNIA since the San Francisco earthquake 83 years earlier. In terms of damage, the earthquake is thought to have been the single most expensive natural disaster in U.S. history up to that time. Estimates of damage are as high as $10 billion. Known as the Loma Prieta earthquake after a mountain near its epicenter, the 1989 disturbance was centered just south of San Jose, in the vicinity of Gilroy, Watsonville, and Santa Cruz, and was caused by movement along the SAN ANDREAS FAULT at a depth of more than 11 miles

(18 km). The rock along the western side of the FAULT moved northward several feet. The earthquake occurred at 5:04 P.M., measured 7.1 on the RICHTER scale, lasted some 15 seconds, and is thought to have been the most powerful earthquake in California since the KERN COUNTY earthquake in 1952.

Although media coverage concentrated on destruction in San Francisco, where structures built on unconsolidated material suffered serious damage, destruction from the Loma Prieta earthquake was greatest within a radius of about 20 miles (32 km) from the EPICENTER. Buildings collapsed in Hollister, Gilroy, and Santa Cruz, and major damage was reported in Watsonville as well. A portion of Highway 101 collapsed over a slough near Watsonville. The earthquake reportedly shifted hundreds of frame homes in Santa Cruz County from their foundations and blocked highway access to the community of Santa Cruz, which was cut off by rockfalls and damaged bridges. Collapsing walls of aged buildings allegedly killed three people in Santa Cruz. Damage was light at the University of California at Santa Cruz but relatively heavy at nearby Stanford University, where destruction was estimated at $165 million. The ground cracked in many locations near the epicenter in the Santa Cruz Mountains, but no signs of surface rupture along the fault itself were found, although one estimate put the length of the underground

rupture from this earthquake at about 30 miles (48 km). One crack near Highway 17 measured more than 600 yards (550 m) long and several feet wide. Sixty-seven people were reported killed by the Loma Prieta earthquake.

Loss of human life was concentrated in western Oakland, on the east side of the bay, where a portion of Interstate 880, some 1.5 miles (2.4 km) long and called the Cypress Viaduct, collapsed and crushed dozens of vehicles and their occupants as the upper level of the two-level highway fell on the lower level. Survivors on the lower level were few; although one man was found alive 89 hours after the collapse, he died soon afterward. Another collapse occurred on the Bay Bridge between San Francisco and Oakland, where a 50-foot (15-m)-long segment of the bridge's upper deck fell onto the lower. Heavy damage was also reported to the Embarcadero Freeway in downtown San Francisco, and damage including landslides and harm to bridges forced the closing of Highways 1 and 101 south of San Francisco.

The earthquake interrupted the third game of the World Series at Candlestick Park. Fires from a ruptured gas main consumed portions of the Marina district, but the fires were brought under control within hours. The Marina district also exhibited widespread damage from liquefaction because the district was built on fill laid down early in the century. Other

Aerial photograph of the collapsed Cypress viaduct of route I-880 in Oakland, California, as a result of the 1989 Loma Prieta earthquake. This collapse caused most of the casualties suffered during this event. *(Courtesy of the USGS)*

The Marina district of San Francisco, California, suffered much worse destruction than the rest of the city during the 1989 Loma Prieta earthquake. The destruction shown here at Beach and Divisadero Streets is a good example. *(Courtesy of the USGS)*

areas of heavy damage included portions of the Mission district and the area south of Market Street. Significant damage also occurred in the Mission, Haight, Sunset, and Tenderloin neighborhoods.

Lomo Poisentepe   *See* APOYO.

Long Beach   *See* CALIFORNIA.

Long Island *caldera, Papua New Guinea*   The Long Island volcanic complex is located on the northern coast of Papua New Guinea and includes two STRATOVOLCANOES, Cerisy Peak and Mount Reaumur, on opposite sides of Lake Wisdom, a CALDERA lake some six miles (9.5 km) wide. Motmot Island, a small island within the lake, was built by volcanic activity within historical times. Powerful seismic and volcanic events appear to have occurred at the caldera some 300 years ago, but records were not kept on a scientific basis. An eruption in 1660 is believed to have been as powerful as the KRAKATOA eruption of 1883. Eruptions occurred in 1933, 1938, 1943, 1953, and 1955 from a VENT or vents under the lake and built up Motmot Island. Discolored water in the lake, accompanied other volcanic activity in 1968, 1974, 1976, and 1993.

long-period wave   A seismic wave with a period of six seconds or greater. The period of a wave is the time it takes for an entire wave to pass a point. If you were standing in the ocean, it would be the time it takes between the wave crests that hit you. Long-period waves mean a slow rise and fall as they pass. These waves tend to be less damaging.

Long Valley *caldera, California, United States*   This CALDERA near Mammoth Lakes appears to have undergone a series of eruptions 500 to 600 years ago that produced RHYOLITE domes in and near the caldera. It is perhaps the largest modern rhyolitic centers in North America. A much earlier eruption (about 3.6 million years ago) is thought to have been so powerful that it sent a PYROCLASTIC FLOW rushing up the flank of the SIERRA NEVADA and over the mountains. The output from this eruption is believed to have flowed down Owens Valley for some 40 miles (64 km) and to have covered almost 600 square miles (1,554 km^2) of land in the present states of CALIFORNIA and NEVADA with thick ASH layers. A famous deposit called the Bishop Tuff appears to date from this eruption. The ASHFALL from this eruption has been traced into what is now NEBRASKA. The eruption is thought to have forced some 150 cubic miles (615 km^3) of MAGMA to the surface in the form of PLINIAN eruptions. This volume of ash makes the Long Valley eruption among the largest documented.

Eruptions in the area continued at Mono-Inyo Craters volcanic field. The most recent eruption was at Mono Lake in 1720 and 1850. Earthquake activity and uplift in the caldera during the 1980s provided evidence that magma was moving underground at that time. In 1994, dying forests were found to be attributed to carbon dioxide concentrations of 30% to 96% within the soil.

Los Alamos *New Mexico, United States*   Los Alamos is built on a volcanic plateau made of ASHFLOW deposits laid down in an eruption that is believed to have resulted in the collapse of the volcano summit and the formation of a CALDERA, now called Valle Grande.

Los Angeles *earthquakes, California, United States*   The Los Angeles area has been the location of numerous earthquakes, some of which in historical times have been powerful and destructive. One of the first earthquakes experienced by European-descended Americans in this part of CALIFORNIA occurred during a visit by Captain Gaspar de Portola, governor of MEXICO, in July 1769. Another series of earthquakes in 1812 caused such widespread damage to Spanish missions in southern California that 1812 came to be known as *"el año de los temblores,"* or "the year of the earthquakes." Other notable earthquakes in southern California history include the FORT TEJON earthquake of 1857; the San Jacinto earthquake of 1899; the IMPERIAL VALLEY earthquakes of 1915 and 1979; the San Jacinto earthquake of 1918; the Santa Barbara earthquake of 1925; the LONG BEACH 1933 earthquake; the Terminal Island earthquakes of 1949, 1951, 1955, and 1961; the KERN COUNTY earthquake of 1952; the SAN FERNANDO earthquake of 1971; the SANTA BARBARA earthquake of 1978; and NORTH RIDGE earthquake of 1994.

It is widely assumed that an earthquake of unprecedented destructive potential is possible in the greater Los Angeles area, although it is difficult to predict when and where it may occur. Projected casualty figures from the Federal Emergency Management Agency (FEMA), in the case of a postulated earthquake of magnitude 8.3 on the SAN ANDREAS FAULT, put deaths at 3,000 to 12,500 and hospitalized injured at 12,000 to 50,000, with 52,000 long-term homeless. The same agency projected figures for another postulated earthquake, of magnitude 7.5 on the Newport-Inglewood Fault Zone, at 4,400 to 21,000 killed and 17,600 to 84,000 hospitalized injured, with 192,000 long-term homeless.

The greater Los Angeles area is highly susceptible to earthquakes because of a complex and unstable geology generated by the collision between northward-moving Baja California and the deeply rooted SIERRA NEVADA range. This collision has produced Southern California's TRANSVERSE RANGES and a network of active faults, some of which have the potential to generate highly destructive earthquakes. Prominent fault zones in the Los Angeles area include the following:

- The Elsinore Fault Zone extends almost 120 miles (192 km) northwest from the Mexican border to the northern boundary of the Santa Ana Mountains.
- The Newport-Inglewood Fault Zone is actually a collection of small faults along a line extending past Newport Beach and Santa Monica all the way into Baja California. The powerful Long Beach Earthquake of 1933 occurred along this fault zone.
- The Palos Verdes Hills Fault lies about five miles (8 km) west of Long Beach.
- The San Andreas Fault Zone is the most famous fault affecting the Los Angeles area and extends approximately along a line connecting Palmdale, San Bernardino, and Banning and then along the southern boundary of the Little San Bernardino Mountains. The San Andreas Fault Zone was the site of the Fort Tejon earthquake of 1857.
- The San Pedro Basin Fault Zone extends approximately from Point Dune down the San Pedro Channel to the east of Catalina Island.
- The small but nonetheless dangerous San Fernando Fault, about 20 miles (32 km) south of the San Andreas Fault, was responsible for the destructive San Fernando earthquake of 1971.

Louisiana *earthquakes, United States* Destructive earthquakes originating within Louisiana itself appear to be infrequent, although the state occupies the lower end of the seismically active Mississippi Valley. A strong earthquake on October 19, 1930, damaged buildings at Napoleonville, and another earthquake on November 19, 1958, caused alarm in Baton Rouge.

Love wave Named after the British physicist A. E. H. Love (1853–1940), Love waves are earthquake waves that are restricted to shallow depths near ground surface. The motion is horizontal and perpendicular to the direction of travel, with no vertical motion involved. They are very destructive waves that cause structures to fall over sideways as the result of their side-to-side movement.

low velocity zone At 37–155 miles (60–250 km) depth beneath Earth's surface, seismic velocities decrease abruptly. At both shallower and deeper levels, velocities increase consistently. The reason that the velocity drops at this depth is because the rocks behave in a DUCTILE manner there. Above it rocks are BRITTLE. Considering that seismic waves are elastic, ductile deformation is not conducive to high speeds. The response of the gum-like material is much slower than the rubber bandlike response of the brittle rocks. In terms of plate tectonics, the low velocity zone corresponds to the ASTHENOSPHERE. Lithospheric plates float on this layer.

Luzon *earthquake, Philippines* At 7:26 A.M. on July 16, 1990, one of the most infamous earthquakes in the history of the PHILIPPINES struck the city of Luzon. The MAIN SHOCK had a magnitude of 7.8 and was followed by numerous AFTERSHOCKS for many months. The death toll was 1,621, with more than 3,100 injuries reported. It could have been far more devastating if it had occurred later in the day. LANDSLIDES, LIQUEFACTION, and abrupt subsidence caused extensive damage to buildings, roads, and utility services. Total damage was estimated at $2 billion.

M

maar Also known as a tuff cone, a maar is a shallow, flat-floored crater that is believed to have formed above diatremes as the result of a violent explosion of gas or steam. Maars are typically 200–6,500 feet (60–1,950 m) across and 30–650 feet (9–195 m) deep. The craters are commonly filled with water, forming lakes. The rims of maars are low and consist of a mixture of fragments of volcanic rock and rocks from the walls of the diatreme.

Macdonald Seamount *Tubuai Islands and Seamounts, South Pacific Ocean* This seamount reaches to within about 100 feet (30 m) of the surface. Its activity was discovered in 1967 when undersea hydrophones picked up noise from it. Bubbles of gas and fragments of BASALT were seen issuing from the seamount in 1987.

mafic A compositional term for MAGMA and LAVA, meaning that they are rich in iron and magnesium. The word derives from *ma* for magnesium and *f* for Fe (iron). Mafic rocks include BASALT for volcanic rocks and GABBRO for PLUTONIC rocks. These rocks are composed of PYROXENE and calcium-rich PLAGIOCLASE but commonly include OLIVINE as well. They are very dark in color. Rocks of this type are formed at very high temperatures relative to other IGNEOUS ROCKS.

magma Molten rock underground is generally known as magma, as distinct from LAVA, which is molten rock that has emerged onto the surface from a volcanic VENT. Magma differs greatly in composition and behavior. Some magmas are high in dissolved gases and, when they reach Earth's surface and the high subsurface pressures on them are removed, they form the frothy rocks PUMICE and SCORIA as dissolved gases bubble out of solution as the rock hardens. Such gas-rich magma tends to produce explosive eruptions because of its high gas content. Alternatively, magma that is relatively low in dissolved gases may emerge from vents as streams or floods of highly fluid lava that may flow across the land for many miles before solidifying and halting. Rocks produced by such eruptions include BASALT. Great flows of basalt from eruptions of this kind may be found in the Pacific Northwest of the United States, especially the COLUMBIA PLATEAU and SNAKE RIVER PLAIN. When magma solidifies underground, it may form PLUTONS such as DIKES (vertical walls of igneous rock created when magma enters a vertical crack in subterranean rock and solidifies there) or SILLS (shelves of IGNEOUS ROCK that form in similar fashion but along strata of COUNTRY ROCK). Masses of magma that solidify at depth are known generally as plutons. When magma in the chimney of a volcano solidifies and remains standing after the external layers of the volcano have been eroded away by wind and water, the resulting columnar structure is known as a VOLCANIC NECK. Where magma reservoirs near the surface heat GROUNDWATER, the result may be an area suitable for GEOTHERMAL ENERGY production. Movement of magma underground is thought to be responsible for minor earthquakes that may signal the imminent eruption of a volcano.

magma chamber The store of MAGMA beneath a volcano that provides the molten rock for LAVA and EJECTA during an eruption. Magma is ever-present in these large chambers up to a kilometer beneath the surface. It can remain there for long periods of time without an eruption taking place. However, if it is charged with additional magma or VOLATILES (especially water) from deeper or even laterally within the earth or it is squeezed by local STRESS, the magma may ascend to the volcano. This process produces an eruption. Pulses of magma may feed the chamber periodically and drive the eruptions. Movement of the magma in the chamber is accompanied by an EARTHQUAKE SWARM as the magma opens FRACTURES. If all the magma is drained from the chamber, the volcano may undergo CALDERA collapse in which massive subsidence of the complex occurs. If the magma crystallizes underground, it forms a PLUTON.

magmatic arc In a CONVERGENT BOUNDARY between OCEANIC CRUST and CONTINENTAL CRUST, an igneous belt develops over a SUBDUCTION ZONE. This igneous belt is analogous to an ISLAND ARC on ocean crust, except that a magmatic arc is on continental crust. These arcs exhibit extensive volcanic activity like island arcs but much more plutonic activity. The prime example of a magmatic arc is the ANDES MOUNTAINS of SOUTH AMERICA. The volcanic rocks are ANDESITES but dominantly DACITES and even rhyodacites in some cases. The plutonic rocks are primarily GRANODIORITES but GRANITES are not uncommon. These extensive PLUTONS coalesce to form huge BATHOLITHS exemplified by the SIERRA NEVADA of CALIFORNIA.

magnitude   A measure of the energy released in an earthquake.
See also SEISMOLOGY.

Maine *United States*   The northernmost state of New England is known for frequent minor seismic activity but also experiences an occasional strong earthquake originating on or near its borders. Maine is located along the highly earthquake-prone SAINT LAWRENCE VALLEY and is sometimes affected by shocks that originate there.

One of the strongest earthquakes in Maine's history occurred on March 21, 1904, and was felt over much of New England and the Canadian provinces of Nova Scotia and New Brunswick. The earthquake apparently was most powerful near Eastport and Calais in Maine but also was felt strongly near the community of Saint Stephen in New Brunswick. The shock was felt in Montreal, 300 miles (483 km) distant, and at points 380 miles (612 km) away in CONNECTICUT. This shock was followed by several minor shocks hours later. An earthquake almost as strong occurred on July 15, 1905, and was felt in central Maine and NEW HAMPSHIRE, a distance of some 100 miles (161 km) westward from the shore. This earthquake, however, affected a much smaller area altogether than the earthquake of 1904.

main shock   The main earthquake in a group of earthquakes is considered the main shock. Designation of main shock from FORESHOCK or AFTERSHOCK usually must be done in hindsight. If there is an earthquake of MAGNITUDE 5, and the next earthquake is a magnitude 3, then the 5 was the main shock. If the next earthquake is a magnitude 7, then the 5 was a foreshock.

Maipo   *See* DIAMANTE.

major earthquake   An earthquake with a magnitude greater than 7 on the RICHTER scale. Major earthquakes are typically restricted to the tectonically active areas of Earth. In the United States, CALIFORNIA and ALASKA commonly experience major earthquakes, and yet throughout the entire rest of the country, they are so rare that historical ones can be counted on one hand. A designation of a major earthquake tells nothing about the damage it causes. For example, the HECTOR MINE earthquake in California was a 7.1 on the Richter scale. It was a major earthquake but in a remote area and therefore there were no casualties. On the other hand, the Armenian earthquake was a 6.9 on the Richter scale, just below major earthquake status, and yet killed more than 25,000 people. It is important to understand that there are strong and major earthquakes all over the world, all of the time. Only when they are in populated areas and cause major destruction do we hear about them. The press does not report on all of the earthquakes in remote areas.

Makalia, Mount   *See* DAKATAUA.

Makian *volcano, Halmahera, Indonesia*   Makian is a STRATOVOLCANO that has had eight historic eruptions. Five of them were large (VEI = 3–4) most of which produced fatalities. The 1646 eruption devastated villages, but records are poor. The 1760 eruption killed some 2,000 people largely from LAHARS. The 1861 eruption caused 326 fatalities from *NUÉE ARDENTE*s and people drowning while trying to escape. The most recent eruption was in 1988. There have been some large TECTONIC earthquakes in the area, notably in 1972.

Makran coast *earthquake, Iran-Pakistan*   At 3:26 A.M. on November 27, 1945, a large earthquake swept the Makran coast on the IRAN-Pakistan border. The quake was an impressive 8.0 on the RICHTER scale, lasting more than 30 seconds. The shaking was felt strongly throughout the region and caused a lot of damage. LIQUEFACTION produced several MUD VOLCANOes off the coast that wound up as islands. One emitted large amounts of natural gas that caught fire, sending flames hundreds of feet into the sky. The real shock of this earthquake, however, was the series of TSUNAMIS that it produced. The typically gentle waves of the area were replaced by tsunamis up to 40 feet (12 m) in height and stretched all of the way to INDIA. Several coastal villages were washed out to sea as a result. The DEATH TOLL from this event was hard to estimate because it involved a tsunami, several countries, and numerous coastal towns and boats. The best estimate was about 4,000 people.

Mammoth Hot Springs *Wyoming, United States*   A terraced natural structure at YELLOWSTONE NATIONAL PARK, Mammoth Hot Springs is a famous example of a formation generated by GEOTHERMAL activity.

Managua *earthquake, Nicaragua*   The Managua earthquake of December 23, 1972, reportedly killed some 11,000 people and injured approximately 20,000 others. About three-fourths of residences in the city were destroyed, and approximately a quarter of a million people were left without homes. Property damage was reported at more than $500 million. This earthquake is believed to have had a minor FORESHOCK approximately two hours before the MAIN SHOCK, with more than 100 AFTERSHOCKS during the following weeks.

Manam *volcanic island, near Papua New Guinea*   A small island off the north coast of Papua New Guinea. The first recorded eruption of Manam was in 1616, and it has erupted at least 30 times since then. Twenty-three of these eruptions were in the 20th century. It has been in constant eruption since 1974 with small, frequent explosions of the

The 1972 Managua earthquake caused the collapse and complete destruction of this three-story, reinforced concrete Customs House office building. *(Courtesy of the USGS)*

STROMBOLIAN type. It had a major eruption in 1994 when a plume shot six miles (10 km) in the air and projectiles 0.62–1.2 miles (1–2 km) above the vent.

Maninjau *caldera, Indonesia*   Maninjau CALDERA is located on the island of Sumatra, along the Semangko Fault, which lies along the southwest shore of the island parallel to the offshore Java Trench. Volcanoes in the vicinity of the caldera have undergone explosive eruptions on numerous occasions since the late 18th century.

mantle   The intermediate layer of Earth's internal structure, between the thin CRUST and the dense CORE. The mantle in recent years has undergone intensive study that has revealed many details of its structure. Once presumed to be homogeneous, the mantle now is known to have considerable heterogeneity. There is an upper mantle that has several layers. The top layer is rigid mantle that is adhered to the base of the crust. The boundary between the crust and the mantle is called the MOHOROVICIC DISCONTINUITY. Below the rigid mantle is the ASTHENOSPHERE or plastic mantle. The asthenosphere gets stiffer with depth and undergoes several transitions through the lower mantle. These transitions involve the collapsing of the mineral structures to denser atomic configurations as a result of the extreme pressures.

See also EARTH, INTERNAL STRUCTURE OF.

mare   On Earth's MOON, a plain of IGNEOUS ROCK produced by LAVA FLOWS following a major meteorite impact is known as a mare (plural, *maria*), from the Latin word for sea. Viewed from Earth, the lunar maria resemble seas, although the effective lack of atmosphere on the moon prevents liquid water from existing there. The lunar maria are thought to be relatively recent additions to the lunar surface because they exhibit few IMPACT STRUCTURES (craters), a sign that their surface is younger than other, more heavily cratered areas. Nonetheless, these seas of BASALT are more than 3 billion years old. Earth may have had a similar history in its early development.

Mars   Although little is known about the geology of Mars compared to that of Earth, volcanism is known to have played a large part in shaping the surface of the red planet. Unmanned probes to Mars have revealed the existence of large volcanoes, such as OLYMPUS MONS and the volcanoes along nearby Tharsis Ridge in the northern hemisphere. Olympus Mons is the largest volcano in the solar system to the best of our knowledge. Although volcanic mountains are evident on its surface, Mars appears to lack the pattern of shifting and colliding crustal plates that characterizes TECTONIC and volcanic activity on Earth.

It has been suggested that material expelled through the volcanoes of Tharsis Ridge came from MAGMA underlying Valles Marinensis, a great valley thought to have been formed by subsidence as molten material underground was removed through eruptions. Volcanism evidently has given the northern hemisphere a GEOMORPHOLOGY differing from that of the southern hemisphere in ways besides the mere existence of volcanic mountains. Impact cratering is much more visible in the southern hemisphere, whereas in the northern hemisphere, TEPHRA from eruptions apparently has covered many impact craters and other such features. The Martian volcanoes affect the planet's meteorology on a small but interesting scale; clouds form in the lee of Olympus Mons and are thought to have been observed through telescopes on Earth at various times before probes reached Mars. Mars, like the moon, appears to be much quieter than Earth from a seismic viewpoint. Evidence of erosive activity on the Martian surface indicates that Mars once possessed substantial amounts of liquid water and a denser atmosphere than at present. Although the Martian atmosphere is only a fraction as dense as that of Earth, occasional dust storms on Mars are capable of transporting large quantities of sediment aerially. These storms are capable of hiding even the giant Martian volcanoes from view.

marsh gas   *See* METHANE.

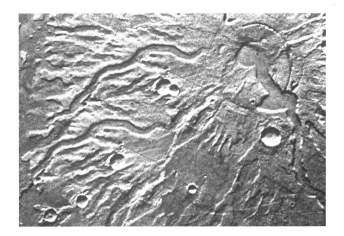

Tyrrhena Patera is an example of a Martian volcano with large valleys on the sides. The valleys are interpreted to have formed as the result of water rather than lava flow. There must have been more water on Mars at one time. *(Courtesy of NASA)*

Marum, Mount   *See* AMBRIM.

Maryland   *United States*   The state of Maryland is not especially noted for earthquake activity, but the state has experienced substantial earthquakes on occasion. The eastern portion of the state has a considerable potential for damage from LIQUEFACTION in any future major earthquakes because much of the urbanized area of eastern Maryland is built atop moist, unconsolidated SOIL that might be expected to lose coherence in a strong earthquake. Maryland sometimes has felt shocks that originated in adjacent VIRGINIA. On April 24, 1758, Annapolis experienced an earthquake (also felt in PENNSYLVANIA) that lasted about a half-minute and followed noises from underground. Shocks in Harford County, Maryland, on March 11–12, 1883, made clocks stop. On January 2, 1885, Frederick County was shaken by an earthquake that also was felt in Virginia and knocked small objects off of shelves and furniture; this earthquake was reportedly silent except for rattling noises from windows. There is a likely IMPACT STRUCTURE in Chesapeake Bay that may localize some of the activity.

Masaya   *volcanic complex, Nicaragua*   The Masaya complex, a CALDERA with three small STRATOVOLCANOES in its center (Masaya, Nindiri, and Santiago), has been reported active since the 16th century. It is the most active volcano in the region with at least 19 eruptive periods since 1524. The most recent eruption was in 1999. Although most Nicaraguan volcanoes have the steep sides characteristic of PYROCLASTIC cones, Masaya has a wide, low configuration (SHIELD VOLCANO), having been built up from BASALTic lava flows without a great amount of explosive eruptions. A lava lake like that of KILAUEA in Hawaii formed in Masaya in the 15th century and again from 1965–1979. Masaya's activity has also included LAVA FLOWS and explosions. It also has a history of SULFUR dioxide emissions.

Masaya, however, is most famous for a huge eruption in 4,550 B.C. It had a VEI of 5 and was one of the largest eruptions in the last 10,000 years.

Mascara   *earthquake, northern Algeria*   On September 18, 1994, an earthquake of MAGNITUDE 5.9 occurred. At least 159 people were killed, 289 were injured, and more than 10,000 were left homeless.

Mashū   *See* KUTCHARO.

Massachusetts   *United States*   The state of Massachusetts is located in New England, one of the most seismically active regions of the United States, and has a history of strong earthquakes dating back to colonial times. Some of the most powerful earthquakes in United States history have occurred in Massachusetts. Eastern Massachusetts in particular has a history of strong earthquakes, often accompanied by remarkable acoustical effects, such as rumbling noises. In addition to earthquakes that originate within Massachusetts, the state sometimes undergoes earthquakes originating in other areas of New England and also along the SAINT LAWRENCE VALLEY between the United States and Canada. Portions of Massachusetts, especially in the east, appear highly susceptible to

damage in any future major earthquake because parts of the large urban area in and around Boston are built on moist, unconsolidated soil that would be vulnerable to LIQUEFACTION as earthquake waves passed through it.

Possibly the greatest earthquake in the history of Massachusetts occurred on November 18, 1755. This earthquake is thought to have had an EPICENTER east of Cape Ann and a MERCALLI intensity of approximately VIII. The earthquake was felt over an area of some 300,000 square miles (776,996 km²), from Nova Scotia to Chesapeake Bay, and some 200 miles (322 km) at sea, where a ship in deep water experienced a shock that made it feel as if the vessel had run aground. (Vessels in harbors along the shore had much the same experience.) Starting with a sound like thunder, the earthquake made it difficult to keep one's footing so that people caught in the earthquake had to hold on to nearby objects to keep from being thrown down. In Boston, chimneys fell, and the ground motion reportedly resembled that of waves at sea. Treetops swayed vigorously, weather vanes fell from buildings, and stone walls were knocked over. In parts of Massachusetts, cracks formed in the earth, and sand moved upward through the cracks to the surface. One curious effect of this earthquake was that it appears to have killed fish in great numbers along the seacoast. Another powerful shock occurred on November 22, 1755, and still more shocks barely a month later on December 19.

mass wasting   Mass wasting encompasses all downhill movement of mass, including soil and rock, in response to gravity. These movements are generally classified based upon composition and velocity. If the dominant material is rock, the movement may either be a ROCKFALL or a ROCKSLIDE, depending upon the mode of movement. Otherwise, mass wasting of debris and soil is generally divided by type of movement, whether fluid-saturated (fluid-type flow) or fluid-bearing (coherent or debris-type movement). The categories of mass movement in each of these types are subdivided by velocity. The slowest coherent movement is CREEP, transitioning to SLUMP, DEBRIS SLIDE, DEBRIS AVALANCHE, and STURZSTROM with increasing velocity. Water-saturated flows are SOLIFLUCTION at its slowest, transitioning to EARTHFLOW and MUDFLOW with increasing velocity. *LANDSLIDE* is the general term for several of these mass movements.

Matsushiro   *caldera, Japan*   The Matsushiro CALDERA is located on the island of Honshū and has not experienced any eruptions within historical times, but earthquake activity is frequent in the vicinity of the caldera and has received extensive study. There are various opinions on what may be happening under the caldera. One view is that intrusion of MAGMA may be involved. Other processes, including HYDROTHERMAL ACTIVITY, have been suggested to account for unrest at Matsushiro.

Mauna Kea   *volcano, Hawaii, United States*   Mauna Kea is located on the island of Hawaii and is the tallest mountain on Earth, more than 30,000 feet (9,144 m) from its base on the seafloor to the summit. Unlike MAUNA LOA and KILAUEA, also on Hawaii, Mauna Kea is not active.

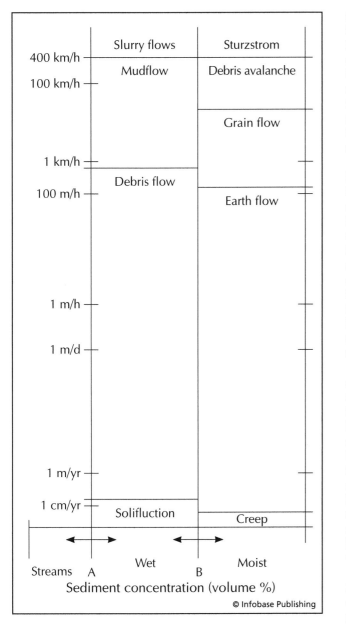

Chart showing the types of mass movements and their conditions of velocity and water content

Mauna Loa *volcano, Hawaii, United States* Mauna Loa, a BASALTIC SHIELD VOLCANO located on the island of Hawaii, is one of the most active volcanoes on Earth but also one of the least dangerous because its LAVA is low in dissolved gases. Therefore, it is not given to producing violent, explosive eruptions. LAVA FLOWS from Mauna Loa occur every several years on the average, tend to last only a few days, and have amounted to more than 3 billion cubic yards (2.4 billion m³) in the past century. Lava emerges from the summit CRATER (called Mokuaweoweo), from RIFT zones that extend from that crater, and from two pit craters, South Pit and North Bay, adjacent to the main CALDERA that was produced by

subsidence. One of five shield volcanoes that make up the island of Hawaii, the largest island in the Hawaiian chain, Mauna Loa has erupted on numerous occasions since records of eruptions began to be kept in the early 19th century. An impressive caldera on the summit of Mauna Loa is approximately two to three miles (3.2 to 4.8 km) in diameter and formed several hundred years ago when collapse accompanied the withdrawal of MAGMA from an underground chamber. The caldera walls stand about 600 feet (183 m) high in some places.

In the late 1800s and the first half of the 20th century, Mauna Loa was very active. The volcano fell quiet in 1950 but became active again in 1975. An eruption in 1984 was associated with precursory earthquakes. Relatively little is known about activity at Mauna Loa compared with the nearby KILAUEA volcano because Mauna Loa was quiet from 1950 to 1975 when methods for studying volcanic activity were advancing. Earthquake activity and other evidence usually provide abundant warning that an eruption is forthcoming, and geologists have had considerable success in predicting the volcanic eruptions. Eruptions of Mauna Loa are notable for "curtains of fire," fountainlike eruptions of extremely fluid lava that may last a day or so before subsiding.

On occasion, lava flows from Mauna Loa have threatened the nearby city of HILO. One flow in 1881 stopped only at the edge of town, and another lava flow approaching Hilo in 1935 made history because of an attempt to divert the lava by aerial bombardment. The U.S. Army Air Corps, on the advice of the director of the Hawaiian Volcano Observatory, bombed the lava flow at selected points in hopes of changing its direction. The theory behind this tactic was that bombing might release gas and molten rock beneath the "crust" of the lava flow and thus stop its advance. The bombing appeared to succeed, for the lava flow halted soon after the bombardment. There is disagreement, however, about whether the bombing actually stopped the lava or whether the lava flow was about to cease anyway.

Mayon *volcano, Philippine Islands* The beautiful STRATOVOLCANO Mayon has a long history of destructive eruptions. It has erupted 47 times since 1616. Of these, 18 produced *NUÉE ARDENTE*s and 12 resulted in fatalities. The 1616 eruption was extremely violent and is said to have destroyed dozens of villages near the volcano. In 1766, another eruption, characterized by flows of hot mud down the volcano flanks, killed some 2,000 people. The 1814 eruption of Mayon caused at least another 2,000 deaths. Major eruptions took place in 1886, 1888, and 1897. In the 1897 eruption, approximately 400 people are believed to have died. Eruptions in 1914 and 1928 were less costly in terms of lives lost. The previous eruption was in 1993. It killed 68 people and prompted the evacuation of 60,000 others. The latest eruption lasted from 1999 to March 2000 and resulted in an evacuation of more than 10,000 but no deaths.

Mayon is steeped in legend. Magayon was the jealous uncle of a beautiful princess. As she escaped with a lover, Magayon chased her. The gods sent a LANDSLIDE that buried Magayon. His anger now wells up as eruptions from Mayon.

Mazama *volcano, Oregon, United States* Mazama is the name given to the volcano that collapsed and thus created the BASAL WRECK occupied by CRATER LAKE today.

Mazar-e Sharif *earthquake, Afghanistan* On September 18, 1994, an earthquake of MAGNITUDE 6.5 occurred. More than 4,000 people were killed, and many thousands were injured.

McDonald Island *volcano, Australia* McDonald Island is a new volcano that was discovered in March 1997. It is believed to have begun erupting in December 1996. It is the first active volcano to be discovered in the Southern Hemisphere in more than 50 years. It is also AUSTRALIA's second active volcano. It is considered very dangerous for anyone on the new island.

Me-Akan *See* AKAN.

Medicine Lake *volcano, California, United States* Located in northern CALIFORNIA near Tule Lake, just south of the OREGON border, Medicine Lake volcano has an area of some 900 square miles (2,331 km²) and rises approximately 4,000 feet (1,219 m) above the surrounding land. Although BASALTIC LAVA from comparatively quiet eruptions is found on the flanks of the volcano, the mountain also has a history of violent eruption from its summit. One of these violent eruptions, believed to have occurred about 1,000 years ago, deposited PUMICE on Mount SHASTA, more than 30 miles (48 km) away. RHYOLITIC lava flowing from Medi-

cine Lake volcano formed Glass Mountain, and a separate flow on the west rim of the CALDERA formed Little Glass Mountain. The volcano has been generally quiet since then, although strong seismic activity in 1910 fractured the surface of the ground and was accompanied by a minor eruption of ASH. Earthquake activity around the volcano in 1988 indicated possible movement of MAGMA underground. Lava Beds National Monument occupies a portion of the Medicine Lake volcano.

Mediterranean Sea The Mediterranean basin is the location of numerous powerful earthquakes and volcanic eruptions and has a complex geology involving both the northward movement of AFRICA toward Europe and the interaction of numerous smaller plates. Volcanoes in the Mediterranean include VESUVIUS, ETNA, and STROMBOLI. The catastrophic eruption of THIRA, comparable in destruction to that resulting from the 1883 eruption of KRAKATOA, is believed to have wiped out the Minoan civilization in 1470 B.C. The Italian Peninsula especially is known for intense seismic activity as well as volcanism, and the presence of MAGMA near the surface in portions of ITALY has enabled that country to satisfy part of its energy needs by drawing on GEOTHERMAL sources.

medium-rate ridge A MID-OCEAN RIDGE with an intermediate two to four inches (5–10 cm) per year rate of spreading. The JUAN DE FUCA Ridge, 200 miles (322 km) west of WASHINGTON State, is the most studied medium-rate ridge in the world. This ridge contributes to the formation of the PACIFIC CRUSTAL PLATE. It has a small axial rift and is crossed by a

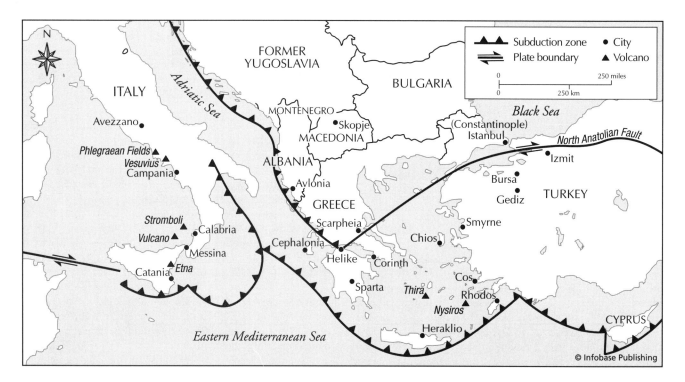

Map of the eastern Mediterranean Sea showing the plate boundaries and locations of volcanoes and several of the epicentral cities for major historical earthquakes. Lines with triangles represent subduction zones, with triangles on the overriding plate.

linear chain of HOT SPOTS. There is an axial volcano atop the ridge where the two intersect.

Medvezhii *caldera, Kuril Islands, Russia* Explosive eruptions at the somma volcano Medvezhii are recorded in 1778 or 1779 and in 1883. The 1883 eruption reportedly involved LAVA FLOWS. PHREATIC activity was recorded in 1946, and activity of undetermined nature was recorded in 1958. Eruptions are relatively small (VEI = 1–2).

megathrust A THRUST FAULT is sometimes simply referred to as a thrust. A megathrust is not simply a large thrust but a thrust associated with the main motion on a SUBDUCTION ZONE. As such, these faults are capable of massive amounts of movement in a single event in terms of length of the rupture, amount of movement, and MAGNITUDE of the earthquake. The BANDA ACEH earthquake was the latest megathrust-type event, though Kamchatca, Chile, and Alaska had such events in 1952, 1960, and 1964, respectively, and CASCADIA off the northwest coast of the United States produced a megathrust-type event (estimated at 9.0 on the Richter scale) in 1700. Because these events are submarine and thrust in nature, there is a strong potential for them to produce TELETSUNAMIS.

megatsunami A megatsunami is a grandiose description of an excessively large TSUNAMI. Sensationalistic television documentaries have cited the huge wave generated in LITUYA BAY, ALASKA, in 1957 as evidence for a megatsunami. This wave, which sloshed some 1,700 feet (510 m) up a slope was not a tsunami but a SEICHE. There is no record of a wave even close to this size having existed in a large ocean basin (such waves tend to be restricted to small basins). An asteroid impact in the open ocean might produce such a wave, but such large impacts are extremely unlikely. Large marine volcanic eruptions in SUBDUCTION ZONES could also generate a huge wave. There is evidence that THIRA generated a more than 200-foot (60-m) wave and that KRAKATOA generated a 135-foot (40-m) wave. Although not as impressive as a seiche, these are still very destructive waves.

meizoseismal region This is the area of particularly strong shaking and significant damage in an earthquake. It is typically located near the EPICENTER. These regions are best displayed in a SHAKEMAP analysis.

Mendeleev *caldera, Kuril Islands, Russia* Mendeleev is thought to have erupted in or around 1880, killing some trees on the island through SOLFATARIC activity. PHREATIC activity may have occurred in 1900, although this appears uncertain. Some earthquakes in recent years may have been associated with a GEOTHERMAL ENERGY facility on the island.

Merapi *volcano, Java, Indonesia* The STRATOVOLCANO Merapi (Mountain of Fire) has a LAVA DOME on its summit and is believed to have erupted on more than 68 occasions since A.D. 1548. It has produced more *NUÉE ARDENTE*s than any volcano on Earth (32 of the 68 eruptions). Merapi is located just 15 miles (24 km) north of the city of Yogyakarta with a population of 3 million, making it a very dangerous volcano. The eruptions are very destructive (VEI = 3). One eruption in 1006 was so destructive that the Hindu rajah of the island moved to nearby Bali and Java fell under Islamic influence. Other large eruptions occurred in 1786, 1822, 1872, and 1930.

The current eruptive phase began in 1987. At one point, it was generating 40 *nuée ardentes* per day. In late 2006, the lava dome collapsed sending PYROCLASTIC FLOWS some five miles (8 km) from the volcano. This event killed 43 people and forced the evacuation of 6,000. In 1998, evacuations of many thousands of people was still a common occurrence.

See also IJEN.

Mercalli, Giuseppe (1815–1914) Italian scientist who made a study of Italian earthquake areas. Mercalli is best known for the Mercalli scale, used to rank earthquakes on the basis of their destructiveness. The Mercalli scale uses Roman numerals, starting with I for the mildest earthquakes and ranging upward to XII for the most destructive. The advantage of the Mercalli scale over the more widely known RICHTER scale of earthquake magnitude is that the Mercalli scale requires no special instrumentation for the user to apply it. The Mercalli scale is based on visual and other noninstrumental observations of the earthquake's effects.

See also SEISMOLOGY.

mesa A flat-crowned hill that is slightly broader than a butte but smaller than a plateau. Mesas are formed in areas with flat-lying sedimentary or volcanic and sedimentary strata. Erosion proceeds through these layers rapidly until it encounters a resistant layer. Erosion eventually breaks through the resistant layer locally and quickly cuts through the underlying layers. The area still containing the remaining resistant layer remains elevated relative to the eroding surroundings. It is a flat-topped, table-like feature because the strata are flat. Mesas are common geomorphic features in the western UNITED STATES, where erosion is more physical than in the wet and temperate east, where chemical weathering dominates.

Mesa de los Hornitos *volcanic formation, Mexico* Associated with the volcano PARICUTÍN, Mesa de los Hornitos (*hornitos* means "little ovens," or SPATTER CONES) was the site of major LAVA FLOWS from 1944 to 1947. One flow in 1944 overwhelmed the village of San Juan Parangaricutiro but allowed enough time to evacuate the community.

Messina *earthquake, Italy* The most destructive earthquake in modern European history struck the port city of Messina in Sicily, ITALY, at 5:25 A.M. on December 28, 1908. The earthquake was estimated to have been 7.5 on the RICHTER scale and came in a series of shocks ranging from 10 to 45 seconds in duration—the MAIN SHOCK lasting for about 30 seconds. The strong SURFACE WAVES flattened everything in a 120-mile (192-km) radius. Strong aftershocks continued for well over one month.

Immediately following this tragic earthquake, a 40–50-foot (12–15-m) TSUNAMI roared across the Messina Straits, destroying many coastal towns and cities. The loss of life from this combined onslaught was staggering. The official death toll was 160,000, but other estimates placed it as high as 250,000. Messina began with a population of 150,000 and was reduced by well over half, with more than 90% of the city flattened. Beautiful architectural treasures, such as the Norman Cathedral of the Annunziata dei Catalani (Annunciation of the Catalans) and the Munizone and Vittorio Emmanuele theaters, were lost. There were so many bodies in the rubble that Messina was dubbed "Citta di Morte" (city of the dead). In the nearby province of Reggio di Calabria, up to 50,000 people were also killed. Although there had been many devastating earthquakes in southern Italy, it was the Messina earthquake that finally spurred the government to institute strict seismic construction codes. Messina was reconstructed under these regulations. Hopefully, they will prevent another tragedy.

metamorphic rock   Rock that has been physically and/or chemically changed through recrystallization under the influence of changes in temperature, fluid content, and chemistry and/or pressure. The changes are mostly mineralogical and textural, but chemical changes are possible as well. There are several types of metamorphism, the most common being regional, contact, and dynamic. CONTACT METAMORPHISM and a subset called pyrometamorphism are most related to igneous activity. The metamorphic rock is the result of contact with hot MAGMA or LAVA which bakes the COUNTRY ROCK. The rocks produced by this process include hornfels and granofels with minerals formed by high temperature processes.

See also IGNEOUS ROCK; SEDIMENTARY ROCK.

methane   A simple hydrocarbon molecule consisting of a carbon atom with four hydrogen atoms attached to it (CH_4), methane is the centerpiece of a hypothesis by physicist Thomas Gold (one of the contributors to the "Steady State" model of the universe) that presents release of methane from underground as an explanation for many curious phenomena associated with earthquakes. Gold's theory, known as the "deep gas" hypothesis, suggests that large quantities of methane (also known as marsh gas), left over from the formation of the planet, still exist inside Earth and are vented to the surface from time to time in outgassing phenomena. The methane alternatively could be from petroleum sources. The aforementioned phenomena, coinciding with earthquakes, might account, for example, for the fish kills and incidents of "boiling" water that have been seen to accompany earthquakes along the shore. Release of methane into the water would give the sea the appearance of boiling, and a sufficient quantity of the gas could asphyxiate fish, especially if the methane contained other gases such as hydrogen sulfide. (Analysis of gas bubbles rising from the ocean floor off Malibu Point following the SAN FERNANDO earthquake of 1971 showed the gas to be 93% methane, with small percentages of nitrogen, carbon dioxide, oxygen, and argon.) Methane outbursts accompanying earthquakes also might account for the strange phenomenon of EARTHQUAKE LIGHT, a glow that has been seen in the night sky during and near the time of earthquakes. Methane gas escaping from the earth and igniting near the surface could burn with enough luminosity

to explain earthquake light, although there appears to be no proof that earthquake light originates in this manner. Methane eruptions from the seabed have even been suggested as a possible mechanism for generating TSUNAMIS. A large emission of methane gas from the seabed, approaching the surface, might lift water above it into a dome, which then would collapse and generate a major disturbance in the sea, with a consequent tsunami.

Metis Shoal *volcano, Tonga Islands*   In June 1995, this normally submerged volcano produced a small island. This is not the first time an island was produced by this volcano. In 1851, it produced another island, but ocean waves quickly eroded it back below the surface. In 1969, another VENT produced an island nearby, but it met with the same fate. This volcano produces huge amounts of PUMICE that form rafts hundreds of kilometers across. People traveling in ships are sometimes awakened by a squealing sound, only to come on deck to see pumice as far as the eye can see. It feels as if the ship is on land. The sound comes from the pumice scraping along the sides of the ship.

Mexico   As part of the westward-moving landmass of NORTH AMERICA, Mexico collides with the OCEANIC CRUST of the PACIFIC CRUSTAL PLATE to the west of the continent. This collision generates numerous earthquakes along the west coast of Mexico, especially in BAJA CALIFORNIA, which is part of a northward-moving block of CRUST that is grinding against CALIFORNIA and generating numerous earthquakes there. A special set of geological conditions in the Mexico City area has made that metropolis highly vulnerable to damage from earthquakes, even those originating along the coastline far to the west. Mexico City's vulnerability to earthquake damage was demonstrated clearly in the great earthquake of September 19, 1985, when an earthquake originating along the Mexican Pacific shore struck Mexico City during rush hour and caused numerous fatalities and extensive property damage. The capital has been constructed atop unconsolidated sediment that once formed a lake bed. Buildings atop this kind of soil are especially susceptible to damage from SURFACE WAVES in earthquakes. Moreover, buildings in Mexico City were designed and built largely without consideration for earthquakes and consequently experienced more damage than newer, quake-resistant buildings such as those required in California. Severe damage and fatalities were reported in four states of Mexico: Colima, Jalisco, Guerrero, and Michoacan. Several vessels at sea were reported missing following the earthquake, and the crew of a trawler reportedly witnessed waves some 100 feet (30 m) high rising from the sea.

Mexico also has some very active volcanoes. The eruption of El CHICHÓN in 1982 had a VEI of 5 and produced beautiful sunsets worldwide for the next year. It is also said to have been responsible for the unusual weather patterns. An eruptive period of PARACUTÍN from 1943 to 1952 also produced some spectacular events and features. COLIMA is another active volcano that produced explosive eruptions in the 1980s and 1990s as is POPOCATÉPETL. Other volcanoes and volcanic features include Iztaccíhuatl, JORULLO, La Reforma, and Pínacaté.

An eight-story frame structure with brick walls was broken away from its foundation and broken in half during the 1985 Mexico City earthquake. *(Courtesy of the USGS)*

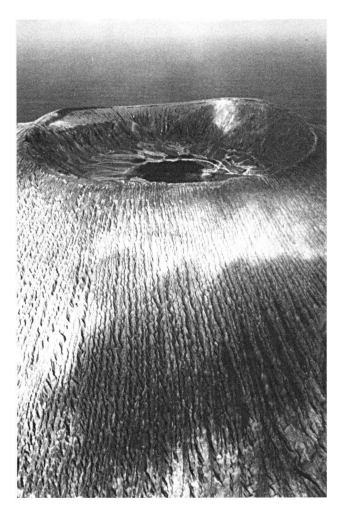

The volcanic crater and slopes of San Benedicto Island along the northern Pacific coast of Mexico. *(Courtesy of the USGS)*

Rayleigh waves caused the crush and collapse of the upper floors of the Lasca Building during the 1985 Mexico City earthquake. Most of the windows were popped out and turned into missiles. *(Courtesy of the USGS)*

Mexico City *earthquake, Mexico*   One of the most famous earthquakes of the 20th century, the Mexico City earthquake had an estimated RICHTER magnitude of 8.1, making it approximately as powerful as the SAN FRANCISCO earthquake of 1906. The earthquake occurred early on the morning of September 19, 1985, and originated along the Pacific coast of Mexico. This timing helped reduce the number of deaths from the earthquake, which killed more than 8,000 people and injured some 30,000. Had the earthquake occurred later in the day, the number of deaths might have been much higher because many public buildings vulnerable to damage would have been occupied. Approximately 500 buildings in Mexico City were destroyed or experienced heavy damage. Damage was estimated at about $4 billion, but serious damage was confined to a small total portion of the city. The character of underlying material did much to determine how much damage buildings in the city experienced from the earthquake. Damage was worst in areas underlain by moist, unconsolidated material deposited in the bed of what once was Lake Texcoco. The lake was drained by Spanish settlers after the conquest of the indigenous Aztec civilization. Construction began afterward atop the sediments of the lake bed. This kind of SOIL is especially susceptible to ground motion in a strong earthquake, and any structure built atop such soil is vulnerable to damage from the ground motion. The earthquake's destructive potential was reduced by the distance (some 200 miles [320 km]) between Mexico City and the earthquake's point of origin. An earlier earthquake in Mexico City, in 1957, produced a pattern of damage similar to that from the 1985 earthquake.

mica A mineral group of sheetlike, smooth, and shiny layers that are adhered together in a stack. The stacks are referred to as books. They include white mica, MUSCOVITE, black mica, biotite, as the main varieties. Biotite can occur in ANDESITE, DACITE, LATITE, and RHYOLITE and their PLUTONIC counterparts. Muscovite is mostly restricted to rhyolite, and even there it is rare. It is more common in GRANITE, the plutonic counterpart. Even if present, mica constitutes a minor component in IGNEOUS ROCKS.

Michigan *United States* Although it generally lies within a zone of minor seismic risk, the state of Michigan is subjected from time to time to strong earthquakes originating in the SAINT LAWRENCE VALLEY. An earthquake at Calumet on July 26, 1905, is thought to have been induced by conditions associated with mining operations. There was reportedly a great explosion. Chimneys toppled, and windows were broken. The earthquake was felt throughout Michigan's Keweenaw Peninsula. On May 26, 1906, some kind of seismic event at a mine on the Keweenaw Peninsula affected an area as wide as 40 miles (64 km) and resembled in its effects a powerful earthquake; subsidence was reported, and rails were twisted.

microearthquake An earthquake with a MAGNITUDE less than 2 on the RICHTER scale. Microearthquakes are common on all ACTIVE FAULTS and can number in the hundreds to thousands per year without foreshadowing any major activity. They are even common in areas of low seismic risk. A radical increase in the number of microearthquakes in an area, however, is commonly considered a warning signal that increased major seismic activity is possible. Microearthquakes also may foretell of volcanic eruptions. As MAGMA forces its way upwards through the crust to the volcano, it does so by cracking the rock in its way. This cracking yields an earthquake. There are swarms of earthquakes present in advance of an eruption in most volcanoes. These earthquakes usually start out as microearthquakes but as the eruption nears, they can be much larger.

microzonation The practice by which a town or county is subdivided into smaller subareas according to the variation in seismic hazards. Depending upon physical features as well as underlying geology, different seismic hazards will be dominant. For example, hills are prone to LANDSLIDES, whereas water may yield TSUNAMIS or SEICHES.

mid-ocean ridge These structures are undersea mountain chains that mark zones where MAGMA rises from the MANTLE and solidifies, generating new OCEAN CRUST. This newly formed rock moves laterally outward from fracture down the middle of the ridge. Midocean ridges are part of

a worldwide system of DIVERGENT BOUNDARIES that is not confined to oceans but may extend into continents as well. Mid-ocean ridges are sites of extensive earthquake and volcanic activity. Not only are there numerous normal faults forming GRABENS along the peak of the ridge, but there are also regularly spaced TRANSFORM FAULTS that break the ridge into offset segments. These transform faults are seismically active and spread several hundred kilometers across the ocean basin thus extending earthquake activity to a much larger area. Large transform faults are FRACTURE ZONES. The rocks produced at mid-ocean ridges are the key evidence that geologists used to prove the existence of PLATE TECTONICS. New ocean crust is constantly being formed at the center of the mid-ocean ridge and moving away toward the edges of the basin like a conveyor belt. The IGNEOUS ROCK that solidifies along mid-ocean ridges preserves a record of the orientation of Earth's magnetic field at the time the rock was formed. Because Earth's magnetic field changes radically on a regular basis, the ocean crust acts like a piece of magnetic recording tape and preserves a history of the magnetic field. The magnetic signatures (patterns) of the ocean crust are exact mirror images across the mid-ocean ridge as are the ages, elevations, and sediment thickness. The theory of how plates are formed and moved are based on this important discovery.

A black smoker at a mid-ocean ridge hydrothermal vent from the Atlantic Ocean. *(Courtesy of NOAA)*

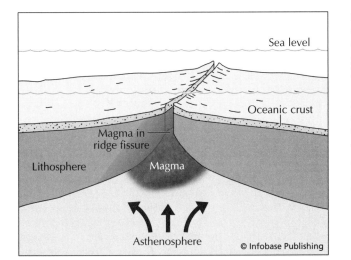

Diagram of a mid-ocean ridge showing newly formed ocean crust moving away from the axis in both directions at the same speed. Circulation of the mantle is shown by the arrows and generates magma beneath the ridge. The magma fills the fissures left as the two plates are pulled apart. The ridge is higher than the surrounding ocean floor both because the mantle is pushing it up and because the new crust is hot and therefore less dense and more buoyant. A mid-ocean ridge is an example of a divergent margin.

Exotic chemical and biological environments are found at certain points along mid-ocean ridges. For example, explorers from the Scripps Institute of Oceanography and the Woods Hole Oceanographic Institution examined the mid-ocean ridge near the GALÁPAGOS ISLANDS in 1977 and discovered a dense concentration of undersea life around volcanic vents on the seabed at depths of about 9,000 feet (2,743 m). Fauna found at this site included giant clams more than a foot long, and huge polychaete worms that lived in tubes. At another site along the mid-ocean ridge off the western coast of MEX-ICO, mineral-laden hot water rising from vents in the seabed has produced peculiar "chimneys" made of minerals precipitated from the heated water. This process of precipitation may have created extensive deposits of rich metal ores on the seabed along certain portions of the mid-ocean ridges, but such deposits do not appear to be commercially exploitable.

Mihara-yama *See* IZU-ŌSHIMA.

Minnesota *United States* Minnesota does not have a long history of strong earthquake activity because the state occupies a zone of minor seismic risk. A strong earthquake reportedly occurred in Minnesota in 1860, but details are unavailable. Another strong earthquake may have occurred in Minnesota between 1860 and 1865, but this is not certain. On September 3, 1917, an earthquake at Staples caused considerable damage to buildings and knocked objects off shelves. Other, minor earthquakes were recorded in 1909, 1925, 1935, and 1968.

Mino-Owari (**Nobi**) *earthquake, Japan* The greatest inland earthquake experienced in Japan occurred in the provinces of Mino and Owari on October 28, 1891, at 9:38 P.M. The esti-

mated RICHTER magnitude for this event was 8.4. The area of damage was over 4,300 square miles (11,150 km²). The EPICENTER was on the island of Honshu, on one of a series of northwest-to-southeast faults (now called the Mino-Owari Fault Zone) that crosses the area, and the FOCUS was less than 9.5 miles (15 km). It was therefore not one of the more common SUBDUCTION ZONE–related earthquakes, which are much deeper. The sense of movement on these faults was a combination of LEFT-LATERAL strike-slip and reverse. The surface ruptures from this event were extraordinary. They could be traced on land continuously for 40–70 miles (24–42 km). They exhibited horizontal offsets up to 13 feet (4 m) and vertical offset up to 20 feet (6 m), producing spectacular scarps. Later geodetic surveys showed widespread uplift of some 28 inches (70 cm) in some areas and subsidence of 12–16 inches (30–40 cm) in others. Over the two years following the main shock, there were some 3,365 aftershocks—10 of which were classified as violent and 97 of which were classified as strong.

In all, some 7,273 people lost their lives in this event, and more than 17,000 were injured. Officially, 142,177 houses were destroyed. In addition to destruction by shaking, some 10,000 LANDSLIDES occurred throughout the area. One blocked a river, causing a flood and producing an earthquake lake. There were also reports of TSUNAMIS along the coast. The 1891 earthquake spurred the government to change building codes to more earthquake-resistant designs.

Mississippi *United States* The state of Mississippi is not known for frequent and strong earthquake activity and lies largely within an area of modest seismic risk. Nonetheless, strong earthquakes have occurred in Mississippi. One of these took place on December 16, 1931, in the northern portion of the state. Walls and foundations were damaged and some chimneys destroyed at CHARLESTON, SOUTH CAROLINA. Damage also was reported at Belzoni, Water Valley, and Tillatoba. This earthquake was felt over an area of about 65,000 square miles (168,349 km²) and was rated at MERCALLI intensity VI–VII.

Missoula, Lake *northwest United States* The landscape of northwest UNITED STATES was reshaped extensively by the floods that resulted from the draining of Lake Missoula, which apparently occupied much of western MONTANA 15,000 to 20,000 years ago. Glacial in origin, Lake Missoula is thought to have begun draining in catastrophic fashion when a dam of ice in the Clark Fork Valley in IDAHO gave way and let several hundred cubic miles of water flow westward, carrying away SOIL and even eroding the underlying BEDROCK. One result of this assault by water is the Channeled Scabland of WASH-INGTON, an area of some 2,000 square miles (5,180 km²) partly denuded by the flood. This same flood is believed to have invaded OREGON's Willamette Valley and reached southward to the location of present-day Eugene. Other floods also originated from Lake Missoula. The floods made the walls of the COLUMBIA RIVER gorge steeper and thus contributed to the conditions that are thought to have resulted in the great LANDSLIDES along that gorge, notably the BONNEVILLE SLIDE, which may have been set off by a powerful earthquake.

Missouri *United States* Missouri and its neighboring states have undergone some of the most powerful earthquakes in the history of the United States. Strong seismic activity in the Mississippi Valley region combines with large areas of unconsolidated, moist SOIL to produce conditions that are highly favorable for LIQUEFACTION. The most famous earthquakes in Missouri history are thought to have been also the most powerful recorded in NORTH AMERICA since European settlement began. These earthquakes occurred between December 1811 and February 1812 in the NEW MADRID area along the Mississippi River and are thought to have affected an area estimated at a minimum of 2 million square miles (5,179,976 km²). The earthquakes are believed to have altered topography over an area of as many as 50,000 square miles (129,499 km²). Although the earthquakes have been estimated at MERCALLI intensity XI, casualties were few because the area was settled sparsely at the time.

The first earthquake in this series began in the early morning of December 16, 1811, shaking houses and knocking down chimneys. This earthquake had dramatic effects on the landscape: It caused LANDSLIDES, generated waves in the river, and made entire islands vanish. Lesser shocks occurred over the following days. Another shock, comparable in many ways to the initial earthquake, occurred on January 23, 1812. A third great shock, apparently even more powerful than the first two, took place on February 7, 1812, and was followed by occasional AFTERSHOCKS during the next two years or longer. Considerable uplift and subsidence occurred in these earthquakes. Some areas were uplifted 15 feet (5 m) or more; one such area, the Tiptonville Dome, is some 15 miles (24 km) long and as much as eight miles (13 km) wide. Submergence at Reelfoot Lake in TENNESSEE occurred to depths of some 20 feet (6 m) or even greater.

"Moho" *See* MOHOROVICIC DISCONTINUITY.

Mohorovicic discontinuity A thin transition zone between Earth's CRUST and uppermost MANTLE constitutes the "Moho" for short. The Moho was first identified by a rapid change in the velocity of seismic waves at a depth ranging from about 22 to 50 miles (35 to 80 km), depending on the elevation of the land. It was later determined that this transition marked the top of the mantle. The Moho occurs midway through the LITHOSPHERE and marks the boundary with the rigid mantle. In continent-continent collisions, sometimes the rigid mantle is stripped off of the crust approximately along the Moho in a process called delamination.

moment magnitude The moment magnitude, Mw, is the measured magnitude of an earthquake using seismic moment. The SEISMIC MOMENT is a measure of earthquakes based on the leverage of forces across the area of rupture on the fault plane. In mathematical terms, it is equal to the rigidity of the rock times the area of faulting times the amount of slip on the fault.

Momotombo *volcano, Nicaragua* A STRATOVOLCANO in western NICARAGUA. Momotombo has erupted at least 15 times since 1524. Its last major eruption was in 1905. Its most famous eruption was in 1605–06, when it destroyed the city of Léon, the former capital of Nicaragua.

Mono Lake *California, United States* Located in CALIFORNIA near the NEVADA border and immediately north of the LONG VALLEY caldera, Mono Lake is considered one of the most likely sites for future volcanic activity in the UNITED STATES and has undergone moderate earthquake activity in recent years. Extending approximately north-south between Mono Lake and Long Valley caldera is a set of volcanic landforms known as the Mono Craters. VENTS in the vicinity of Mono Lake appear to have been active within the past two millennia and to have become active as a series of vents along a zone several miles in length, rather than erupted at a single point, as is the case with many volcanoes. Two major series of eruptions at Mono Lake are thought to have occurred in the 14th century. This activity is believed to have begun when molten rock moved along a FRACTURE to the surface, came in contact with GROUNDWATER there, and generated steam that caused a powerful explosion. A string of new vents some four miles (6.4 km) long formed during this eruption, which expelled large amounts of ASH over what now is central California and western Nevada. PYROCLASTIC FLOWS followed this initial eruptive activity but affected only the area immediately around Mono Lake. Next highly viscous lava emerged from the ground and formed domes and COULEES, including Panum Dome. Soon after these eruptive events, another series of explosions commenced at Inyo Craters, immediately to the south of Mono Lake, under circumstances apparently similar to those which led to the first eruptive cycle. (These two outbursts may have been separated by only a few weeks.) This eruption also produced large quantities of ash and steam explosions, followed by formation of domes and flows of OBSIDIAN. A series of earthquakes estimated up to 6.0 on the RICHTER scale of MAGNITUDE occurred in 1980, and earthquakes of comparable magnitude happened again in 1986 along the nearby White Mountain fault, reminding observers that the area around Mono Lake maintains a considerable potential for seismic and volcanic activity. Depending on the intensity of a future eruption at Mono Lake, as well as other factors including wind direction and velocity, ASHFALLS might affect areas hundreds of miles downwind.

Monowai Seamount *Kermadec Islands, Pacific Ocean* It is a SUBMARINE VOLCANO that is located about 932 miles (1,500 km) northeast of the north island of NEW ZEALAND. It has erupted at least eight times in the past 20 years including a 1977 eruption that lasted for 18 months.

Montana *United States* The state of Montana is moderately active from a seismic standpoint, with much of the earthquake activity concentrated in the southwestern portion of the state. Earthquakes centered in Montana have affected vast areas of the western UNITED STATES.

One of the greatest earthquakes in Montana's history, estimated at MERCALLI intensity VII and RICHTER magnitude 6.75, occurred on June 27, 1925, east of Helena. This earthquake was felt very strongly over an area of some 600 square miles (1,554 km²), and the total area affected was more than 300,000 square miles (776,996 km²). On an east-west axis, the earthquake was felt from the border of NORTH DAKOTA to WASHINGTON State, and on a north-south axis from the border of CANADA to the central portion of WYOMING. AFTER-SHOCKS continued for months following the earthquake. Damage was concentrated at Logan, Lombard, Manhattan, and Three Forks. A large school building at Manhattan was demolished. Chimneys were seen to fall, not in one prevailing direction but rather in many different directions; this phenomenon led one observer to speculate that some kind of "twist" was involved in this particular earthquake. Railroad tracks shifted, and rockfalls interfered with service on the Northern Pacific Railroad and the Chicago, Milwaukee, and St. Paul. A powerful roar accompanied the earthquake at Bozeman. A church and a schoolhouse were damaged at Three Forks.

The earthquake of August 17, 1959, near HEBGEN LAKE, Montana, was of Mercalli INTENSITY X and was estimated at 7.1 on the Richter scale. This earthquake affected an area of some 600,000 square miles (1.535 million km²) and was felt at locations as distant as Seattle on the west, North Dakota on the east, UTAH in the south, and Banff, Alberta, CANADA to the north. Twenty-eight people were killed, most of them in LANDSLIDES. One spectacular landslide occurred along the south wall of Madison River Canyon, where a huge mass of earth, rock, and trees fell and blocked the Madison River. Within several weeks, this landslide had created a lake with a depth of almost 175 feet (53 m).

Hadly Rille, a sinuous channel that was probably formed from flowing lava. Astronaut James Irwin from Apollo 15, 1971, with the lunar rover, samples the rock in the area. *(Courtesy of NASA)*

Monte Negro *earthquake, Croatia* At 7:00 A.M. on April 6, 1667, the Monte Negro area of Croatia along the Adriatic coast was rocked by a powerful earthquake. The estimated MAGNITUDE of the quake was 7.2, and the INTENSITY was estimated as high as IX in the town of Budva. It was reported that the shock was accompanied by strong winds in the same east-west direction. The sea withdrew four times that day, and TSUNAMIS struck in many areas. It was said that the town of Ragusa (Dubrovnik) lost 5,000 people, but there was also near–total destruction on the island of Mozzo as well as in the towns of Kastel Novo, Budva, and Kataro. The DEATH TOLL for the event was likely much higher.

Monte Pelato *See* CAMPO BIANCO.

Montserrat *volcanic island, Caribbean* Until recently, there had been no eruptions at Montserrat within historical times. However, the recent and current eruption of SOUFRIÈRE HILLS changed that. Previously, the island exhibited considerable earthquake and SOLFATARIC activity related to intrusions of MAGMA under the Soufrière Hills at the southern end of the island.

Moon Instruments placed on the Moon have reported large numbers of seismic events, but the pattern of lunar seismicity differs from that of Earth in important ways. Although several thousand moonquakes are believed to occur each year, the Moon appears to release much less energy in such quakes annually than Earth does. Total energy released in terrestrial quakes each year is thought to be more than 1 billion times greater than on the Moon. Some evidence of moonquakes is visible to the unaided eye, as where seismic events have dislodged boulders from positions on slopes and caused the rocks to roll downhill. Lunar quakes also are believed to

occur at greater depths than similar events on Earth. These differences reflect the fact that Earth is more active geologically than the Moon, both on the surface and at depth. The ever-changing pattern of plate TECTONIC activity, with its characteristic distribution of earthquake foci and volcanic eruptions, is absent from the Moon. The character of lunar rock also reflects the difference between processes at work on the moon and on Earth. Lunar rock is igneous (BASALTS in the maria, or "seas," and a GABBRO-like IGNEOUS ROCK called anorthosite in the highlands), whereas terrestrial rock may be igneous, SEDIMENTARY, or METAMORPHIC, as a result of erosion, sedimentation, and other processes related to the action of wind and water on Earth. IMPACT STRUCTURES are prominent on the lunar surface. The maria are believed to have formed when large meteorites struck the Moon and caused great flows of molten rock. Some of the rills, channel-like features on the lunar landscape, are thought to have formed through flows of LAVA.

See also MARE.

Moro Gulf *earthquake, Philippines* On August 17, 1976, at a few minutes after midnight, a strong earthquake struck the island of Mindanao in the Philippines, producing a great TSUNAMI and destroyed 438 miles (700 km) of coastline along the Moro Gulf in the North Celebes Sea. The RICHTER magnitude of this event was 7.9. AFTERSHOCKS were reported to have occurred at the rate of 140 per day over the next few days. The source of the earthquake was the Cotabato Trench, part of the SUBDUCTION ZONE system in the Philippines. Within minutes after the main shock, the sea began receding from the shore for about 1.8 miles (3 km) in typical pre-tsunami fashion, 10 minutes later, it came roaring back with waves up to 50-feet (15-m) high at speeds approaching 40 miles (64 km) per hour.

The resulting tsunami from the 1976 Moro Gulf earthquake produced extensive damage such as this at Lebak, Mindanao. *(Courtesy of the USGS)*

The tsunami wreaked havoc, especially in Padigan City and Cotabato City. Total deaths were 4,791, with 2,288 missing and presumed dead. There were 9,930 injured, and over 93,000 left homeless. More than 85% of all deaths were the result of the tsunami. Properties along the coast were washed away, and bodies were left scattered all along the shore.

Motagua Fault *earthquake, Guatemala* The Motagua Fault is a very active TRANSFORM FAULT that separates the CARIBBEAN SEA plate from the NORTH AMERICAN CRUSTAL PLATE. It also connects the Central American SUBDUCTION ZONE to the Caribbean subduction zone and has a LEFT-LATERAL offset of 0.9 inch (2.2 cm) per year. On February 4, 1976, the Motagua Fault produced its latest major earthquake. The quake measured 7.5 on the RICHTER scale. It killed more than 23,000 people and caused $1.1 billion in property damage. It was felt over 38,610 square miles (100,000 km²). Extensive LIQUEFACTION and LANDSLIDES accompanied the earthquake. The observed surface rupture showed up to 11.5 feet (3.4 m) left-lateral STRIKE-SLIP displacement along 199 miles (320 km) of the fault. This earthquake was the largest strike-slip event since the 1906 SAN FRANCISCO earthquake in North America.

mudflow This is one type of MASS WASTING, with high fluidity and very high flow velocity. Mudflows are fast-moving rivers of watery mud that can inundate whole towns, causing widespread death and destruction. Mudflows are reported in the news on a regular basis because they occur in many areas as the result of heavy rain. Mudflows are also common with volcanic eruptions. A LAHAR is a volcanic mudflow composed

of hot ASH and melted alpine glacial ice. The fine size of ash makes it susceptible to forming a mudflow with heavy rains as well.

mud volcano A mud volcano is not, strictly speaking, a volcano at all but rather a hot spring whose waters mix with sediment to form mud, which behaves in a manner roughly comparable to that of LAVA. Mud volcanoes look like SPATTER CONES and tend to be only a few feet high. They are commonly the products of LIQUEFACTION during seismic events. They can also be called sand volcanoes, depending on the sediment in the shallow subsurface. They form from sand or mud "blows" in which a slurry of sand or mud is shot out of the ground up to tens of feet high.

See also SAND BOIL.

Muradiye *earthquake, Turkey* On November 24, 1976, at 12:22 P.M., a strong earthquake struck Turkey. The EPICENTER was located in the town of Muradiye on Mount Ararat, and the FOCUS was at 21.5 miles (36 km) in depth. The

En echelon cracks, or "mole tracks," in a soccer field at Gualan, Guatemala, showing the trace of the strike-slip Motagua Fault that produced a devastating earthquake in 1976. *(Courtesy of the USGS)*

The town of Tecpan, Guatemala, was totally destroyed during the 1976 Motagua earthquake. *(Courtesy of the USGS)*

MAGNITUDE of the event was 7.3–7.9 on the RICHTER scale and its intensity was X on the modified MERCALLI scale. The area was inhabited by Kurdish peasants who were embroiled in conflict with the Turkish government. They resisted the efforts of relief workers and refused to be relocated. Numerous AFTERSHOCKS and blizzards exacerbated the situation. In all, some 5,000 people perished, and the cost of the damage was estimated at nearly $60 million.

muscovite   A mineral otherwise known as white mica. Muscovite occurs as if it were clear shiny stacked sheets of paper. These sheets are flexible like plastic and can be peeled away from the rock. Muscovite can be found in GRANITE and its volcanic counterpart, RHYOLITE.

Muzaffarabad *earthquake, Pakistan*   The third devastating earthquake in consecutive years (BAM, IRAN, in 2003, and BANDA ACEH, INDONESIA, in 2004) struck the KASHMIR-Pir Panjal Mountains near the city of Muzaffarabad, Pakistan, in 2005. On October 8, 2005, at 9:20 A.M., an earthquake of MAGNITUDE 7.6 rocked the Indian subcontinent. The FOCUS of the tremor was a shallow six miles (10 km) on a THRUST FAULT caused by the collision of India with Asia. The duration was approximately two minutes, but that was all it took

to become the deadliest event in this earthquake-prone region. The strong SURFACE WAVES not only destroyed 90% of buildings in an 8,000-square-mile (20,720-km³) area around the EPICENTER, but they also released LANDSLIDES and ROCKFALLS

Sand boil or sand volcano produced by liquefaction during the El Centro earthquake, California, on October 15, 1979. *(Courtesy of NOAA)*

throughout the region, cutting all utilities and blocking roads. AFTERSHOCKS continue with 22 in the first 24 hours—including a magnitude 6.4 at 4:16 P.M. that afternoon—and at least 12 events greater than 5.5 in the first three weeks.

The current DEATH TOLL is approximately 86,000 people. At least 40,000 perished in the area around the epicenter, and upward of 38,000 died in the North West Frontier Province of Pakistan. Some 1,360 died in INDIA, and a few died in Afghanistan. Children were especially vulnerable, as many were in poorly constructed schools that collapsed during the event. At least 2.5 million people were left homeless by the event—a problem exacerbated by the severe winter typical of the Kashmir Mountains. Despite a tremendous international outpouring of support, the destruction of the roads hindered rescue and relief efforts from reaching hard-hit areas.

This event illustrated that the ever-increasing population densities throughout the world will only increase the impact of such natural disasters.

mylonite FAULT rocks from deeper than where earthquakes are produced. Earthquakes are produced by brittle DEFORMATION of rocks in fault zones; in other words, rocks break like breaking a glass bottle. The cracking sound is equivalent to an earthquake. It too sends shock waves throughout the bottle. They are just too small to observe. However, rocks below about six to nine miles (10 to 15 km) in depth are DUCTILE; in other words, they do not crack, but they stretch like silly putty. The fault rocks are not like broken glass but instead are stretched, flattened out, and fine-grained. These are mylonites.

Myozin-Syo *volcano, Bonin Islands, Pacific Ocean* Myozin-Syo volcano, about 300 miles (483 km) south of JAPAN, emerged from the Pacific in a 1952 eruption. Two ships arrived at the site to watch the eruption, but one ship was destroyed, with no survivors, when it strayed over the top of a vent.

See also BAYONNAISE.

N

Nakadake *See* ASO.

Nakajima *See* KUTCHARO.

Napier *earthquake, New Zealand* A large earthquake struck the Hawke's Bay area of New Zealand of February 3, 1931, at 10:46 A.M. and devastated the city of Napier. The EPICENTER was located nine to 12 miles (15–20 km) north of Napier, and the FOCUS was interpreted to have been on a BLIND thrust fault. The focus was shallow in contrast to the typical New Zealand earthquakes, which are deep and related to the SUBDUCTION ZONE. The RICHTER magnitude was 7.8, and the duration was two and one-half minutes, with a 30-second lull in the middle. Some believed that it was actually two closely spaced temblors. The Napier area was uplifted up to seven feet (2 m), creating 5,575 acres of new coastal land. Similarly, the Ahuriri Lagoon was raised and emptied by the earthquake, creating 9,000 acres of dry land.

The earthquake caused 258 deaths, primarily in Napier. Fires raged out of control and spread to the neighboring city of Hastings. Some 525 AFTERSHOCKS were felt in the region over the next two weeks. Napier was rebuilt using newly instituted earthquake-resistant design standards that persist today.

Naples *earthquake and volcano, Italy* The area around Naples is noted for earthquake and volcanic activity. Located near Naples is the famous volcano VESUVIUS, whose eruptions have caused repeated and extensive damage within historical times, notably the destruction of POMPEII AND HERCULANEUM in A.D. 79. Major earthquakes have occurred in the vicinity of Naples, notably in A.D. 63 and 1456.

Nasu *volcanic group, Honshū, Japan* A group of overlapping STRATOVOLCANOes that have produced nine eruptions between 1397 and 1963 from the Chausu-dake vent. A 1410 eruption killed 180 people. The recent eruptions (1953, 1960, and 1963) were small (VEI = 1) and PHREATIC.

The nurses' home at Napier Hospital was a pile of rubble after the 1931 Napier (Hawke's Bay) earthquake. *(Courtesy of the USGS)*

natural frequency Objects will vibrate at a specific frequency if they are struck by a single force or impulse and not damped. If a string on a guitar is plucked or a tuning fork struck, the noise that is produced is the natural frequency, A problem arises if the interaction of the natural frequency of a building constructively interacts with the frequency of the

Sailors were brought in to search the ruins of Napier for victims and survivors of New Zealand's worst earthquake, the 1931 Hawke's Bay event. *(Courtesy of the USGS)*

waves of an earthquake. A series of vibration nodes will form, which can cause the building to literally shake itself apart.

Navajo Volcanic Field   *See* SHIP ROCK.

Nazca crustal plate   The Nazca crustal plate is one of the smaller major plates of Earth's CRUST and is located under the PACIFIC OCEAN immediately to the west of SOUTH AMERICA. It is composed completely of OCEANIC CRUST. The ongoing collision between the South American plate and the Nazca plate is believed to account for much of the belt of strong seismic and volcanic activity along the ANDES MOUNTAINS of western South America, through BOLIVIA, CHILE, ECUADOR, and PERU. The Nazca plate is bounded on the north by the smaller COCOS PLATE, on the west by a MID-OCEAN RIDGE, and on the south by the Antarctic plate.

See also PLATE TECTONICS; "RING OF FIRE."

Nebraska   *United States*   Earthquake activity in Nebraska appears to have been minor since the territory was settled by Americans of European descent. A strong earthquake was reported in eastern Nebraska on November 15, 1877, and was felt over a wide area of MINNESOTA. Two shocks 45 minutes apart were reported. Some damage occurred to property, and the earthquake was felt over an area of about 300 by 600 miles (483 by 966 km).

neck   The remnants of a volcanic pipe after erosion has removed the volcanic cone. It is an erosional remnant. DEVILS TOWER is an example of a volcanic neck.

neotectonics   Current seismic and TECTONIC activity in a region. Neotectonics includes earthquakes, uplift and/or subsidence, and active faulting and/or folding. This current activity typically results in LANDSLIDES and strongly altered drainage patterns. Rivers can shift their course and lakes may appear in the areas of subsidence. Neotectonism only occurs in active areas.

Nepal The small, landlocked country of Nepal sits high in the Himalayas within the zone of the most intense continental collision on Earth—India with southern Asia. India is attempting to dive beneath Asia, but because the CRUST is so buoyant, the collision is more like a car accident, with both continents crumpling. There are no active volcanoes in this area, and there have not been any for tens of millions of years, but there are numerous instances of earthquakes. The first reliable report of an earthquake was from A.D. 1255. It had an estimated modified Mercalli intensity of X, and it killed one-third to one-fourth of the population of the Katmandu Valley. Other catastrophic earthquakes in Nepal were reported in 1408, 1681, and 1810. In 1833, the PAPHLU earthquake, with a magnitude of 7.8, devastated the Katmandu Valley. It produced a surface rupture more than 42 miles (70 km) long that is located about 30 miles (50 km) north Katmandu. Two large FORESHOCKS warned the residents, and many evacuated their homes before the MAIN SHOCK struck. As a result, there were only 500 casualties. The most devastating earthquake to have stuck Nepal in modern times was the 1934 BIHAR event, which killed more than 10,000 Nepalese and produced a surface rupture some 120 miles (200 km) long.

Nevada *United States* The state of Nevada is located in a seismically active region of the country and has been the site of several notable earthquakes in the 20th century. The earthquake of October 2, 1915, had an estimated RICHTER magnitude of 7.7 and generated effects of MERCALLI intensity X at its epicenter in Pleasant Valley near Winnemucca. This earthquake was felt over a vast area of the western United States, from San Diego, CALIFORNIA, to OREGON but did little damage because the affected area was settled so lightly. The earthquake wrecked adobe houses, tossed persons from bed, and caused mine tunnels to collapse. A FAULT with a new vertical scarp 22 miles (35 km) long and up to 15 feet (5 m) high appeared along the base of the Sonoma Mountains. Western Nevada experienced another strong earthquake, this one of magnitude 7.3 with effects of Mercalli intensity X, on December 20, 1932; again, there was little damage because the region affected was so underpopulated. Considerable faulting occurred in a valley near Mina between the Pilot and Cedar Mountain ranges; several dozen FISSURES, with lengths up to four miles (6.4 km), were reported. The shock was felt over much of California, from SAN FRANCISCO to San Diego. The Dixie Valley earthquake on December 16, 1954, affected an area of some 200,000 square miles (517,998 km²), had a Richter magnitude of 7.1, and produced effects of Mercalli intensity X.

Nevado del Ruiz *volcano, Colombia* Nevado del Ruiz was the site of a deadly LAHAR in 1985. An eruption of VEI = 3 produced MUDFLOWS that traveled from the volcano down the Lagunillas River and killed approximately 23,000 people in the town of Armero. Mudflows are common during eruptions of Nevado del Ruiz. The reason is that it is a high mountain 17,453 feet (5,320 m) and therefore covered with ice and snow. During eruptions, the snow and ice is melted and mixes with ASH to form mudflows that can travel 10 miles (16 km) from the mountain. Similar eruptions occurred in 1595 and 1845, but the lahars did not hit inhabited areas.

Nevados de Chilian *caldera, Chile* Located in southern central Chile, Nevados de Chilian has a history of explosive eruptions dating back to the 17th century. Although the nearby city of Chilian stands some 60 miles (97 km) from the shore, saltwater and mud erupted in large quantities after a strong earthquake in 1835. Numerous new hot springs formed at this time. (It is not certain whether this activity occurred at Chilian itself or at Nevados de Chilian.) There may have been a minor eruption at Nevados de Chilian in 1906, several days before a powerful earthquake that wrecked Valparaiso. A new volcano, Volcan Nuevo, formed in this eruption. Activity, including an incident of dome formation, continued in the 1970s and 1980s.

Nevados Huascarán *earthquake, Peru* On May 31, 1970, an earthquake of RICHTER magnitude 7.7 occurred beneath the PACIFIC OCEAN some 84 miles (135 km) away from Nevados Huascarán. Shaking lasted 45 seconds and loosed a giant part of west-facing slope of the north peak of Nevados Huascarán. When it failed, there was a loud explosion and a cloud of dust obscured the mountain. The side of the mountain between elevations 18,000 and 21,000 feet (5,486 and 6,401 m) including a 100-feet (330-m)-thick glacier fell away and down the mountain. The mass was about 100 million square meters (258,998,810 km²) of rock, ice, and water. It started as a free fall for some 1,300 to 3,000 feet (400–914 m) before landing and scooping up more rock and ice. It then moved down the Rio Santa Valley as a DEBRIS FLOW or AVALANCHE for 30 miles (48 km). It moved at speeds of 175 to 210 miles (282 to 338 km) per hour and destroyed the towns

The path of lahar leading down the valley from its origin at the eruption of Nevado del Ruiz, Colombia, in 1985. More than 23,000 people died in this disaster. *(Courtesy of the USGS)*

Blocks such as this 700-ton example came hurtling down the mountain at speeds of up to 210 miles (350 km) per hour in Nevados Huascarán, Peru, in 1970. Houses literally exploded into toothpick-size fragments upon impact. More than 18,000 people were killed. *(Courtesy of the USGS)*

Aerial view of the Andes mountain range, including the highest peak in Peru, Nevados Huascarán. The arrow shows where the 3,000-foot (1,000-m)-wide mass of glacial ice and rock broke loose during the 1970 earthquake. The 22-mile (37-km) travel path extends to the white deposits at Yungay and Ranrahirca. *(Courtesy of the USGS)*

of Yungay and Matacoto. More than 18,000 people were killed, and Yungay was buried 100 feet (30 m) deep in debris. The boulders were up to several tons apiece. They left bomb-like craters along their path, and houses literally exploded when hit by them.

New Britain Island   *See* PAPUA NEW GUINEA.

New Hampshire   *United States*   New Hampshire is located in one of the most seismically active areas of the United States, in northern New England along the SAINT LAWRENCE VALLEY. Earthquakes, some of them powerful, have shaken the state frequently within historical times. The earthquake of November 29, 1783, was felt from New Hampshire to PENNSYLVANIA. On November 9, 1810, an earthquake at Exeter, New Hampshire, involved a peculiar noise like that of a strong explosion underfoot. The earthquake lasted a minute at Portsmouth, New Hampshire, and was strongest between Portsmouth and Haverhill, MASSACHUSETTS. A ship in Portsmouth Harbor appeared to hit the bottom, and the earthquake broke windows in Portsmouth. The earthquake was felt also in MAINE.

Ancient volcanism in the White Mountains of New Hampshire created a curious set of circular and nearly circular structures called RING DIKES that formed when molten rock oozed up around the edges of cylindrical masses of rock overlying volcanic chambers far underground. These circular structures can be found at the Ossipee Mountains, Mount Lafayette, the Pliny Range, and the Crescent Range.

New Jersey   *United States*   New Jersey is located in a region of generally minor seismic risk but has experienced strong earthquakes from time to time. In 1884, an earthquake of an estimated 5.2 on the RICHTER scale occurred. Church bells rang in CONNECTICUT as a result. Faults in the area have historically produced such earthquakes on a 140-year recurrence interval. The Ramapo Fault is likely the most dangerous seismic risk in the New York metropolitan area. On September 1, 1895, an earthquake in Hunterdon County, estimated at MERCALLI intensity VI, affected an area of some 35,000 square miles (90,650 km²); this earthquake knocked articles off shelves in Hunterdon County and was felt throughout the city of Newark. Strong shocks were reported also at Burlington and Camden, and windows were broken in Philadelphia. A rumbling sound accompanied an earthquake on January 26, 1921, in Moorestown and Riverton. The coast of New Jersey between Toms River and Sandy Hook experienced three shocks on June 1, 1927; between Long Branch and Asbury Park, chimneys and plaster walls were damaged, and objects were knocked off shelves. On December 19, 1933, an earthquake near Trenton was felt sharply and knocked pictures off walls in Lakehurst.

The Salem County earthquake of November 14, 1939, was estimated at Mercalli intensity V and was felt over an area of some 6,000 square miles (15,540 km²), from Trenton to Baltimore, MARYLAND, and from the Philadelphia area to Cape May. Western central New Jersey experienced an earthquake on March 23, 1957, that caused considerable damage and alarm in the area of Hamden, Lebanon, and Long Valley. An earthquake on December 27, 1961, along the border of New Jersey and Pennsylvania was estimated at Mercalli intensity V and caused much fright in the affected area, which included Bristol and the northeast Philadelphia area, as well as Langhorne and Levittown. The December 10, 1968, earthquake in southern New Jersey broke windows and caused considerable alarm and was felt also in portions of Pennsylvania and DELAWARE; this earthquake was estimated at Richter magnitude 2.5, and effects of Mercalli intensity V were reported at some locations. Small earthquakes occur at the rate of several per year in New Jersey, mostly in the New Jersey Highlands and near the shore.

New Madrid Fault Zone *Missouri, United States*   The
New Madrid Fault Zone along the Mississippi River was the center of what are believed to be among the most powerful earthquakes in the history of the United States, in the winter of 1811–12. These earthquakes are believed to have had magnitudes between 8.4 and 8.8 on the RICHTER scale and were felt from New Orleans to CANADA and all along the Eastern Seaboard. Strongly affected were areas in the states of ILLINOIS, MISSOURI, ARKANSAS, TENNESSEE, and KENTUCKY. It was so powerful that pedestrians in Richmond, VIRGINIA, allegedly had trouble standing on their feet as the vibrations from the earthquake passed, and in Kentucky the earthquake was felt so strongly as to leave people feeling dizzy and disoriented. The New Madrid area was sparsely settled at the time, and casualties were few, compared with other, similar earthquakes in urban communities; only 11 men, women, and children were formally reported as lost in the New Madrid earthquakes, although the total is thought to have been greater.

Even in 1904, lakes formed by the 1811–12 New Madrid earthquake could be seen flooding forests in Varney River, Missouri. *(Courtesy of the USGS)*

Skeletons of cypress trees killed by submergence in Lake St. Francis, Arkansas, in 1904 resulting from the 1811–12 New Madrid earthquakes. *(Courtesy of the USGS)*

The earthquakes laid waste to some 3,000 to 5,000 square miles (7,770 to 12,950 km²) of land. The most spectacular effects were to the Mississippi River and its tributaries. LANDSLIDES fell from the bluffs and banks all along the river, ruining channels and producing large and damaging waves. The course of the river was so shifted in areas that swamps and lakes were formed in areas that were formerly forests. One hundred years later, these shifts and floodings were still plainly visible. LIQUEFACTION in the unconsolidated bank sediments formed MUD VOLCANOes in which the water was reportedly shot some 10–15 feet (3–5 m) in the air.

Detailed records of the earthquakes and their effects have been compiled from personal journals, letters, and newspaper accounts. The earthquakes caused the banks of the river to collapse into the water, carrying anglers with them. Submerged trees were dislodged from the riverbed and floated to the surface, and one report said the current in the river was actually reversed briefly because of elevation of the riverbed. There were numerous observations of strange behavior on the part of animals (both wild and domestic) before and during the earthquakes. Naturalist John James Audubon noted that the horse he was riding stopped and began to groan at the time of the earthquake. A naturalist visiting the New Madrid area indicated later that birds behaved abnormally. Waterfowl congregated on boats, and on land small birds settled on houses. In some cases, birds were even seen to alight on humans. Panthers, foxes, and wolves were said to have been seen next to wild deer.

Peculiar atmospheric phenomena appear to have accompanied the New Madrid earthquakes. One correspondent for the *Louisiana Gazette* mentioned that fog obscured the sky at the time of one shock and that as another shock occurred, observers noticed what looked like frost covering houses and fences. On close examination, however, the "frost" turned out to be a vapor that lacked the chill of frost. Another interesting phenomenon reported during the New Madrid earthquakes was EARTHQUAKE LIGHT in the form of either bright flashes of light or a dull glow in the sky.

The sociological and psychological effects of the New Madrid earthquakes have been reported in detail, specifically a religious revival that occurred among settlers in the area. Church membership increased dramatically in the weeks and months following the earthquakes, but apparently the revival was short-lived.

Another strong earthquake struck the New Madrid area again in January 1843. A report from Little Rock, Arkansas, indicates the shaking lasted for about one minute and caused windows and cupboards to rattle. Hundreds of residents of Memphis, Tennessee, ran into the streets when they felt the earthquake, and some deaths allegedly occurred at auction houses where people were crushed and trampled by crowds struggling to escape the buildings. The noise of the earthquake was compared to that of hundreds of heavily laden wagons moving down a street. In 1895, another earthquake centered near New Madrid was felt strongly in a zone about 100 miles (161 km) long and was felt as far away as GEORGIA and WEST VIRGINIA. The shocks lasted almost one minute, cracked walls, and shook down brick chimneys. As in 1811–12, a strange phenomenon resembling earthquake light was reported as a luminous "streak" that appeared in the sky just prior to the shock. Yet another earthquake, in 1968, startled residents but caused little damage.

If and when another earthquake comparable in magnitude to those of 1811–12 strikes the New Madrid Fault Zone, the results may well be disastrous for metropolitan areas in the Mississippi Valley, notably St. Louis, Missouri, and Memphis, Tennessee. Located on the edge of the New Madrid Fault Zone, Memphis may expect severe ground motion, and large areas of the city stand to suffer greatly from liquefaction, which also poses a threat to St. Louis in the event of a major earthquake there. As one U.S. Geological Survey report describes the situation in St. Louis: "Much of the old part of the city of St. Louis, and particularly the modern highway network, is built on uncontrolled fill. This fill is generally [found] in stream valleys, but there are many rubbish-filled pits in the old portion of the city. All this rubbish and fill is prone to large differential settlements in an earthquake."

The report adds that "parts of the downtown area are underlain by open, underground mines, where clay was mined long ago for making tile. Their locations are generally . . . not known."

Landslides are another danger facing St. Louis, the report points out: "Landslides [in the St. Louis region] are likely to be commonplace on many natural and highway slopes. The natural slopes most prone to landslides in uplands are thick, silt-rich bess [soils] on steep slopes. . . . Many highway cuts [will] fail and cause very serious problems to highway traffic." Even today, there are significant landslides in the cuts along Interstate Highways I–44, I–244, I–270, and U.S. 67. Some highway fills [material in the roadbed underlying the asphalt] would probably fail, especially on the lake sediments near the airport and on flood plains of the major rivers. The following is an eyewitness account of the event, reported by F. A. Sampson:

Accompanying the noise, the whole land was moved and waved like waves of the sea, violently enough to throw persons off their feet, the waves attaining a height of several feet, and at the highest point would burst, throwing up large volumes of sand, water and in some cases a black bituminous shale, these being thrown to a considerable height, the extreme statements being forty feet, and to the tops of trees.

With the explosions and bursting of the ground there were flashes, such as the result from the explosion of gas, or from the passage of the electric fluid from one cloud to another, but no burning flames; there were also sulphuretted gasses, which made the water unfit for use, and darkened the heavens, giving some the impression of its being steam, and so dense that no sunbeam could find its way through. With the bursting of the waves, large fissures were formed, some of which closed again immediately, while others were of various widths, as much as thirty feet, and of various lengths. These fissures were generally parallel to each other, nearly north and south but not all.

In some cases instead of fissures extending for a considerable distance there were circular chasms, from five to thirty feet in diameter, around which were left sand and bituminous shale, which latter would burn with a disagreeable sulphurous smell. . . .

No more startling change of scenery could well be imagined than that at old man Culberson's, who lived with his family in a bend of the Pemiscot River, ten miles below Little Prairie. There was about "an acre of ground" between his house and the river, and on it was his smoke-house and well. On the morning of the 16th, Mrs. Culberson went out to get some meat from the smoke-house, but no well or smoke-house was to be seen. Upon search, they were both found on the other side of the river. A fissure across the bend had been so large that the river flowed through it, and the great pressure on the isolated spot forced it to the opposite side of the river when the next earthwave occurred. . . .

The persons who experienced the shocks generally did not theorize as to the cause, but Bradbury found a man near the Lower Chicksaw Bluffs who gave his theory. It was that a comet, which had appeared a few months before, had two horns, over one of which the earth had rolled, and was then lodged between them. The shocks were occasioned by the attempts to surmount the other horn. If this should be accomplished all would be well; otherwise inevitable destruction of the world would follow.

New Mexico *United States* Central New Mexico has a long history of earthquakes, mostly localized but strong. Earthquakes in July and November 1906 affected an area of about 100,000 square miles (259,000 km²). New Mexico also has numerous volcanic landforms.

See also BANDELIER TUFF; RIO GRANDE RIFT; VALLES.

Newport-Inglewood Fault Zone *California, United States* A fault zone that directly underlies the LOS ANGELES, CALIFORNIA, area, the Newport-Inglewood Fault Zone appears to pose a serious threat to the city in the event of any major seismic events along the fault zone in the future.

New York *United States* New York State borders on the seismically active SAINT LAWRENCE VALLEY between the United States and CANADA, and earthquake activity is commonplace in upstate New York, including occasional powerful shocks. Even the New York City area, ordinarily not known for seismic activity, has undergone strong earthquakes over the past three centuries; on December 18, 1737, for example, an earthquake estimated at magnitude VII on the MERCALLI scale struck the New York City area. Other substantial earthquakes to hit the vicinity of New York City within historical times include: December 10, 1874 (Westchester and Rockland counties, New York, and Bergen County, NEW JERSEY; Mercalli intensity VI); August 10, 1884 (Bergen County, intensity VII); March 9, 1893 (magnitude V); June 8, 1916 (intensity IV–V); May 11, 1926 (New Rochelle; intensity V), and November 22, 1967 (Westchester County; intensity V). The 1884 earthquake was especially strong and affected a broad area of the northeastern United States, from Portsmouth, NEW HAMPSHIRE, and Burlington, VERMONT, to points as far south as Baltimore, MARYLAND, and Atlantic City, New Jersey. Maximum damage occurred in Amityville and Jamaica, New York, where large cracks formed in walls. The 1893 earthquake had dramatic effects on New York City itself; the EPICENTER was located between 10th and 8th streets, and the earthquake distressed animals at the city zoo and interfered with the playing of billiards.

The frequency of small to moderate earthquakes in New York State has made it a convenient natural laboratory for studying techniques of earthquake prediction, and some success has been reported in predicting seismic activity there.

New Zealand New Zealand has a complex geology and a long history of strong volcanic and earthquake activity. New Zealand and the Macquarie Islands to the southwest are believed to account for more than 1% of the seismic activity on Earth. One area of intense seismic activity is Fiordland on South Island.

Notable earthquakes in New Zealand's history have occurred in 1848 at Awatere, 1855 at Wellington, 1881

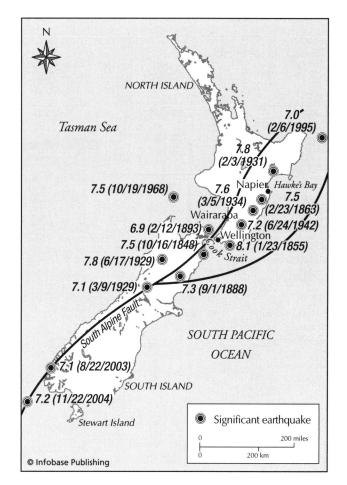

Map of New Zealand showing the locations of the major faults and the epicenters of the major historical earthquakes, including the magnitude and the date (in parentheses) of the event

at Amuri, 1901 at Cheviot, 1929 at Murchison, 1931 at Hawke's Bay, 1932 at Wairoa, 1942 at Wairarapa, and 1953 at Bay of Plenty. The Awatere earthquake in 1848 was felt over much of New Zealand and caused serious damage on both islands, although damage reportedly was worst around Cook Strait, notably in Wellington. The 1931 Hawke's Bay earthquake was even more severe and is thought to have had a RICHTER magnitude of 7.8. This earthquake stands out in the history of seismology because data from it indicates to geologists that Earth has an INNER CORE. North Island exhibits extensive volcanic activity. Its volcanoes include NGAURUHOE, RUAPEHU, TARAWERA, and White Island. WHITE TERRACE and Pink Terrace, two beautiful formations deposited by HYDROTHERMAL ACTIVITY, were destroyed by a volcanic eruption in the 19th century. New Zealand has GEOTHERMAL ENERGY facilities developed after World War II on North Island in a thermal area near Ngauruhoe and Ruapehu. This area features many manifestations of geothermal activity, including, but not limited to, GEYSERS, hot springs, and FUMAROLES.

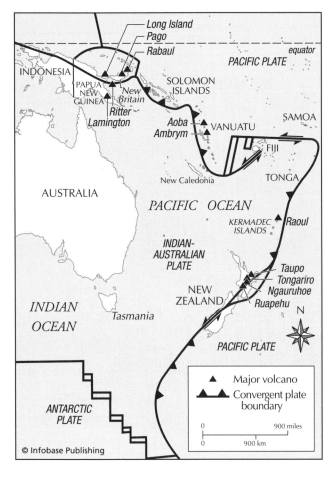

Map of the New Zealand–Australia area showing the plate tectonic relations and several of the major volcanoes

Nicaragua is also noted for having active volcanoes. The eruption of Cosigüina in 1835 was one of the most violent of the 19th century. A powerful volcanic eruption occurred in Nicaragua on November 14, 1867, when a mountain east of the city of Lion erupted for more than two weeks and cast out great amounts of black sand, which was carried as far as the Pacific shore some 50 miles (80 km) away. INCANDESCENT rocks were reportedly ejected from the crater to a height estimated at 3,000 feet (914 m). The city of Léon, the former capital of Nicaragua was earlier destroyed by the 1605–06 eruption of Momotombo. An eruption of yet another dangerous volcano, Masaya in 4550 B.C. was one of the largest on Earth in the last 10,000 years. Other volcanoes include Concepción, the spectacular Cerro Negro, La Madera, Las Pilas, San Cristobal, and Telica.

Seismic and volcanic activity in Nicaragua played an important part in formulating plans to build a canal through Central America to connect the Atlantic and Pacific Oceans. Historian Charles Morris explained in *The Volcano's Deadly*

The near-perfect stratovolcano of Ngauruhoe is one of the most active in New Zealand, with 61 eruptions since its discovery in 1839. *(Courtesy of NOAA)*

Ngaurahoe *volcano, North Island, New Zealand* The beautiful STRATOVOLCANO Ngauruhoe is NEW ZEALAND's most active volcano. It has erupted at least 61 times since 1839 and has produced LAVA FOUNTAINS, and minor *NUÉE ARDENTE*s on some occasions. The most recent eruption was in 1977. Ngauruhoe is adjacent to two other volcanoes, Tongariro and RUAPEHU.

Ngorongoro *volcano, Tanzania* Ngorongoro is part of a set of several large volcanoes in the LAND OF THE GIANT CRATERS, along Africa's GREAT RIFT VALLEY. The subsidence CRATER in the extinct Ngorongoro volcano is some 35 miles (56 km) in circumference and once was home to large herds of wild animals, including wildebeest, hippopotamus, rhinoceros, giraffe, impala, and elephant. Hunting reduced the wildlife population of Ngorongoro, but in 1959 a conservation area was established.

Nicaragua Nicaragua has a history of highly destructive earthquakes, notably one that destroyed much of MANAGUA on December 23, 1972. That earthquake was associated with Nicaragua's Tiscapa Fault.

A collapsed bridge after the June 16, 1964, earthquake in Niigata, Japan. The earthquake of magnitude 7.4 killed 26, injured 447, and damaged 9,750 structures in Niigata. Two piers on this bridge collapsed, and successive spans were pulled away from piers as the bridge was pulled apart. *(Courtesy of NOAA)*

Work, published in 1902, when deliberation on the route of the canal was continuing:

> There are three volcanoes in Lake Nicaragua itself, which body of water it was proposed to make the summit level of the projected canal on this line. Indeed, the evidence of geology is that Lake Nicaragua was once an arm of the Pacific, and that the central plateau of Nicaragua was formerly much nearer to the Caribbean coast than at present. The forces which effected so vast a change in the configuration of the land are still active. The eruption in 1835 of Cosigüina, which lies but sixty miles from the proposed Nicaragua canal route, was of extraordinary violence, So tremendous was this explosion, and so great was the storm of dust and ashes, that absolute darkness prevailed for thirty-five miles in every direction. . . . Cosigüina could have filled up ten times in one hour a canal prism which the contractors, with all their . . . labor-saving devices and the employment of tens of thousands of hands, would require eight years to excavate. . . .
>
> The danger of such convulsions at Panama is far less. We are told . . . that in Panama there is, within a distance of one hundred and eighty miles

from the canal, no volcano. The Isthmus there . . . lies in an "angle of stability," so called by seismographers. Except for rare and not very violent seismic vibrations, originating at distant centers, the Isthmus of Panama has never been affected by volcanic disturbances. One earthquake of some violence, indeed, has occurred there during the historic period, in 1621, when the greater part of Panama City was shaken down. Aside from this the most destructive earthquake known in the history of Panama was that of September 7, 1882. It lasted only a minute, but in that time shook down the courthouse and ruined the front of the old cathedral. Yet it may be affirmed that no paroxysmal convulsions have remodeled the geographical features of the Isthmus, as is the case with Nicaragua, and that its hills are nearly if not quite as stable as those of the Appalachian system. . . .

The canal route eventually was decided in favor of Panama.

Niigata *earthquake, Japan* On June 16, 1964, an earthquake of MAGNITUDE 7.4 occurred. It killed 26 people and moderately to severely damaged more than 9,750 houses. The Niigata is one of the best examples of LIQUEFACTION and resulting SLUMPING and unequal settling. About one-third of the city subsided some seven feet (2 m) as the result of compaction. The earthquake is famous for a group of apartment buildings that were built to be earthquake-proof. They withstood the severe shaking by the seismic waves, but the footings around their foundations could not withstand the severe liquefaction, and they fell over on their sides, still completely intact.

Nisyros   *See* NYSIROS.

Niuafo'ou Island *Tonga*   Niuafo'ou Island is also known as Tin Can Island because of the way in which mail used to be delivered to the island. Because the island had no harbor, mail would be sealed in a metal container and dropped over the side of a ship. A swimmer then would carry the can of mail to shore. This method is said to have been discontinued after a shark attacked a swimmer who was retrieving the mail. The island is located in the northern portion of the Tonga Archipelago and consists of a LAVA shield surmounted by a composite volcanic cone. A large lake, called simply Big Lake, occupies much of a CALDERA on the summit. A second, smaller lake known as Little Lake is also located in the caldera. Eruptions within historical times are recorded as early as 1814, when an explosive eruption apparently was accompanied by PHREATIC activity and LAVA FLOWS. This eruption occurred in the caldera itself. Details of a reported eruption in 1840 are unavailable. Eruptions in 1853 and 1867 involved FISSURES on the flanks of the volcano.

Observers of an eruption in 1886 left detailed accounts of the event, which was heralded by earthquakes and an apparent dramatic uplift in the seabed. The captain of a ship anchored at Niuafo'ou discovered, after an earthquake on August 12, that his ship was floating in only 48 feet (8 fathoms) of water, compared to some 120 feet (20 fathoms) the previous day. A very loud noise was heard (compared afterward to the sound of a heavy gun), and some kind of discharge occurred in the lake. Earthquakes stopped at this time. Some subsidence appears to have occurred around the caldera during this eruption.

A curious detail was noted in eruptions in 1919 and 1929: Coral was expelled along with lava, even though there appeared to be no recently formed coral on the island. Very little seismic activity was reported before the 1929 eruption. An eruption in 1943 required the evacuation of the island, and the inhabitants did not return for more than a decade. Another eruption in 1946 was preceded by a very strong earthquake just more than an hour before the eruption began. Considerable uplift was reported in this eruption. One account says that a reef on the northern side of the island was raised 30 to 40 feet (9 to 12 m) out of the water in places. Moderately strong earthquakes occurred in 1985.

normal fault   *See* FAULT.

North America   The North American continent is characterized by localized earthquake activity and active volcanism. According to the PLATE TECTONIC model and its dynamics, North America occupies a plate of Earth's CRUST that is moving slowly westward as new crust is formed along the MID-OCEAN RIDGE that lies in the middle of the ATLANTIC OCEAN. The ongoing collision between the North American mainland and the OCEANIC CRUST underlying the PACIFIC OCEAN has generated a complex geology along the Pacific coast of the continent. It is marked by major and minor FAULTS with associated seismic activity, although there are other earthquake-prone areas in North America. Much of North America's seismic activity is concentrated in the state of CALIFORNIA, where the SAN ANDREAS FAULT, which lies along the coast, has been associated with numerous strong earthquakes, such as FORT TEJON in 1857, that which destroyed much of SAN FRANCISCO in 1906, and the less powerful but also highly destructive SAN FERNANDO earthquake of 1971, the LOMA PRIETA earthquake of 1989, and NORTH RIDGE earthquake of 1994.

In both northern and southern California, but especially in the LOS ANGELES area, urban development has created a tremendous potential for destruction in the event of another major earthquake comparable to that of 1906. Similar dangers exist in other parts of the continent, such as the cities surrounding PUGET SOUND. Other areas of North America with a history of strong earthquake activity in the continental United States include New England, where earthquakes have caused widespread damage to Boston, MASSACHUSETTS, within historical times. In addition, LOUISIANA's Gulf coastal plain; upstate NEW YORK along the SAINT LAWRENCE VALLEY; the Appalachian Mountains, especially the southern portion of the chain, from VIRGINIA through ALABAMA; UTAH, which occupies part of a belt of pronounced earthquake activity extending northward from ARIZONA to the Canadian border; coastal SOUTH CAROLINA, site of the devastating CHARLESTON earthquake of 1886; and the Mississippi River valley in the vicinity of NEW MADRID, MISSOURI, where the most powerful earthquakes in the history of the continental United States are believed to have occurred during the winter of 1811–12. Virtually every part of the United States has experienced seismic activity at one time or another since settlement by Europeans began, although certain portions of North America, such as the stable rocks of the Canadian Shield, are all but free of substantial earthquakes.

The earthquake potential of western North America is generally better understood than that of the eastern portion of the continent, partly because earthquakes occur more often in the west and partly because active faults in the east tend to lie under thick layers of sediment that have hindered efforts to study the seismology. Earthquake activity is also intense in ALASKA, the site of one of the greatest earthquakes of modern times, the GOOD FRIDAY EARTHQUAKE of 1964, which was accompanied by a powerful TSUNAMI that caused great damage as far away as CRESCENT CITY, California. Earthquake activity in Alaska concentrates along the ALEUTIAN ISLANDS and extends into the central portion of the state.

MEXICO, too, has a long history of strong seismic activity. Earthquakes have caused widespread destruction in Mexico over the centuries, notably in the MEXICO CITY earthquake of 1985. Western Mexico shares with the western United States and CANADA a potential for devastating earthquakes, as smaller blocks of crust along the coast are squeezed between the westward-moving continent and the oceanic crust to the west. Especially susceptible to earthquakes are the northwest corner of Mexico, which is subjected to the same set of geological conditions that make southern California susceptible to major earthquakes, and the extreme southern Pacific shores of Mexico, where earthquake epicenters are clustered almost as densely as in California. The depth of earthquake foci differs greatly from one portion of North America to another. Earthquakes in Alaska, for example, extend much more deeply than those in southern California and NEVADA.

Volcanism has done much to shape the landforms of North America. The western states of the United States exhibit numerous signs of past and current volcanic activity. Great flows of BASALT in the northwestern states of the United States testify to past volcanism in the Columbia plateau. A SUBDUCTION ZONE off the coast of the Pacific Northwest of the United States and the Canadian province of British Columbia has given rise to the volcanoes of the CASCADE MOUNTAINS, including Mount HOOD, Mount BAKER, Mount RAINIER, LASSEN PEAK, and the recently active Mount SAINT HELENS, as well as the photogenic CALDERA of CRATER LAKE (MAZAMA) in Oregon. Volcanism in Mexico has a long and colorful history; numerous volcanic mountains stand near Mexico City and southward along the CENTRAL AMERICAN arc. The volcano PARICUTÍN formed in a spectacular eruption in a Mexican farmer's field earlier in this century. Others include El CHICHÓN, whose massive eruption in the early 1980s affected the weather patterns, JORULLO, and COLIMA. Alaskan volcanoes have erupted on many occasions in this century, sometimes with great violence, as in the Mount KATMAI-Novarupta eruption of 1912, which laid down a plain of FUMAROLES, the VALLEY OF TEN THOUSAND SMOKES. Alaskan volcanism is tied to a subduction zone along the Aleutian Island arc and includes many active volcanoes.

North American volcanism sometimes is extended to cover that of the Hawaiian Islands because of their inclusion in the United States but only for political reasons. Hawaiian volcanoes include KILAUEA and MAUNA LOA, known for their frequent but generally nonviolent and harmless eruptions. Tsunamis pose a particular threat to the Hawaiian Islands because of their position in mid-Pacific; the tsunami that struck HILO in 1946, for example, caused extensive damage and loss of life.

The presence of large amounts of still-hot magma near the surface in western North America has given rise to abundant geothermal activity and HYDROTHERMAL ACTIVITY in certain locations, such as YELLOWSTONE NATIONAL PARK, where GEYSERS have become tourist attractions. GEOTHERMAL ENERGY has been exploited to generate electricity on a large scale in northern California. There is also much evidence of past volcanism like that which resulted from the separation of North America and AFRICA. The WATCHUNG BASALTS of NEW JERSEY and Palisades Sill are a record of it.

North American crustal plate
The North American crustal plate generally underlies the continent of NORTH AMERICA and includes CANADA and the 48 contiguous UNITED STATES in addition to ALASKA and MEXICO. The North American plate is thought to be moving westward

relative to the PACIFIC CRUSTAL PLATE, with which it shares a boundary along the western shore of North America. This boundary between the North American and Pacific plate has been the location of some of the most famous and destructive earthquakes and volcanic eruptions in history. The Pacific and North American plates are not the only ones involved in generating seismic and volcanic activity there. The subduction of the oceanic JUAN DE FUCA CRUSTAL PLATE, for example, is the source for earthquakes and volcanism along the CASCADE MOUNTAINS of BRITISH COLUMBIA and the northwest United States, including portions of the states of CALIFORNIA, OREGON, and WASHINGTON. On the east, the North American plate is bounded by the MID-OCEAN RIDGE that traverses the middle of the ATLANTIC OCEAN. Here, new crust for the North American plate is created as molten rock rises from beneath and solidifies. To the southeast, the North American crustal plate adjoins the plate underlying the CARIBBEAN SEA, along the borders of which have occurred some of the most destructive eruptions and earthquakes in the recent history of the Western Hemisphere. The Caribbean plate is bounded to the north by a TRANSFORM FAULT and to the east by a SUBDUCTION ZONE. The small COCOS PLATE, off the Pacific coast of Central America, also shares a boundary with the North American crustal plate. The boundary is a subduction zone that yields the Central American arc.

See also PLATE TECTONICS.

North Carolina
United States North Carolina is located in a region of moderate seismic activity in the southeastern UNITED STATES. Although severe earthquakes have been rare in North Carolina's history, the state has a record of mild seismic activity extending back into colonial times. An earthquake on March 9, 1828, probably centered in VIRGINIA, was felt as a severe shock at Raleigh, North Carolina, and was associated with noises like thunder at Hillsborough. Another earthquake on April 29, 1852, again probably centered in Virginia, was felt in Raleigh, Greensboro, Hillsborough, and Milton, North Carolina.

A notable series of earthquakes occurred between February 10 and April 17, 1874, in McDowell County. Strong shocks were followed by a curious rumbling sound. The earthquakes shook buildings vigorously. Between 50 and 75 shocks are thought to have occurred altogether, each one associated with rumbling. On some occasions, a noise like artillery fire was also reported. The earthquakes affected an area approximately 25 miles (40 km) wide.

North Dakota
United States Although characterized by low seismic activity, North Dakota is affected from time to time by earthquakes in nearby states. Notable effects were reported in North Dakota, for example, from the MONTANA earthquake of 1959.

North Pagan
caldera, Mariana Islands Located on Pagan Island, North Pagan has erupted on several occasions since the late 17th century. Details of eruptions are few, but a large eruption appears to have occurred in 1872, and earthquakes reportedly accompanied other eruptions in 1917 and 1923. A huge eruption occurred in 1981 (VEI = 4) in which

A *nuée ardente* was suddenly produced from a lateral blast in the previously dormant Lamington volcano, Papua New Guinea in 1951. The hot gas cloud sped down the mountain at 65 miles (100 km) per hour, and killed some 3,000 people. *(Courtesy of NOAA)*

the eruptive column was 11 to 12 miles (18 to 20 km) high. The CALDERA appears to have been quiet until 1984, when a small eruption followed an earthquake. Another minor eruption occurred in 1987, and there is evidence of further eruptive activity in 1988.

North Ridge *earthquake, southern California*   On January 17, 1994, an earthquake of MAGNITUDE 6.7 occurred. Some 60 people were killed, more than 7,000 people were injured, and 20,000 were left homeless. More than 40,000 buildings were damaged or destroyed in LOS ANGELES, Orange, Ventura, and San Bernadino Counties. Services were disrupted in many areas, and collapsed overpasses closed several major freeways in the area. Total damage is estimated at $20 billion dollars. ROCKFALLS and LIQUEFACTION occurred throughout the affected area. The fault that is responsible for the earthquake is a BLIND FAULT that was previously undetected. Major AFTERSHOCKS continued for nearly one year in the Los Angeles area.

The famous points of destruction in this earthquake included the Northridge Fashion Mall, which was essentially destroyed. Many homes and apartment buildings had first-floor carports and garages; during the earthquake, these bottom floors collapsed and the buildings crushed down onto the cars underneath. The California State University parking structure was another spectacular sight as it bent over and collapsed in on itself. The I–10 and I–5 Freeway collapses and damage to the famous Fillmore Theater were also spectacular.

nuée ardente   A term from the French meaning "fiery cloud," *nuée ardente* refers to a violent and potentially highly destructive phenomenon that occurs when a VOLCANIC DOME collapses or a LATERAL BLAST occurs. A PYROCLASTIC FLOW emanates from the volcano along with an overlying cloud of ASH. The resulting cloud of ash, rock, and superheated

gas (800°C) flows along the ground at high velocity 40 to 70 miles per hour (64 to 113 km) and incinerates or melts virtually anything in its path. A *nuée ardente* can devastate the area around a volcano almost as effectively as a nuclear explosion. Perhaps the most dramatic example of a *nuée ardente* in the 20th century is the one that accompanied the eruption of Mount PELÉE in the CARIBBEAN SEA. Mount SAINT HELENS also produced a spectacular *nuée ardente*.

Nyamuragira *volcano, Zaire*   Nyamuragira is a SHIELD VOLCANO and has erupted frequently in the 20th century. Since 1882, it has erupted 34 times. Only an eruption in 1912–13 caused fatalities. The latest eruption was in 1994–95. A lava lake was active in the 1920s and 1930s.

Nyiragongo *volcano, Zaire*   A STRATOVOLCANO, Nyiragongo has erupted often in the late 19th and the 20th centuries. Some 100 people were killed by fast-moving 40 miles (64 km) per hour LAVA FLOWS from the volcano in 1977. In 1958, an expedition to the volcano succeeded in measuring surface temperatures within the lava lake of Nyiragongo. Measurements at the surface indicated temperatures between 1,000° and 1,200°C. The volcano was active in 1982 and again in 1994. Both produced FIRE FOUNTAINS and PHREATIC ERUPTIONS.

Nyos *toxic gas, Cameroon*   The village of Lower Nyos was overwhelmed on August 21, 1986, by a cloud of toxic gas that erupted from the bed of Lake Nyos and asphyxiated some 1,200 people in that community, plus some 500 others in the vicinity. Some 3,000 domestic animals also died. The cloud consisted largely of carbon dioxide that had accumulated at or near the bottom of the lake. This phenomenon is thought to be common in volcanic lakes, although most of those lakes undergo sufficient mixing (that is, overturn of warmer upper waters with colder, gas-laden bottom waters) that prevents excessive buildup of toxic gas in deep waters. It is not known exactly what caused the gases to rise from the bottom of Lake Nyos. According to one explanation, the lake had gone so long without overturning that toxic gas had built up to tremendous concentrations in bottom waters; then, when something finally initiated mixing, the asphyxiating gases rose to the surface and were released. An earthquake has been suggested as the possible agent of this overturn; and indeed, a rumbling noise reportedly was heard from the vicinity of Lake Nyos just before the catastrophe. It also seems possible that a minor volcanic eruption expelled the gas from the lake bed, but this hypothesis has not been confirmed.

Nysiros *caldera, Greece*   The volcanic island Nysiros lies in the Aegean Sea near TURKEY. The caldera of Nysiros is about 2.5 miles (4 km) wide and occupies what was the summit of an andesitic STRATOVOLCANO. The date of collapse is related to one of two explosive phases that occurred about 25,000 years ago. Several RHYOLITIC domes have formed inside the CALDERA. Unrest has been reported at the caldera on several occasions since the early 15th century. In 1422 there was a report of an eruption. Detonations and roaring noises, along with emissions of vapor and hot water, reportedly occurred at the caldera in 1830. In 1871 earthquakes were reported, accompanied by "flames" and detonations; ash and LAPILLI from this eruption are said to have wiped out gardens of fruit on the floor of the caldera. A CRATER some 20 feet (6 m) wide formed in 1873 following an earthquake; again on this occasion, lapilli and ASH were cast out from the volcano. A lake formed in the caldera from hot, salty water, and eruptions of dark mud lasted for several days. Also in 1873, the caldera ejected lapilli on many occasions, along with saltwater and mud. Eruptive activity had formed a crater approximately 75 feet (23 m) wide by the autumn of 1888. SOLFATARIC activity was reported in 1956.

O

O-Akan *See* AKAN.

oblique slip fault *See* FAULT.

obsidian A dark, glassy volcanic rock, obsidian is really a supercooled liquid with no crystalline structure. Although it is commonly black, obsidian is commonly of RHYOLITIC composition. If the quickly cooling LAVA gets drops of water in it, small alteration spots formed by the devitrification process, form white spots in the obsidian. This texture is called snowflake obsidian because the white spots look like snowflakes. Obsidian was used as material for arrowheads by Native Americans because it can hold a very sharp edge.

oceanic crust The CRUST that makes up the ocean floor. Oceanic crust has a distinct three-layer pattern. The bottom layer is composed of PERIDOTITE, which is really uppermost MANTLE rather than crust. The next layer is dominated by closely spaced sheeted GABBRO dikes that acted as feeders to the volcanoes. The top layer is composed of PILLOW basalts overlain by deep-sea sediments. They are mostly mud and chert but closer to the continents other sediments are possible. The three-layered sequence is called Steinman's Trinity and coined for pieces of ocean crust that had been broken and lifted onto CONTINENTAL CRUST through PLATE TECTONIC collisions. Ocean crust on continental crust is called an ophiolite.

Ocean crust is produced at the MID-OCEAN RIDGE. The splitting apart of the ocean ridge is filled in with MAFIC magma from the mantle. That forms the sheeted DIKES. The dikes feed underwater volcanoes that produce the top layer. Therefore, new crust is constantly being formed at the mid-ocean ridge. As the newly formed crust is in turn split, it is pulled in both directions away from the ridge. Therefore, the youngest oceanic crust is at the ridge, and it becomes progressively older as it is pulled away. It is like a conveyor belt. Because the ocean crust is eventually destroyed at a SUBDUCTION ZONE, it is all relatively young. The oldest ocean crust is 200 million years and the vast majority is much younger. In contrast, continental crust can be 3.4 billion years old.

offset The amount of movement on a FAULT. Offset is a quantitative measure of how far a fault has moved. It is used for all fault types. There can be total offset for how much a fault has ever moved or offset for the amount of movement during a single earthquake. Offset can be both in the vertical and horizontal sense. On a DIP-SLIP FAULT, the vertical offset is called the throw, and the horizontal offset is called the heave.

Ohio *United States* Ohio undergoes significant earthquakes on occasion. Several examples are listed here. The June 18, 1875, earthquake was felt strongly in Chicago, ILLINOIS, Columbus, and Cincinnati and was strongest at Urbana and Sicily, Illinois. On September 19, 1884, an earthquake near Columbus caused slight damage. This earthquake was felt at the top of the then unfinished Washington Monument in Washington, D.C. A September 20, 1931, earthquake at Anna caused minor damage but was felt at great distances, in ALABAMA, ARKANSAS, and TENNESSEE. In March 1937, a series of strong earthquakes was reported. An earthquake on March 2 near Anna and Sidney caused alarm and slight damage. On March 3, minor damage was reported from an earthquake that was felt at Anna, Botkins, Jackson Center, and Sidney. The March 8 earthquake in western Ohio caused minor damage and was felt on upper stories of high buildings in Chicago, Illinois; Milwaukee, WISCONSIN; and Toronto, CANADA.

Oklahoma *United States* Oklahoma is not one of the more seismically active states of the UNITED STATES, but strong earthquakes do affect the state on occasion. A strong shock on April 9, 1952, at El Reno, for example, was felt over an area of 140,000 square miles (362,598 km^2).

Okmok *caldera, Aleutian Islands, Alaska, United States* The Okmok CALDERA is located on Umnak Island and has two summit calderas overlapping each other. The Tulik volcano is located several miles from the caldera. Eruptions have been reported here since the early 19th century. A major explosive eruption in 1817 deposited several inches of ASH over the surrounding area, and large rocks were cast out, some landing more than 30 miles (48 km) away. Earthquake activity and a TSUNAMI were reported in 1878, although it is not certain how this activity may have been related to the Okmok caldera. Tulik emitted black smoke in 1931, and an eruption took place on the island several weeks later. The island was relatively quiet until 1944, when activity resumed. In 1945 an earthquake was felt at the army base on the island, and several days later pilots reported seeing an eruptive column. Later investigations showed that ash and a LAVA FLOW had emanated from a cinder cone in the caldera. Airborne observations revealed another lava flow in 1958. From 1960 to 1961, explosive eruptions occurred, and a minor ash eruption was observed in 1983. Several small eruptions of ash took place between 1986 and 1988.

Okushiri *earthquake, Japan*   On July 12, 1993, an earthquake of magnitude 7.8 occurred in the Sea of Japan just to the north of the small island of Okushiri. A mere two to five minutes after the earthquake, a TSUNAMI with vertical run-up of 50 to 100 feet (15 to 30 m) struck the island. The coastal areas were devastated by these enormous waves. The EPICENTER was too close for warnings and evacuations, so people were taken completely off guard. Some 120 people were killed, and there was more than $600 million in property damage.

Old Faithful *geyser, Yellowstone National Park, United States*   The GEYSER Old Faithful has erupted approximately once every 40 to 80 minutes since the 19th century. Water from the geyser reaches a height up to 150 feet (46 m). Each eruption is thought to release some 100,000 pounds of water.

olivine   A common mineral in MAFIC and ULTRAMAFIC igneous rocks. Olivine is a glassy green to brown mineral that in gem form is called peridot. It is a silicate mineral that is formed at very high temperatures (1,400°C). It is commonly found as crystals in BASALT. On one beach in HAWAII, the olivine has been preferentially weathered out of the basalt and makes up the sand on the beach. It is bright green. There are impressive olivine-rich layers in the Palisades Sill along the Hudson River in NEW JERSEY.

Olympus Mons *volcano, Mars*   The largest-known shield volcano in the solar system, Mars's Olympus Mons stands some 75,000 feet (22,860 m) high. Olympus Mons is one of several large volcanoes in the Tharsis region of the northern hemisphere. The Martian volcanoes appear to have played a major role in shaping the topography of that hemisphere in relatively recent times, although the existence of older volcanoes elsewhere on Mars indicates that volcanism has been active on Mars for much of the planet history. The major Martian volcanoes are larger by far than any on Earth. The evident absence of PLATE TECTONICS on Mars may account for the impressive size of its volcanoes. Unlike Earth, where

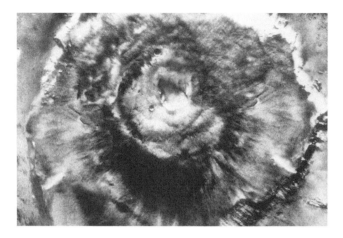

The Martian volcano Olympus Mons is one of the largest in the solar system. The volcano is more than 360 miles (600 km) across and 16 miles (27 km) high. *(Courtesy of NASA)*

horizontal motion of crustal plates can prevent a volcano from growing past a certain height because the volcano can move out from its source, Mars appears to have a fixed, comparatively immobile crust. This lack of movement gives a volcano time to reach tremendous size by terrestrial standards. Because the processes that create volcanoes along crustal plate margins on Earth are apparently lacking on Mars, it is believed that radionuclides underground may supply the heat required for volcanism there.

Onikobe *caldera, Japan*   The Onikobe volcanic center is located in the northern portion of the island of Honshū. The volcano Kurikoma stands several miles northeast of the rim of the CALDERA. The Narugo caldera and a group of VOLCANIC DOMES are situated just southeast of the Onikobe cal-

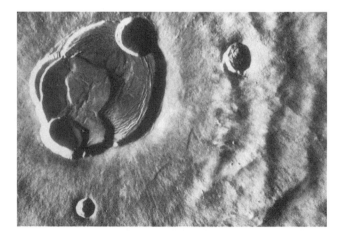

The summit of the Olympus Mons, Mars. It is composed of several large collapse features that indicate a long period of activity. There are also wrinkle ridges, and grabens suggest a complex tectonic history. *(Courtesy of NASA)*

dera. Although the Onikobe caldera itself appears to have been quiet in historical times, eruptions of Kurikoma were recorded in 1946 and 1950. These eruptions do not appear to have released large amounts of material, although a MUD-FLOW was associated with the 1950 eruption. Some unrest was reported at the volcano in 1957, but there is no indication of an actual eruption.

On-Take *volcano, Japan* The On-Take volcano is located on the island of Honshū, some 50 miles (80 km) northeast of Nagoya. The volcano rests within a CALDERA and has a summit caldera more than one mile (1.6 km) wide. The volcano's first eruption within historical times occurred in 1979 following several hours of seismic activity in the vicinity. A strong earthquake in 1984 caused extensive LANDSLIDES. FUMA-ROLES in the area of the 1979 vent appeared unaffected by the earthquake, but changes were noted in a fumarole on the summit of the volcano. Although it is not one of the more active volcanoes in JAPAN, On-Take has received particular attention because of its eruption in 1979. The eruption occurred after many hundreds of years of quiescence, and the eruption is thought to be associated with EARTHQUAKE SWARMS in the area between 1976 and 1984 and with the strong earthquake near the volcano in 1984.

Opala *volcano, Kamchatka, Russia* The Opala volcano has been generally quiet over the last few centuries, although an explosive eruption is reported for the year 1776. The CALDERA was formed between 31,000 and 39,000 years ago. About 1,500 years ago, explosions are believed to have formed a CRATER about one mile (1.6 km) wide on the average within the Opala caldera, on the northern rim of which the Opala volcano is located. Intense SOLFATARIC activity was reported in 1827, 1854, and 1984 but no actual eruptions.

Öræfajökull *volcano, Iceland* Although not one of ICE-LAND's more active volcanoes, Öræfajökull has erupted on two notable occasions within historical times. The first was in 1362, when TEPHRA and JÖKULHLAUPS caused considerable damage in the vicinity of the volcano. Powerful earthquakes are thought to have occurred just before this eruption. The next eruption occurred in 1727 and was apparently similar to the eruption of 1362.

Oran *earthquake, Algeria* On October 9, 1790, at 1:15 A.M., a major earthquake struck the northern part of Algeria. The epicenter was in the town of Oran, which was completely destroyed. The INTENSITY was estimated at X on the modified MERCALLI scale, and the maximum ground ACCEL-ERATION was calculated to have been 0.42 g. The recurrence interval for earthquakes of this size is estimated at 500 years. The source of the earthquake was the Oran Fault, which lies east-west across the area with THRUST and STRIKE-SLIP movement. The DEATH TOLL for this earthquake was estimated at 3,000 people.

Oregon *United States* Located in the seismically active Pacific Northwest of the UNITED STATES, Oregon nonetheless has not had a major earthquake since the territory was settled by Americans of European descent in the 19th century. This historical lack of seismicity does not mean that a powerful earthquake in Oregon is impossible. However, parts of Oregon would stand in risk of severe damage in any such event, especially where numerous buildings are constructed of unreinforced masonry, a kind of construction that is extremely vulnerable to damage from seismic events. Some notable earthquakes in Oregon history are listed below.

On February 3, 1892, a strong shock caused buildings to sway in Portland, and frightened residents of the city ran into the streets. The earthquake was also felt strongly at other locations, including Salem and Astoria. An earthquake at McMinnville on April 2, 1896, involved three shocks and awakened sleepers. The principal shock was felt also in Portland and in Salem. The earthquake of January 10, 1923, was felt in Lakeview and Klamath Falls and probably was very strong in an underpopulated area; minor damage occurred in CALIFORNIA.

Oregon shows abundant evidence of volcanic activity in the recent geologic past, notably the peaks of several volcanoes in the CASCADE MOUNTAINS, which extend north-south through the western and central portions of the state. Volcanism in Oregon is thought to be the result of the gradual and ongoing destruction of the JUAN DE FUCA CRUSTAL PLATE in the CASCADIA SUBDUCTION ZONE beneath the PACIFIC OCEAN immediately to the west of Oregon. As molten rock from the fusion of the descending crustal plate rises back to the surface along the Cascade range, the rising MAGMA carries with it a heavy load of dissolved gases thought to be derived from ocean water carried downward into Earth's MANTLE with sediments from the seafloor. As the magma approaches the surface and pressure on the molten rock is relieved, the gases pass out of solution and into gaseous form, thus giving the Cascade volcanoes the potential for highly explosive eruptions similar to that of Mount SAINT HELENS in WASH-INGTON in 1980. Volcanoes in Oregon include Mount HOOD, Three Sisters and CRATER LAKE, the water-filled CALDERA left from the collapse of Mount Mazama in prehistoric times.

origin time This is the exact time that seismic waves are initiated at a seismic source. This time is calculated from the arrival times of P-WAVES and S-WAVES at a SEISMOGRAPH. Knowing the velocities of these two waves, SEISMOLOGISTS backtrack the waves to their FOCUS to determine the exact second that the rupture occurred. This origin time is most commonly reported in universal time (UTC) rather than local time.

Osaka *earthquake, Japan* Two closely spaced submarine earthquakes struck Japan and caused great destruction in December 1854. The EPICENTER of the first tremor was off Totomi Sea, and the second was off Shikoku. The first of the earthquakes struck Osaka at 8:00 A.M. on December 23, 1854, and was felt strongly but produced only a moderate TSUNAMI. People were nonetheless nervous, and many decided to spend the night outside their homes or on boats. The second event struck at 5:00 P.M. on December 24, 1854, but the tsunami did not arrive for two hours. It arrived in the outer harbor at a height of 10 feet (3 m) but was 20 feet (6 m) high by the time it reached the port. It damaged or completely wrecked upward of 1,500 boats, some of which were swept along by the wave for a great distance, damaging at least 25 bridges. Between the SURFACE WAVE damage and the tsunamis, 60,000 houses were destroyed as well. It was fortunate that only about 3,000 people lost their lives because it could have been far worse. The first moderate event likely forewarned people sufficiently that they were better prepared for the second.

See also ANSEI TOKAI.

Öskjuvatn *See* ASKJA.

outer core Earth's core is composite. It contains a solid iron-nickel INNER CORE and a liquid iron-nickel outer core. We can tell that the outer core is liquid because S-WAVES will not pass through it. S-waves will not pass through a liquid. There is a shadow zone on the opposite side of Earth from the earthquake where no S-waves arrive. Only P-WAVES arrive. By comparing the records from many earthquakes, we can tell the exact size of the liquid core. The interaction of the liquid core with the solid core is the reason for our strong magnetic field.

Owens Valley *earthquake, California, United States* The Owens Valley earthquake on March 26, 1872, is believed to have been about 8.0 on the RICHTER scale and to have killed from 50 to 60 people. The Owens Valley Fault along the eastern face of the SIERRA NEVADA exhibited surface rupture for more than 100 miles (161 km). Vertical displacement in this earthquake was dramatic, especially near Independence and Lone Pine; total relative displacement along the fault exceeded 20 feet (6 m). The community of Lone Pine was virtually leveled, and 23 of its inhabitants were killed. The initial shock of the earthquake is said to have been the most powerful. It is said to have created a large wave in Owens Lake and dried up the Owens River for hours. Observers at one bridge saw fish thrown out of the water and onto the riverbank by the earthquake. Earthquakes continued for eight weeks. Witnesses to the earthquake included naturalist John Muir, who was asleep in his cabin near Sentinel Rock when the earthquake occurred. He ran outside to see what was happening and was forced to take shelter behind a tree as boulders dislodged by the shock rolled downhill to the floor of the valley. Muir improvised a primitive earthquake detector (a bucket of water set on a table) to indicate when shocks were occurring. The Owens Valley earthquake provided dramatic proof of how poorly adobe structures withstood earthquakes, notably in the devastated community of Lone Pine.

P

Pacaya *volcano, Guatemala* A volcanic complex with an older ANDESITIC and younger BASALTic STRATOVOLCANO. The complex appears to have formed in the last 23,000 years. It lies within the Cerro Grande–Pacaya–Cerro Chino volcanic complex. Pacaya has erupted 23 times since it was discovered in 1565. There is little information about early eruptions because of its remote location. Its current phase of activity began in 1965. Eruptions are highly variable and include STROMBOLIAN and Vulcanian explosions, minor summit flows from the CRATER, and larger flows from FLANK ERUPTIONS. It is rarely quiet and as a result has become a tourist attraction.

See also AMATITLÁN.

Pacific crustal plate The biggest plate of Earth's CRUST in terms of area, the Pacific plate underlies much of the PACIFIC OCEAN and is colliding with the Americas on the east and with the Asian landmass on the west. The boundaries of the Pacific crustal plate include much of the so-called RING OF FIRE, a belt of intense earthquake and volcanic activity surrounding the Pacific basin. Numerous SUBDUCTION ZONES, associated with volcanic activity, are found along the boundaries of the Pacific crustal plate, such as the subduction zone found along the ALEUTIAN ISLANDS arc in ALASKA. It is also bounded by several MID-OCEAN RIDGES especially to the east where it bounds the NAZCA, COCOS, and JUAN DE FUCA CRUSTAL PLATES. There are also many small plates in the western Pacific. Although the Pacific plate is almost entirely oceanic, it contains one notable piece of continental crust. The BAJA CALIFORNIA and all areas west of the SAN ANDREAS FAULT, including LOS ANGELES, California are actually not part of NORTH AMERICA but instead part of the Pacific plate. The direction of movement of the Pacific plate can be seen in the track of volcanoes produced by a HOT SPOT over which it passed. Currently, the hot spot underlies the HAWAIIAN ISLANDS, which show the plate to be moving northwest. However, the Hawaiian chain connects to the EMPEROR SEAMOUNTS, which show that the Pacific plate was

Small ash eruption from the volcano Pacaya in the Central American arc in Guatemala *(Courtesy of the CVO-USGS)*

moving nearly due north until it shifted abruptly to its new course about 42 million years ago.

See also PLATE TECTONICS.

Pacific Ocean The Pacific Ocean basin is surrounded by a ring of intense earthquake and volcanic activity known as the "RING OF FIRE," produced by ongoing collisions between the Pacific and related OCEANIC CRUSTAL plates and adjacent plates of Earth's crust. Other, smaller plates of crustal rock are also involved. HOT SPOTS of volcanic activity within the Pacific plate have produced some large volcanic islands, of which the HAWAIIAN ISLANDS are familiar examples.

See also PLATE TECTONICS.

Pacific Tsunami Warning System (PTWS) The rim of the Pacific Ocean is by far the most seismically and volcanically active zone on Earth. It is called the "RING OF FIRE" because of the huge number of volcanoes and earthquakes.

This activity has also made it more prone to TSUNAMIS than anywhere else on Earth. Over the millennia, residents of the Pacific rim have lived with constant threats and regular devastation. After the great CHILEAN 1960 EARTHQUAKE and the GOOD FRIDAY EARTHQUAKE of 1964, both of which produced killer tsunamis, an international group decided to take action. Thus was born the idea for an early warning system for the Pacific. Over the next few years, not only would the technology be perfected to sense tsunamis, but also coordination centers and a system of warning and evacuation would be established. The technology is the DART system of sensors, buoys, and transmitters as well as the Advanced National Seismic System (ANSS), National Earthquake Information Center (NEIC), and other seismic networks. There are automatic alerts issued to the centers for all seismic events with MAGNITUDES greater than 6.5 and warning bulletins issued for all events greater than 7.5 (7.0 for ALASKA). Local authorities are also notified in this latter case. The DART system is monitored closely at this point, and public warnings on radio and television are issued upon sighting of a tsunami. Evacuation routes are well marked and known by residents if necessary. The United States has three centers, the Pacific Tsunami Warning Center in Hawaii, the Alaska Tsunami Warning Center, and the West Coast Tsunami Warning Center in California. After the BANDA ACEH disaster, plans were made for an Indian Ocean Tsunami Warning System.

Pagan *volcano, Mariana Islands* The STRATOVOLCANO Pagan has erupted 19 times in history, on several occasions in the 20th century. Four minor eruptions occurred between 1909 and 1925. A powerful explosive eruption in 1981 produced a cloud of ASH that rose some 12 miles (19 km) above the volcano. Lesser eruptions have taken place since then in 1987, 1988, 1992, and 1993–94.

Pago   *See* WITORI.

pahoehoe   LAVA having a "ropy" surface, as distinct from the broken, blocky surface of AA lava. *Pahoehoe* is a Hawaiian word named for the lava that is very fluid and flows at high speeds. The ropy texture of the surface results from surface flow DEFORMATION coupled with surface tension.

paleoseismicity   Evidence of past earthquakes. Because seismic waves are ELASTIC, there is no direct record of them. Therefore, only the damage is left. Surface damage like fallen trees and broken buildings is typically not preserved in the geologic record. Preserved evidence might include LANDSLIDE deposits or intrusions of sand and mud into SOIL and sediments. The intrusions are from LIQUEFACTION and are distinctive. The landslides must be more carefully evaluated because there can be many reasons for them.

paleovolcanic   Evidence for past volcanoes. Old blast zones, LAHARS, and LANDSLIDES can indicate prior eruptions in addition to the physical evidence of volcanic flows and related deposits. However, as the volcano ages, discerning even that there was volcanic activity becomes more difficult. Much volcanic material is chemically unstable and weathers away quickly. ASH and PUMICE are so permeable that groundwater can flow through them unhindered. They are quickly

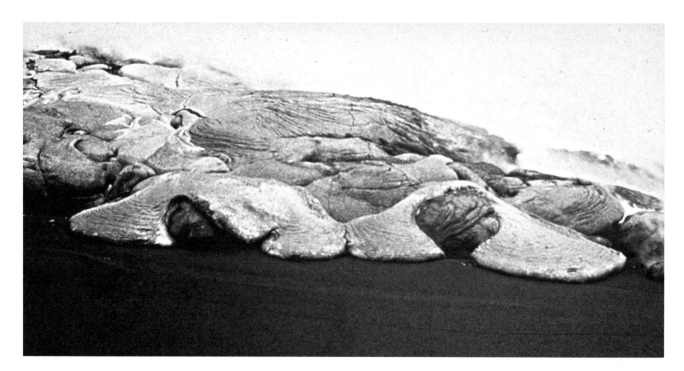

The toe of an advancing pahoehoe flow on the black sand at Kaena Point, Hawaii, on October 11, 1972. The sand is composed of particles of basalt. The toe of the flow advances by solidifying the tip, and the lava spills over from the back. *(Courtesy of NOAA)*

altered to clays and mud and washed away. VOLCANIC CONES are thereby destroyed and removed. The volcanic necks like that at SHIP ROCK remain behind for a while. The old LAVA FLOWS and feeder DIKES can last for millions to billions of years below ground. Therefore, depending on the age of the volcano, different features must be sought.

Paleozoic A geologic age ranging from about 540 to 245 million years ago. *Paleozoic* means "old life," so it covers the time when there was only primitive life on the planet. It begins with the first hard-shelled invertebrate animals in history. Through the Paleozoic Age, vertebrate fish and later amphibians also evolved. Plants and animals moved from ocean to land, and the barren soil and rock landscape was transformed to one of vegetation and activity. However, very few of the species that existed then are still around today. The supercontinent PANGAEA was built during this time through a series of continental collisions. These collisions produced vast volcanic provinces and FAULTS that record heavy earthquake activity. PLATE TECTONIC activity was intense during this time. The MID-OCEAN RIDGES were spreading so fast that they inflated and expanded. They displaced so much ocean water that it flooded the continents. Most of the continental United States was covered with a shallow sea throughout most of Paleozoic Age.

palisade A palisade is any clifflike, fencelike formation of rocks, of which there are many (Pacific Palisades in CALIFORNIA, for example). The cliffs along the Hudson River facing New York City are probably the most famous palisades for their great height and prominent columnar structure. They are the edge of a huge SILL of diabase (a type of GABBRO) that was formed 200 million years ago during the opening of the ATLANTIC OCEAN and breakup of the supercontinent of PANGAEA.

Palmdale *uplift, California, United States* Located approximately where the Garlock Fault and the SAN ANDREAS FAULT meet, Palmdale was the site of an uplift called the Palmdale bulge in the 1970s. The Palmdale bulge drew considerable attention as a potential generator of destructive earthquakes.

Pangaea A supercontinent that existed about 250 million years ago. Pangaea was constructed throughout the PALEOZOIC Age by colliding continents to build a single huge mass. In the Appalachians, there were at least three separate collisions known as the Taconian, Acadian, and Alleghanian Orogenies. There was the Ouachita Orogeny in the south and the Antler Orogeny in the southwest UNITED STATES. Worldwide, the Scandian Orogeny in Scandinavia, Caledonian in England, Hercynian and Variscan Orogenies in Europe, and Uralian Orogeny in Russia are examples of other Pangaean events. One continent meant one ocean, and it was called Panthalassa. Pangaea broke apart in the continents we see today beginning about 200 million years ago. If you notice how some continents appear to fit together like puzzle pieces, it is because they were once together when they were in Pangaea.

The Earth about 250 million years ago when all of the continents were joined forming the supercontinent Pangaea

Papandayan (**Papandajan**) *volcano, Java, Indonesia* Also known as Papandayang, and located in western Java, Papandayan is a STRATOVOLCANO. A 1772 eruption killed some 3,000 people and destroyed several dozen villages through explosions and LANDSLIDES. An eruption in 1883 accompanied the destruction of nearby KRAKATOA. Another eruption occurred in 1925 and again in 1942.

Paphlu *earthquake, Nepal* A major earthquake struck eastern Nepal at the town of Paphlu on August 26, 1833, at 11:56 P.M. The estimated RICHTER magnitude for the quake was between 7.5 and 7.9. The entire area of eastern Nepal and surrounding India was destroyed by this massive event. There were LANDSLIDES throughout the area and LIQUEFACTION, causing SAND BOILS and foundation failures. This earthquake has been compared to the BIHAR earthquake of 1934, which killed about 11,000 people. The Paphlu earthquake took the lives of less than 500 people. The reason for the relatively low death toll was that there were two strong FORESHOCKS several hours before the MAIN SHOCK. After the second event, the vast majority of the population decided to heed these warnings and evacuate their homes. The main shock devastated the region and would have cause the deaths of many thousands of people had it not been for these precursors and the quick thinking of the people.

Papua New Guinea The PLATE TECTONICS of Papua New Guinea are complex with a small MID-OCEAN RIDGE, a major TRANSFORM FAULT, and the large and several smaller SUBDUCTION ZONES that dominate area. The result is a highly active area with both seismic and volcanic activity on several islands of various sizes, including New Britain.

Some of the most famous volcanic activity in history has occurred in Papua New Guinea. The most notable eruption was from Mount LAMINGTON in 1951 which had a VEI of 4 and killed some 3,000 people. The eruption of RABAUL in 1937 caused 441 fatalities from PYROCLASTIC FLOWS. Ritter Island produced an AVALANCHE during its 1888 eruption that spawned a 39–49-foot (12–15-m)-high TSUNAMI that killed more than 3,000 people. The eruption of LONG ISLAND in 1660 was comparable in intensity to the eruption of KRAKATOA in 1883. KARKAR, Pago, and MANAM, among others, have also been active in historical times.

parasitic cone   A conical buildup of material around a VENT that erupts near the base or on the flank of an existing volcano. A parasitic cone may grow to become a large volcano itself.

parental magma   MAGMA that chemically and isotopically reflects its source in the MANTLE. It is uncontaminated by melted crustal rock that it encountered during ascent to the surface or a MAGMA CHAMBER. Typically, the earliest LAVA from a volcanic vent is the parental magma. In ISLAND ARCS, early lava is typically MAFIC (BASALT) with a distinctive composition. Later lavas are INTERMEDIATE, starting with ANDESITE and finally DACITE as the amount of crustal contamination increases with time. In the ANDES, some lavas are so contaminated that they produce primary metamorphic minerals (andalusite). Generally, the thicker the CRUST that the magma must pass through, the more contaminated the lava. The speed of ascent, however, is also an important factor.

Paricutín *volcano, Mexico*   In one of the most spectacular eruptions of the 20th century, the volcano Paricutín arose beginning in February 1943 from a farmer's field in Michoacan, MEXICO, in the Sierra Occidental on the western edge of the Central Plateau. There are stories of the farmer plowing his field when the eruption began and his hat, which had been on the ground, being shot up into the air. The volcano

Summit eruption and eruption column of Paricutín from Tipacua, Mexico, October 30, 1947. The lake in the foreground was formed because a lava flow from the eruption crossed a river and dammed it. The tree shows that the flooding was recent. *(Courtesy of the USGS)*

attained a height of about 1,200–1,300 feet (366–396 m) before the eruptions ceased in 1952. It has been quiet ever since. There was a particularly violent eruption on June 10, 1944.

passive margin   In contrast to an active margin, where volcanoes and earthquakes are commonplace, a passive margin is quiet and relatively safe. The term *active margin* refers to a plate tectonic setting of active convergence (SUBDUCTION ZONE) or divergence. A passive margin is not a plate margin but an ocean margin (coast) in which the plate margin is somewhere else. An example is the Atlantic margin of the United States. The plate margin is more than 1,000 miles (1,610 km) to the east at the Mid-Atlantic Ridge. There are no volcanoes and few damaging earthquakes along the east coast, thus it is passive.

Patan *earthquake, Pakistan*   Pakistan has a long history of devastating earthquakes. One such event struck a 70-mile (112-km) stretch of the Indus Valley and especially the town of Patan (EPICENTER) on December 28, 1974, at 12:12 in the early afternoon. The quake had a RICHTER magnitude of 6.3

A spectacular nighttime eruption of Paricutín volcano in Mexico. *(Courtesy of the USGS)*

and a modified MERCALLI damage of VII. The FOCUS was at a depth of 12 miles (20 km). Although this was not a terribly strong earthquake, it struck a region that had used very poor construction methods. Mud and brick houses collapsed, crushing inhabitants, and ROCKFALLS after the earthquake crushed many houses that remained. The death toll was nearly 5,300, with more than 15,000 injuries. Helicopters were used to evacuate survivors and air-drop supplies to towns that were otherwise cut off. Those who could not be brought to safety were forced to sleep outside in subzero temperatures. The total cost of the event was estimated at $3.25 million.

Pavlof *volcano, Alaska, United States* The STRATO-VOLCANO Pavlof is located on the Alaskan Peninsula, next to the volcano Pavlof's Sister. Pavlof is probably the most consistently active volcano in the ALEUTIAN ISLANDS. It has undergone at least 40 eruptions since 1790. Most of the eruptions are STROMBOLIAN in nature. There is evidence that one of the two volcanoes (probably Pavlof's Sister) was active between 1762–86. A very strong earthquake occurred at about the time of an eruption in 1844, although it is not known what precise relationship the earthquake may have had to the eruption. A FISSURE on the side of Pavlof erupted for a single day in 1846. An unusually strong earthquake struck the area again in 1866, several months after an eruption in March 1866. Pavlof's biggest eruption in historical times occurred in December 1911. Earthquakes accompanied this eruption. LAVA flowed from a fissure on the north flank, and large blocks were cast out of the volcano. Earthquakes attributed to explosions became more numerous at Pavlof in July 1983. An eruption preceded by several weeks of earthquakes may have occurred on November 14, 1983, when an aircraft pilot reported sighting a plume, and a glow was seen. Earthquake activity declined after this event and remained at a generally low level. In 1986 Pavlof entered a new period of eruption. LAVA FLOWS have occurred in 1846, 1958, 1960–63, 1966, and during the 1970s and 1980s. Lava may also have flowed from the volcano between 1936 and 1948. Pavlof was active in 1996–97 and produced 25,000-foot (7,620-m) plumes and lava flows. It is constantly monitored.

pegmatite A very coarse-grained PLUTONIC rock that is most commonly FELSIC in composition. *Pegmatite* is really a textural term meaning very coarse-grained. However, pegmatites never form large PLUTONS and by far most commonly form DIKES. Pegmatites form from water-rich melts. The water is so abundant that it inhibits nucleation of new

The snow-covered volcano Mount Pavlov at the western end of the Alaskan Peninsula *(Courtesy of NOAA)*

crystal grains. Therefore, the grains that do form have a huge space to grow without bumping into other grains. Grains are typically six inches (15 cm) long or more, but there are pegmatites where the grains are up to 39 feet (12 m) long. Water-rich melts are typically the last liquid left over in a plutonic event. The water concentrates because there is no place for it in the minerals as they crystallize. Many other elements also fall into this category and are termed incompatible elements because they don't readily combine with the common elements in minerals. They can therefore be found in high concentrations in pegmatites. Elements such as lithium, boron, gold, and uranium are concentrated in pegmatites. Some of them can combine with regular elements to form odd minerals. Such minerals include beryl, garnet, emerald, tourmaline, indicolite, and other gem stones. That is why pegmatites are common exploration targets.

Pelée, Mount *volcano, Martinique* The volcano Pelée's eruption on May 8, 1902, was one of the most destructive in the history of the CARIBBEAN SEA. The eruption produced a *NUÉE ARDENTE* that wiped out the city of Saint Pierre and is thought to have killed more than 29,000 people; only six individuals from the city survived. The first signs of eruption occurred late in April. A cloud of smoke began to rise from the volcano on April 23, accompanied by occasional emissions of ASH. On May 5, the eruption intensified and sent LAVA and mud spilling over the rim of the CRATER and down the valley of the River Blanche; a sugar mill was destroyed, and more than 20 people were killed.

Many firsthand accounts of the destruction of Saint Pierre and vicinity were recorded, despite the virtually complete loss of life on shore. Some of those reports are worth quoting at length, for they convey perhaps more clearly than any other documentation of volcanic eruptions the power and destructive effect of such events. One of the most comprehensive and vivid accounts of the catastrophe comes from Comte de Fitz-James, a French traveler, who with his companion Baron Fontenilliat witnessed the destruction of Saint Pierre from the relative safety of a boat in the harbor. He is quoted in Charles Morris's 1902 book, *The Volcano's Deadly Work*:

> From the depths of the earth came rumblings, an awful music which cannot be described. I called my companion's name, and my voice echoed back at me from a score of angles. All the air was filled with the acrid vapors that had belched from the mouth of the volcano
>
> From a boat in the roadstead . . . I witnessed the cataclysm that came upon the city. We saw the shipping destroyed by a breath of fire. We saw the cable ship *Grappler* keel over under the whirlwind, and sink as though drawn down into the waters of the harbor by some force from below. The *Roraima* was overcome and burned at anchor. The *Roddam* . . . was able to escape like a stricken moth which crawls from a flame that has burned its wings
>
> When we got ashore we called aloud, and only the echo of our voices answered us. Our fear was great, but we did not know which way to turn, and had it been our one thought to escape we would not have known how to do so. It was about one o'clock in the afternoon when we reached shore. Our weariness was beyond description. Sleep was the one thing that I wanted, but I overcame the desire and, with Baron de Fontenilliat, set off to make our way to St. Pierre, hoping that we might still render some assistance to the injured. . . .
>
> We saw great stones that seemed to be marvels of strength, but when touched with the toe of a boot they crumbled into impalpable dust. I picked up a bar of iron. It was about an inch and a half thick and three feet long. It had been manufactured square and then twisted so as to give it greater strength. The fire that came down from Mont Pelée had taken from the iron all of its strength and had left it so that when I twisted it, it fell into filaments, like so much broom straw. . . .
>
> I know that the explosion of Mont Pelée was not accompanied by anything like an earthquake, for . . . when we entered St. Pierre we found the fountains all flowing, just as though nothing had happened. They continued to flow, and are flowing still, unless destroyed by the later explosions.
>
> There was no flow of lava. It was all ashes, dust, gas and mud.

The sole survivor of the catastrophe at Saint Pierre was a murderer named Joseph Surtout. He survived because he was locked in a cell so far underground that it protected him from the worst effects of the explosion. He was trapped in his cell for four days after the eruption without food and water, though somehow he got enough air to survive. Because his cell was windowless, he had no way of knowing exactly what had happened. He called for help and eventually was rescued. Workers disposing of bodies found many curious sights among the devastation left by the catastrophe. The charred body of a woman was found holding a silk handkerchief unburned to her mouth. Carbonized bodies were found with their shoes undamaged. Severely burned bodies lay adjacent to others that the fire appeared to have touched only slightly. Articles such as purses were discovered virtually intact. The May eruption of Pelée destroyed approximately eight square miles (21 km^2). The area of destruction was enlarged slightly by another eruption on August 30. Yet another eruption occurred in December.

In October, a LAVA DOME, or great mass of solidified lava, arose in the caldera and grew to a height of some 800 feet (244 m) by the December 1. This formation was called the Tower of Pelée and stood more than 1,000 feet (305 m) tall at its maximum height. Later activity at the volcano shook down the Tower of Pelée, however, and only a small portion of the dome remained by the end of 1903. Mount Pelée became active again in 1929; this eruption was less intense than that of 1902 but lasted more than three years. On this occasion, Saint Pierre and other communities on the island were evacuated. The eruption lasted until 1932.

As fluid lava is shot out of a volcano, it is stretched and strung out. The drops leading the stretched out lava are Pele's tears and the threads of stretched lava that trail behind are called Pele's hair. *(Courtesy of the USGS)*

Pele's hair *See* PELE'S TEARS.

Pele's tears Specific shapes of LAPILLI from Hawaiian volcanoes. As pure molten rock is shot to great heights above an active volcano, it forms droplets. These droplets fall back to the earth like rain. However, they are very hot molten. They form raindrop shapes, but because the air is so cool, they solidify into volcanic glass. While still molten they fall through the air dragging tails of glass strands that lag behind the drops. The drops are called Pele's tears because they look like teardrops, and the strands are called Pele's hair. Pele is the Hawaiian goddess of the volcano.

Pelileo *earthquake, Ecuador* An earthquake on August 5, 1949, was among the deadliest in the history of ECUADOR. The EPICENTER of the quake was located in the village of Pelileo, and the FOCUS was at a depth of 25 miles (40 km). The earthquake was caused by SUBDUCTION of the NAZCA CRUSTAL PLATE beneath SOUTH AMERICA. The RICHTER mag-

nitude of the earthquake was 7.5. The quake affected an area of over 1,500 square miles (3,900 km²).

The DEATH TOLL from this earthquake was over 6,000 people, with more than 20,000 injured. Besides Pelileo, five towns virtually disappeared. A huge FISSURE about one mile long swallowed the community of La Libertad. During rescue and relief operations, Ecuador had to divert some of its troops to protect the impacted area from Salasaca Indians, who attacked relief columns and looted villages. Some 100,000 people were left homeless, and damage to the Ecuadorian infrastructure was over $66 million. The enormous price tag destabilized the Ecuadorian economy.

Pennsylvania *United States* Pennsylvania is located in a region of minor seismic risk but has undergone strong earthquakes on occasion. Very strong shocks were reported at Philadelphia on March 17 and November 29, 1800, and again on November 11 and 14, 1840; this latter earthquake was accompanied by a large swell on the Delaware River. There was a sizeable earthquake in DELAWARE in 1974 that affected Pennsylvania. There is also a region where regular small earthquakes occur called the Lancaster seismic zone.

peridotite A plutonic ULTRAMAFIC rock, peridotite is a general name for a whole group of ultramafic rocks. They are exceedingly high in iron and magnesium. These rocks are composed only of FERROMAGNESIAN minerals such as PYROXENE, OLIVINE, and, less common, HORNBLENDE. The group includes rocks such as dunite, lherzolite, websterite, and wehrlite. Locally, these rocks can contain significant quantities of chromite and platinum, in addition to other strategic minerals.

period oscillations There is a natural period of a building based on its dimensions and characteristics. There is also a natural period of earthquake waves. If the two match, build-

Ruins of a church at Santa Rosa as the result of the Pelileo, Ecuador, earthquake of 1949 *(Courtesy of the USGS)*

ings can swing wildly or oscillate. This swinging can cause severe structural damage to the building. The real problem happens when two adjacent buildings undergo period oscillations. They can swing into or "hammer" each other repeatedly, thereby destroying both buildings. Several famous examples of these occurred during the 1985 MEXICO CITY earthquake.

Peru   Located on the Pacific shore of SOUTH AMERICA, Peru suffers from earthquakes as a result of the ongoing collision of the South American crustal plate with the NAZCA CRUSTAL PLATE to its west. Earthquakes in Peru and ECUADOR on August 13–15, 1863, are believed to have killed some 25,000 people. Another major earthquake in Peru in 1892 killed an estimated 25,000. The Lima earthquake on May 25, 1940, is believed to have killed 200 to 300 and caused some 5,000 injuries in the port of Callao. The ANCASH earthquake of November 10–13, 1946, caused some 700 fatalities. More than 100 were reported killed in the Cuzco earthquake on May 21, 1950. Strong earthquakes in Peru, BOLIVIA, and CHILE on January 13, 1960, killed more than 2,000 people and caused hundreds of millions of dollars in damage. An earthquake on May 30, 1970, at Yungay killed more than 66,000 and left more than a half-million homeless after the inundation of a resort. It was the most devastating earthquake in South America. On May 31, 1970, in NEVADOS HUASCARÁN, Peru, a tremendous LANDSLIDE killed 18,000 people. An unusual case of a landslide apparently causing an earthquake, rather than vice versa, was observed on April 25, 1974, along the Mantaro River in Peru. A great landslide there produced seismic effects comparable to those of an earthquake of about 4.5 on the RICHTER scale of magnitude.

The STRATOVOLCANO El Misti is located in Peru near the city of Arequipa. The volcano is famed for its beauty more than for its eruptions; in historical times, El Misti has been characterized by mere steaming and occasional minor explosive eruptions. A new volcano, Sabancaya, formed in 1986–88.

petrology   The study of rocks. Subdivisions of petrology deal with IGNEOUS, SEDIMENTARY, and METAMORPHIC ROCKS. Petrology deals with the full classification of rocks, including texture and chemical composition. It also deals with the processes of formation.

phaneritic   A term describing IGNEOUS ROCKS whose individual minerals are large enough to see with the naked eye. Basically, these are rocks that were formed in PLUTONS, where they cooled slowly and had enough time to form large crystals. Contrast this texture with the small grains found in volcanic rocks making up an APHANITIC texture.

Philippine Islands   Volcanic activity has been frequent and destructive in the Philippine Islands within historical times. The stratovolcano MAYON has undergone explosive eruptions on more than 40 occasions since 1616 and has generated both LAVA FLOWS and *NUÉE ARDENTES*. The volcano TAAL, located in the central Philippines, has erupted on more than 30 occasions since 1572, and its eruptions are

Massive destruction of adobe houses in the town of Huarez during the 1970 earthquake, by far Peru's worst natural disaster *(Courtesy of the USGS)*

sometimes accompanied by TSUNAMIS in the lake from which the volcano rises. The island of LUZON has many volcanic mountains including Bulusan, which has erupted 13 times between 1886 and 1988. The volcano Malaspina is located on the island of Negros, and the volcano Camaguin erupted on an island approximately 90 miles (145 km) southeast of Negros in 1876. *Nuée ardentes* from an eruption of HIBOK-HIBOK in 1861 killed 326 people there. In 1831, Babuyon Claro volcano produced a large eruption (VEI = 4). Volcanoes on the island of Mindanao include Ragang, Apo, and Cottobato. Hot springs and SULFUR deposits in the Philippines are further evidence of volcanic and HYDROTHERMAL ACTIVITY there. The 1991 eruption of Mount PINATUBO was one of the most powerful eruptions of the 20th century and forced the UNITED STATES to abandon an air base near the volcano. The eruption was well covered by scientists and the media. The impact on human life was minimized by this coverage because evacuations were timely. Unfortunately, a typhoon swept in right after the eruption and turned the fresh ASH into raging torrents of boiling mud, causing many fatalities.

The Philippine Islands are also vulnerable to earthquakes, of which perhaps the most famous occurred on July 3, 1863, in Manila. The earthquake destroyed many government buildings, hospitals, and churches and is thought to have killed some 1,000 persons. Damage to the tobacco industry was estimated at more than $2 million. Another earthquake in Manila in 1880 caused widespread destruction but reportedly no loss of life. More recent eruptions at Luzon and MORO GULF were also destructive.

Phlegraean Fields *volcanic area, Italy*   This volcanic area, several miles west of Naples and VESUVIUS, has a high concentration of FUMAROLES and SOLFATARAS. Indeed, Solfatara, one of the volcanoes located in the Phlegraean Fields, provided the name for that phenomenon. Nineteen CRATERS occur within an area of about 25 square miles (65 km²) here. Classical literature is full of references to the Phlegraean Fields, including a mention in Virgil's *Aeneid* and they are said to have provided imagery for the *Inferno* of Dante.

The field is made up of explosion craters from PHREATIC ERUPTIONS, and cinder cones inside a collapsed CALDERA some eight miles (13 km) wide. The caldera is thought to have formed in prehistoric times either in or following the eruption of vast volumes of TUFF. A large portion of the volcanism at the Phlegraean Fields is thought to have occurred between 3,000 and 5,000 years ago.

The migration of VENTS at the volcanic field has followed a pattern: Vents have tended to migrate from the rim of the caldera toward the center and to diminish in the volumes of their eruptions. Another interesting pattern is that early activity at the Phlegraean Fields appears to have taken place underwater, whereas later activity moved onto the land. The patterns of activity at the Phlegraean Fields indicate that a reservoir of MAGMA near the surface is crystallizing slowly and causing activity at the surface to move inward toward the center of the caldera as crystallization proceeds underground. The magma reservoir is believed to be within two–four miles (3–6 km) of the surface. The boundaries of the volcanic field, however, are indistinct. The Phlegraean Fields are noted for their history of dramatic uplift and subsidence. The motions may be slow by everyday human standards, but on the geological time scale they are swift.

A case in point is the marketplace at Pozzuoli, which once was underwater, as shown by the borings of mollusks in the ancient columns. The marketplace sank by perhaps 30 feet (9 m) or more by the year A.D. 1000 and then rose again by about six feet (1.8 m) between 1000 and 1198 when the volcano Solfatara erupted. (Outstanding examples of uplift and subsidence are also recorded at other points in the vicinity. One pier built in the second century A.D. reportedly dropped almost 20 feet (6 m) by the 18th century.) Following the eruption in 1198, uplift continued. The uplift was accompanied by earthquakes, including a strong tectonic earthquake under the Campanian Apennines that occurred in 1456 and apparently caused severe damage to Pozzuoli. Other powerful local earthquakes shook the area around Pozzuoli in 1488. The rate of uplift increased around 1500. A considerable amount of new shoreline had emerged from the waters by 1503 and was taken over by a local school. Between 1000 and 1503, the rate of uplift was perhaps an inch per year on the average, for a total of approximately 36 feet (11 m). Most of that uplift took place after 1500, as the rate of uplift increased to about seven inches per year.

Earthquakes occurred frequently in 1534 and between 1536 and 1538 and apparently originated along the southern and eastern edge of the caldera. (A colorful story is associated with one of these earthquakes, in 1534. The earthquake occurred during a worship service just before Easter, as the gospel account was read of the earthquake that took place at the resurrection of Christ. This coincidence made a tremendous impression on the churchgoers.) Fumaroles became more active between 1536 and 1538 in the vicinity of Sudatoio, slightly to the west of Pozzuoli.

In this area, a new volcano, Monte Nuovo, was about to appear. Numerous earthquakes preceded this eruption, as did a spectacular display of uplift. On September 26–27, 1538, the ground at Pozzuoli rose perhaps 15 feet (5 m), and the shoreline retreated some 1,200 feet (366 m). Water started to gush from FISSURES on September 28, and subsidence lowered ground level by approximately 12 feet (3.7 m) the following day. On September 29, uplift resumed near what would be the site of the Monte Nuovo eruption. The eruption began in the evening with great noise, and great quantities of ASH and PUMICE mixed with water emerged from the ground. Evidently uplift had caused a considerable retreat of the shoreline because one account of the eruption mentions townspeople carrying fish that they had picked up along the shore. New springs of water, one hot and the other cold, also reportedly appeared at this time. Fire was reported from this eruption and is thought to have been burning gas from fumaroles. An explosive eruption near the community of Trepergule cast out pumice. Eruptive activity lasted five days. On October 6, 24 persons who had climbed the cone of the volcano were killed in an explosion and in PYROCLASTIC surges. This outburst knocked down trees three miles (4.8 km) from the volcano. After the 1538 eruption, subsidence proceeded at an average rate of less than an inch per year, with only a short period of uplift between 1951 and 1952.

Several earthquakes in the late 16th century took place around Pozzuoli. Another strong earthquake occurred in 1832 and appears to have been centered near Monte Nuovo. Later though weaker earthquakes occurred between 1887 and 1892. Several earthquakes in the early 20th century are thought to have occurred around the borders of the caldera. A mud volcano at Solfatara erupted with great energy after an earthquake about 30 miles (48 km) east of the Phlegraean Fields in 1930. In 1969, uplift resumed in the caldera, with accompanying mild earthquake activity. Seismic activity and uplift diminished in 1972. A strong earthquake about 60 miles (97 km) southeast of the Phlegraean Fields in 1980 preceded a renewal of uplift. Earthquake activity started to increase in 1983. A measuring device at Pozzuoli showed uplift of approximately four feet (1.2 m) between mid-1982 and late 1984 and approximately 10 feet (3 m) between 1969 and 1985. Earthquakes released much more energy during this later period of uplift than in 1969 to 1972. During the 1982–84 episode, much of the earthquake activity was focused around the Solfatara volcano and around a location immediately north of Pozzuoli. Uplift had ceased by the end of 1984, and a very gradual deflation began in January 1985.

Uplift, volcanism, and earthquake activity at the Phlegraean Fields are thought to reflect the presence of a small, shallow MAGMA CHAMBER a couple of miles below the surface. The on-and-off pattern of uplift at the Phlegraean Fields, along with the history of phreatic activity there, also indicates that groundwater heated by a subterranean body of magma plays an important role there.

phreatic eruption Generally speaking, a nonincandescent explosive volcanic eruption. A phreatic eruption involves mud, steam, or other nonincandescent material and results from rapid heating of GROUNDWATER by igneous material

Hydra explosion clouds are emitted from a phreatic explosion on Hawaii on June 15, 1960. The clouds are pure white because they are completely steam (water) and devoid of ash. *(Courtesy of the USGS)*

underground to the flash point. The rapid expansion of water from liquid to gaseous state fuels the explosion.

pillow lava   A distinctive form of LAVA FLOW in which the solidified rock forms rounded masses with a glassy exterior. Pillow lava is found where LAVA flows into the ocean or a lake and quickly cools. The liquid lava forms a sphere when it enters the cold water as a result of surface tension. Drop hot wax into water and the same phenomena will occur. These hot balls of quickly solidifying lava settle to the ground quickly. As they settle, they flatten and become pillowlike in form. As more and more pillows settle to the seafloor, the weight flattens them even more. Pillow lavas therefore form piles around the vent or entry point of a lava flow. They make up much of the OCEANIC CRUST. Pillow lava also is found in TABLE MOUNTAINS, curious volcanic formations that originated in ICELAND where eruptions occurred beneath thick layers of glacial ice. Herdubreid is an example of such a mountain.

Pinatubo, Mount   *Philippine Islands*   The June 1991 eruption of Mount Pinatubo, after more than 600 years of dormancy, killed at least 847 people (mostly as the result of the typhoon that closely followed) and forced the abandonment

of the United States's Clark Air Force base in the PHILIPPINES. More than 1 million people were left homeless, and damage was estimated at several hundred million dollars. The eruption was closely monitored by scientists, who evacuated

Pillow lavas on the slope of Hawaii from a submarine eruption *(Courtesy of NOAA)*

most areas prior to the main eruption or the DEATH TOLL would have been much higher. Working from Clark, the team watched as the eruption grew; they barely escaped with their lives as the peak of eruption overwhelmed the base. The eruption column was more than 19 miles (30 km) high and some 11 miles (18 km) wide at its base. More than 172 billion cubic feet (5 billion m³) of ASH and EJECTA were shot into the atmosphere. PYROCLASTIC FLOWS and LAHARS devastated the agricultural heartland of Central LUZON. In the hardest hit provinces, more than 212,500 acres (86,000 ha) of farmland and fish farms were destroyed, forcing more than 650,000 people out of business. The devastation of the eruption was severely compounded by a major typhoon that struck the island just days after the peak of the eruption. Heavy rains mobilized the newly fallen and unconsolidated and very hot ash into rivers of mud. Much of the death and destruction was caused by the added effects of the typhoon. Even today, the thick layers of ash blanket and consequently conserve the heat at depth. Any particularly heavy erosional event, such as another typhoon, will excavate this hot material and again produce a boiling mudflow even after all of these years.

The eruption cast between 15 million and 20 million tons of SULFUR dioxide into the atmosphere, where the sulfur dioxide combined with water to form an estimated 30 million to 40 million tons of sulfuric acid particles. The climatic effects of the Mount Pinatubo eruption, combined with those of a lesser eruption of CHILE's Mount Hudson volcano, are believed to have contributed to global cooling the following year. In the lower and middle atmosphere, temperature measurements reportedly showed a change of almost 1°F between 1991 and 1992. The cooling effect might have been greater but for the warming effect of El Niño, a periodic increase in temperature in the PACIFIC OCEAN that influences weather.

Mount Pinatubo experienced a minor eruption again in 1994 for about four hours, but other than that, it has been quiet ever since.

pipe   A volcanic pipe is a conduit for MAGMA to produce SUMMIT ERUPTIONS in a volcano. When the pipe is clear, large summit eruptions occur. If the pipe is clogged with sticky magma or hardened LAVA, lateral eruptions are possible. The magma in the pipe hardens and forms a VOLCANIC NECK. It remains behind after the volcano is eroded away. DEVILS TOWER is such a feature.

Piton de la Fournaise *volcano, Réunion Island*   An active SHIELD VOLCANO, Piton de la Fournaise has a summit CALDERA surrounded by a set of curving FAULTS. As a rule, Piton de la Fournaise does not exhibit dramatic precursory phenomena such as strong earthquakes before eruptions. Nonetheless,

The eruption column from the 1991 Mount Pinatubo eruption as seen from the Clark Air Force Base *(Courtesy of the USGS)*

Ashfall deposits buried this straw hut at the base of Mount Pinatubo after the 1991 eruption. Note the odd landscape in the background resulting from erosion caused by the typhoon the following day. *(Courtesy of the CVO-USGS)*

The mudflow generated by the typhoon that struck the Philippines the day after the eruption of Mount Pinatubo was far more disastrous than the eruption itself. *(Courtesy of the USGS)*

eruptions do tend to follow minor but notable changes, such as small earthquakes and the formation of FISSURES in the caldera. Eruptions were recorded in 1708 and 1800 and every several years thereafter up to the late 1980s. Most eruptions have been explosive with associated LAVA FLOWS. Piton de la Fournaise is thought to be fed by BASALTic MAGMA. Geologists believe that the magma is stored in a central reservoir near the surface before erupting through the summit and through RIFTS on the volcano's flanks. Apparently, inflation occurs at the summit for some weeks or months before an eruption. Reinflation, however, does not appear to start soon after each new eruption. Piton de la Fournaise is presumed to be a HOT SPOT volcano like Hawaii. A volcanic observatory has been set up there.

plagioclase   Probably the most common igneous mineral, plagioclase is a type of FELDSPAR, a SILICATE with a framework structure. At high temperature, plagioclase is calcium rich. It is common in BASALT and GABBRO. At low temperatures, plagioclase is sodium-rich. it occurs in GRANITE and RHYOLITE. In INTERMEDIATE rocks, the plagioclase is half-calcium and half-sodium. It is most common in intermediate rocks.

plastic deformation   *See* DUCTILE.

plateau basalt   *See* FLOOD BASALTS.

plate tectonics   One of the foundations of modern geology, the theory of plate tectonics has been described as "the glue that holds geology together." It describes the interactions among several large, and a number of smaller, rigid plates of Earth's LITHOSPHERE. The lithosphere is composed of Earth's CRUST, which forms the top layer adhered to a bottom layer of rigid MANTLE rock. The interactions among the plates are complex, but for the purpose of understanding earthquakes and volcanic eruptions, plate tectonic theory may be described as follows. Lithospheric plates float, so to speak, atop the denser rocks of the ASTHENOSPHERE (soft upper mantle) by a process known as ISOSTASY. These plates are not stable but instead move around the planet. Thermally induced CONVECTION CURRENT in the mantle drive this motion or continental drift.

The plates interact with each other along their margins. There are three different types of margins: CONVERGENT where plates are pushed together, DIVERGENT where plates are pulled apart, and TRANSFORM where plates slide laterally past each other. There are three different kinds of convergent margins, depending upon the types of crust involved. They are ocean-ocean, ocean-continent, and continent-continent collisions. The ocean-ocean and ocean-continent collisions form SUBDUCTION ZONES in which the downgoing ocean crust is consumed in the mantle. In ocean-continent collisions, also called Andean margins after the type example, the ocean crust is subducted and melted to form a MAGMATIC ARC. Magmatic arcs are seismically and volcanically active and very dangerous. Ocean-ocean collisions form ISLAND ARCS, where ocean crust is subducted beneath other ocean crust in a deep ocean trench. Examples such as the ALEUTIAN ISLANDS are equally

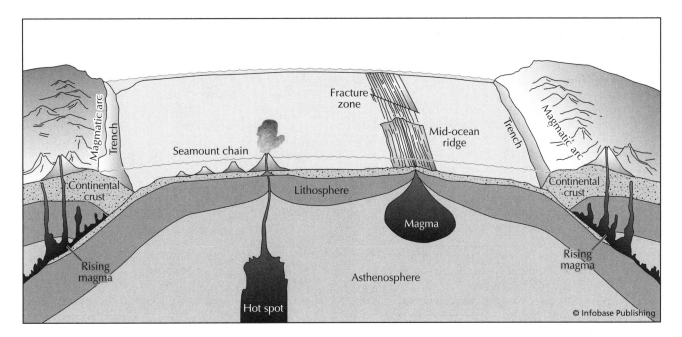

Diagram of plate boundary types, magmatic arcs above subduction zones (convergent boundary), and a mid-ocean ridge comprising a divergent boundary. Fracture zones form transform boundaries. A hot spot is also shown.

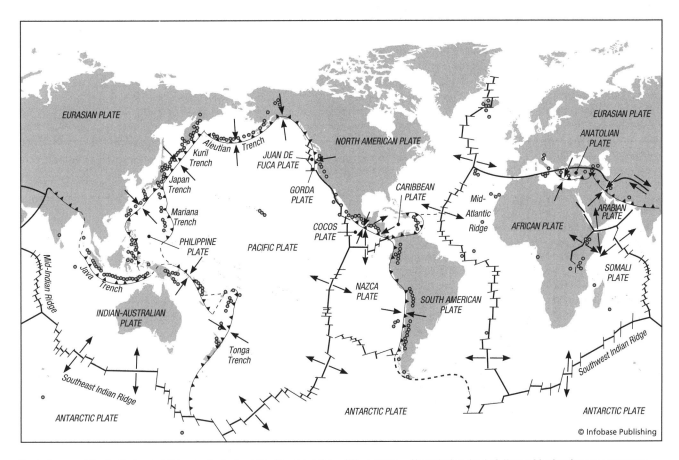

Tectonic map of the Earth showing the major tectonic plates. There are three different types of boundaries: the dark lines with triangles are convergent boundaries (subduction zones), the double lines are mid-ocean ridges, and the single lines are transform faults. *(Courtesy of the USGS)*

active and dangerous. Continent-continent collisions are more like car accidents: One plate plows under the other, but most of the deformation is taken up in crushing and elevating the rock. This scenario is called a Himalaya margin after the type example. There is intense seismic activity at the margins but no volcanic activity.

DIVERGENT BOUNDARIES form more in stages. On continental crust, the margin begins with uplift followed by normal faulting and GRABEN formation. Seismic activity is common in this phase. Next, extensive volcanic activity occurs including FLOOD BASALTS and possible RHYOLITE volcanoes as in the BASIN AND RANGE PROVINCE. The crust is then pulled apart, and ocean crust forms between, making a narrow ocean basin like the Red Sea. Finally, a full ocean basin with a MID-OCEAN RIDGE forms including extensive seismicity and volcanism.

Transform margins are large STRIKE-SLIP FAULTS that separate lithospheric plates. Transform margins must connect other types of plate margins. Therefore, there are divergent-divergent, divergent-convergent, and convergent-convergent transform margins. About 99% of transform margins are the divergent-divergent type. They form small offsets all along every mid-ocean ridge of various sizes and are almost exclusively on ocean crust. A major exception to this rule is the SAN ANDREAS FAULT of CALIFORNIA, which is on continental crust. Like the San Andreas Fault, these other transform faults are also highly seismically active. Regardless of the type of transform margin, they are all similar in their seismic activity and general lack of volcanic activity. Larger transform faults are called FRACTURE zones. Other examples of transform faults include the South Alpine Fault in NEW ZEALAND and the MOTAGUA FAULT in GUATEMALA, both of which have produced devastating earthquakes.

New ocean crust is formed along mid-ocean ridges, where molten rock rising from the asthenosphere solidifies. The newly formed rock moves outward to either side of the ridge in a pattern much like that of objects on a moving conveyer belt. Volcanic activity is therefore commonplace along the ridges and may build up whole islands, of which ICELAND is a spectacular example. As much ocean crust is destroyed in subduction zones as is produced at the mid-ocean ridges to preserve the size of Earth. As the subducted plate descends into the asthenosphere, the plate melts, and its lighter components rise back toward the surface as molten rock, or magma. This magma finds its way to the surface through openings in the crust and produces volcanoes. (Not all volcanic activity is linked to subduction zones; HOT SPOTS, or points where ascending plumes of molten rock are thought to approach Earth's surface, may be responsible for volcanic activity, as in the case of the HAWAIIAN ISLANDS.)

The source of energy for the movement of crustal plates has been a subject of debate since plate tectonics was developed from the earlier theory of continental drift. In recent years, it has been widely believed that much of the energy behind the motions of crustal plates is derived from convection cells, or circulation patterns produced within the mantle by convective activity.

See also CONVECTION CURRENT; EARTH, INTERNAL STRUCTURE OF; WEGENER, ALFRED.

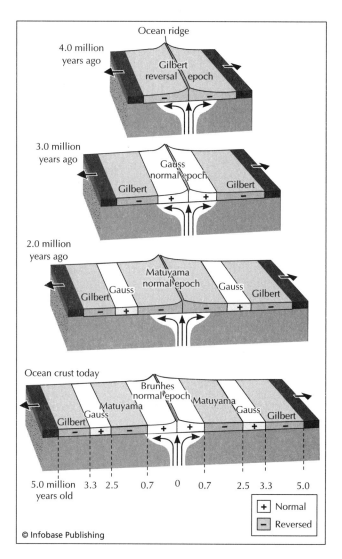

Production of new ocean crust at a mid-ocean ridge during a reversing magnetic field of the Earth. The stripes on the ocean floor show normal and reversed magnetic polarity over the past 5 million years.

Plinian eruption A type of volcanic eruption characterized by incredible vertical explosions that shoot gas, ASH, and EJECTA many miles into the atmosphere. These eruptions are the most violent of the VESUVIAN events. They are commonly the final phase in an eruptive cycle after the pipe has been cleared by earlier eruptive types. They have VEIs of 3–8.

plug A volcanic neck.

pluton A body of intrusive, typically coarse-grained IGNEOUS ROCK. Plutons intrude forcibly or passively into the preexisting COUNTRY ROCK of an area and cool underground. Through slow cooling, the minerals grow large and form a CRYSTALLINE, phaneritic texture. Tabular DIKES and SILLS are less circular than STOCKS and BATHOLITHS. Laccoliths and lopoliths are subvolcanic intrusions. They are typically forcible intrusions in contrast to tabular intrusions which are typically passive.

plutonism The formation of PLUTONS and related processes.

Poás *volcano, Costa Rica* Located northeast of the communities of Naranjo, Grecia, and San Pedro, Poás volcano has arisen within a pair of nested CALDERAS and is suspected of having a lake rich in SULFUR, underlain by a layer of molten sulfur, in its main CRATER. Molten sulfur has sometimes been expelled in eruptions of Poás. Most of its eruptions, which have been recorded on numerous occasions in the 19th and 20th centuries, have been small and either magmatic or PHRE-ATIC in character. It has erupted at least 39 times since 1828, but the activity has been nearly continuous. Notable eruptions in 1828, 1834, and 1880 deposited ASH on nearby communities. Other significant eruptions occurred in 1910 and 1953. The 1910 event began with the eruption of a GEYSER, and a plume rose to an altitude of more than four miles (6.4 km). Mud covered the whole top portion of the volcano. Numerous strong earthquakes occurred later that year, including one especially powerful earthquake that caused widespread damage in nearby communities. A 1953 eruption eliminated a crater lake and dropped ash 30 miles (48 km) or more away from the volcano. An ERUPTION column rose about two miles (3 km) over the volcano in 1976, and geysers erupted in 1979, 1980, and 1988. The crater lake gradually drained in 1989 and was replaced by a six-foot (1.8-m)-deep pool of liquid sulfur. This was the first observation of a sulfur lake on Earth, although liquid sulfur is common on Io, Jupiter's nearest moon. Summit FUMAROLE activity continued through 1995.

Pompeii and Herculaneum *volcanoes, Italy* Two of the most famous cities in the history of volcanology, the Roman communities of Pompeii and Herculaneum were destroyed in the eruption of VESUVIUS in A.D. 79. The destruction of the two cities has provided a popular theme for authors, notably Edward Bulwer-Lytton in his novel *The Last Days of Pompeii.* Bulwer-Lytton was apparently influenced by the historian Dion Cassius, who reported that the fallout from Vesuvius buried the people of Pompeii as they sat in the theater. This melodramatic scenario appears to have little basis in fact, and it is thought that Dion Cassius was merely drawing on traditional (and exaggerated) accounts of the catastrophe. Other of Dion Cassius's statements sound improbable, to say the least, especially his report that a great number of giants appeared in the vicinity of the volcano during the eruption. An eyewitness account of the event was made by Pliny the Younger (*see* Appendix B).

Excavations at Pompeii have revealed much about life in that coastal city some 2,000 years ago. One visitor, quoted in Charles Morris's study of volcanoes, *The Volcano's Deadly Work,* commented on "the weirdness of the scene" at Pompeii and the remarkable preservation of details of life there:

We have before us the narrow lanes, paved with tufa, in which Roman wagon wheels have worn deep ruts. We cross streets on stepping-stones which sandaled feet long ago polished. We see the wine shops with empty jars, counters stained with liquor, stone mills where the wheat was ground, and the very ovens in which bread was baked more than eigh-

teen centuries ago. "Welcome" is offered us at one silent, broken doorway; at another we are warned to "Beware of the dog!" The painted figures, some of them so artistic and rich in colors that pictures of them are disbelieved, the mosaic pavements, the empty fountains . . . the marble pillars and the small gardens are there just as the owners left them. Some of the walls are scribbled over by the small boys of Pompeii in strange characters which mock modern erudition. In places we read the advertisements of gladiatorial shows, never to come off, the names of candidates for legislative office who were never to sit. There is nothing like this elsewhere.

The streets of Pompeii must have had a charm unapproached by those of any city now in existence. The stores, indeed, were wretched little dens. Two or three of them commonly occupied the front of a house on either side of the entrance, the ostium; but when the door lay open, as was usually the case, a passerby could look into the atrium, prettily decorated and hung with rich stuffs. The sunshine entered through an aperture in the roof.

As the life of the Pompeiians was all outdoors, their pretty homes stood open always. There was indeed a curtain betwixt the atrium and the penstyle, but it was drawn only when the master gave a banquet. Thus a wayfarer in the street could see, beyond the hall described and its busy servants, the white columns of the peristyle, with creepers trained around them, flowers all around, and jets of water playing through pipes which are still in place. In many cases the garden itself could be observed between the pillars of the further gallery, and rich paintings on the wall beyond that.

But how far removed those little palaces of Pompeii were from our notion of well-being is scarcely to be understood by one who has not seen them. It is a

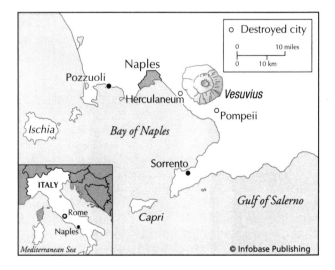

Map of the Mount Vesuvius (Naples) area of Italy showing the relation of the volcano to the two destroyed cities of Pompeii and Herculaneum.

question strange in all points of view where the family slept in the houses, nearly all of which had no second story. In the most graceful villas the three to five sleeping chambers round the atrium and four round the peristyle were rather ornamental cupboards than aught else. One did not differ from another, and if these were devoted to the household the slaves, male and female, must have slept on the floor outside. The master, his family and his guest used these small, dark rooms which were apparently without such common luxuries as we expect in the humblest home. All their furniture could hardly have been more than a bed and a footstool ... The kitchen of each villa certainly was not furnished with such ingenuity, expense or thought as the stories of Roman gourmandising would have led us to expect. In [one] house ... the cook seems to have been employed in frying eggs at the moment when increasing danger put him to flight. His range, four partitions of brick, was very small; a knife, a strainer, a pan lay by the fire just as they fell from the slave's hand. Pompeii was buried by falling tephra, whereas Herculaneum was covered by mudflows.

Popocatépetl *volcano, Mexico* The word *Popocatépetl* means "smoking mountain" and was given to the volcano by the Aztec. The STRATOVOLCANO stands south of MEXICO CITY and has undergone explosive eruptions, most of them small, on more than 36 occasions in recorded history, 15 since the arrival of the Spanish in 1519. There was a major eruption in 1720. Popocatépetl was one of the first volcanoes into which European explorers actually made a descent, to collect SULFUR for gunpowder. The most recent eruption was from 1995 to 1999 and forced the evacuation of 75,000 people. It started to erupt again in the spring of 2000. Some 20 million people live close enough to the volcano to be threatened by eruptions.

porphyritic A textural term for volcanic rocks meaning that there are two grain-size populations. Large grains (usually one-eighth to one-half inch [2 mm–1 cm]) of a mineral form perfect crystals floating in a fine-grained matrix. This texture reflects a two-stage history. The large crystals were formed while the MAGMA was cooling in the MAGMA CHAMBER. Because they were surrounded by pure liquid, they could grow without bumping into other grains. That is why they form crystals. The minerals are varieties that form at higher temperatures than the rest of the rock. Otherwise, they would also be crystallizing as well. Later, the crystals and liquid are erupted from a volcano during an active phase. The liquid cools quickly during the eruption producing very small mineral grains and/or glass to fill between the crystals. In BASALT, the large crystals could be OLIVINE or PLAGIOCLASE and the matrix is PYROXENE and plagioclase. In ANDESITES and DACITES, the large crystals are plagioclase and possibly HORNBLENDE, pyroxene, or BIOTITE. The fine minerals are QUARTZ and plagioclase. In RHYOLITE, the large grains are K-FELDSPAR and possibly quartz or biotite. The matrix is quartz, K-feldspar, and plagioclase. These relations are determined by the crystallization sequence as shown by BOWENS REACTION SERIES.

Port Royal *earthquake, Jamaica* The earthquake that hit Port Royal on June 7, 1692, struck about 11:43 A.M. (In 1959, divers brought up the remains of a watch that apparently stopped at that moment.) The earthquake caused the waterfront built on sandy soil to fall into the sea. A tavern, a warehouse, and other buildings slid seaward and crumbled into the waters. Approximately half the town was wiped out.

Precambrian All time from the beginning of Earth 4.6 billion years ago until the beginning of the PALEOZOIC Age about 535 million years ago. The Precambrian can be subdivided into the early Archean and the later PROTEROZOIC. The division between the two was 2.5 billion years ago. *Billion years* is written *Ga* in scientific terminology and it means "giga annum," where *annum* means "years."

precursor Any number of changes in the geological conditions that foreshadows the coming of an earthquake on a FAULT. These changes can include several small earthquakes, local uplift or subsidence along a fault, changes in water levels in wells, and increased radon emissions, among others. Most times it is difficult to recognize and properly evaluate these precursors otherwise earthquakes would not be as deadly.

pressure ridge Common topographic ridges resulting both in earthquakes and volcanoes. The ridges are formed through lateral transverse shortening of the land surface. In volcanic flows, the crust on the top of the flow can fold into a ridge as the result of pressure generated from flowing subsurface LAVA.

primary effects The direct hazards from a volcanic eruption. These include LAVA FLOWS, *NUÉE ARDENTES*, LAHARS, explosions, ASH fallout, and any other direct effects. In contrast, secondary effects might include flooding from dammed rivers from lava flows, polluted water, ruined crops and famine, disease-resulting pollution, interrupted transportation routes and services, and any other hazard that is not directly caused by the eruption.

Proterozoic The later part of Precambrian.

pseudotachylite Rock within a FAULT that is melted by movement during an earthquake. It is then hardened in the fault. In shallow rocks near the surface, there is a lot of friction on a fault during an earthquake. The friction can heat up the rock and melt it. The newly formed liquid fills in the fault plane or cracks around the fault. The solidified melt is pseudotachylite. Tachylite is volcanic glass and *pseudo* means "false."

Puerto Rico Puerto Rico experiences numerous earthquakes, but highly destructive ones are rare. However, the CARIBBEAN SEA, in which Puerto Rico is located, is an area of high-intensity seismic and volcanic activity.

On October 11, 1918, an earthquake of RICHTER magnitude 7.5, with MERCALLI intensities in the range of VIII–IX, caused property damage estimated at about $4 million and killed more than 100 people. The earthquake was centered in the northwestern Mona Passage and was accompanied by a TSUNAMI that caused widespread destruction and loss of life.

The sea withdrew before the tsunami's arrival, then returned, and reached heights possibly greater than 19 feet (5.8 m). A church in Aguada was destroyed, and all brick buildings were ruined in Anasco. Structures of wood or reinforced concrete reportedly came through the earthquake with little damage where building materials were sound. At Aguadilla, the earthquake caused serious damage to buildings constructed on alluvium. The collapse of a factory at Mayagüez killed several people. Damage to railways, bridges, pipelines, and chimneys was widespread, and two cable links in the Mona Passage were broken. Strong AFTERSHOCKS followed the October 11 earthquake on October 24 and November 12.

Puget Sound *Washington, United States* An inlet of the PACIFIC OCEAN between the Olympic Peninsula to the west and the mainland to the east, Puget Sound is an area of intense seismic activity.

See also CASCADIA SUBDUCTION ZONE; CASCADE MOUNTAINS; WASHINGTON.

pumice An extremely lightweight, "foamy" volcanic rock, pumice forms when MAGMA bearing a high concentration of dissolved gases solidifies as glass just as the gases are bubbling out of solution. It forms on top of a magma or LAVA pool just as foams such as whipped cream would. Pumice is cast out in large quantities during the eruptions of some volcanoes and is so lightweight that it may float on water. In eruptions of certain volcanoes on islands or near seacoasts, pumice may accumulate on the water and form great floating "rafts" capable of supporting a man's weight. Such rafts of pumice may impede navigation. They are also the vehicle by which certain species of plants and animals migrate from island to island across the PACIFIC OCEAN.

Puracé *volcano, Colombia* The DACITIC shield volcano capped by an ANDESITIC stratovolcano Puracé has undergone explosive eruptions on more than two dozen occasions since 1827, 12 of which were in the 20th century. Two of these eruptions caused fatalities.

Puu Oo *volcanic cone, Hawaii, United States* This cone, whose name means "hill of the Oo Bird," was produced by an eruption of KILAUEA that began in 1983 and came to be known as the Puu Oo eruption. Puu Oo underwent several episodes of eruption in the 1980s and reached a height of more than 750 feet (229 m). Although episodes of eruption became generally shorter, the extrusion rate of LAVA increased, and FIRE FOUNTAINS rose higher. Some fire fountains rose approximately 1,200 feet (366 m) into the air, approaching the height of the Empire State Building in New York City.

P-waves See SEISMOLOGY.

pyroclastic Made up of fragmented pieces of rock that were produced in an explosive volcanic eruption. The EJECTA are transported, deposited, and lithified just as a clastic rock. The *pyro* part of the word means "fire," so it is a clastic rock produced by fire. An ignimbrite, for example, is composed of pyroclastic material.

This pyroclastic flow of hot ash and debris is moving down the slopes of Mount Saint Helens. In the second photo, a geologist studies the larger material deposited by the flow. *(Courtesy of the CVO-USGS)*

pyroclastic flow More commonly known as an ASH FLOW, a pyroclastic flow is made up of clastic material included and carried along in a flow of ash. The pyroclastic flow behaves as a fluid, hence its name. An ignimbrite is an example of a pyroclastic flow. Such a flow may extend outward many miles from the volcano that produced it. A *NUÉE ARDENTE* is a fast-moving cloud of gas released in a volcanic eruption and containing pyroclastic material.

pyrometamorphism Rock that has undergone extreme metamorphism in LAVA or MAGMA. As magma intrudes into preexisting rock, pieces of the cool rock are broken off into the extremely hot liquid. The broken rock is immediately cooked at very high temperatures. The minerals and tex-ture change accordingly. The same process happens when extremely hot lava runs over the ground surface. The soil beneath is immediately cooked and new minerals and textures are produced. *Pyro* means "fire."

pyroxene A common igneous mineral in MAFIC, ULTRA-MAFIC, and, less commonly, INTERMEDIATE rocks. It includes clinopyroxene and orthopyroxene. Pyroxene is a FERROMAGNESIAN mineral. It is a SILICATE with a chain structure.

quake *See* EARTHQUAKE.

quartz The most common mineral in Earth's CRUST, quartz is found in many FELSIC and INTERMEDIATE igneous rocks. It is especially common in GRANITE and PEGMATITE, where it can make up 25% or more of the rock. It can also be common in RHYOLITE, but quartz takes on two different structures at high temperatures that are called cristobalite and tridymite. Because the grain size is so small in RHYOLITE, commonly the quartz grains cannot be seen. Quartz is nearly pure silicon dioxide (SiO_2) and has a framework type SILICATE structure. Impurities give quartz colors forming smokey quartz (gray), rose quartz (pink), amethyst (purple), or any of the hydrated (water-bearing) family such as agates and opal. Quartz can also be a replacement mineral in any volcanic rock and commonly fills amygdules and cavities.

See also AMYGDALOIDAL.

quartz diorite A plutonic rock of INTERMEDIATE composition. It contains PLAGIOCLASE and typically contains PYROXENE, HORNBLENDE, and minor amounts of QUARTZ. The volcanic equivalents of quartz diorite are ANDESITES and TRACHYTES. It is therefore a common rock in magmatic arcs and even ISLAND ARCS.

Quchan *earthquake, Iran* Quchan has had more than its share of historical seismicity. A more recent event occurred on January 17, 1895, just before noon. The event lasted for about one minute, but strong AFTERSHOCKS lasted for several months. Buildings were leveled throughout the city by the MAIN SHOCK, and the few that remained were rendered unsafe because of the aftershocks. Survivors were forced to live outside in tents or in makeshift shelters. The problem was that it was an intensely cold winter, and exposure caused great suffering and death. It also prevented the arrival of medical and relief staff. It is unclear whether the earthquake or the weather caused the greater death toll. In all, some 11,000 people were said to have perished, although some estimates put the number closer to 8,000. As a result of this earthquake, Quchan was relocated some 7.8 miles (13 km) east of the former location. Other modern destructive earthquakes in Quchan include a MAGNITUDE 5.6 event on December 23, 1871, that killed 2,000 people; a magnitude 7.0 event on January 6, 1872, that killed over 4,000 people, and a major event on November 17, 1898, of magnitude 6.6 that had a 6.6-mile (11-km) focal depth and a death toll of 18,000.

Quetta *earthquake, Pakistan* A devastating earthquake struck the town of Quetta and surrounding area of Balochistan, Pakistan, on May 31, 1935, at 2:33 A.M. The MAIN SHOCK had a RICHTER magnitude of 7.7 and a duration of three minutes. The shaking was felt as far away as 600 miles (1,000 km) to the east and 360 miles (600 km) to the south. AFTERSHOCKS were reported to have continued until October, with the strongest (MAGNITUDE 6.0) on June 2, 1935. The apparent source for this event was the north-south-oriented Ghazaband Fault Zone, which underwent LEFT-LATERAL strike-slip movement.

Approximately 35,000 people were killed in this event. Over 26,000 of the casualties were from Quetta alone, which had a total population of 40,000 at the time. EARTHQUAKE LIGHT was reported to the west of Quetta during the event. Ground LIQUEFACTION followed the earthquake and included the reactivation of a MUD VOLCANO for a reported nine hours. LANDSLIDES and ROCKFALLS accompanied both the main shock and main aftershock, which reportedly raised a cloud of dust to a height of 1,650 feet (500 m). The area was under British control at this time. After the event, 12,000 troops stationed in the area participated in rescue and relief operations. They sealed the town to prevent looting and set up refugee camps to care for the survivors.

Quito *earthquake, Ecuador* On February 4, 1797, most of the population of Quito, ECUADOR died in a massive earthquake. The earthquake was generated by STRESS from the underlying Andean SUBDUCTION ZONE. Quito sits on a plateau at an elevation of 10,000 feet (6,000 m) in the ANDES MOUNTAINS. The earthquake dislodged rock, soil, and ice, causing massive AVALANCHES, ROCKFALLS, and LANDSLIDES. Immediately after the earthquake, the volcanoes COTOPAXI and Chimborazo reactivated and showered the town of Ambato with VOLCANIC BOMBS and debris, wreaking widespread damage there. The combined DEATH TOLL for this tragedy was estimated at 40,000 people, making it one of the largest in SOUTH AMERICAn history.

Another tragic earthquake struck the unfortunate city on March 22, 1859, at about 8:30 A.M. The tremors lasted for six minutes and destroyed the city. The death toll from this event was estimated at 5,000 people.

Q-wave   *See* LOVE WAVE.

R

Rabaul *volcanic cluster, Papua New Guinea* Located at the end of the Bismarck volcanic arc, Rabaul is thought to lie near the meeting point of three plates of Earth's CRUST: the Bismarck, Solomon Sea, and PACIFIC CRUSTAL PLATES. Two collapse events are thought to have occurred at Rabaul within the last 3,500 years. Earthquakes occur often in the vicinity of Rabaul, and the CALDERA has a history of eruptive activity starting in the 18th century. An eruption was recorded in approximately 1767, probably either at Tavurvur on the east side of the harbor or at Rabalankaia, about two miles (3.2 km) north of Tavurvur. Another eruption in 1791 occurred at Tavurvur. In 1850, an eruption at Sulfur Creek at the north end of the harbor followed a time of powerful earthquakes and extensive uplift. During the next 28 years, uplift in the vicinity of Vulcan on the western side of the harbor raised reefs out of the water. This period of uplift preceded a series of powerful earthquakes that started early in 1878 and were accompanied by TSUNAMIS (on February 4) and dramatic uplift. Along the coast near Tavurvur, the ground reportedly was elevated by about 20 feet (6 m) in some places. At the same time, subsidence occurred at Davapia Rocks north of Vulcan and put some homes underwater. Steam and PUMICE rose to the surface of the harbor near Vulcan at a spot that soon would become the site of Vulcan Island. Several hours after this eruption of pumice and steam, Tavurvur underwent a phreato-magmatic eruption. Rabaul remained quiet for the next several years, but unrest resumed in 1910 when a strong earthquake shook the area and was accompanied by a smell of SULFUR. In 1916, another powerful earthquake occurred, along with considerable subsidence. Uplift during an earthquake in 1919 again raised some of the land that had subsided in 1916, and in 1919 FUMAROLES became highly active at Tavurvur and at Matupit Island, about two miles from Tavurvur.

After this eruption, a causeway connected Matupit Island with the mainland. Earthquakes resumed in 1937, including one strong event on May 28 that destroyed at least two buildings and set off landslides and possibly a tsunami as well. Dramatic uplift was observed at this time. Earthquakes continued, especially near Vulcan, on the following day. In between the stronger earthquakes, a steady vibration, possibly volcanic rather than TECTONIC in origin, was felt. On May 29, an eruption began. Vulcan Island was uplifted about six feet (1.8 m), and cracks appeared on Matupit Island. Just prior to the eruption, rumbling sounds were heard, and gas was seen rising to the surface of the waters. An area at the south end of Matupit Island was uplifted several feet, exposing a reef. Another reef near Vulcan Island was lifted out of the water, submerged again and reexposed within 10 minutes. Some 500 people were killed by PYROCLASTIC FLOWS from this eruption. Another, less powerful eruption occurred at Tavurvur. This eruption was PHREATIC and generated destructive rains of mud. Great amounts of carbon dioxide emanated from vents along the east side of the caldera and killed animals there for months after the eruption. Earthquakes stopped with the onset of the eruption.

In 1938 emissions of sulfur dioxide and hydrogen chloride increased at Tavurvur and then increased again in 1940. Minor steam explosions occurred at Tavurvur in 1940, and a powerful tectonic earthquake shook Rabaul that same year. Late in 1940, fumaroles at Tavurvur showed a marked increase in temperature, and in January 1941, a major earthquake occurred in the vicinity of the caldera. Fumarole temperatures increased sharply just before the earthquake and continued rising afterward. Meanwhile, hydrogen chloride and sulfur dioxide output also increased. These changes apparently were restricted to Tavurvur because fumaroles on the opposite side of the harbor at Matupit Harbor were reportedly unaffected. Earthquakes occurred often at Rabaul in the latter half of 1941, early in 1942, and again in late 1943. Strong tectonic earthquakes occurred east of Rabaul again in 1967, but no changes were observed in local volcanoes. In 1971, earthquake swarms resumed, and the caldera floor started to rise. Earthquakes became more frequent (increasing from approximately two per day to 10 per day in some periods) between 1971 and 1983. After a strong tectonic earthquake about 120 miles (193 km) east of Rabaul in 1983, earthquake activity diminished in 1984 and 1985. Emissions of carbon dioxide from a CRATER on Tavurvur killed birds in 1981, and fresh activity from fumaroles destroyed plants at a site slightly more than one mile northwest of Tavurvur in 1983.

In view of the recent and historic unrest at Rabaul, there is a chance that new eruptions will occur there in the near future. Calderas apparently were formed by eruptions there some 3,500 years and 1,400 years ago, and eruptions of comparable magnitude were considered possibilities for the years ahead. With such events in mind, scientists and public officials prepared an emergency plan that would respond in several stages, up to and including evacuation of the local population, to a threatened eruption at Rabaul. On September 19, 1994, both Vulcan and Tavurvur began to erupt simultaneously and without warning. The eruption was powerful (VEI = 4) producing heavy amounts of ASH (up to 30 inches [75 cm] deep) and forcing the evacuation of 50,000 people. Because of the plan, only one person was killed. The eruption ended in 1995.

See also PLATE TECTONICS.

radioactivity   The loss of particles and energy from the nucleus of an atom, resulting in a more stable product. Unstable versions of elements undergo spontaneous decay to other elements and give off radiation. The unstable (radioactive) atoms are called isotopes, although only about 25% of isotopes are unstable. Isotopes differ from normal atoms in that they have different numbers of neutrons in the nucleus (i.e., not equal to the number of protons). Decay occurs through three kinds of radiation: alpha, beta, and gamma. Gamma is like X-rays and does not change the element, but both alpha and beta actually change one element into another. The radioactive element is called the parent, and the element it changes into is called the daughter. The daughter may also be radioactive and in turn produce another daughter. There can be many steps from an unstable parent to a final stable daughter forming a decay series. One of the most commonly studied decay series is from radioactive uranium to stable lead. There are more than 30 steps from the starting parent to the final daughter, including the decay to and from both radium and radon, which are environmental concerns. The rate at which the change from a parent to a daughter takes place is constant. Geologists measure parent relative to daughter to calculate the age of a rock.

radionuclides   Radioactive atoms that generate heat during their decay. Radionuclides are believed to be an important source of internal heat for the Earth. They include isotopes of such elements as uranium and thorium and are commonly found in IGNEOUS ROCKS such as GRANITE.

See also RADIOACTIVITY.

Rahoum   *See* AMBRIM.

Rainier, Mount   *Washington, United States*   One of the most famous volcanoes on Earth, Mount Rainier stands near Seattle and is part of the CASCADE MOUNTAINS, a volcanic

The sleeping giant Mount Rainier overlooks an encroaching population of western Washington. It poses one of the biggest volcanic threats in the United States. *(Courtesy of the CVO-USGS)*

range associated with the CASCADIA SUBDUCTION ZONE off the Pacific coast of northwest UNITED STATES. Mount Rainier is a STRATOVOLCANO and has been dormant since the mid-1800s, when a minor eruption reportedly took place.

The most recent major eruption of Mount Rainier is thought to have occurred approximately 2,000 years ago and laid down a thick layer of TEPHRA east of the mountain. This eruption also is believed to have produced a LAVA cone at the summit of the volcano, as well as large MUDFLOWS. One mudflow, estimated to have occurred almost 3,000 years ago, traveled down the South Puyallup River and Tahoma Creek valleys as a mass of mud hundreds of feet thick; this same flow is thought to have exposed the core of the volcano. In any future eruptions, mudflows may recur on a large scale, through melting of glacial ice on the peak, and may present a threat to nearby communities. The potential for devastating AVALANCHES from Mount Rainier is considerable because internal heat is believed to have turned the mountain core into soft material that might fail easily during a powerful earthquake (a distinct possibility in the seismically active Pacific Northwest), releasing overlying material to fall downslope in vast amounts. Such an avalanche, in turn, might allow an explosion comparable to the one that destroyed the area around Mount SAINT HELENS during the eruptions of 1980. The White River valley is believed to be especially vulnerable to major avalanches from Mount Rainier, should conditions ever permit them to occur. ROCKFALLS said to have exceeded 10 million cubic yards (7.9 million m³) occurred in the White River Valley in 1963.

Mount Rainier is noted for its steam caves, a network of ice tunnels and chambers about 1.5 miles (2.4 km) long altogether. The steam caves are produced by heat from the volcano acting on firn, a moderately dense form of ice that results when snow accumulating on the volcano is compressed to a density not quite that of the crystalline ice found in glaciers. Heat from the volcano melts spaces between the ice and the rock of the mountain. Warm air circulating in these spaces enlarges them. The result is a network of passageways within and beneath the ice. The steam caves represent a delicate equilibrium between FUMAROLE activity and the accumulation of snow and ice on the volcano. The caves were discovered in 1870 when climbers took shelter in them, although there is evidence that Native Americans knew of the caves' existence long before European-descended settlers arrived in the Northwest. Most of the tunnel network exists in the East Crater atop Mount Rainier, although the steam caves extend also to the adjacent West Crater. Similar systems of steam caves are found at nearby Mount Baker and also at WRANGELL in ALASKA.

Ranau *caldera, Sumatra, Indonesia* Located on the Semangko Fault Zone, the Ranau caldera is occupied in part by Lake Ranau. The CALDERA has undergone some unrest but no major eruptions during the late 19th and early 20th centuries. On two occasions, in 1887 and 1888, large numbers of dead fish were discovered on Lake Ranau and on its beaches. The lake water was discolored and had an odor of SULFUR. The deaths of the fish were attributed to volcanic activity in the lake. A powerful earthquake in 1895 reportedly produced SEICHES in Lake Ranau; dead fish were noted again on the surface of the lake in December 1903. This fish kill was accompanied by discoloration of the lake waters, as in 1887 and 1888, and a smell of sulfur was detected near the shore.

Rasshua *volcano, Kuril Islands, Russia* The Rasshua volcano has been active on an intermittent basis since the early 19th century. There appears to have been some unrest at the site in 1810, but details are unavailable. In 1846, there was a moderate to large eruption (VEI = 3). Increased steaming activity at Rasshua in 1946 preceded a similar increase in activity at nearby Sarichev volcano, which erupted five days later. FUMAROLES at Rasshua became considerably more active in 1957, and small explosions were reported. Young LAVA FLOWS extend down the eastern, northwestern, and southern flanks of the volcano.

Raung *See* IJEN.

Rayleigh waves These earthquake waves occur only along Earth's surface (i.e., SURFACE WAVES). They make the ground

Crushing of the base of a column in an overpass during the 1971 San Fernando earthquake. The damage done to the column is characteristic of that done by a Rayleigh wave. The motion of the ground is like an ocean wave. The return upward motion causes the crushing because it is held in place by the overpass. *(Courtesy of the USGS)*

surface move like the waves on the ocean. This motion is technically referred to as retrograde, which means that they orbit in an elliptical pattern in the direction opposite to that of vibrations from the earthquake EPICENTER. Such motion is highly destructive but fortunately only in the area of the epicenter. Energy of the waves decreases rapidly away from the source. Throw a pebble in a lake, and watch the rings form. They model Rayleigh waves. Rayleigh waves, also known as ground roll, are named after John William Strutt, Lord Rayleigh (1842–1919), British mathematician and physicist.

Raymond Fault *California, United States* The Raymond Fault is about five miles (8 km) long and is located several miles southeast of Pasadena. The FAULT has exhibited surface rupture within historical times and is suspected of being the source of the LOS ANGELES earthquake of 1855.

Reaumur, Mount *See* LONG ISLAND.

recurrence interval A recurrence interval is the average amount of time between earthquakes of a given MAGNITUDE for a given area. There are different recurrence intervals for various-size earthquakes in an area and different recurrence intervals for the same-size earthquakes between areas. A recurrence interval for a given city might be 50 years for a magnitude 7.0 earthquake. Recurrence intervals can be among the best tools for earthquake prediction because the tectonic STRESS buildup for a fault is consistent and the strength of the rocks does not vary. As long as the historical records are old enough and complete, a recurrence interval can be determined. The problem is that they are only accurate to 15 or 20 years. This kind of uncertainty is usually not acceptable to most people.

Redoubt *volcano, Alaska, United States* The STRATOVOL-CANO Redoubt, located some 110 miles (173 km) southwest of Anchorage, is one of the easternmost in the ALEUTIAN ISLANDS. It underwent major eruptions several times in the 20th century including four moderate to large ones (VEI = 3) since 1966. It has erupted at least 30 times in the last 10,000 years. Major explosive eruptions also have occurred in this century at nearby Spurr and AUGUSTINE volcanoes. Redoubt's eruption of December 14–19, 1989, produced columns of ASH that rose to altitudes exceeding 40,000 feet (12,192 m) and had serious effects on airline travel in the vicinity of the volcano. The ASHFALLS from this series of explosive incidents extended over the heavily populated Cook Inlet area and inland to the vicinity of Fairbanks. A period of LAVA DOME building activity followed these ashfalls and produced more than a dozen such domes over four months. Each of the domes was destroyed by explosive eruptions that produced ash clouds, although the clouds did less to disrupt air travel than the clouds from the early phase of the volcano's eruption. More than 20 significant falls of TEPHRA were observed at Redoubt between December 14, 1989, and the end of April 1990. A fall of tephra on December 15 included coarse-grained deposits rich in PUMICE, but subsequent tephra falls were generally characterized by ash-size material. Damage and loss of revenue from ash and debris flows totaled some $160 million, making this eruption second only to Mount SAINT HELENS in cost in U.S. history.

Recent eruptions of Redoubt have allowed few direct observations because of poor weather conditions and short periods of sunlight in the winter. Studies of these eruptions have yielded considerable information, however, about lightning in the clouds surrounding the volcano. Ordinarily, lightning is rare in the vicinity of Redoubt, but lightning discharges have been observed during most of the eruptions of Redoubt that released ash. A commercial lightning detection system deployed by the Alaska Volcano Observatory detected lightning in 11 of 12 eruptions and located the lightning strikes in nine of those eruptions, following the activation of the system early in 1990. Most of the strikes between clouds and the ground occurred close to the mountain, and the number of cloud-to-ground discharges was found to be related to the amount of ash released. Lightning discharges in the early stages of the eruption had a negative charge, whereas discharges later in the eruption were positive. This pattern indicates that coarse particles of ash, expelled early in the eruption, carry a negative charge, and finer particles expelled in later stages of the eruption carry a positive charge. "Intra-cloud" strikes of lightning, within the volcanic cloud itself, tend to be more numerous than cloud-to-ground strikes and to occur even at great distances from the volcano. Intracloud discharges during the February 15, 1990, eruption of Redoubt occurred up to about 80 miles (128 km) from the mountain.

Intracloud lightning may prove useful in developing systems capable of tracking airborne volcanic clouds. Satellites have been used to track ejecta from Redoubt in more than 20 eruptions since late 1989. Environmental observation satellites in geostationary and polar orbits have monitored emissions of ash and volcanic debris. Data provided by the advanced very high resolution radiometer (AVHRR) on two U.S. weather satellites in polar orbit were used to detect a large MUDFLOW along the Drift River near Redoubt as the event occurred during the February 15, 1990, eruption, and to provide warnings to Anchorage. Studies of AVHRR images of eruptions of Redoubt and Augustine indicate that volcanic clouds can be detected under various weather conditions, whether during day or night or over water or land. Multiple infrared channels are useful here, but there is as yet no single algorithm capable of detecting all eruption clouds.

See also AVIATION AND VOLCANOES.

reef A reef is a barely submerged expanse of rock or coral, often found in shallow water near a continent or island. Coral reefs are often associated with volcanic islands and SEA-MOUNTS, when a SUBMARINE VOLCANO's summit rises close enough to the surface for light to penetrate and allow the growth of coral. This scenario is only possible between 30° north and south latitude because coral can only grow in the tropics. When coral builds up to the surface from the summit of a seamount and forms a ring of dry land, the resulting formation is called an atoll. The ocean water in these areas becomes supersaturated with calcium carbonate. Limestone is formed in these areas. Reefs can serve as useful indicators of uplift or subsidence at a volcanic site. On some occasions, reefs may be uplifted, dropped, and then lifted again very rapidly. At RABAUL, for example, a reef on one occasion was seen to appear and disappear and then reemerge from the water, within only about 10 minutes.

reflected wave Earthquake waves that bounce off an interface between different rock units. The reflected waves can return to the surface and be picked up by SEISMOGRAPHS. These stray waves contribute to the noise registered on seismograms. Generating synthetic seismic waves can be used for exploration.

refracted wave Earthquake wave that bends at an interface between different rock units. Just as light bends as it passes from air to water, seismic waves bend as they pass from one rock type to another. The waves bend sharply across each of the major layers in the Earth. That is the way seismologists can tell the internal architecture of Earth. There are sharp seismic velocity contrasts across the INNER and OUTER CORE, the MANTLE, and the CRUST.

resonance Objects resonate when the proper sound or vibration is imposed upon them. The same idea applies to seismic waves passing through a material or object. Technically, resonance is the largest vibration of a geological system (like a soil layer) due to energy enhancement at a certain frequency that characterizes that system.

Réunion Island   *See* PITON DE LA FOURNAISE.

Reventador *volcano, Ecuador*   A STRATOVOLCANO, Reventador has erupted at least 24 times since 1541. Its most recent reported eruptions were in 1972, 1973, and 1976. Little is known about the eruptions because of the remote location and altitude.

reverse fault   A contractional fault. A DIP-SLIP FAULT in which the block on one side of the fault moves upward. The fault forms a slope underground. The rock units on the top of the fault move up the slope. The overall effect is one of shortening the strata in an area. They occur as the result of compression. Shallow angle reverse faults are called thrust faults.

A dip-slip fault in which the rock on top of the fault plane (hanging wall) moves up relative to the rock under the fault (footwall). Movement is shown by offset of the rock layers across the fault. *(Courtesy of the USGS)*

Rhode Island *United States*   Rhode Island has not been the site of numerous strong earthquakes. It belongs to seismically active New England, however, and has been affected on occasion by earthquakes elsewhere in the region. Two notable earthquakes did occur in the Narragansett Bay area in the 1960s. The earthquake of December 7, 1965, rattled doors and windows and caused widespread alarm in Warwick; this earthquake was felt also in MASSACHUSETTS. The earthquake of February 2, 1967, registered 2.4 on the RICHTER scale of earthquake magnitude and frightened residents of communities including Newport and North Kingstown.

Rhodos *earthquake, Greece*   A massive earthquake struck the eastern MEDITERRANEAN region on August 8, 1303. The EPICENTER was on Rhodos, which was completely destroyed, but extensive damage also occurred on Crete and Cyprus and even in CAIRO and ALEXANDRIA in EGYPT. The RICHTER magnitude of this earthquake was estimated at 8.0, and the intensity was X on the modified MERCALLI scale on Rhodos. There were detailed descriptions of damage in the town of Candia (Heraklio), Crete, where 26 churches in the area were destroyed and some 4,000 people lost their lives. In Alexandria and Cairo, walls, towers, houses, and government buildings either collapsed due to the SURFACE WAVES or were washed away by a large TSUNAMI that reached as far inland as Bab al Bahr. In all, 10,000 people lost their lives in Egypt. The DEATH TOLL for the whole event is a matter of debate, but it is likely to have exceeded 25,000 people.

rhyolite   A category of FELSIC volcanic rock, rhyolites are commonly tan or light gray in color. OBSIDIAN, or volcanic glass, a widespread volcanic rock of rhyolitic composition, is dark brown or black. Rhyolite is APHANITIC (fine-grained) rock with large grains of QUARTZ and/or K-FELDSPAR, if any.

Richter, Charles (1900–1985)   A U.S. geologist and physicist, Richter's name is applied to a widely used scale of earthquake MAGNITUDE, indicating energy released in an earthquake. The logarithmic Richter scale was first proposed in 1927, and Richter refined it with the help of Beno GUTENBERG. The formula for Richter magnitude is $M = \log a + b$, where a = AMPLITUDE, and b = ATTENUATION factor that depends upon the rock. The log scale means that each number yields 10 times as much energy as the next lower. In other words, a magnitude 5 is 10 times as powerful as a magnitude 4 and 100 times as powerful as a magnitude 3. Richter magnitude is expressed in Arabic numerals and decimal (for example, 5.5). The Richter scale is different from but complements the MERCALLI scale of earthquake intensity.
    See also SEISMOLOGY.

rift   A breach or "crack" between two masses of crustal rock that once were joined together. The process by which CONTINENTAL CRUST is pulled apart and a divergent margin is created. Rifting includes normal faulting (HORSTS and GRABENS as well) and resulting earthquakes. It can lead to BASALT volcanism (FLOOD BASALTS) and RHYOLITIC volcanism in some cases. Eventually, it will create an ocean basin with a MID-OCEAN RIDGE.

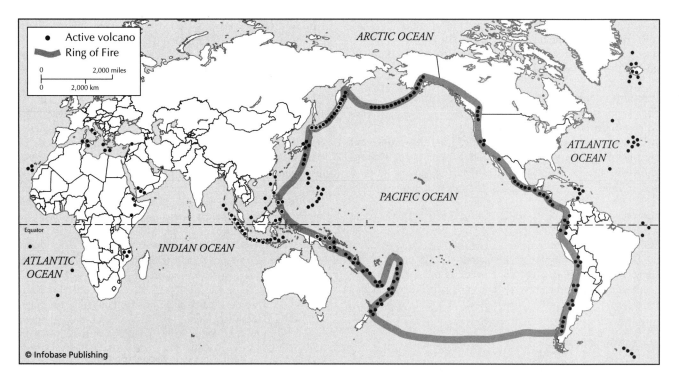

Map of the world showing the "Ring of Fire" that encircles the Pacific Ocean basin as well as the distribution of volcanoes in the ring. Other major volcanic areas of the world are also shown.

rift valley   A long, linear crack in Earth's CRUST that occurs where MAGMA rising from below reaches the surface as FLOOD BASALTS and RHYOLITIC cones. They are also areas of intense seismic activity related to movement on normal faults. These FAULTS form SCARPS along the valley walls. Drainage into the basin leads to thick sedimentary deposits that commonly contain petroleum deposits. The GREAT RIFT VALLEY of Africa is a good example of a rift valley and is the early phase in the development of a DIVERGENT margin.

right-lateral fault   A type of STRIKE-SLIP FAULT in which the rock on the opposite side of the fault from the observer moves to the right. They are also called dextral strike-slip faults. The SAN ANDREAS FAULT is a right-lateral strike-slip fault.

Rincón de la Vieja *volcano, Costa Rica*   Rincón de la Vieja, "Nook where the Old One lives," is the biggest volcano in the northern part of COSTA RICA and appears to be located inside nested CALDERAS. The volcano is made up of several volcanic centers arrayed along a ridge. It has erupted at least 16 times since 1851, most of which are PHREATIC. Explosive eruptions were recorded between 1860 and 1863, but details are unavailable. Explosive activity was observed again at the volcano in April 1922, together with evidence of a very powerful recent eruption. A strong eruption occurred on June 4, 1922, preceded by SULFUR dioxide emissions and loud noises from FUMAROLES. In 1963, the volcano was reported to be steaming and giving off sulfur gases. Vapor emissions from the volcano in 1965 preceded an eruption

of ASH in 1966. On a single day (February 23, 1968), more than 20 eruptions occurred in a half hour. These eruptions required that villages in the area be evacuated. Steam blasts from the volcano occurred in 1983 that affected locations more than a mile from the CRATER. Another eruption took place in 1984, and more eruptive activity is thought to have occurred in the autumn of 1985. Minor explosions were recorded in 1986–87, 1992, 1995, and 1998. The 1995 eruption forced the evacuation of 300 families.

ring dikes   Circular DIKES around a volcano. If MAGMA is removed from the MAGMA CHAMBER beneath a volcano, it can undergo CALDERA collapse. The volcano and the area around it collapse because they are no longer being held up by the liquid. A series of concentric FAULTS and cracks develop around the collapsing volcano. As they do, magma will squeeze up along the cracks and faults forming ring dikes.

"Ring of Fire"   This is the popular name for a narrow belt of intense earthquake and volcanic activity that follows approximately the borders of the PACIFIC OCEAN basin and includes most of the active volcanoes on Earth. The Ring of Fire extends from Tierra del Fuego, at the southern extremity of SOUTH AMERICA, northward along the ANDES MOUNTAINS of South America, through CENTRAL AMERICA, along the CASCADE MOUNTAINS in the Pacific Northwestern UNITED STATES and southwestern CANADA, down the ALEUTIAN ISLANDS in ALASKA, westward to volcano-filled KAMCHATKA PENINSULA, and southward to JAPAN and eastern CHINA, the PHILIPPINE ISLANDS, INDONESIA, NEW GUINEA, and NEW

ZEALAND. Altogether, the Ring of Fire measures some 30,000 miles (48,280 km) in length and encloses an area of approximately 70 million square miles (181,299,167 km²).

Because the Ring of Fire contains some of the largest cities in the world, notably LOS ANGELES and TOKYO, it has been the site of numerous fatalities from earthquakes and volcanic ERUPTIONS over the centuries, and the potential for loss of life in future earthquakes and eruptions is great despite all precautions that may be taken. Adding to the potential for damage from earthquakes and eruptions around the Ring of Fire is the danger of TSUNAMIS, which can carry destruction all the way across the Pacific Ocean and even farther. The following selected listings of major earthquakes and volcanic eruptions along the Ring of Fire for the past several centuries provide some idea of the destructive potential of seismic and volcanic activity along it:

- *China, 1556.* Although little historical information is available about this earthquake, it is thought to have affected three provinces and killed more than 800,000 people.
- *Japan, 1596.* An earthquake offshore reportedly generated a tsunami that destroyed the island of Uryu-Jima completely and caused more than 4,000 deaths.
- *Philippine Islands, 1616.* An eruption of the volcano MAYON buried numerous villages near the mountain under ASH, although casualty figures are unavailable. A later eruption in 1766 is said to have killed some 2,000 people and produced large flows of hot mud, possibly from lakes that spilled out of the CALDERA of the volcano. An additional 2,000 people were reported killed in an 1814 eruption of Mayon. The volcano has erupted frequently since then, but casualty figures have been comparatively low.
- *Japan, 1737.* A tsunami said to have been more than 200 feet (61 m) high struck the north shores of Japan and also Russia's Kamchatka Peninsula.
- *Chile, 1757.* A major earthquake occurred at CONCEPCIÓN, allegedly accompanied by a tsunami. Some 5,000 people perished, and 10,000 were injured.
- *Mexico, 1759.* The emergence of JORULLO volcano is thought to have killed some 200 people. Jorullo's activity continued for approximately 40 years.
- *Java, Indonesia, 1772.* Powerful earthquakes at PAPANDAJAN generated a huge depression, approximately six miles (10 km) wide and 15 miles (24 km) long. An entire town was destroyed, and some 2,000 people were killed.
- *Japan, 1793.* The volcano UNZEN exploded, killed some 50,000 people, and deposited PUMICE on the sea in layers thick enough to support a person.
- *Venezuela, 1812.* An earthquake at CARACAS killed about 10,000 people there and thousands more in nearby communities.
- *Indonesia, 1815.* The eruption of TAMBORA is estimated to have cast out more than 30 cubic miles of solid material in a single week and killed 92,000 people both directly and by resulting disease and famine.
- *Java, Indonesia, 1822.* The volcano GALUNG GUNG erupted, killing some 4,000 people.
- *Chile, 1822.* An earthquake at Valparaiso killed some 10,000 people and uplifted the shoreline several feet, exposing shipwrecks on the ocean floor.

- *Chile, 1835.* One of the most calamitous earthquakes in history destroyed Concepción and Santiago on February 20 and was accompanied by a tsunami that caused widespread destruction along the shore.
- *New Zealand, 1855.* A very strong earthquake in southeast NEW ZEALAND uplifted the Rimutaka mountain range by several feet.
- *Japan, 1857.* An earthquake and subsequent fire destroyed Tokyo and killed an estimated 100,000 people.
- *Ecuador, 1877.* The most violent recorded eruption of the volcano COTOPAXI killed some 1,000 people and destroyed much of the mountain's summit.
- *Indonesia, 1883.* The eruption of the volcanic island KRAKATOA in the Sunda Strait generated a huge (120-foot [37-m]-high) tsunami that killed 36,000 people along shorelines near the volcano. The tsunami circled the globe several times before diminishing completely. Ash ejected into the upper atmosphere from this eruption made sunsets redder in the following year.
- *Alaska, 1891.* A series of extremely powerful earthquakes near Yakutat Bay raised a nearby mountain range almost 50 feet (15 m) and affected an area of some 200,000 square miles (517,998 km²).
- *Guatemala, 1902.* An earthquake and subsequent fire destroyed Guatemala City and killed more than 12,000 people.
- *Formosa (Taiwan), 1906.* An earthquake reportedly destroyed more than 6,000 buildings and took some 1,300 lives.
- *California, United States, 1906.* The SAN FRANCISCO earthquake of this year is perhaps the most famous seismic event in United States history.
- *Chile, 1906.* An earthquake struck Valparaiso and killed approximately 1,500 people.
- *Java, Indonesia, 1919.* Kelud volcano erupted and generated a vast flow of hot water and mud that is thought to have killed more than 5,000 people.
- *China, 1920.* GANSU Province experienced a powerful earthquake that reportedly killed about 200,000 people.
- *Japan, 1923.* The "great KANTO earthquake" destroyed much of Tokyo, killing approximately 143,000 people and leaving a half-million more homeless.
- *Japan, 1927.* An earthquake almost as powerful as the Kanto earthquake struck the TANGO PENINSULA on Japan's western shore, killing some 3,000 people and destroying approximately 14,000 buildings.
- *Netherlands East Indies (now Indonesia), 1928.* The volcano Rokotinda erupted and killed more than 200 persons with LANDSLIDES and falling rocks.
- *Java, Indonesia, 1931.* MERAPI volcano's eruption lasted three weeks and took more than 1,000 lives.
- *California, United States, 1933.* The LONG BEACH earthquake killed more than 100 people and caused extensive damage.
- *Chile, 1939.* An earthquake destroyed Concepción, killing some 50,000 people and leaving perhaps 750,000 more without shelter.
- *Japan, 1946.* An undersea earthquake generated large tsunamis in the Inland Sea that obliterated some 50 communities

along the shore. About 2,000 people are thought to have died in the tsunami, and possibly 500,000 were left homeless after the waves had passed.

- *Ecuador, 1949.* An earthquake in central Ecuador killed some 6,000 people and injured perhaps 20,000 others.
- *Japan, 1952.* The eruption of a submarine volcano at MYOZIN-SYO in the Bonin Islands destroyed a ship, leaving no survivors, when the vessel passed over a vent during the eruption.
- Hawaii, United States, 1960. A tsunami that accompanied an earthquake in CHILE caused extensive destruction in Hawaii. Some 60 people died in Hilo City. More than 400 other people were killed when the wave reached Japan and the Philippine Islands, in addition to the 2,000 people killed in Chile.
- *Alaska, United States, 1964.* The GOOD FRIDAY EARTHQUAKE of 1964, with its accompanying tsunami, killed more than 100 people and carried destruction down the Pacific coast to CRESCENT CITY, CALIFORNIA.
- *Chile, 1968.* Almost 500 people died in an earthquake that struck central Chile. The earthquake itself was responsible for only a few of these deaths. Most resulted from the failure of dams that released their impounded water and buried communities in mud.
- *Peru, 1970.* A huge earthquake set off a devastating AVALANCHE where house-size blocks traveled at speeds up to 270 miles (435 km) per hour. A total of 66,000 people were killed in all.
- *California, United States, 1971.* The SAN FERNANDO earthquake was small by some standards (fewer than 100 persons killed) but caused heavy damage to property and became one of the most intensively studied earthquakes in history.
- *China, 1976.* The TANGSHAN earthquake by official reports caused 250,000 fatalities. However, it is suspected that more than 655,000 actually perished.
- *Guatemala, 1976.* A devastating earthquake on the MOTAGUA FAULT killed more than 23,000 people.
- *Mexico, 1985.* The devastating MEXICO CITY earthquake demonstrated that city's vulnerability to seismic events. Much of the city is built on unconsolidated sediment, which reacted to passing seismic waves in such a way that numerous buildings collapsed. The earthquake originated not in the Mexico City area but rather along the Pacific coast of Mexico.
- *California, United States, 1989.* The LOMA PRIETA earthquake that hit San Francisco damaged the Bay Bridge between San Francisco and Oakland and killed motorists whose vehicles were crushed under a collapsing segment of highway in the East Bay region.
- *Philippines, 1991.* The eruption of the long sleeping giant, Mount PINATUBO killed more than 3,000 people and forced the closure of Clark Air Force Base. It had a VEI of 5.
- *California, United States, 1994.* The NORTH RIDGE earthquake devastated the Los Angeles area, killing only 60 people but causing more than $20 billion in property damage.
- *Japan, 1995.* The KOBE earthquake killed more than 5,500 people and reduced much of the city to rubble.
- *Taiwan, 1999.* The Tai-Ching earthquake killed some 2,400 people and caused $14 billion in property damage.

Rio Grande Rift *New Mexico, United States* The Rio Grande Rift began its activity about 13 million years ago. It borders the COLORADO PLATEAU and extends through the San Luis Valley between the Sangre de Cristo Mountains and the San Juan Mountains. The northern part of the rift lies near Taos and Albuquerque, NEW MEXICO. The lower portion of the rift is characterized by flat territory, such as the Plains of Saint Augustin. Signs of volcanic activity are abundant along the Rio Grande Rift. An example is the VALLES caldera in the Jemez Mountains near LOS ALAMOS. This is one of the largest CALDERAS known and is believed to have formed through the eruption and eventual collapse of a volcano whose output of TEPHRA filled in large portions of the rift in the vicinity of Los Alamos about 1.12 million years ago. Cinder cones in the Valles Caldera area indicate that volcanism was active here only a few thousand years ago. Numerous LAVA FLOWS are visible along the rift, notably the Malpais lava beds near Carriozo. This lava flow may be seen at Valley of Fires State Park. Molten rock may still underlie the rift; there is evidence of a pool of MAGMA between 10 and 15 miles (16 and 24 km) deep beneath the rift near Socorro. The Rio Grande Rift has been investigated as a possible source of GEOTHERMAL energy.

Riviera *earthquake, France and Italy* An earthquake took the Riviera at the French-Italian border by surprise at 6 A.M. on February 23, 1887. The seismic waves were felt strongly in the cities of San Remo and Genoa, ITALY; Monte Carlo, Monaco; and Nice, France. Earthquakes are not common in this part of the MEDITERRANEAN basin, so preparation was poor. In addition, the quake struck on the morning of Ash Wednesday in towns that were crowded with tourists there to celebrate Mardi Gras. The earthquake was estimated at a 6–6.2 on the RICHTER scale and an X on the modified MERCALLI scale. This earthquake marked the first use of the Cecchi instrument, an early SEISMOGRAPH. The sea level was said to have dropped three feet (1 m) after the first shock and then rose six feet (2 m), suggesting that a TSUNAMI or SEICHE accompanied the event. In all, about 2,000 people lost their lives. Equally striking was the mass exodus of people—especially tourists—immediately after the event. Some 50,000 people choked the rail systems, boats, and other modes of transportation in a mad attempt to escape.

Roccamonfina *volcano, Italy* The STRATOVOLCANO Roccamonfina is located some 30 miles (48 km) north of VESUVIUS on the western shore of the Italian Peninsula. The volcano is about 11 miles (17.6 km) in diameter and has two CALDERAS. The main caldera is approximately four miles (6.4 km) in diameter and occupies the summit of the mountain, whereas a second, smaller caldera is located on the north side of the volcano. Roccamonfina has many PARASITIC CONES. The volcano has generally been quiet within historical times but has had one recorded period of unrest, around 270 B.C., when a flame reportedly rose up and burned for several days. What exactly happened in this case is uncertain. It may or may not have been an actual eruption.

rock avalanche A general term for rock masses moving downhill at high speed. They include rock and debris falls and

DEBRIS FLOWS. Rocks and debris can break loose as the result of an earthquake among other causes and race downhill. In rugged terrains and high elevation, these masses can attain speeds of well over 200 miles (322 km) per hour. They will devastate anything in their way and only stop at the bottom of a valley. The most spectacular example of a rock or debris flow AVALANCHE was in NEVADOS HUASCARÁN in 1970.

rockfall   In contrast to a ROCKSLIDE, which stays in contact with the surface throughout its movement, a rockfall is a mass of rocks that breaks loose from the BEDROCK and falls through the air to the slope below. Most of the travel is a free fall, but it may include bouncing as well. Rockfalls typically break away from the bedrock along steeply inclined joints. Undercutting of slopes by rivers, wind, or human activities also hastens rockfalls. Toppling is a particular type of rockfall that involves the rotation of the rock mass away from the exposure around a fixed fulcrum. Whatever the mechanism, repeated rockfalls or rockslides produce a pile of rock rubble at the foot of a cliff or slope. This rubble is called TALUS. Earthquakes in mountainous regions produce numerous rockfalls.

rockslide   A mass of solid rock that slides quickly down an inclined discontinuity. The most common discontinuities are bedding and fractures. In sedimentary rocks, bedding is usually the primary discontinuity. With bedding, a rigid layer such as limestone or sandstone typically breaks free and slides down an underlying, softer layer such as shale. In crystalline rocks (igneous and metamorphic), the most common discontinuities are joints and FAULTs. The rock masses may break loose along steeply inclined joints through frost-wedging processes, but they slide along shallower joints and faults. Rockslides may be generated by a number of processes, but they are especially common during earthquakes. The SURFACE WAVES shake loose otherwise stable rock masses, and they slide into the valleys below with disastrous results.

Rodinia   There was another supercontinent that preceded PANGAEA, and it has been named Rodinia. It was constructed during the middle to late PROTEROZOIC and was finally assembled about 1 billion years ago. The continents were in completely different positions and have been shuffled around during later PLATE TECTONIC interactions. For example, the

Talus

Block diagram of a cliff showing a rockfall with talus slope at the base

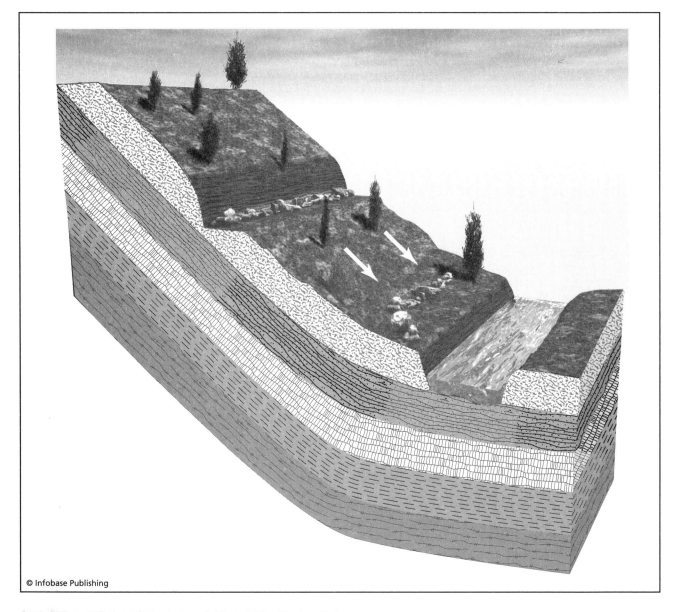

© Infobase Publishing

Block diagram of tilted rock layers and a rockslide or debris slide along the layers

east coast of the United States appears to have faced onto South America. The building of this continent included significant earthquake and volcanic activity along active belts. The Blue Ridge, NEW JERSEY–Hudson Highlands, Adirondacks, Green Mountains, and Long Mountains of eastern NORTH AMERICA contain such a mobile belt. However, evidence for the building of this supercontinent is abundant worldwide. Rodinia broke apart during the late Proterozoic and into the Cambrian.

Rostaq *earthquake, Afghanistan*   On February 4, 1998, an earthquake of MAGNITUDE 5.9 occurred in the Afghanistan-Tajikstan border region. At least 2,323 deaths were reported as were 818 injuries. At least 8,094 houses were destroyed, and 6,725 livestock were killed.

Rotorua *caldera, North Island, New Zealand*   The Rotorua caldera is located in the TAUPO volcanic zone and is noted for its variable HYDROTHERMAL ACTIVITY. In 1886, there apparently was a failure for several weeks in a hot-water supply to baths at Rotorua, although this change may or may not have been connected with activity at the CALDERA. In 1932 a strong hydrothermal explosion took place in Lake Rotorua.

Ruapehu *volcano, North Island, New Zealand*   Adjacent to the volcanoes NGAURUHOE and Tongariro, Ruapehu has a CRATER lake at its summit whose ERUPTIONS cause occasional MUDFLOWS. It has erupted about 50 times since 1861. Most of the eruptions are PHREATIC. A mudflow in 1953 destroyed a railroad bridge and brought about the deaths of more than 150 people in a train wreck. According to studies of the vol-

Rockslide blocks destroy a highway after the April 24, 1984, earthquake in Morgan Hill, California. The earthquake had a Richter magnitude of 6.2 and caused $30 million in property damage. *(Courtesy of NOAA)*

cano that span the last 75,000 years, Ruapehu has been in a phase of low-emission-volume, low-magnitude eruptions. It has been in this phase for the past 1,800 years. It erupted in 1997 and again in 1999. The 1997 eruption was the largest in 400 years. It included explosions, a six-mile (10-km)-high eruptive column, and a LAHAR.

Ruby Seamount *Mariana Islands* It is a SUBMARINE VOL-CANO located about 25 miles (40 km) northwest of Saipan, Guam. The last major eruption was in 1966, but there was a minor eruption in 1995 as well. TSUNAMI alerts were posted during this eruption.

run-up The run-up height is the elevation of the water level in feet or meters caused by the arrival of a TSUNAMI that is above sea level at the expected tidal elevation. The run-up height is the added height of sea level caused by a tsunami. In the vast majority of tsunamis, run-up heights are on the order of centimeters. There are historical accounts of tsunamis that claim run-ups of greater than 200 feet (60 m). These would be extremely rare if they ever existed.

rupture Slippage along a FAULT PLANE that produces break-age or rupturing of rock. After STRESS exceeds the strength of the fault rock, rupture occurs, producing an earthquake. Sur-face rupture does not necessarily coincide with the EPICENTER of an earthquake. It can be related to slumping and shifting of fault blocks in the subsurface.

See also SEISMOLOGY.

rupture front As a FAULT breaks to produce an earth-quake, the broken area expands along the fault to produce the full rupture. There is a boundary between this broken zone in the fault and the unbroken but soon to be broken zone. This is the rupture front, and it expands outward at the FOCUS at a very high velocity called the rupture velocity.

Ryon-dake *See* DAISETSU-TOKACHI.

S

sag A sag is a long and narrow topographic depression that lies along a STRIKE-SLIP FAULT. May sags are filled with water. In this case, they are called a sag pond.

sag pond Pond common on an ACTIVE FAULT. A depression can form on an active fault by subsidence or disrupted drainage patterns. The rock is broken up and weakened in the faulting process. Erosion can proceed much faster in these zones, leading to the formation of depressions. Juxtaposing streams against other topographic features through faulting disrupts normal drainage. The small ponds form all along the faults. The SAN ANDREAS FAULT has several good examples of sag ponds.

Saint Helens, Mount *Washington, United States* The 1980 eruption of Mount Saint Helens caused $1.5 billion or more in total damage, destroyed more than $100 million in crops, and demolished some 150 square miles (389 km²) of timber. More than 100 people were reported dead or missing following the eruption. The first indications of an impending eruption that year began on March 20 when a seismograph at the University of Washington in Seattle detected an earthquake approximately 20 miles (32 km) north of Mount Saint Helens. Earthquake activity near the volcano increased over the following several days. On March 27, the volcano emitted a loud explosion along with a plume of ASH and vapor that rose to more than 20,000 feet (6,000 m) in altitude. This eruption left a CRATER some 250 feet (76 m) wide on the summit. Another cloud of vapor and ash two days later reached Bend, OREGON, some 150 miles (241 km) south of Mount Saint Helens.

By April 7, tremors indicated that magma was flowing beneath the volcano. An eruption on April 8 lasted more than five hours. Soon afterward, the north flank of the mountain was noticed bulging outward, in much the same manner as a blister on an overinflated tire. The blister indicated that MAGMA and gas released from it were building up pressure within the peak. Governor Dixy Lee Ray declared a state of emergency on April 9, and the National Guard took steps to prevent onlookers from approaching the volcano. The bulge remained intact for more than another month, despite a violent eruption of May 7. Eventually, the bulge grew to be about 2,000 feet (610 m) long and 500 feet (152 m) high. The bulge disintegrated at about 8:30 A.M. on May 18, 1980, when an air-blast explosion from Mount Saint Helens created a blast wave that moved at an estimated velocity of 200 miles (322 km) per hour and knocked down trees in "blowdown" some 20 miles (32 km) away. A plume of ash rose from the mountain to an altitude of about 12 miles (19 km). The mountains north flank, which had been destroyed in the explosion, became an AVALANCHE that flowed into the south fork of the nearby Toutle River and Spirit Lake, displacing water from the lake and sending it into the north fork of the Toutle River. The resulting flood of debris-laden water carried away virtually everything in its path, including a railroad

The 1980 eruption of Mount Saint Helens made this volcano in Washington one of the most intensively studied in the world. *(Courtesy of the USGS)*

"Blowdown" of trees from the lateral blast of the explosion of Mount St. Helens, 1980 *(Courtesy of the USGS)*

A tree trunk splintered at its base by the tremendous force generated by the lateral eruption of Mount Saint Helens in 1980. Thousands of acres of trees suffered the same fate. *(Courtesy of the CVO-USGS)*

bridge. Much of the fine sediment from the avalanche and ash cloud made its way into the Columbia River, where it interfered with navigation.

One of the personal stories associated with the eruption concerned one Harry Truman, proprietor of a lodge at Spirit Lake near the volcano. Truman refused to leave his lodge as earthquakes and eruptions became more frequent. He claimed the mountain "didn't dare" harm him. Truman and his lodge were buried under volcanic debris during the eruption of May 18.

Trees flattened by the blowdown of Mount Saint Helens carried along by the lahar are plastered against a house at the base of the mountain. *(Courtesy of the CVO-USGS)*

Mud line 25-feet (7.6-m) high on trees shows how deep the lahar was that passed through this area during the eruption of Mount Saint Helens in 1980. *(Courtesy of the USGS)*

The 1980 eruptions of Mount Saint Helens made the mountain one of the most intensively studied volcanoes. Mount Saint Helens has added to the knowledge of many phenomena of volcanology, notably air-blast explosions.

Mount Saint Helens was built about 40,000 years ago and was widely intermittently eruptive until 2,500 years ago. In 1855 B.C., Mount Saint Helens produced a huge (VEI = 6) eruption, one of the largest in NORTH AMERICA. In the new eruptive phase, it has gone DORMANT for two to seven centuries between eruptive intervals. The modern volcano was built between 2,200 and 1,600 years ago. The last eruptive phase prior to the 1980 eruption is termed the Goat Rocks eruptive period. It lasted from about 1800 to 1857 and produced numerous minor eruptions. There are also reported minor eruptions in 1898, 1903, and 1921.

Saint Lawrence Valley *Canada and United States* The Saint Lawrence River valley forms the boundary between a

portion of NEW YORK State and eastern CANADA and has been the location of numerous earthquakes, some of them severe, over the past several centuries. It would be impractical in this space to provide a complete listing of earthquake activity in the Saint Lawrence Valley since settlement of the area by Europeans began, but descriptions of several notable earthquakes may convey a sense of the seismic potential of this region.

An earthquake on June 11, 1638, for example, is believed to have occurred in the Saint Lawrence Valley. This very strong earthquake affected Trois-Rivières (Three Rivers), Quebec, and was felt so strongly in New England that people in MASSACHUSETTS had trouble standing upright and that portions of chimneys were shaken down in the vicinity of Lynn, Plymouth, and Salem, Massachusetts. CONNECTICUT and Narragansett, RHODE ISLAND were also affected by the earthquake, as were ships off the coast. Lesser shocks occurred for some three weeks after this earthquake.

Another powerful earthquake on September 16, 1732, caused substantial damage in Montreal and reportedly killed seven people. Houses in NEW HAMPSHIRE were damaged, and the earthquake was perceived in Boston and as far away as Annapolis, MARYLAND. On October 20, 1870, a powerful earthquake was felt strongly at Montreal and Quebec and was reported over a vast area of the eastern UNITED STATES, as well in such widely separated states as IOWA, MICHIGAN, and VIRGINIA. (There is some question, however, whether certain reports, from Virginia and Iowa, are accurate.) The shock was strong enough in MAINE to break windows and was felt along the northeast coast of the United States from Portland, Maine, to New York. The vicinity of Lake George in eastern New York State was also affected. The earthquake lasted one minute and was accompanied by a rumbling sound at Albany, New York. Clocks reportedly stopped in Cleveland, OHIO.

The 20th century has also seen some notable earthquakes in the Saint Lawrence Valley. An earthquake on February 28, 1925, was centered near Laurentides Provincial Park, between La Malbaie (Murray Bay) and Chicoutimi. The exact EPICENTER was difficult to determine because instrumentation in the region at the time was inadequate for that task. Although this was a powerful earthquake, measuring about VIII on the MERCALLI scale in the vicinity of its epicenter, and was perceived over a great area of east Canada and the northeast United States (it was felt as far away as Boston, Massachusetts, and New York City), it appears to have caused little significant damage, and that was confined to a narrow zone on either side of the Saint Lawrence River. More than a hundred AFTERSHOCKS were experienced at Chicoutimi in the week following the earthquake. For some months, shocks continued, including strong shocks on April 10 at La Malbaie and April 25 at Chicoutimi. (There also appears to have been a FORESHOCK on September 30, 1924, felt in Maine and VERMONT and estimated at magnitude 6.1 on the RICHTER scale.) Reportedly, no one was killed directly by the earthquake, although several deaths from shock were attributed to its effects. Among the intriguing effects of this earthquake was the rotation of monuments. Statues and furniture at La Malbaie rotated clockwise during the earthquake.

Map of the northeastern United States and southeastern Canada showing the seismically active St. Lawrence Valley and earthquake epicenters, the date of the earthquake, and the approximate Richter magnitude

This earthquake distributed its energy in ground motion of moderate intensity over a very wide area, with considerable variations in intensity at separate locations equally distant from the epicenter in different directions. The earthquake had interesting effects on buildings in Quebec. In some cases one structure was damaged while another standing nearby was relatively untouched. For example, sheds and elevators for handling grain underwent significant damage at one location, but a nearby hotel built on rock rode out the earthquake so easily that there was hardly any sign an earthquake was occurring. Near the epicenter, there was apparently little damage to buildings because there were few structures of any great size except for churches. A church at Saint-Urbain underwent heavy damage, and the bells were shaken out of place and the organ pipes were bent at another church at Rivière-Quelle. As a rule, buildings situated atop rock came through the earthquake well. Vertical ground motion near the epicenter must have been considerable; at Rivière-Quelle, a 200-pound rock was reportedly thrown off its foundation, and a chimney was destroyed by vertical thrust from below. Cracks formed along the south shore in a grid pattern. One such crack near Rivière-Quelle was several inches wide and deeper than two feet (0.61 m). In some areas, frozen Earth near the surface appears to have moved atop the unfrozen portion and cracked, producing FISSURES through which wet sand and water made their way to the surface.

The GRAND BANKS earthquake of November 18, 1929, estimated at magnitude 7.2 on the Richter scale, was felt throughout New England and parts of Canada south of the

Saint Lawrence River and the Strait of Belle Isle. The earthquake and subsequent undersea LANDSLIDE (TURBIDITY CURRENT) broke submarine communications cables that traversed the area of the epicenter. A TSUNAMI associated with the earthquake was responsible for loss of life and property damage at Placentia Bay, Newfoundland. In Maine, objects were knocked from shelves, and clocks stopped. Ships at sea also felt the shock strongly.

Saint Vincent *See* SOUFRIÈRE.

Sakhalin *earthquake, Russia* Sakhalin is a large, remote island on the south side of the Sea of Okhotsk north of JAPAN in the northwest PACIFIC OCEAN. It was literally flattened on May 28, 1995, by a huge earthquake with a MAGNITUDE of 7.6. The FAULT that produced the earthquake lies northeast across the area and ruptured several tens of miles along its length. Huge FISSURES opened, and many residents fell into them only to be crushed whey they closed. LIQUEFACTION produced GEYSERS of sand and water that shot up several feet into the air. The town of Neftegorsk, with a population of 3,000, was demolished. In all, some 1,989 people were killed in this event.

Sakurajima *volcano, Japan* Located on the island of Kyūshū on Kagoshima Bay and on the south rim of the AIRA caldera, Sakurajima (or Sakura-zima) is one of the most active volcanoes on Earth. Since its first documented historical ERUPTION in A.D. 708, it has been in nearly constant eruption. It is typically STROMBOLIAN in character, but it erupted with great violence in 1476, 1779, and 1914. An account of the 1779 event reports that strong earthquakes of varying duration preceded the eruption, which began when the volcano's summit exploded and expelled a great white cloud. Hot rocks fell from the volcano and reportedly set houses on fire. Some of the rocks were so large that they were compared to birds in flight even when viewed from a distance of several miles. Darkness covered the area around the volcano, accompanied by thunder and lightning. PUMICE reportedly covered the waters to such a depth that one could walk on it, and two men allegedly used this route to escape from the volcano. Estimates placed the number of dead at approximately 150. One group of refugees hid in a cave and perished there when an AVALANCHE buried the entrance. Snow-white TEPHRA accumulated up to five feet (1.5 m) deep in places. Following the eruption, authorities ordered everyone, regardless of wealth or social station, to work on cleanup.

The 1914 eruption was preceded by earthquake activity and several violent eruptions from KIRISHIMA, a volcano about 30 miles (48 km) away. Settlements around Sakurajima were evacuated, and not one life was reported lost as a direct consequence of the eruption. Thirty-five people were reported killed and more than 100 injured, but as a result of earthquakes, not the eruption. The 1914 eruption started on the morning of January 12. A large cloud burst from a FISSURE on the volcano's west flank shortly after 10 A.M. and rose about five miles (8 km). A powerful earthquake in the early evening generated a TSUNAMI that damaged the waterfront area. A spectacular display of gas-rich lava emerged from the volcano. LAVA FLOWS from this eruption covered some 10 square miles (26 km²) and joined the volcanic island to the mainland, thus converting the island into a peninsula.

The current eruptive phase of Sakurajima began in 1955. The annual number of explosive eruptions ranged from 88 to 282 between 1991 and 1994, and ash emissions ranged from 2.9 to 16.8 million tons during the same period. About 7,000 people live at the foot of the volcano, and about one-half million live 10 kilometers west in Kagoshima City, making the volcano of great concern to Japan.

Sakurajima has been the subject of intensive research into predicting volcanic eruptions. Japanese researchers report that explosions at the CRATER on the summit can be predicted, using data on tilt and strain at the summit and on EARTHQUAKE SWARMS occurring at shallow depths below the crater.

San Andreas Fault *California, United States* Perhaps the most famous FAULT in the world, the San Andreas Fault lies along the CALIFORNIA coast about 500 miles (805 km) in length and marks the approximate boundary between the NORTH AMERICAN CRUSTAL PLATE and the PACIFIC CRUSTAL PLATE. The San Andreas fault is a TRANSFORM margin that exhibits RIGHT-LATERAL strike-slip motion. It is one of the few transform margins on land, about 99+% are undersea. Because SAN FRANCISCO is east of the fault and on the North American plate and LOS ANGELES is west of the fault and on the Pacific plate, the two cities are actually getting closer together all of the time. In fact, in 8 million years, they will be twin cities.

The San Andreas Fault is not a single fault but rather a system of faults joined together in a braidlike pattern. The fault has different character in the north than in the south. The northern half of the San Andreas Fault, which lies in the San Francisco area, has REVERSE FAULTS and is generally mountainous. The southern portion of the fault, which traverses the Los Angeles area, has more normal faults and basins associated with it (Los Angeles basin, Salton Sea). There are many faults of various orientations that come together to the south of Los Angeles (the Garlock Fault, for example). The faults generally wind around the SIERRA NEVADA massif. This meeting of crustal blocks, faults, and mountain ranges has given the Los Angeles area a complex geology. The San Andreas Fault in southern California has numerous tributaries, including the Garlock Fault, Elsinore Fault, White Wolf Fault, and San Jacinto Fault. The NEWPORT-INGLEWOOD FAULT ZONE (actually a string of several shorter faults) also extends through greater Los Angeles. The 1857 FORT TEJON earthquake occurred where the Garlock Fault meets the San Andreas Fault. Even though many of the earthquakes in California are not on the San Andreas Fault, they are all the result of deformation around the fault. The NORTH RIDGE earthquake of 1994, for example was on a BLIND FAULT near Los Angeles. That fault was produced to compensate for the tremendous stress exerted by the nearby San Andreas Fault.

Despite its fearsome reputation, the San Andreas Fault is difficult to see in many locations because development has erased its traces on the surface. Near San Francisco, for example, the fault is clearly visible only in a few places, such as Lake San Andreas, which lies parallel to the fault just south of the city, and Mussel Rock, where the fault extends out to sea several miles away. The San Andreas Fault was named not for a saint of the Roman Catholic Church but rather in honor of Andrew Lawson, a geologist whose study of the 1906 San Francisco earthquakes is considered one of the classics in the literature of geology.

See also PLATE TECTONICS.

San Cristobal *volcano, Nicaragua* A STRATOVOLCANO with a BASALTic cone composed of alternating layers of LAVA and TEPHRA. It is in the San Cristobal volcanic complex, which consists of five volcanoes. The volcano was DORMANT from 1685 to 1971 when it became active. Since then, it erupted eight times in quick succession, the last one in 1977. That eruption lasted one day and had a VEI of 2 making it moderate in size. A new period of activity began in 1997 and continued through the last report in late 1999. There are also unconfirmed eruptions in 1985 and 1987.

sand boil Produced by LIQUEFACTION, a sand boil is the equivalent of a MUD VOLCANO. The two terms are typically interchangeable, though sand boils supposedly contain more sand than mud.

San Fernando *earthquake, California, United States* The earthquake that struck CALIFORNIA's San Fernando Valley near LOS ANGELES on February 9, 1971, is one of the most famous and most intensively studied earthquakes in history. The earthquake, also known as El CENTRO, registered about 6.9 on the RICHTER scale, killed 64 people, and injured about 2,400 others. Most of the fatalities (47) occurred at the Veterans Hospital by Pacoima Canyon, much of which collapsed completely. At another hospital, Olive View, only three people were killed, one by falling debris and two others by

This and other ambulances, crushed by the collapse of the Olive View Hospital, Los Angeles, were rendered useless for rescue operations after the 1971 San Fernando earthquake in southern California. *(Courtesy of the USGS)*

a power failure that cut off their life-sustaining equipment. The Olive View facility was damaged so badly, however, that it was almost a total loss. The extent of damage was remarkable because the building was said to be earthquake resistant. Because the earthquake occurred at 6:01 A.M., injuries and fatalities were few compared to what might have happened if the quake had struck several hours later when more people would be expected at the hospital.

The earthquake came within a narrow margin of destroying the Lower San Fernando Dam and releasing a torrent of water on some 80,000 residents of the valley downstream. Part of the dam collapsed into the reservoir and reduced the freeboard, or the distance between the reservoir level and the top of the dam, to only a couple of feet. Had the earthquake lasted a few seconds longer, the dam might have failed. A curious effect of the earthquake was seen near the intersection of Rajah and Wallaby Streets in Sylmar. Here the ground motion was so intense that it plowed grass under, so to speak, and brought up to the surface fresh subsoil from below. After the earthquake had done its work, only 10% of the surface soil at this location was made up of grassy soil clods. The other 90% consisted of fresh subsoil churned up from underground. Surface ruptures occurred in a zone roughly 10 miles (16 km) long, between San Fernando and Big Tujunga Canyon. Displacement involved slip of more than six feet (1.8 m) in some places along the San Fernando Fault Zone.

San Francisco *California, United States* Widely known as "Earthquake City" or "City That Waits to Die," San Francisco sits beside the SAN ANDREAS FAULT and close to other active faults, including the HAYWARD FAULT just across the San Francisco Bay. Numerous earthquakes have shaken San Francisco since its founding. Nonreinforced adobe houses

A classic picture of a picket fence offset across strike-slip fault movement during the 1906 San Francisco earthquake, Marin County, California. Notice that the segment of fence on the far side of the fault has moved to the right relative to the closer fence segment. The total offset is 8.5 feet (2.6 m). *(Courtesy of the USGS)*

Even geologists were not safe from the 1906 San Francisco earthquake. This photo shows the geology building at Stanford University. During the earthquake, a statue of Louis Agassiz, a famous glacial geologist, fell from its perch on the roof of the building and landed head first on the plaza. *(Courtesy of the USGS)*

reportedly suffered heavy damage from a series of earthquakes that hit the city in June and July 1808. Mission Santa Clara and Mission San Jose were damaged by a strong earthquake in 1822, and two powerful earthquakes struck San Francisco in 1836 and 1838. A big quake in 1851 damaged some buildings severely, and another earthquake the following year is said to have cracked the ground so severely that the waters of Lake Merced drained into the sea. An 1856 earthquake was felt as far away as Stockton and disturbed the waters of the San Francisco Bay. In 1864, an earthquake broke store windows in San Francisco, and another in 1865 destroyed numerous buildings and created a FISSURE two blocks long in the earth along Howard Street. An earthquake in 1868, centered near San Leandro on the east shore of the bay, damaged every building in adjacent Hayward and was felt more than 150 miles (241 km) away. Mark Twain experienced this earthquake while walking down a street in San Francisco. He described the event as "terrific" and noted an interesting detail: Many people suffered motion sickness from the earthquake, and some remained incapacitated for hours and even days later.

The destruction of San Francisco in 1906 by earthquake and subsequent fire is one of the most widely known catastrophes in history, although earthquakes elsewhere have caused greater loss of life. Fire consumed almost 500 blocks (or about 2,800 acres) of the city. Roughly 30,000 buildings were destroyed, about 3,000 of those by fire. The surface rupture from this earthquake extended almost 300 miles (483 km). Thirty schools and 80 churches and convents were listed as destroyed, and approximately a quarter of a million men, women, and children were left homeless. How many died in the earthquake and subsequent fire is still uncertain. Initial reports put the number of dead at 315, plus 352 missing and unaccounted for, but the total of casualties has risen to more than 2,000 as information continues to accumulate about victims of the earthquake. The earthquake, estimated at 8.2 on the RICHTER scale, struck San Francisco at 5:13 A.M. on April 17, 1906. The first tremor lasted some 40 seconds and was followed by another shock lasting a minute and a half. Falling chimneys damaged or destroyed numerous wood-frame homes. As many as 80 guests were reported killed at one particular hotel, though not all of them by the earthquake itself; many of them were said to have drowned when a ruptured water main flooded the first floor. Reports indicate that a cooking fire in the area south of Market Street burned out of control, and the fire spread quickly through the city, which was built without close attention to fire safety and therefore was soon ablaze. The army tried to slow or halt the fires spread by using dynamite to blow up buildings and create firebreaks, but the powder they used may itself have ignited fires and contributed to the conflagration. The army also used artillery to generate firebreaks by shelling houses to the ground, but this measure proved ineffective. San Francisco was rebuilt on much the same layout as before, but this time planners and builders incorporated into the city an extensive system of reservoirs, secondary water mains, and cisterns to provide an adequate water supply for fighting future fires.

Sangay *volcano, Ecuador* Sangay is the southernmost STRATOVOLCANO in the northern volcanic zone of the ANDES. The first documented eruption of Sangay was in 1628. It has been in nearly constant eruption since 1934. The eruptions are generally STROMBOLIAN with mild explosions. In 1976, an expeditionary team was caught in an explosion and two people died.

San Jacinto Fault Zone *California, United States* The San Jacinto Fault Zone includes a number of FAULTS in the greater LOS ANGELES area and is responsible for most of the strong earthquakes there. The fault zone extends for some 200 miles (322 km) along the CALIFORNIA coast and includes the San Jacinto, Glen Helen, Lytle Creek, Claremont, Casa Loma, Hot Springs, and Clark faults. The potential for powerful and highly destructive earthquakes appears to be greater, however, along the nearby SAN ANDREAS FAULT because the fault segments in the San Jacinto Fault Zone are comparatively short (about 50 miles [80 km] long or less) and separated from one another.

San Marcelino *See* COATEPEQUE.

San Pedro *See* ATITLÁN.

Sanriku *earthquake, Japan* During the late morning of June 15, 1896, a large earthquake struck the Tuscarora Deep in the SUBDUCTION ZONE off the east coast of JAPAN. The tremors were slow and undulating and felt from the Hokkaido to the KWANTO region but with no real damage. The disaster came about 20 minutes later when a TSUNAMI struck the coast. In most areas, this wall of water was 10–20 feet (3–6 m) high. In several others, it was 75 feet (22.5 m) high and even 100 feet (30 m) high in one location. In Sanikru, people were celebrating the Boys Festival; thousands of people gathered along the seashore were subsequently swept out to sea. The DEATH TOLL was a staggering 27,122, making the event one of the worst disasters in Japanese history. There were 106,170 houses destroyed. Deepwater fisherman returned at the end of the day to find the harbors strewn with bodies and the wreckage of houses. The tsunami was recorded on tidal gauges at Sausalito, CALIFORNIA, the next day.

Santa Ana *volcano, El Salvador* The first recorded eruption of this STRATOVOLCANO was in 1520. Since then, it has erupted at least 12 times. The most recent eruption was in 1920, but there are few details available on it. There is still fumarolic activity in the CRATER.
See also COATEPEQUE; FUMAROLE.

Santa Barbara *See* CALIFORNIA.

Santa María *volcano, Guatemala* The ANDESITIC stratovolcano Santa María ("The Naked Volcano") has a CRATER that contains a small active LAVA DOME called Santiaguito that began to form in 1922. Santa María is famous for its first historic eruption in 1902. It was the second-largest volcanic eruption of the 20th century (VEI = 6). The eruption lasted 19 days and produced 1.3 cubic miles (5.3 km³) of PYROCLASTIC emissions. One PLINIAN ERUPTION produced an eruption column that was 16 miles (25.6 km) high. More than 5,000 people were killed during this event, and ASH reached as far away as SAN FRANCISCO. The activity continued in the crater until 1913. The lava dome Santiaguito has been nearly continuously active since 1922 but with six main periods of rapid extrusion to date. In 1929, the most damaging eruption produced a large pyroclastic flow in a partial collapse of the dome. From several hundred to perhaps as many as 5,000 people were killed as the flow traveled more than six miles (10 km). From 1972 to 1975, a particularly vigorous phase of activity was recorded. The volcano is currently active.

Santorini *caldera, Greece* Located on the island of THIRA (formerly known as Thera), Santorini was the site of the catastrophic explosion that destroyed the Minoan civilization and much of the island around 1470 B.C. and is thought to have given rise to the myth of ATLANTIS. During some 2,000 years following that ERUPTION, a dome complex emerged in the center of the underwater CALDERA and became the Kameni Islands, Micra Kameni and Palaea Kameni. Palaea Kameni may have appeared in 197 B.C. and was enlarged in A.D. 46 and 726, and the dome of Micra Kameni appeared between 1570 and 1573. Although there were reports of an eruption in 1457, it appears that a cliff on Palaea Kameni merely collapsed.

Seismic activity in 1649 required residents of Santorini to evacuate, and very powerful earthquakes occurred in 1650,

An old engraving of the 1866 eruption of Nea Kameni, Santorini. The eruption of Santorini in 1,650 B.C. was probably the largest on Earth in the past 10,000 years. It destroyed the Minoan civilization and may explain how in the Bible Moses parted the Red Sea. The Island of Thira was destroyed by the explosion. It may have been the lost continent of Atlantis. *(Courtesy of NOAA)*

followed several days later by an eruption that formed a cone. The cone was eroded by wave action and became Colombo Banks. Apparently, subsidence followed this eruption because the sea reportedly moved inland in at least two locations on Thira, the main island. In May 1707, earthquakes preceded the emergence of a shoal some 20 feet (6 m) high, called Nea Kameni, off the northeast shore of Micra Kameni. One account says that mariners, curious to see what was happening, went to examine the shoal and came back to report that a new shoal was rising from the seafloor. The following day, more investigators set out for the shoal and managed to land on it, even though the shoal was growing perceptibly under them. The explorers brought back souvenirs, including some large oysters that were found clinging to a rock and were said to have an excellent flavor. In July 1708, the water grew hotter and more turbulent at the site during the next several weeks, and the smell of SULFUR intensified. During an explosive event that month, Nea Kameni continued rising, while Thira and Micra Kameni reportedly subsided about six feet (1.8 m). The new island rose to a height of about 30 feet (9 m) and a diameter of approximately three miles (4.8 km) by November 1708 and attained a height of almost 250 feet (76 m) by 1711. Micra Kameni subsided during this period of Nea Kameni's growth.

In the early 19th century, copper-bottomed ships took advantage of the submarine volcanism at Nea Kameni to clean their hulls. The ships simply anchored in the bay for a while and let the sulfur-laden waters clean and polish the metal sheathing on their hulls. Late in January 1866, FISSURES appeared between Nea Kameni and Micra Kameni. Earthquakes occurred, the water grew hotter, and gas rose through the bay water south of Nea Kameni. The south side of Nea Kameni subsided dramatically. An eruption began around February 1, and a cone perhaps 15 feet (5 m) high reportedly formed on the seafloor. Earthquakes continued all through the eruption. The LAVA DOME rose, while other nearby areas subsided; at one place along the shore of Nea Kameni, a pier sank almost six feet (1.8 m) between May 1866 and March 1867. Several new domes had grown on the south side of Nea Kameni by 1870.

The temperature rose in the waters around Nea Kameni in 1925, and local earthquakes were reported. At one point, the heat and the turbulence were so great that navigation ceased in the area. An explosive eruption in August 1925 led to the formation of lava domes and LAVA FLOWS. Seismic and eruptive activity continued in the late 1920s. An eruption occurred in 1939, and a dome formed in 1950.

satellite observations Earth satellites have proven to be useful tools for monitoring volcanic ERUPTIONS. Satellites cover large areas of the globe in a single field of view and can provide multispectral images that can reveal images invisible in a single band of the spectrum. A famous example of satellite monitoring of a volcanic eruption occurred in 1980 when the U.S. Geostationary Operational Environmental Satellite (GOES) photographed the ongoing eruption of Mount SAINT HELENS. In the UNITED STATES, satellite data have also been used to monitor eruptions of AUGUSTINE and REDOUBT volcanoes in ALASKA. Weather satellites have monitored ash clouds from other eruptions, including those of GALUNG GUNG in 1982, MAYON in 1984, and KELUT in 1990. There are limits to what satellite imagery can reveal. It can be difficult for satellites to distinguish between the cloud from a volcanic eruption and ordinary clouds in the surrounding atmosphere. Satellites can, however, help monitor ASH clouds from major eruptions that pose hazards to aircraft. Australia's Bureau of Meteorology uses images from the Japanese Geostationary Meteorological Satellite to help in providing warnings to aviators about clouds of volcanic ash over AUSTRALIA and in areas to the north, where volcanic eruptions are commonplace. The Australians base their warnings also on reports from aircraft and from information supplied by the country in which the eruption occurs.

scarp *See* ESCARPMENT.

scoria A lightweight, BASALTic volcanic rock with a dense concentration of VESICLES, scoria differs from pumice in the density of the gas bubbles and the lack of a glassy texture. Scoria is composed of identifiable volcanic rock. Scoria will not float on water as will PUMICE. It may accumulate around a volcano in quantities great enough to form a cone several hundred feet high.

seamount A seamount is a volcano that rises from the ocean floor. Some seamounts reach above the ocean surface and form islands, of which the HAWAIIAN ISLANDS are good examples, or atolls, where the summits are covered with coral reefs. If the seamount does not quite reach the surface and has a flattened crown indicative of erosion by wave action in the past, the mountain is called a guyot. Seamounts may occur in chains or clusters, like the EMPEROR SEAMOUNTS that connect to the Hawaiian chain. Seamounts spaced very near to one another are known as aseismic ridges.

secular Secular changes are those that occur slowly, on a long-term basis. Examples of secular changes include the slow growth of a mountain chain or the slow buildup of STRESS in a FAULT zone between earthquakes.

sediment *See* SEDIMENTARY ROCK; SEDIMENTATION.

sedimentary rock Rock that was deposited in the form of particles, such as sand, silt, or clay or that was precipitated from a solution and then transformed into rock. The material making up sedimentary rock was originally part of another rock, IGNEOUS, sedimentary, or METAMORPHIC, but was either physically removed (eroded) or chemically removed (dissolved). The particles or solution was then transported and finally deposited by wind, water, or glaciers. Sedimentary rocks vary greatly in composition and texture. They are generally subdivided into either clastics (deposited as particles) or precipitants (from solution). Examples of sedimentary rock are sandstone, shale, and limestone.

sedimentation The process by which clastic or dissolved material is laid down to form sedimentary rock. Sedimentation may involve deposition by air and/or water, liquid, or ice and generally produces horizontal layers of sediment, although sloping beds of sediment may occur in some situations.

A volcanic rock that is shot full of vesicles, holes left by escaping gas. If the volcanic rock has only a few holes, it is termed *vesicular.* Foam textured rock composed completely of glass and holes is termed *pumice. (Courtesy of the USGS)*

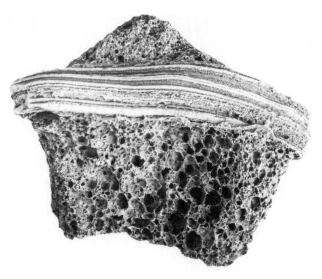

Ribbon scoria is a mixed volcanic rock showing a two-stage history of flow and explosion. *(Courtesy of the USGS)*

Segara Anak *caldera, Lombok, Indonesia* Located on the island of Lombok, this CALDERA includes Gunung Rinjani, a STRATOVOLCANO on its east rim, and Gunung Baru, a cone within the caldera. A strong earthquake in 1884 preceded an eruption of Gunung Rinjani, which is said to have expelled smoke and fire for several days. Another earthquake in 1898 was suspected of having originated at Gunung Rinjani. Two strong earthquakes occurred under Gunung Rinjani in November 1903, and the volcano was reported smoking in November 1915. Several weeks after an earthquake early in 1966, a climber reported seeing a flow of LAVA from Gunung Baru and detecting a strong smell of SULFUR. The most recent eruption was in 1994 when a LAHAR killed 30 people. The eruption produced ASH clouds but was relatively mild.

segmentation Many FAULTS are not a single continental-scale plane of weakness but a series of small fault pieces or segments that are stuck together. The slip on a fault to produce an earthquake typically occurs along a single segment. Segmentation may form from bends in fault planes, intersections with other faults, or lateral changes in rock units.

seiche A periodic oscillation in an enclosed, or nearly enclosed, body of water. Basically, seiches are large waves that slosh back and forth in an enclosed basin. These waves can be large (up to 100+ feet [30.8 m]) and cause considerable destruction. Earthquakes may contribute to producing seiches.

seismic gap This expression refers to a recently "quiet" portion of an active earthquake-generating FAULT. This section forms a gap on a map of recent earthquake EPICENTERS along the fault and may be a likely spot for strong earthquakes in the future because the forces that generate earthquakes may be building up along the gap and "waiting" to be released.

seismic moment The seismic moment of an earthquake is the measure of its strength based upon the area of the FAULT rupture, the amount of slip on the fault in the event, and the force required to overcome the friction or strength of the rock opposing the movement. Seismic moment is usually calculated from the AMPLITUDE spectra of seismic waves on a SEISMOGRAM.

seismic sea wave *See* TSUNAMI.

seismic velocity The speed at which earthquake waves travel through Earth. The speed is a function of the rock through which the waves travel. It depends upon the elastic moduli of the rock. There are measures of incompressibility and resistance to shear and torsion as well as density that determine the speed. Basically, seismic waves travel faster in hard, tight rock than in SEDIMENTS or soft (DUCTILE) rocks or liquids.

seismogram A SEISMOGRAPH records the arrivals of incoming seismic waves through time. It prints those records on a seismogram. The piece of paper showing the wiggly lines of an earthquake's, P-waves, S-WAVES, or SURFACE WAVES is a seismogram.

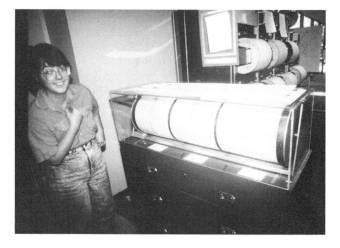

The USGS seismic station at Menlo Park, California. The three recorders show vibrations in the three dimensions, records shown are those of the 1989 Loma Prieta earthquake. *(Courtesy of the USGS)*

seismograph A device for detecting and recording the vibrations of earthquakes.
See also SEISMOLOGY.

seismologist A GEOPHYSICIST who studies earthquakes and earthquake phenomena. RICHTER scales, MERCALLI intensities, FAULT PLANE SOLUTIONS, and FOCUS and EPICENTER locations as well as a full analysis of the seismic waves and damage are conducted by seismologists.
See also SEISMOLOGY.

seismology The scientific study of earthquakes. Earthquakes generate several kinds of waves. These are sometimes

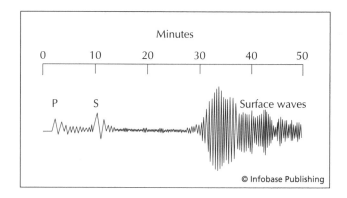

Typical seismogram that is registered on a seismograph. The wiggly line is what appears on a seismograph as an earthquake occurs. Each wiggle represents a different kind of wave: P-wave, S-wave, or surface waves (both Rayleigh and Love waves). The great delay between each arrival means that the focus of the earthquake is very far away because each wave travels at a different velocity. If the focus was closer, the arrivals would be closer together. If the seismograph was at the focus, all waves would be grouped together in one set of wiggles in which no specific wave type could be discerned. Notice that the surface waves give higher amplitudes and longer durations. They cause the most damage.

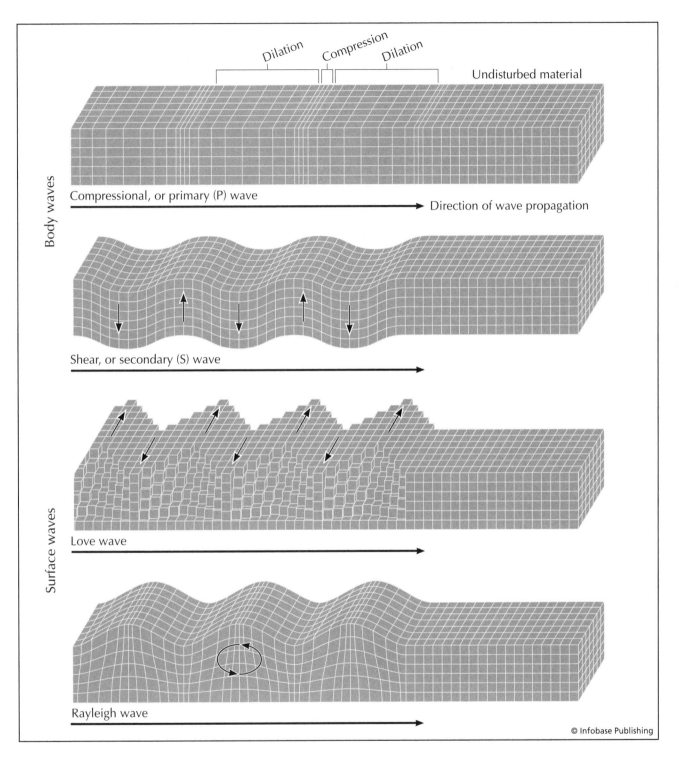

Diagram showing the response of rock as different types of seismic waves pass through. No permanent deformation of the rocks occurs from the passage of these waves. Body waves include P- and S-waves. A P-wave is like a low speed accident, where the bumper is compressed at impact but expands (dilation) and recovers after the impact. An S-wave is like whipping a jump rope. In surface waves, the ground surface moves like an ocean wave in a Rayleigh wave. In Love waves, the surface moves from side to side as the wave passes through. It causes structures to fall over sideways.

classified in two categories: SURFACE WAVES and BODY WAVES. Surface waves, are confined to the surface and shallow sub-surface of Earth like ocean waves are confined to the surface and near surface of the water. They are very slow. There are two kinds of surface waves, named LOVE WAVES and RAY-LEIGH WAVES after their discoverers. Love waves move like a snake. They exhibit horizontal shear, meaning a side-to-side or transverse shaking motion as they go forward. They cause buildings and other structures to fall over sideways. Rayleigh waves make the surface of Earth move like an ocean wave. They involve an elliptical motion of the ground that is actually retrograde, or opposite to the direction of travel. In other words, for a wave moving to the right, the movement of the rock will be counterclockwise. Body waves are also called P-waves and S-WAVES, named for the arrival times: P is the primary or first arrival, and S is the secondary arrival. P-waves are the fastest of the seismic waves. They move by compression and subsequent rarefaction of the rock media through which they move, in much the same manner as sound waves traveling through the air or vibrations from a hammer strike on a rock. S-waves move in a transverse motion like the wave generated by whipping a jump rope. They also have a similar motion to a Love wave and are slower than P-waves.

FACTORS AFFECTING DESTRUCTION FROM EARTHQUAKES

The extent of destruction from an earthquake depends on several factors:

Magnitude and Intensity

Earthquake magnitude refers to how much energy the earthquake released. Magnitude is not always proportional to damage. Under some conditions, such as high population density in the affected area, a relatively minor earthquake may do far more damage than a more powerful quake in a sparsely populated area.

There are two scales that are most commonly used to measure earthquakes: the RICHTER scale and the MERCALLI scale. The Richter scale, named after Dr. Charles Richter of the California Institute of Technology, is a logarithmic scale (expressed in powers of 10—*see also* MAGNITUDE) that reflects the AMPLITUDE of ground movement. Several steps are involved in establishing an earthquakes magnitude on the Richter scale. First, the location of the quake must be determined. It is done by noting the time a wave passed two widely separated locations and inserting the data into a set of equations to determine how far each station was from the origin of the quake. The next step is to estimate the intensity of ground motion at the earthquake's origin by using a correction factor that introduces some error into the calculation of

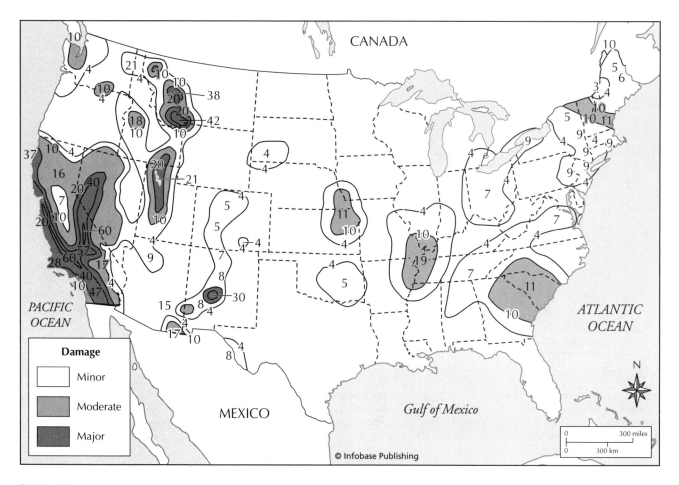

Seismic risk map for the United States. Based on historical records of earthquakes, both intensity and recurrence interval, the country is divided into zones of earthquake risk. Certain areas will be relatively low-risk for earthquakes over a 50-year interval, whereas others are high.

Energy Released and Equivalent Energy of Richter Magnitudes

Magnitude	Energy Released (ergs)	Energy Equivalent
-2	6×10^8	100 Watt-light left on for one week
-1	2×10^{10}	
0	6×10^{11}	Seismic waves from one pound of explosive
1	2×10^{13}	A 2-ton truck traveling at 75 mph
2	6×10^{14}	
3	2×10^{16}	Typical lightning bolt
4	6×10^{17}	Seismic waves from 100 tons of explosives
5	2×10^{19}	Average tornado
6	6×10^{20}	Hiroshima 1945 atomic bomb
7	2×10^{22}	One-megaton nuclear explosion
8	6×10^{23}	Average kinetic energy of a hurricane
9	2×10^{25}	Average annual seismic energy released on Earth
10	6×10^{26}	U.S. annual energy consumption
11	2×10^{28}	Earth's annual internal heat flow
12	6×10^{29}	Earth's daily receipt of solar energy

magnitude. Measuring magnitude on the Richter scale does not yield precise results, but within certain limitations, it does provide a useful estimate of the strength of the earthquake. The Richter scale starts at zero and is open ended. An earthquake of magnitude 1 on the Richter scale probably would go unnoticed by anyone but an observer with a SEISMOGRAPH. Magnitude 4 earthquakes are mild but still perceptible, and earthquakes of magnitude 8 and above are devastating if they strike densely populated areas. The SAN FRANCISCO earthquake of 1906 is estimated to have been about 8.2 on the Richter scale, and two other earthquakes of the 20th century—one beneath the PACIFIC OCEAN near JAPAN in 1933 and the other off the coast of ECUADOR in 1933—are thought to have ranked around 8.6.

The second common scale for measuring earthquake INTENSITY is the Mercalli scale and modified Mercalli scale. It uses Roman numerals to avoid confusion with the Richter scale and is based on easily observable effects, such as destruction of buildings.

Seismic moment is another measure of the size of an earthquake. Seismic moment is the product of the average amount of slip, the area of the rupture and the shear modulus, or strength, of the rocks affected. An expression of magnitude related to seismic moment is moment magnitude, which is approximately the same as Richter magnitude in range 3–7.

Duration

Some earthquakes last only a few seconds, but others have been known to last for a minute or two. As a rule, damage increases with duration.

Local Geology

The character of underlying material does much to determine the extent of damage from an earthquake. The safest material on which to build, generally speaking, is solid rock, which transmits vibrations from an earthquake but maintains its structural integrity. More dangerous is unconsolidated sediment, which may exhibit LIQUEFACTION and extensive consequent damage where GROUNDWATER rises close to the surface. Fractured rock, as found in many parts of the western UNITED

Mercalli Intensity	Observed Effects
I	Not felt at all
II	Felt by only a few individuals
III	Felt indoors by many persons, but not necessarily identified as an earthquake
IV	Felt both indoors and outdoors and comparable to vibrations produced by a passing train or heavy truck
V	Strong enough to awaken a sleeping person; small objects fall off shelves, and beverages may splash out of cups or glasses
VI	Perceptible to everyone; pictures fall off walls, and plaster may fall from ceilings.
VII	Difficult to stand upright; ornamental masonry falls from buildings; waves may occur in ponds and swimming pools.
VIII	Mass panic may occur; chimneys and smokestacks may fall; unsecured frame houses slide off their foundations
IX	Heavy damage to masonry structures and underground pipes; large cracks open in ground; panic is general
X	Numerous buildings collapse; water splashes over riverbanks
XI–XII	Virtually complete destruction

STATES, may absorb energy from earthquake waves and thus reduce the geographical extent of damage from a major earthquake. By contrast, in areas where geological formations stretch unbroken for hundreds of miles, as in portions of the eastern United States, vibrations may travel great distances with much of their initial destructive potential intact. The 1906 San Francisco earthquake, for example, was barely felt at all outside CALIFORNIA, but the NEW MADRID, MISSOURI, earthquakes approximately a century earlier were perceived over much of the United States east of the Mississippi River. The irony here is that the same processes that predispose a given part of Earth's CRUST to major earthquakes may also confine damage from earthquakes through extensive fracturing, whereas comparatively earthquake-free regions may harbor much greater potential for destruction because of the relatively undisturbed, contiguous rock that underlies them.

Time of Day

This is an important factor in determining the number of casualties from an earthquake. Deaths and injuries from collapsing buildings in rural areas, for example, are likely to be more numerous if an earthquake occurs at night or in the early morning when the population is indoors sleeping or having breakfast. In cities, on the other hand, deaths and injuries may be more likely during the working day, when large numbers of pedestrians are exposed to harm from falling walls and ornamental masonry.

Architecture

Wood-frame buildings have a reputation for standing up well to earthquakes because wood can bend and sway under earthquake vibrations, thus improving a building's chance of survival. Buildings of brick and stone construction, by contrast, are more likely to crumble and collapse because they lack wood's resiliency, although careful design and construction may increase the survivability of a non-wooden structure. Possibly the worst building material in regard to earthquake resistance is adobe, which was responsible for numerous fatalities during earthquakes in the early years of Euro-American settlement of the western United States. *See also* FORT TEJON earthquake for an illustration of the dangers of adobe construction.

Predicting Earthquakes

Earthquake prediction has been a goal of scientists for centuries, and efforts in this direction have often been colorful, even when unsuccessful. Charles Mackay, in his 1841 book *Extraordinary Popular Delusions and the Madness of Crowds*, tells how one man named Bell apparently used a primitive mathematical model to predict the date of an earthquake in England during the 18th century:

> In the year 1761, the citizens of London were alarmed by two shocks of an earthquake, and the prophecy of a third which [supposedly] was to destroy them altogether. The first shock was felt on the eighth of February and threw down several chimneys in the neighborhood of Limehouse and Poplar; the second happened on the eighth of March and was chiefly felt in the north of London. . . . It soon became the subject of general remark, that there was an interval of exactly a month between the shocks; and a . . . fellow named Bell, a soldier in the Life Guards, was so impressed with the idea that there would be a third in another month that he lost his senses altogether and ran about the streets predicting the destruction of London on the fifth of April [that is, four weeks after the previous earthquake].

Mackay tells how Bell's prediction affected the public:

> There were not wanting thousands who confidently believed the prediction and took measures to transport themselves and their families from the scene of the impending calamity. As the awful day approached, the excitement became intense, and great numbers of credulous people [migrated] to all the villages within 20 miles (32 km), awaiting the doom of London. Islington, Highgate, Hampstead, Harrow, and Blackheath were crowded with panic-stricken fugitives who paid exorbitant prices for accommodation to the housekeepers of these secure retreats. Such as could not afford to pay for lodgings at any of these places remained in London until two or three days before the time [of the prediction], then encamped in the surrounding fields, awaiting the tremendous shock which was to lay their high city all level with the dust. . . .
>
> [The] fear became contagious, and hundreds who had laughed at the prediction a week before packed up their goods when they saw others doing so and hastened away. The [Thames] river was thought to be a place of great security, and all the merchant vessels in the port were filled with people who passed the night between the fourth and fifth on board, expecting every instant to see St. Paul's totter and the towers of Westminster Abbey rock . . . and fall amid a cloud of dust.

Even when the predicted earthquake failed to occur on April 5, the public's fears took days to subside completely:

> The greater part of the fugitives returned on the following day . . . but many judged it more prudent to allow a week to elapse before they trusted their dear limbs in London. . . . [Bell] lost all credit in a short time and was looked upon by even the most credulous as a mere madman. He tried some other prophecies, but no one was deceived by them; and . . . a few months afterwards, he was confined in a lunatic asylum.

Such "earthquake prophets" have remained active from Bell's day to ours, but only in recent years has knowledge of earthquakes and their origins advanced to the point where accurate prediction of earthquakes, or some quakes at least, has become a possibility.

Determining the potential for future rupture along a fault can be difficult because faults differ in their behaviors. One fault may show no activity for many centuries and then suddenly move 30 feet (9 m) or more, whereas another fault may exhibit virtually continuous, gradual movement and many small earthquakes. This variability in behavior means that geologists cannot make accurate predictions of a fault's future activity merely by establishing that the fault is active. Two selected faults, though both active, may move at greatly different intervals and exhibit equally great differences in rates of slip. Also, the rate of slip along an individual fault may change. The slip rate along California's SAN JACINTO FAULT ZONE, for example, is believed to have fluctuated by more than 1,200%.

Another source of uncertainty in determining slip rate and history of activity is difficulty in estimating the ages of offset deposits or other features along a fault. In the Los Angeles area, for example, such estimates are unreliable in many or most locations where faults are active. Yet another problem in determining the true slip rate for a fault is the partial nature of information on components of slip. Data may be restricted to either the vertical or horizontal component. Such limited data may yield an unreliable estimate of slip along a fault, although the horizontal or vertical component alone may prove useful if the ratio of vertical to horizontal motion along a fault is known already. (In situations involving a STRIKE-SLIP FAULT, where motion along a fault is primarily horizontal, the horizontal component alone may provide a reasonably accurate measure of the slip rate. This is the case along the SAN ANDREAS FAULT in California. Vertical component data reportedly have provided an approximation of true slip rates along the reverse-slip faults in California's TRANSVERSE RANGES.) The best information that can be supplied, in many cases, is an average figure for slip rate over a long period of time. As a rule, however, the higher the average slip rate, the more active the fault and the more closely it bears watching as a potential source of future earthquakes. Only in a few areas—namely, boundaries between major plates of Earth's crust, as along the San Andreas Fault—do faults show very high average slip rates, perhaps a half-inch or more per year. Active faults in other areas generally show less slip.

The future behavior of a fault may be inferred, to some extent at least, from evidence including the rate of slip, the size of earthquakes and intervals between them and the amount of slip in each incidence of movement. No one set of criteria exists, however, for determining how active a fault may be in the future.

Complicating such analysis further is the fact that some active faults do not extend to the surface of Earth and, on the surface, may display only faint evidence of recurring seismic activity. A highly damaging earthquake that struck Coalinga, California, in 1983 involved such a BLIND FAULT similar to the NORTH RIDGE earthquake of 1994. Moreover, simply establishing that a fault is active does not allow geologists to make accurate predictions of its future activity. Ongoing measurements of seismic activity are useful tools for estimating the likelihood of earthquakes along a given fault in the future. Seismic data may be misleading, however, because ongoing earthquake activity, or the absence of it, along a fault does not necessarily reflect the potential of that fault for generating destructive earthquakes. An active fault may have very little potential to cause destructive earthquakes because it gradually releases its energy through CREEP, without actually generating earthquakes. Also, data on seismic activity along a particular fault may span too short a time to provide a useful means of predicting the behavior of earthquake-generating faults over a long period. Even when historical records are more extensive, as in parts of Asia, great uncertainties remain in reconstructing the seismic history of a given locality or region.

Despite these limitations, certain methods exist for estimating the size and frequency of possible future earthquakes along a fault. These methods involve, but are not limited to, analyzing the earthquake history of a region to find the biggest seismic event linked to a given fault and comparing a given fault's history of earthquake activity with that of other faults similar in structure and tectonic characteristics. Another approach is to use empirically established relations between earthquake magnitude and length of faults. It is assumed that the greater the dimensions of the fault surface involved in an earthquake, the greater magnitude the earthquake will have. One widely used method for estimating the most powerful earthquake that is likely to occur on a given fault rests on the assumption that half the total length of the fault may rupture in a particular earthquake. This approach is not fully reliable, however, because experience has shown that a major earthquake may involve rupture of anywhere from only a small percentage to almost the whole extent of a fault surface that existed prior to the earthquake. Another drawback to this method is difficulty in measuring the length of a given fault accurately. Much of a particular fault may lie hidden under sediment and water. In greater Los Angeles, the 1857 Fort Tejon earthquake is thought to represent about the most powerful earthquake that might affect that area, although an earthquake of still greater magnitude is possible in theory. It has been suggested that the whole San Andreas Fault might rupture in southern California, producing a single gigantic earthquake; but there are questions about whether or not the geology of the region would allow such a single catastrophic event.

Semeru *volcano, Java, Indonesia* One of the world's most active volcanoes, Semeru has erupted at least 55 times since 1818. The typical eruption has a VEI = 2–3 and produce LAVA FLOWS and *NUÉE ARDENTE*s on several occasions. Ten of the eruptions produced fatalities including some 600 death from LAHARs during the 1909 and 1981 eruptions. The current eruptive period began in 1967 and typically produces VULCANIAN-type eruptions. To date nearly 500 people have been killed by PYROCLASTIC FLOWS and AVALANCHES.

Semisopochnoi *volcanic island, Alaska, United States* Semisopochnoi Island is located in the western ALEUTIAN ISLAND arc and consists largely of a single volcano, Pochnoi. It is the youngest volcanic island in the western Aleutians. It is composed of an older SHIELD VOLCANO. The central portion collapsed and formed a CALDERA following the eruption of a great volume of PUMICE, although the date of the eruption and collapse is uncertain. After the caldera formed, several new cones formed. Accounts of eruptive activity from the 18th and early 19th centuries are spotty, but there is reason to think the volcano was active in or around 1772, 1790, 1792, and 1830. An eruption appears to have occurred in 1873. A plume arose from the island in 1987. Several days after this event, a pilot reported seeing snow darkened by ASH. That eruption was explosive with a VEI = 2.

Serravalle di Chienti *earthquake, central Italy* On September 26, 1997, an earthquake of MAGNITUDE 5.7 of occurred. A total of 11 people were killed, 100 people were injured, and 80,000 buildings were damaged or destroyed. The earthquake received extensive press coverage because of the damage it did to the Basilica of St. Francis at Assisi. Footage of a collapsing roof in a church was on the evening news throughout the world.

Sete Cidades *caldera, Azores* The Sete Cidades CALDERA on San Miguel Island has undergone eruptions on at least eight occasions within historical times. An ERUPTION occurred around the year 1444 from a vent on the southwestern floor of the caldera. In 1638 very strong earthquakes in May and early June were followed in July by an underwater eruption from a vent to the west of the caldera. A SUBMARINE ERUPTION in 1682 followed strong earthquake activity, and an EARTHQUAKE SWARM in 1713 may have accompanied an underwater eruption west of Sete Cidades. Strong earthquakes in 1810 and early 1811, including one that caused widespread destruction on June 24, 1810, were followed in 1811 by a submarine eruption several hundred yards southeast of the site of the underwater eruption of 1638. This 1811 eruption produced an islet about a mile and a half (2.4 km) off the western shore of the island. The eruption subsided in early February but resumed in June, when powerful earthquakes shook the island and a strong odor of SULFUR was detected. An underwater eruption began on June 14, 1811, near the site of the eruption in February. Further submarine eruptions were recorded in 1861 and 1880. The caldera contains two crater lakes.

shadow zone On the opposite side of Earth from an earthquake, there are no S-WAVE arrivals. S-waves move with a shearing motion. This shearing motion cannot pass through liquid. Therefore S-waves cannot pass through the liquid core. Seismic waves that must pass through the liquid core to reach the SEISMOGRAPH do not include S-waves. Only P-waves arrive, and their path is so bent (REFRACTED) that there is even a small zone where they do not arrive. The shadow zones are where there are no arrivals at all or no S-wave arrivals. This phenomenon is the main way we can tell that the OUTER CORE is liquid.

ShakeMap ShakeMap is a new tool developed for earthquake response in urban areas that is only available in CALIFORNIA at this point. When an earthquake strikes, emergency management personnel must quickly decide where to concentrate their meager resources to best address the disaster. An overwhelming amount of information on damage

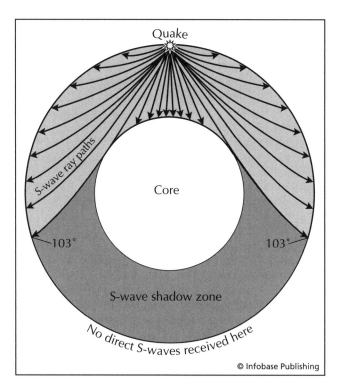

Schematic diagram of the Earth showing travel of S-waves from the earthquake focus to the surface, where they are detected by seismographs. Because S-waves cannot pass through liquid, they are absorbed by the liquid outer core. Seismographs on the opposite side of the Earth from the earthquake do not receive any S-wave arrivals and therefore a shadow zone is formed. A less extensive shadow zone for P-waves is also formed based upon refraction of the waves.

is received and can only be processed in a cursory manner. Because ground shaking causes the most damage to human-made structures, knowing where the ground will shake the most will determine the areas of highest damage. ShakeMap is a map showing the distribution of relative shaking intensity and thus the highest risk for damage. It is constructed using geological bases as well as experience from previous earthquakes. As the seismic waves are still passing through, rescue teams can be mobilizing to the areas of highest damage potential to shave precious minutes or to save the otherwise victims of the disaster.

Shalla *See* ASAWA.

shallow-focus *earthquakes* Earthquakes with foci of 44 miles (70 km) depth. By far and away, these are the most common depth for earthquakes. Earthquakes are only produced during BRITTLE deformation and brittle deformation mainly occurs at depths of less than six to nine miles (10–15 km) in the CRUST. The most common DEEP-FOCUS earthquakes occur in SUBDUCTION ZONES.

Shasta, Mount *Cascades, California, United States* The second tallest of the Cascade STRATOVOLCANOes includes a second cone, Shastina on its shoulder. Mount Shasta is an active volcano. It has erupted 11 times in the past 3,400 years including three events in the past 750 years. The last eruption was in 1786. There is a 300,000-year-old AVALANCHE deposits that extend 27 miles (43.5 km) from the base. This event deposited eight times the debris of the 1980 Mount SAINT HELENS event. The explosion must have been huge. There is pressure to build ever closer to Mount Shasta. Is it worth the risk?

shear   Shear is a type of stress in which force is transmitted by sliding, like a hand sliding on a table, rather than simply by pressing against the object. This motion cannot be transferred to liquid or gas. In SEISMOLOGY, it is defined as motion at right angles to the direction in which an S-WAVE or SURFACE WAVE, from an earthquake is advancing (propagation direction). Shear waves are also known as transverse waves.

shear wave (S-wave)   An S-wave is named as such because it arrives second after the primary wave. *S* stands for secondary, not shear. The S-wave, however, exhibits shear motion as well. If you slide your hand across a table, you are exerting shear on it. You can move a book on the table in this manner or even the table if it is small enough. You are exerting shear stress on it. This is the way shear waves travel. If you substitute liquid or gas for the table, none of the shear stress will be transferred. Your hand will just slide across the top. In the same way, shear waves (S-waves) cannot pass through a liquid or a gas.

Sheveluch *volcano, Kamchatka, Russia*   A large STRATOVOLCANO in north-central KAMCHATKA PENINSULA. It has a small internal cone with a large CALDERA where the recent eruptions occur. After 150 years of DORMANCY, in 1854, Sheveluch erupted violently. The explosion created a 1.25-mile (2-km)-wide crater from which huge PYROCLASTIC FLOWS were emitted. Some of the flows extend an unbelievable 65 miles (105 km). This eruption was the largest in historic times. It has erupted at least 28 times since then. The two major eruptions were in 1944–50 and 1964. The 1964 eruption was the larger of the two and produced a huge

CRATER, an 11-mile (18-km) long and locally 160-foot (49-m)-thick pyroclastic flow. It also produced a cloud with impressive electrical output (thunder and lightning). Minor eruptions occurred in 1879–83, 1897, 1905, 1928–29, 1980, 1984, 1985, 1986, 1988, 1989, 1990, 1991, 1993, and 1996.

shield volcano   A volcano that is shaped like an ancient shield. It is a relatively flat mound with gradual slopes, formed by eruptions of BASALTIC LAVA. The basalt has such low viscosity and lacking in EJECTA that it cannot stack up a steep slope. OLYMPUS MONS on MARS and the HAWAIIAN ISLANDS are examples of shield volcanoes.

Shikotsu *caldera, Japan*   The Shikotsu CALDERA appears to have been the site of a great flow of ASH and fall of TEPHRA in prehistoric times. About 30,000 years ago, the modern volcano was formed in a huge eruption that produced approximately 30.5 cubic miles (125 km³) of ash (VEI = 6–7). Two of the three prehistoric eruptions were also large. The eruptions in 6950 and 800 B.C. had VEI = 5. In the modern volcano, Lake Shikotsu occupies much of the interior of the caldera. The volcano Tarumai (called Tarumae-San) is located on the caldera's southeast edge. In historical times, the caldera has been active on at least 34 occasions. Major eruptions occurred in 1667 and 1739 at Tarumai and left a caldera approximately one mile (1.6 km) wide on the volcano summit. A vigorous eruption of Tarumai in 1874 preceded by about three weeks a powerful TECTONIC earthquake along the northwestern coast of Hokkaidō, the northernmost major island in the Japanese archipelago. Small emissions of ash occurred in 1894. Tarumai underwent a strong eruption in 1909, approximately a year after a major tectonic earthquake off the north coast of Hokkaidō. This 1909 eruption began in late April with loud detonations, shaking of the ground, and a tremendous column of dark smoke. Eruptive activity occurred again in May, apparently with great emissions of smoke and vigorous seismic activity. Minor eruptions occurred every few months from 1917 to 1923, and eruptive activity continued on an intermittent basis through 1936. A powerful explosion occurred in 1920. Another eruption,

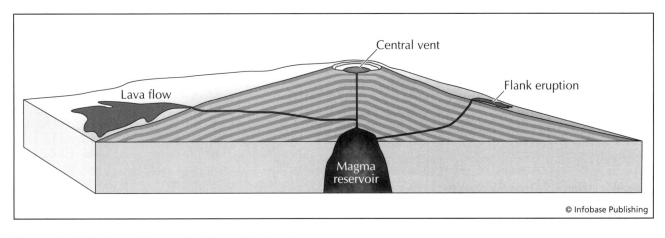

© Infobase Publishing

Diagram of a shield volcano with a summit caldera and flank eruptions

on December 1, 1933, preceded a powerful earthquake on December 5 and a renewal of volcanic activity on December 11. Explosive eruptions, and possible PHREATIC ERUPTIONS as well, occurred between 1944 and 1955. EARTHQUAKE SWARMS started to increase in 1967, and minor swarms of earthquakes were reported again in 1975, 1978, 1979, and 1981. Eruptions resumed in 1978, and small eruptions continued through 1981. A minor eruption in February 1981 marked the end of earthquake swarm activity that had been occurring since 1967, although some earthquakes took place after this eruption. Inflation appears to have occurred at the caldera during the 1980s.

Ship Rock *New Mexico, United States* One of America's most spectacular and famous landforms, Ship Rock is a VOL-CANIC NECK with DIKES radiating from it. The ancient volcanic cone that once surrounded Ship Rock has been removed by erosion. The dikes acted as feeders and probably fed flank eruptions. Ship Rock is part of the Navajo Volcanic Field, which lies in the Four Corners area (intersection of UTAH, COLORADO, ARIZONA, and NEW MEXICO) covering about 7,810 squares miles (20,000 km²). The igneous bodies have an unusual composition and occur as either diatremes (KIM-BERLITES) or TUFF pipes. The activity occurred from about 30 to 25 million years ago, and erosion has proceeded ever since.

Shiraz *earthquake, Iran* During the night of May 4, 1853, a devastating earthquake struck the city of Shiraz, IRAN, and the surrounding towns and villages. The RICHTER magnitude of this event was estimated at 6.5, and the damage was estimated at VIII on the modified MERCALLI scale. The earthquake caused extensive settling of the ground surface and FISSURES and SLUMP in the mountains surrounding the city. There were strong AFTERSHOCKS for months after the event. Only the stronger masonry buildings survived the shaking; all else was laid to waste. In all, some 12,000 people were estimated to have perished in the event, although some sources placed the number closer to 10,000.

Shishaldin *volcano, Alaska, United States* A beautifully symmetric, snow-covered STRATOVOLCANO in the eastern ALEUTIAN

The snow-covered stratovolcano Shishaldin on Unimak Island in the Aleutian Islands, Alaska. This volcano has been compared with Mount Fuji of Japan as one of the most perfect volcanic cones in the world. Only volcanoes in polar regions are completely ice-covered. *(Courtesy of the USGS)*

ISLANDS (Unimak Island), Shishaldin is one of the most active volcanoes in the Aleutians. It has erupted at least 29 times since 1775. The major eruptive periods were 1775–78, 1827–30, 1897–99, 1927–29, 1946–48, 1975–76, and 1986–87. The most recent full eruption was in 1999, but there are constant low-frequency rumblings and steam plumes from the volcano, and it is constantly monitored. An ERUPTION column 45,000 feet (13,716 m) high and an ash cloud that extended 500 miles (805 km) was produced during this last eruption.

Sierra Negra *volcano, Galápagos Islands, Ecuador* A large SHIELD VOLCANO, Sierra Negra occupies Isabela (Albemarle) Island and has a shallow, elliptical CALDERA on its summit. The volcano is less than 4,500 years old. An explosive eruption was reported in 1813, and other eruptive activity of unknown character was recorded in 1817. A LAVA FLOW occurred during an eruption in 1844, and an explosive eruption was reported in 1860. Some kind of unrest appears to have occurred at Sierra Negra in 1911 or 1912, but details are unavailable. An explosive eruption and associated flow or flows of LAVA were reported in 1948 and again in 1953–54. Unrest of unknown character was reported in 1957. Explosive eruptions with lava flows occurred in 1963 and again in 1979–80. An eruption in 1979 injected large amounts of SUL-FUR dioxide, detectable on a global scale, into the atmosphere.

Sierra Nevada *California, United States* Approximately 350 miles (563 km) long and 60 miles (97 km) wide, the Sierra Nevada range lies along the western edge of NORTH AMERICA'S GREAT BASIN and is bordered on the north by the CASCADE MOUNTAINS. The Sierra Nevada comprises a huge BATHOLITH of granitic rock with FAULTS on the east side and which is tilted westward. In some places, MAGMA has made its way to the surface and formed volcanoes atop the PLUTON. The Sierra Nevada batholith forms the roots of an ancient MAGMATIC ARC. A SUBDUCTION ZONE existed under CALIFORNIA until about 25 million years ago when the TEC-

Ship Rock is located in New Mexico. The main part is a volcanic neck that remained after the rest of the volcano eroded away. The radial lines are dikes that fed magma to the volcano. Ship Rock is nearly 1,400 feet (470 m) high, and the dikes form large walls across the desert. *(Courtesy of NOAA)*

TONICS shifted around to the SAN ANDREAS transform margin that exists today. Previously, California looked more like the ANDES with a high range of volcanic mountains.

The Sierra Nevada range is noted for seismic activity, especially along its southern border, where a northward-moving block of crustal rock encounters the deep roots of the mountain range. This interaction generates numerous earthquakes in southern California as the block of crust grinds against, and fragments smaller pieces of crust around the southwest edge of the Sierra Nevada. The TRANSVERSE RANGES have arisen along the zone of collision. Veins of GOLD were deposited in the rocks along the Sierra Nevada and became known as the mother lode. Fragments of gold eroded from the mother lode were laid down with sediments in placer deposits that, when discovered during the early settlement of California by Americans of European descent, led to the gold rush of 1849. Alpine glaciation has done much to sculpt the topography of the Sierra Nevada. The range exhibits a curious inverted topography in places, where LAVA FLOWS that once filled canyons now form ridges because erosion has worn away the granitic rocks on either side of the lava flows, leaving the former valley (topped with lava) in separate streams. The basins parallel to the Sierra Nevada on the west side of the range contain thick deposits of sediment eroded from the mountains. They formed as the basins sank relative to the mountains, giving sediment an opportunity to accumulate to great thickness.

silica Tetrahedra (pyramid-shaped) molecules of silicon and oxygen. Silica is an important component of many minerals, especially the SILICATES, QUARTZ crystals are familiar examples of silica in its pure form.

silicates Silicate minerals incorporate a characteristic molecule, which consists of a silicon atom and four oxygen atoms arranged in a tetrahedral (pyramid) structure. Other atoms such as potassium, iron, and magnesium may occupy spaces within tetrahedra. Silicate tetrahedra may be joined together in various arrangements. In minerals where tetrahedra occur in isolated form, tetrahedra are packed together closely; garnet and OLIVINE are examples of minerals with this structure. An example of a mineral having rings of six tetrahedra is beryl, commonly found in the form of six-sided crystals. Tetrahedra arranged in single chains (PYROXENES, for example) and double chains (amphiboles—HORNBLENDE) are moderately hard and tend to separate in a regular fashion (along cleavage planes), parallel to the chains. In sheet form, tetrahedra are linked together to form a planar structure; mica (MUSCOVITE and BIOTITE) is a familiar example of a mineral with this structure. In frameworks, where tetrahedra join to form a three-dimensional lattice comparable to the framework of a building, the best examples are FELDSPARS (PLAGIOCLASE and K-FELDSPAR), the most common mineral group, and quartz, the most common mineral variety.

silicic Rich in silica.

sill A sill is a tabular body of intrusive igneous rock that lies parallel to and along the layering of the COUNTRY ROCK strata. PLUTONS that lie parallel to strata are said to be conformable

This photo is of a volcanic sill near Edinburgh, Scotland. The magma squeezed along the layering of the existing rock strata before solidifying. The Palisades of New York and New Jersey formed by the same process. *(Courtesy of NOAA)*

in contrast to plutons that cut across strata and are disconformable. DIKES are disconformable but otherwise appear similar to sills. A popular example of a sill is the Palisades Sill of NEW JERSEY and NEW YORK.

Sinano *earthquake, Japan* On May 8, 1847, a strong earthquake struck the Sinano area of JAPAN. The Sinano area is relatively rugged, and the earthquake shook loose literally thousands of LANDSLIDES that crushed and buried the villages located in the valleys below. The largest of the landslides blocked and dammed the Saikawa River, producing an earthquake lake 17 miles (27.2 km) long by 2.5 miles (4 km) at its widest point. The lake filled for 19 days and then the landslide dam gave way, producing a huge flood that swept away all settlements along the banks downstream, including 4,800 houses. In all, 34,000 houses were lost in this earthquake, and the DEATH TOLL was in excess of 12,000 people.

Sissano *earthquake, New Guinea* On July 17, 1998, an earthquake of magnitude 7.1 occurred near the north coast of New Guinea. At least 2,183 people were killed, but at least 500 were reported missing. Thousands were injured and many thousands were left homeless. Much of the damage was caused by TSUNAMIS which reached maximum heights of 32 feet (10 m).

Skaptarjökull *volcano, Iceland* The eruption of Skaptarjökull (also known as Lakagigar or LAKI) in 1783 was one of the most powerful eruptions in history. The eruption started on June 11 and followed a long series of earthquakes that became very violent just before the eruption. LAVA issued from FISSURES in the mountain at numerous locations (although volcanic gases had been vented starting several days earlier), and the separate streams of lava united to form a stream of molten rock that blocked the gorge of a nearby river, which was some 200 feet (61 m) wide and 400 to 600 feet (122 to 183 m) deep. The lava not only filled this gorge but also overflowed it on either side. The lava also filled the bed of a lake along the river. Continuing on its way, the LAVA FLOW encountered an area of ancient volcanic rocks honeycombed with caverns. The lava flowed into these caverns and sent fragments of old lava flying some 150 feet (46 m) into the air.

On June 18, another flow of lava advanced rapidly over the hardened surface of the first flow. On August 3, more lava flowed from the mountain. The fresh lava moved southeastward and spread out over the plain. Eruptions of lava continued intermittently for two years and, by one estimate, amounted to a mass as great as that of Mont Blanc in France. Individual flows averaged some 100 feet (30 m) deep, but in some places depths of 600 feet (183 m) were attained. A total of 3,600 cubic miles (14,731 km^3) of BASALT were erupted making it the second-largest basalt emission in history (Eldgja, A.D. 935 is the largest). At peak flow FIRE FOUNTAINS reached heights of 800 to 1,400 feet (244 to 427 m) and discharge rates reached 303,706 cubic feet (8,600 m^3) per second (the Amazon River discharge is 353,147 cubic feet [10,000 m^3] per second, and it is the largest river in the world). Vapors were still rising from these lavas as late as 1794. Partly because of melting snow and ice and partly

because of blocked rivers, great floods of water occurred and caused extensive damage. Lava overwhelmed 20 villages. Ashes covered the entire island and the waters offshore. Winds carried ashes from the eruption over the European continent on several occasions, creating effects similar to those following the eruption of KRAKATOA. Of the 50,000 people inhabiting Iceland at this time, more than 9,000 are thought to have been killed, along with more than 11,000 head of cattle, some 28,000 horses and more than 190,000 sheep. This loss of life resulted from streams of lava, from vapors, from floods of water, from destruction of plants by volcanic fluorine contamination and acid rainfall, and from a shortage of fish, which disappeared from the coastal waters during the eruption. Previously, fish had supplied much of the food of the people.

In 1784, Benjamin Franklin noted that the year 1783 had been unusually cool and that sunlight reaching the ground seemed diffused. He observed that when the rays of the Sun "were collected in the focus of a burning glass, they would scarcely kindle brown paper." He also reported that a strange "dry fog" hung over the land in the summer of 1783 and that this fog appeared to cut off some incoming sunlight. Franklin suggested that the "vast quantity of smoke" from Skaptarjökull's eruption in 1783 had created the cloud, blocked part of the incoming solar radiation and thus given the northern latitudes a colder than usual year. The eruption of the volcano ASAMAYAMA in Japan that same year also may have contributed to the effects observed by Franklin.

Skopje *earthquake, Macedonia* Two strong shocks rocked the city of Skopje, Macedonia (Yugoslavia), on Friday, July 26, 1963. The first earthquake struck at 5:14 A.M. and the second just three minutes later at 5:17 A.M. The MAGNITUDE was 6.0 on the RICHTER scale, and the EPICENTER was right in the town of Skopje, with the focus at 7.8 miles (13 km) in depth. The damage at the epicenter registered X on the modified MERCALLI scale. The source of the earthquake was the Vardar Zone, the most seismically active structure in the region.

The DEATH TOLL from this earthquake was 1,066 people. More than 3,300 people were injured and 170,000 people, or 75% of the population of Skopje, were left homeless. About 77% of the buildings were damaged or destroyed. Direct economic losses were estimated at $1 billion, which was 15% of the entire Yugoslavian economy. The Yugoslavian army carried out the rescue and relief operations, and 80 countries sent aid for the survivors.

slip model This is a modeled description of the KINEMATICS that portrays the amount, distribution, and timing of slip on a FAULT plane associated with a particular earthquake.

slump The process of slumping is the downhill slipping of a mass of rock or unconsolidated SOIL and/or rocky material along a concave-upward plane of failure, similar to a listric normal fault and usually involving a backward rotation of the material during movement. Slump is one type of MASS WASTING that occurs when a slope exceeds the ANGLE OF REPOSE. It may happen for any number of reasons, including heavy rain, melting of snow and ice, removal of vegetation by fire,

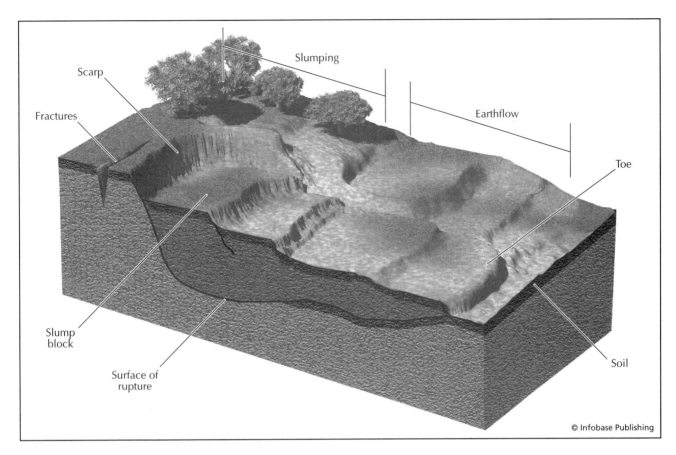

Block diagram of slump, a type of landslide and mass wasting

disease or human activity, or by shaking and/or LIQUEFAC-TION associated with seismic activity. Slumping typically produces a SCARP along the top margin and FISSURES along the upper part of the slumped material. Most fissures produced in earthquakes are from slumping.

Smyrna *See* IZMIR.

Snake River Plain *Idaho, United States* The Snake River Plain in southern IDAHO is part of the Columbia–Snake River Plateau, an expanse of LAVA FLOWS overlying great RHYO-LITE ash deposits extending through much of the states of Idaho, WASHINGTON, and OREGON, with portions in UTAH, NEVADA, CALIFORNIA, and WYOMING. The Snake River Plain stands about 3,000 feet (914 m) high at the west and rises to about one mile (1.6 km) in elevation in the east. The plain exhibits many volcanic cones, particularly at CRATERS OF THE MOON MONUMENT, about 50 miles (80 km) north of American Falls. Two impressive volcanic cones, Menan Buttes, may be seen near Idaho Falls. The plain is not noted for frequent or strong earthquakes but borders on areas of considerable seismic activity. Earthquakes around the edges of the Snake River Plain indicate the plain is descending with respect to the areas around it. The rate of descent is believed to be a few inches per century on the average.

The formation of the Snake River Plain and the COLUM-BIA PLATEAU both had their origin as FLOOD BASALTS, beginning with the Columbia Plateau. The rhyolitic ash deposits underlying the basalts of the Snake River Plain are believed to have been formed when granitic continental crust moving over the HOT SPOT melted and generated vapor-rich MAGMA. The magma emerged in explosive eruptions, creating a series of CALDERAS that became a series of basins stretching through the plain. The overlying basalts presumably did not form during these explosive eruptions but instead were deposited later from FISSURES. The postulated hot spot now lies beneath YELLOWSTONE NATIONAL PARK. It has been suggested that this hot spot is also responsible for the rifting that gave rise to the northern BASIN AND RANGE PROVINCE of the western United States. Dramatic evidence of rifting activity may be seen at Great Rift National Landmark near American Falls.

See also PLATE TECTONICS.

Socorro *caldera, New Mexico, United States* The Socorro caldera is associated with the RIO GRANDE RIFT and has been the site of considerable unrest, although no actual ERUPTIONS, since the mid-1800s. Earthquake swarms were reported at the CALDERA in 1849–50, 1904, and 1906–07 and from the early 1960s on. Earthquakes at Socorro occur within several miles of the surface. Uplift also has been noticed at Socorro. The uplift and seismic activity are thought to be linked to a large body of MAGMA that is located at a depth of perhaps 10 miles (16 km) and causing uplift in an oscillating or episodic pattern.

soil Soil is a naturally occurring layer of unconsolidated mineral and/or organic matter that mantles the surface of the BEDROCK of the Earth. Soils are formed through processes that give them a distinct physical, chemical, mineralogical, and morphological character. They may be formed in place through the weathering and alteration of underlying bedrock, or they may be transported into place from another area. These exotic soils may be glacial in origin but do not reflect the underlying bedrock. Soils are one of our most valuable resources in that most of our food comes from them. On the other hand, seismic waves are usually much more destructive in soil than in bedrock.

soil amplification Seismic waves grow in strength and intensity as they pass from rigid BEDROCK into much less rigid SOIL in virtually all cases. This soil amplification of seismic waves is a danger. As a result, an earthquake which appears to be causing little damage can suddenly cause great damage in an area through this process.

solfatara A volcanic FISSURE or VENT that emits only vapor, notably sulfurous gases. Named for a volcano in the PHLEGRAEAN FIELDS, ITALY, thick yellow sulfurous rinds coat the surficial rocks in these areas. Mining operations typically follow. These mines create a dangerous situation. When the volcano erupts, many SULFUR miners can be killed.

solifluction By definition, solifluction is the slow flowing of water-saturated soil and entrained debris down hill. It is a wet version of CREEP and another of the processes of MASS WASTING. Flow rates are imperceptibly slow, typically on the order of one or two centimeters per week. Solifluction is common in cold climates, where the top few centimeters of soil mantling hills can be thawed and saturated with meltwater. The soil mass moves like a viscous fluid (consistency of cement) but faster toward its center, forming lobes and tongues on the hillslopes.

Sonoran earthquake On May 3, 1887, the Pitaycachi Fault, MEXICO, ruptured and produced the great Sonoran earthquake with magnitude 7.2. It was the largest historic earthquake in the BASIN AND RANGE PROVINCE. The recurrence interval for such large earthquakes on the Pitaycachi fault is 100,000 years.

Soputan *See* TONDANO.

Soufrière 1 *volcano, Saint Vincent* A volcano on the island of Saint Vincent in the CARIBBEAN SEA, south of Jamaica. The volcano's name means "sulfur" or "brimstone." The volcano erupted in 1718 and again in 1812, but its most famous eruption took place in 1902. Earthquakes began to shake the island in April 1902, and in May the peak began to emit steam. The CRATER lake started bubbling on May 5, and large

amounts of steam rose from it. On the afternoon of May 7, the volcano released a *NUÉE ARDENTE,* a mixture of ASH and superheated gases. The cloud was heavier than air and spilled downslope, killing some 1,650 people on the island. Some escapes from death were remarkable: About a hundred people, for example, are said to have avoided destruction by hiding in a rum cellar.

The eruption laid waste about a third of the island and removed more than 700 feet (213 m) from the summit of the volcano. The earlier eruption in 1718 is said to have released LAVA, and the violent eruption in 1812 reportedly lasted three days and caused extensive loss of life. A new crater about a half-mile (0.8 km) wide and 500 feet (152 m) deep formed in the 1812 eruption, northeast of the original crater (which was about the same size). Barbados, some 95 miles (153 km) distant, received an ASHFALL of several inches.

Before the 1812 eruption, a conical hill stood in the middle of the crater, beside which were two lakes, one sulfurous and the other taste-free. Vegetation made the site beautiful, and white smoke, with an occasional touch of light blue flame, emanated from FISSURES on the cone. The eruption of April 27, 1812, began with a strong earthquake, accompanied by what one historian describes as a "tremulous" noise as well as by a column of dense black smoke. The first signs of the eruption were almost comical. A boy who was herding cattle on the mountainside noticed a stone fall near him, then another. He supposed that some other boys were tossing rocks at him from the cliff above, and he began to throw stones upward in reply. He soon saw that an eruption was starting, however, and he ran for his life. The eruption continued for three days and nights. Then, on the 30th, lava flowed from the crater's rim downward to the sea before the eruption stopped. This eruption appeared to be the culmination of almost two years of earthquake activity, especially in the vicinity of CARACAS, VENEZUELA. On March 26, 1812, a great earthquake struck Caracas as large numbers of residents were gathered in churches. The earthquake killed some 10,000 people in the collapse of churches and homes and was felt up to 180 miles (290 km) away. Just more than one month later, the 1812 eruption of Soufrière occurred. On April 30, just as the eruption was ending, a loud underground noise resembling cannon fire was heard at Caracas and other locations, although no shock was reported. The noise was said to be as loud along the shore as it was hundreds of miles inland, and at Caracas preparations were made to defend the city against what sounded like invading troops with heavy guns.

In the 1902 eruption, Soufrière erupted simultaneously with Mount PELÉE. Soufrière's eruption occurred at the north end of the island of Saint Vincent and reportedly wiped out most of the native CARIBBEAN population of the island. In April, unrest was noted at Soufrière, and on May 5, the crater lake started to bubble and to release great clouds of steam. Earthquakes followed, some of them very strong. At noon on Wednesday, May 7, 1902, Soufrière suddenly sent six separate streams of lava down its sides. On the night of May 7, Soufrière's eruption was visible from Saint Lucia. The eruption lasted all night, and the day and night afterward. On Thursday morning, a tremendous black column rose to an estimated height of eight miles (12.8 km). Ashes and rock fell from the column for miles around. On Thursday night, a steamship on the way to Kingstown encountered a floating mass of ashes and was beset for three hours by a cloud of sulfurous gas. When the ship reached Kingstown at dawn, the streets were covered to a depth of two inches (5 cm) with ash and rock. Lava flowed down the side of Soufrière.

The eruption abated slightly on Friday. Showers of rocks ceased falling, although lava continued to flow from the volcano. Corpses of humans and livestock added to the pollution generated by the eruption; hundreds of bodies tragically remained buried. In one ravine, 87 bodies were found together. Nearby lay the carcasses of hundreds of cattle. The bodies were destroyed by quicklime, and the cattle were burned.

Soufrière was active in 1971–72, and again in 1979. The 1979 eruption was VULCANIAN and sent an eruption column 12 miles (19 km) high. Timely evacuations prevented the loss of human life.

Soufrière 2 *volcano, Guadeloupe* The first historical eruption of a volcano in the CARIBBEAN was of Soufrière on Guadeloupe Island in 1660. The volcano has erupted eight times since then, half of which were explosive. The last eruption was PHREATIC and in 1976–77. There were 26 separate explosions and significant seismic activity. The eruption led to the evacuation of thousands of people.

Soufrière Hills *volcano, Montserrat, West Indies* This STRATOVOLCANO began to erupt on July 18, 1995, for the first time in historic times. The volcano grew continuously thereafter. By August 1997, the PYROCLASTIC FLOWS and ASHFALLS had consumed much of the island. The flows reached the capital city of Plymouth and began to engulf it in flames. The British government began action to evacuate all residents of the island permanently. By August 25, 4,000 of the original 11,000 inhabitants remained on the island under the threat of a "massive and cataclysmic" eruption. They were offered $4,000 per adult and $1,700 per child to compensate for their losses if they evacuated. By September 9, all remaining residents were crowded to the northern tip of the island where the British government was considering a five-year rebuilding plan. On September 23, a large benefit concert was held in London in hopes of raising $1.6 million for relocation and new housing; such notables as Elton John, Paul McCartney, Eric Clapton, Sting, and Phil Collins were in performance. There was much media attention on Montserrat at this time. The eruption remained intense and continues today. The few residents who remain are under constant threat of not only the pyroclastic flows and ashfalls but also intense seismic activity, noxious gas accumulations, and regular ROCKFALLS. This one time island paradise is now a wasteland.

South America A continent of the Southern Hemisphere that is noted for its volcanic and seismic activity. Very old CONTINENTAL CRUST of interior South America has been overlain by the huge and active ANDES MOUNTAINS. The Andes form a MAGMATIC ARC over a SUBDUCTION ZONE that bounds the entire west coast of South America. This subduc-

tion zone consumes OCEANIC CRUST of the NAZCA CRUSTAL PLATE under the PACIFIC OCEAN. Volcanic activity is discontinuous because an old ridge is currently being subducted at the center of the range. Nonetheless, the highest (elevation) volcanoes on Earth occur in the Andes. Most of the current seismic and volcanic activity is concentrated to COLOMBIA, PERU, and CHILE, although BOLIVIA, Venezuela, and Ecuador also see periodic activity. The South American plate is also bounded to the east by the mid-Atlantic ridge and to the south by the Scotia arc.

Some of the most devastating disasters have occurred in South America. The city of CONCEPCIÓN, Chile has been ravaged by earthquakes on several occasions as early as 1835 and notably in 1939 when some 50,000 people were killed. The Chilean 1960 earthquake was the largest earthquake ever recorded and sent devastating TSUNAMIS all over the Earth. An earthquake in YUNGAY, Peru, killed some 66,000 people in 1970. This earthquake shook loose an AVALANCHE in NEVADOS HUASCARÁN in which house-sized blocks attained speeds of more than 200 miles (322 km) per hour and cut a 22-mile (35-km) path of destruction. More than 18,000 people were killed. In 1985, a LAHAR generated by the eruption of the volcano NEVADO DEL RUIZ, Chile, killed some 23,000 people.

South Carolina *United States* South Carolina's reputation as a high-intensity earthquake area is due largely to the great CHARLESTON earthquake of 1886. This is not, however, the only strong earthquake in the history of South Carolina. A powerful earthquake accompanied by a loud roar occurred in Pickens County on October 20, 1924, for example. Coastal areas of South Carolina are highly susceptible to LIQUEFACTION and resultant damage in the event of future strong earthquakes.

South Dakota *United States* South Dakota does not have an outstanding history of strong earthquakes, but some such events have occurred since the state was settled by Americans of European descent. The James River Valley earthquake of June 2, 1911, for example, was felt in IOWA and NEBRASKA as well as in South Dakota.

South Sandwich Islands *southern Atlantic Ocean* These islands are the ISLAND ARC above the Scotia Arc SUBDUCTION ZONE. The PLATE TECTONIC geometry is complex. The arc is formed as the South American plate is driven beneath the tiny Sandwich plate on its east side. A small MID-OCEAN RIDGE separates the Sandwich plate from the Scotia plate to the west. East-west trending TRANSFORM FAULTS separate these small plates from the South American plate on the north and the Antarctic plate on the south.

These islands were first discovered in 1775 but not explored until 1818. The islands include Thule, Bristol, Montagu, Saunders, Candlemass and Vindication, Leskov, Visokoi, Zavodovski, and Protector Shoal. Because these islands are uninhabited, shrouded in clouds most of the time, and very cold, little is known about them. Protector Shoal is a SUBMARINE VOLCANO that had a spectacular eruption in 1962. A RHYOLITIC PUMICE raft was formed that covered at least 2,000 square miles (5,180 km²). Bristol Island is made of several overlapping STRATOVOLCANOES that last erupted in 1956. Other confirmed eruptions occurred in 1823, 1935, 1936, and 1950. Candlemass is an active ANDESITE-dacite stratovolcano. It was active in 1775 and had an eruption in 1953–54. GEYSERS and FUMAROLES are in constant activity.

Sparta *earthquake, Greece* An earthquake that changed the course of ancient Greek history took place in 464 B.C. The estimated RICHTER magnitude of this event was 7.2, and the destruction was estimated at X on the modified MERCALLI scale as the result of studying historical records. Using satellite images and field work, geologists identified the probable FAULT for the event. It forms a 12-mile (20-km)-long normal fault SCARP that lies a few kilometer east of Sparta with a north-south orientation. The earthquake was estimated to have killed some 20,000 people in Sparta. The Spartans had recently established themselves as the most formidable army in the region by defeating a massive invading Persian force. The earthquake so weakened the Spartans that they agreed to allow Athens to send troops to help with the rescue and relief operations. When the Athenians learned that the Spartans had at least partially enslaved the Helots (forced them into an agrarian existence to supply the Spartans), they protested based on an agreement that all Greeks should be free and were immediately invited out of Sparta. Removing the Athenians did not end their troubles. The Messenians aided the Helots in a revolt against the weakened Spartans that lasted for many years. Resentment from the events associated with this earthquake eventually led to the Peloponnesian War.

spatter cone A spatter cone is a volcanic structure resembling a giant anthill, standing several feet high, and formed by eruptions of highly fluid LAVA. The eruptions must be relatively nonenergetic and of small volume. Regular spatters of lava are emitted a few feet high into the air and land right around the vent. They slowly form a small mound that resembles a volcanic cone. If the activity continues a long time, the cone shape will be lost to more of a chimney shape and the features are then called HORNITOS, which means "little oven."

spectral acceleration In contrast to peak ACCELERATION, which models the movement of a particle in the ground during an earthquake, spectral acceleration approximates the response of a building during an earthquake. The building is modeled as a particle on a massless vertical rod with a natural period of vibration (natural frequency) of a building.

spectrum A spectrum shows how much of each type of shaking there is from a particular earthquake. Technically, it is an analysis of the vibrations on a SEISMOGRAM that shows the AMPLITUDE and phase of the waves as a function of frequency or period.

splay A FAULT splits or branches so that part of the offset is taken up on the branch or splay of the fault. A splay is therefore a small fault that connects to a larger fault. Splay faults

A hornito on Kilauea volcano, Hawaii in 1970. A hornito is a tall spatter cone. The texture on the hornito is caused by blobs of lava landing and cooling on the outside of the spatter cone. *(Courtesy of the USGS)*

develop because the larger fault can't easily move through a certain rock. The motion must be taken up on both faults.

Srinagar *earthquake, India* A major earthquake struck the city of Srinagar and its surroundings in northern INDIA on May 30, 1885, at 2:45 A.M. The quake had an estimated RICHTER magnitude of 7.0, and the intense damage covered some 750–1,000 square miles (1,945–2,600 km²). Large FISSURES were reported to have formed, and LIQUEFACTION produced sand blows some 24–60 inches (60–150 cm) across. A huge LANDSLIDE at Larridar was reported to have caused many deaths. In all, the DEATH TOLL was reported as 3,200 people, but there is debate on the exact number.

station A station is simply the geographic location of a geophysical instrument to analyze earthquakes.

stick-slip A description for that way rocks deform in a FAULT that produces regular earthquakes. STRESS builds up in the fault as the TECTONIC plates or other large-scale pressure affects an area. Nothing happens until the stress exceeds the strength of the rock. Then an earthquake occurs and relieves much of the stress on the fault rocks. With time, the pressure builds up again until once again there is an earthquake to relieve the stress. The sticking of the fault while the stress builds up and the slip of the fault during the earthquake describes the process.

stock A small, approximately circular igneous INTRUSION. These PLUTONS are spherical to an inverted teardrop shape; however, the view from the surface makes them look circular. They must be less than 100 square miles (259 km^2) in that surface view. Plutons greater than 100 square miles (259 km^2) in the surface view are called BATHOLITHS.

strain Quantitative strain is a ratio of an initial size to a final size, so it is either unitless or expressed as a percent. Strain can be compressional, tensional, extensional, or the result of SHEAR. It takes STRESS to make strain.
See also DEFORMATION.

strain rate The rate at which a material takes on STRAIN when subjected to a given STRESS. Strain is a ratio of an initial size to a final size, so it is either unitless or expressed as a percent. The unit of strain rate is therefore no unit over time, or sec^{-1}. Typical geological strain rates (plate tectonic) are 10^{-14} sec^{-1}, which is so slow that it cannot be modeled in a laboratory. On the other hand, strain rates related to MASS WASTING, volcanoes, or earthquakes can be very fast.

stratovolcano A steep-sided VOLCANIC CONE formed by repeated eruptions of LAVA and PYROCLASTIC material. As SUMMIT ERUPTIONS occur one after another, layer upon layer

of pyroclastic debris is dropped around the vent from ASH-FALLS and ASH FLOWS. These ash layers are thin near the summit and thicken downslope toward the base of the volcano. LAVA FLOWS are sporadically interlayered with the ash, but it is volumetrically minor in comparison. These volcanoes are therefore almost exclusively RHYOLITE, DACITE, and ANDESITE. Because of the great ash content, these cones are highly susceptible to weathering and are removed relatively quickly. All that remains is the VOLCANIC NECK after several million years depending upon the climate.
See also VOLCANO.

stress Stress is the action that causes STRAIN or DEFORMATION. It is an applied force per unit area of a material. If stress is equal on all sides of a material, it is called pressure, like air pressure or water pressure. In all but a few cases, to cause strain in rocks, stress must be different in at least two directions, which is termed differential stress. There is a maximum stress direction in response to which objects are shortened and a minimum stress direction in which objects may extend.

stress drop STRESS may build up in a material if it is resistant to STRAIN. If and when the stress overcomes the strength of the material, it will break abruptly, and the stress of the material will drop. This is the idea behind an earthquake. Stress builds up in a FAULT for a period of time until it overcomes the frictional forces preventing the fault from slipping. The earthquake releases the built-up stress, and there is a stress drop in the fault, before to after. The stress drop of the fault is the energy released as seismic waves in the earthquake.

strike-slip fault A FAULT with purely lateral motion and no vertical motion. The motion on these faults must be viewed in a map sense. The land on one side of the fault slides sideways without producing a topographic expression. The way to determine the type of strike-slip fault is to stand on one side

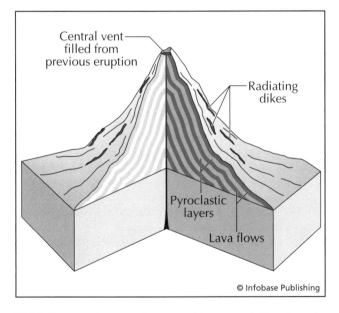

Model of a composite stratovolcano cone. It is composed of interlayered pyroclastic deposits and lava flows. The pyroclastic deposits are mainly ash-flow tuff with lesser amounts of ash-fall tuff. The central vent emits the ejecta through a summit caldera, but the radiating dikes can also feed a lateral vent as well.

Offset rows of lettuce from movement on a strike-slip fault in El Centro, California. The shift of the rows is to the left, making this a left-lateral or sinistral strike-slip fault. The earthquake occurred on October 15, 1979, and had a magnitude of 6.9. *(Courtesy of NOAA)*

The stratovolcano Stromboli, in Italy, has been in almost continuous incandescent eruption for nearly 2,000 years. It has been named the "Lighthouse of the Mediterranean" as a result. *(Courtesy of NOAA)*

of the fault and look across to the other. If the land moved to the left, it is a LEFT-LATERAL or sinistral fault. If it moved to the right, then it is a RIGHT-LATERAL or dextral fault.

Stromboli *volcano, Tyrrhenian Sea, near Sicily* Known as the Lighthouse of the Mediterranean, the volcanic island of Stromboli rises slightly more than 3,000 feet (914 m) above sea level. It is one of the most active volcanoes on Earth, having been in near constant eruption for 2,000 years (some say 5,000 years). The modern volcano developed about 15,000 years ago. Stromboli's eruptions have varied in output and intensity from mild activity of FUMAROLES to violent outbursts. In those of 1907, explosions shattered windows in villages on the island, and INCANDESCENT material was cast out of the CRATER. The 1919 eruption killed four people and destroyed 12 homes. In 1930, two strong explosions on September 11 followed an earthquake that produced a minor TSUNAMI. Three people were killed in an AVALANCHE during the 1930 eruption. In 1986, a biologist was killed on the crater rim. Most of the eruptions, however, consist of small gas explosions that shoot incandescent BOMBS of LAVA in a fireworkslike display. This type of eruption is the basis for the Strombolian eruption.

Strombolian eruption *See* ERUPTION.

sturzstrom If a large rock mass slips down a steep slope, the rough terrain may shatter the mass into large numbers of fragments moving at enormous speeds. These flows can move at speeds greater than 335 feet (100 m) per second, making them the fastest form of MASS WASTING. They are super DEBRIS AVALANCHES. They may contain water, but the mode of transportation is more commonly the result of the debris particles colliding internal to the mass. These internal collisions cause the mass to move as a nearly frictionless cloud. It has been found that they hit the substrate of the chute so that they can cause it to melt locally, forming a glassy rock informally called frictionite because it is gener-

ated by friction. The impacts can also produce shatter cones and even shocked QUARTZ, which is universally accepted as evidence for a meteorite impact. Clearly, the force of impact generated by a sturzstrom is remarkable and the damage truly devastating.

subduction zone An area where the OCEANIC CRUST of a plate is being overridden by another plate and is moving downward into the ASTHENOSPHERE. The overriding plate can be either CONTINENTAL CRUST as is the case in the ANDES or oceanic crust as is the case in the ALEUTIAN ISLANDS. Subduction zones are characterized by deep ocean trenches with curved or "arcuate" shapes, as well as by volcanic activity and earthquakes. The descending plate is marked by earthquakes that originate ever deeper into the subduction zone (*see* BENIOFF ZONE). Eventually the downgoing crust melts and forms MAGMA of INTERMEDIATE composition that rises back toward the surface to source the large BATHOLITHS and voluminous volcanoes that characterize ISLAND ARCS and MAGMATIC ARCS.

submarine earthquake Also known as a seaquake, submarine earthquakes are common. If there is ground surface movement during the earthquake, a TSUNAMI will be produced. Otherwise, they are typically less dangerous than a land-based earthquake.

submarine eruption Eruptions that occur beneath the sea may be modest in scale, but in some cases they are much more powerful and have provided some of the most spectacular events in the history of volcanology. The eruption of SURTSEY near ICELAND and MYOZIN-SYO near JAPAN are recent examples of such eruptions. Iceland has a vivid history of submarine eruptions. Located on a MID-OCEAN RIDGE, volcanic activity is abundant. The name *Reykjanes* means "smoky cape," a reference to submarine eruptions near that Icelandic cape. Several islets were formed here during an eruption in 1240; most of them vanished later, but others reappeared to take their places. In 1783, about a month before the volcano SKAPTARJÖKULL erupted, a volcanic island appeared, but it too was soon reduced to a mere reef by wave action. Yet another submarine eruption in 1830 produced an island. (This eruption was notable for apparently destroying the skerries known as the Geirfuglaska and with them the great auks that had bred on them up to that time.) Another island emerged from the sea approximately 10 miles (16 km) off Reykjanes in July of 1884.

The ALEUTIAN ISLANDS of ALASKA are monuments to submarine volcanism, having arisen from the ocean floor in ISLAND ARC eruptions that continue into the present. These eruptions have been observed frequently since Americans of European descent began to explore and settle Alaska. In 1856, for example, one Captain Newell of the whaling bark *Alice Fraser,* together with men from several other vessels, watched a spectacular submarine eruption in the Aleutians. No island appeared during this eruption, but huge jets of water rose up from the sea, and the waters were strongly agitated. Volcanic ASH and stones soon emanated from the eruption site, and PUMICE was strewn over the sea for great distances. Loud noises accompanied the eruption, as did shocks like those of earthquakes.

An eruption that produces large amounts of PYROCLASTIC materials may build up an island in time, but as described above, wave action quickly reduces such an island to sea level unless lava flows occur to give the new island greater resistance to erosion. When magma is relatively gas-poor and flows freely, as in the case of the great SHIELD VOLCANOES of the HAWAIIAN ISLANDS, lava flows occurring over many years can build up a huge island. These submarine eruptions are produced by HOT-SPOT volcanism and are almost exclusively BASALT.

Volcanoes that exhibit more explosive eruptions may also emit enough LAVA to produce large islands, but the explosive character of their eruptions may destroy those islands soon after they are formed; the island may simply blow itself to pieces, as in the famous case of KRAKATOA. Much of the explosive character is produced by interaction with seawater. The flash boiling of water increases its volume more than 22.4 times instantaneously. These explosions produce a huge amount of steam and are called PHREATIC EXPLOSIONS.

An undersea volcano that produces a mountain whose top has not yet reached the surface is called a SEAMOUNT. A seamount with a flattened top, indicative of wave action and subsequent submergence, is known as a GUYOT. Submarine eruptions occur frequently, though not always with spectacular results, along mid-ocean ridges, where MAGMA rises from Earth's interior to the surface. This process creates new crustal rock, which moves outward on either side, away from the ridge, in somewhat the same manner as goods on a conveyer belt. Not all undersea volcanism, however, occurs along such ridges.

Sukaria *caldera, Flores Island, Indonesia* The boundaries of the Sukaria caldera are uncertain, but it may be as much as 11 miles (17.5 km) in diameter. Likewise, the geologic history of the CALDERA is largely unknown. Several lakes are located at the caldera. The color of one or more of these lakes changed in 1937–38 because of SULFUR and sulfuric acid in the water. A fountain of water some 300 feet (91 m) high arose from one of the lakes in 1968. On April 27, 1986, one CRATER lake showed an increase in gas bubbling, and an earthquake occurred the following day. Evidently some PHREATIC explosive activity occurred at the caldera in the 1860s, but details are unavailable.

sulfur A chemical element by definition but sulfur is also mineral commonly associated with volcanic eruptions. Sulfur in its pure, native form is a yellow mineral that may be found in crystals or in irregular masses. Volcanoes with SOLFATARA are areas with thick sulfur deposits. It can even occur as a molten liquid as on Jupiter's moon Io. Sulfur may join with various metals to produce sulfides, some of which are widely mined as ores. Galena (lead sulfide) and pyrite (iron sulfide) are familiar examples of such ores although native sulfur is more commonly mined.

summit eruption A volcanic eruption from the top of the volcano. The classic volcanic eruption is a summit eruption in which LAVA and/or ASH and gas explodes out of the top of the volcano. In contrast, in a lateral eruption, the side of the volcano is blown out. LATERAL BLASTS are far more dangerous than summit eruptions.

Sunda *caldera, Java, Indonesia* The Sunda caldera on the island of Java may be only about 2,000 years old, according to one estimate. The diameter of the CRATER is not known with certainty but is thought to be between 3.5 and five miles (5.6 and 8 km). The STRATOVOLCANO Tangkubanparahu occupies part of the CALDERA. Although a stratovolcano, it is BASALTic and has a structure like that of a SHIELD VOLCANO. There are several large craters—Kawah Baru, Kawah Ratu, Kawah Upas—and a number of lesser craters on the summit of the volcano.

Tangkubanparahu has erupted about 17 times since 1826. Within historical times, eruptions have tended to be minor and PHREATIC. This mildness is fortunate because the city of Bandung with a population of 1.6 million is a mere 18 miles (29 km) to the south. Carbon dioxide and hydrogen sulfide gases may gather in craters in concentrations high enough to endanger humans and other animals. An apparent eruption was reported in 1829 after several days of "violent" noise. An eruption in 1846 was stronger than most at this volcano and is thought to have involved a PYROCLASTIC surge that destroyed a whole forest and broke off trees several feet above the ground. There is also a report of a flow of ASH and mud from this eruption that moved eastward from the volcano and destroyed trees and shrubbery. In 1896 a renewal of activity at Kawah Upas involved minor explosions and the formation of a new crater, Kawah Batu. The volcano apparently emitted rumbling noises on several occasions in 1908, but no sign of activity was observed in the craters. Underground noises and an ash column from Kawah Ratu were reported in 1910, and a strong earthquake occurred on April 22, followed by a less powerful earthquake on April 30. Ash eruptions were observed in May.

In 1913, SOLFATARAS became more active; subterranean rumbling was reported, along with vigorous steaming from vents involved in the 1896 and 1910 eruptions. In 1926, phreatic eruptions produced a new crater some 150 feet (46 m) wide, Kawah Ecoima, inside Kawah Ratu. Solfataric activity intensified in 1967, and a solfatara cast out mud in May of that year. Small eruptions of ash also occurred in 1967, and certain solfataras showed a sharp increase in temperature in September of the same year. Phreatic eruptions in 1969 deposited a small layer of ash on the volcano. In 1971, the volcano emitted a small quantity of mud, and thermal activity intensified.

Seismic activity started to increase in June 1983 and showed a dramatic increase later that year, from several earthquakes per day in June to more than 1,000 per day on some dates in September. Most of these earthquakes occurred near the surface and in the vicinity of the Baru FUMAROLE field, although only a very slight increase in fumarole temperatures was recorded. Temperatures at fumaroles at Kawah Baru increased sharply in later 1985. Slight earthquake activity was reported in 1986. A MAGMA reservoir approximately one mile under Kawah Upas is thought to have increased by more than 70 million cubic feet (1,982,179 m³) of magma in the 1980s, and measurements taken at Tangkubanparahu indicate that some inflation occurred at the volcano in the early 1980s.

Suoh depression *Sumatra, Indonesia* The Suoh depression is located along the Semangko Fault Zone on the island of Sumatra. There is disagreement about the possible origin of the Suoh depression. It has been argued that the depression is purely TECTONIC in origin and, alternatively, that it formed by a combination of tectonic and volcanic activity. Many hot springs occur within the depression. A strong earthquake in 1933 was followed a few hours later by increased steaming in the Suoh depression. Two weeks later, a large PHREATIC ERUPTION took place in the depression, forming two explosion craters and making a noise that was heard approximately 400 miles (643 km) away. A PYROCLASTIC surge may have been associated with this event, but this is not certain. Another strong earthquake occurred in 1985, but there was no sign of increased volcanic activity.

supervolcano *Supervolcano* is not a scientific term, it was invented by producers of a British television program. It refers to a volcano that has produced massive effects, largely through an explosion and the ejection of immense quantities of ASH and other PYROCLASTICS. Supervolcanoes have the potential to greatly change the climate of the Earth for a temporary but significant period of time. Humans have been very lucky that we have not experienced the eruption of a real supervolcano during modern times or else we may not have been so successful. Even relatively small eruptions from volcanoes such as TAMBORA and LAKI have disrupted modern life. Volcanic eruptions such as those from TOBA and YELLOWSTONE could shift the climate so strongly that worldwide famines could completely destabilize the political systems of the world.

surface faulting Also known as surface rupturing, surface faulting is a slip along a FAULT during an earthquake that reaches the surface of the Earth. In other words, it is fault movement at the surface. The phenomenon is restricted to shallow earthquakes with foci less than 12 miles (20 km) in depth and commonly much less. It also requires significant slip on the fault to propagate all the way to the surface. Surface faulting of a DIP-SLIP FAULT leaves a SCARP, whereas a STRIKE-SLIP FAULT does not. Aseismic CREEP can also produce surface faulting.

surface wave Seismic waves that move only along the surface of Earth. Surface waves include two types, RAYLEIGH WAVES and LOVE WAVES. Ocean waves are a type of surface wave in that they move only along the top 50 feet (15 m) of the ocean water in most cases.

Surtsey *volcanic island, near Iceland* The volcano Surtsey emerged from the North ATLANTIC OCEAN south of ICELAND in 1963. The crew of a fishing boat noticed a cloud rising

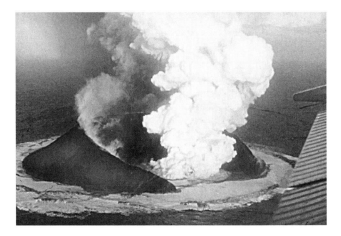

The building of the island of Surtsey, Iceland in 1963. This volcano began 400 feet (130 m) below the ocean surface and during the eruption built an island one-half mile across. *(Courtesy of NOAA)*

from the sea on November 14 and recognized the cloud as coming from a volcanic eruption. Apparently the ERUPTION had been under way beneath the ocean surface for some time. It went unnoticed except for a few small indications such as minor earthquakes and venting of hydrogen sulfide gas, which produced a characteristic odor of rotten eggs, perceptible on the Icelandic shore nearby. Within 10 days of its discovery, the eruption had produced an island about a half-mile (0.8 km) long, almost as wide, and about 300 feet (91 m) high at the rim of the principal CRATER. Made up of fragments of volcanic rock, the island grew in size despite the erosive effects of waves and measured almost a mile wide at maximum by early February. Steam explosions sent volcanic BOMBS flying out from the volcano. As steam explosions diminished, LAVA outflows from the volcano, after cooling and hardening, made the island more secure against erosion. Surtsey's eruption continued intermittently for more than three years before finally subsiding in 1967. Volcanoes in shallow water, of which Surtsey is an example, tend to erupt explosively because of lava contact with relatively cold seawater. The eruption becomes less explosive when the summit rises above sea level and direct contact between sea and MAGMA is reduced.

The eruption produced about 0.25 cubic miles (1 km³) of lava and ASH, but only 9% of it was above sea level. Between 1967 and 1991, Surtsey subsided some 3.6 feet (1.1 m), likely as the result of compaction of seafloor sediments and/or ISOSTASY.

S-waves Seismic waves with a lateral shearing motion that travel throughout the solid Earth from the EPICENTER of an earthquake. They are *S* waves because they are the second waves to arrive at the SEISMOGRAPH. S-waves cannot travel through liquid and therefore do not penetrate the liquid INNER CORE of Earth. There is consequently a SHADOW ZONE on the opposite side of Earth from the earthquake epicenter.

See also SEISMOLOGY.

T

Taal *volcano, Luzon, Philippines*   The volcano Taal, on an island inside Lake Bombon, a CALDERA lake, has a history of ERUPTIONS dating from 1572. It has erupted at least 34 times since then. Records of eruptions in the 16th and 17th centuries indicate that destruction was widespread. An eruption in 1707 is said to have brought about a spectacular display of lightning but caused no damage to nearby communities. In another eruption, in 1719, eruptive activity is said to have progressed down the side of the volcano and into the lake; on this occasion TSUNAMIS appear to have occurred and caused many fatalities, and fish reportedly were killed by heat. Evidently, few people were killed in several subsequent eruptions in the early 18th century. A vigorous eruption in 1754 that expelled great amounts of material and crushed a church under the weight of the ASHFALL is said to have taken only a dozen lives, although it seems likely that many other deaths from the eruption went unreported. Eruptions continued at intervals through the 19th century, but no great loss of life appears to have occurred. The most destructive eruption of Taal occurred in 1911 when activity thought to have centered around Green Lake, one of two CRATER lakes, resulted in an outburst of mud that killed most inhabitants of the island. The waters of Green Lake were said to be highly acidic, and the acid content of the resulting mud is believed to have been responsible for many of the deaths that occurred. At other locations, asphyxiating gases appear to have been responsible for numerous deaths. In addition, a tsunami was reported. More than 1,300 people were reported killed in this eruption. Another, major eruption in 1965 had a VEI = 4 and killed approximately 500 people from a *NUÉE ARDENTE*. Since 1991 Taal has entered another phase of unrest with EARTHQUAKE SWARMS and steam vents and as a result is being closely monitored.

Taal has been designated as one of the 15 "Decade Volcanoes" that present especially large potential hazards to population centers. The densely populated city of Manila is a mere 30 miles (48 km) north.

The stratovolcano Taal sits in the middle of a large caldera lake (eight by 11 miles [13 by 19 km]). This volcano is responsible for more than 1,300 deaths, mostly as a result of tsunamis it generated in the lake during eruptions. *(Courtesy of NOAA)*

Tabas *earthquake, Iran*   In contrast to the more common earthquakes in northern and western IRAN, this earthquake was in the far eastern part of the country near AFGHANISTAN. On September 16, 1978, the small agricultural city of Tabas, right on the edge of the desert, was totally demolished. Of the 17,000 residents, a shocking 15,000 were killed in the earthquake. In all, some 25,000 Iranians lost their lives in this disaster. The RICHTER magnitude of this earthquake was 7.7, and the EPICENTER was right in the center of Tabas. The 40 villages within a 60-mile (100-km) radius of the epicenter were leveled. The quake could be felt as far away as Tehran, some 400 miles (640 km) to the northwest. It was felt through 70% of Iran.

Damage to a building in Tabas, Iran, from an earthquake on September 16, 1978. *(Courtesy of the USGS)*

table mountain   This curious tabular formation in ICELAND represents a volcano that erupted under cover of glacial ice, forming a mountain topped with LAVA FLOWS that exhibit gentle slopes. Table mountains exhibit some of the same features as SUBMARINE ERUPTIONS, such as PILLOW LAVA formed when molten rock encounters cold water or ice. Herdubreid in Iceland is a good example of a table mountain.

Tabriz *earthquake, Iran*   The first report of an earthquake destroying Tabriz is from A.D. 858 but offers little information. The first well-documented earthquake in Tabriz occurred on November 11, 1042. It had an estimated RICHTER magnitude of 7.6 and was felt as far away as AFRICA. Buildings and city walls were demolished, and there were reports of SURFACE RUPTURING and FISSURES. In all, between 40,000 and 50,000 people were said to have lost their lives, making this event one of the worst in IRAN's history. There were apparently strong FORESHOCKS for months prior to the MAIN SHOCK and strong AFTERSHOCKS for months afterward, making this a very seismically unstable period in Tabriz. An interesting anecdote involves a local astronomer, Abu Taher, who was apparently monitoring the foreshock activity. He must have noticed a change and warned the people of Tabriz about the impending doom the day before. Many people, including the governor, evacuated the city and were spared.

The next major earthquake disaster to strike Tabriz occurred on April 26, 1721. Even though it was more recent than the 1042 event, reports are equally vague. There are reports of devastating LANDSLIDES and ROCKFALLS during this event as well as volcanic emissions of poison gas that killed livestock. The number of casualties is even more debatable. At least 8,000 people were killed in this event, but some estimates place the number as high as 250,000. The most careful work places the death toll at about 80,000.

Just six years later, on November 18, 1727, Tabriz was again impacted by a major earthquake. The U.S. Geological Survey lists this earthquake as among the top 15 most deadly of all time, with a reported death toll of 77,000, though there is little additional information available. It is possible that the

1721 and 1727 are the same event. Tabriz is again mentioned in reports of the earthquake of June 7, 1755, as is Kashan. The reported death toll from this event was 40,000, but there is some doubt as to the location. Kashan and Tabriz are some 420 miles (700 km) apart, making a single earthquake an unlikely culprit for destroying both cities. Furthermore, it was reported by European rather than Iranian sources.

Another major earthquake struck Tabriz on February 28, 1780, one hour after sunset. After the main shock, there were 40 strong aftershocks in one day. FISSURES six miles (10 km) long and 6.7 feet (2 m) wide, and some 21 miles (35 km) long, opened near the mountains, and surface ruptures were also reported. Whole villages were reported to have been swallowed. The main damage zone extended 36 miles (60 km) from Tabriz in which virtually everything was leveled. The death toll is again a subject of controversy. Some sources place it at 50,000–60,000, while others claim up to 200,000 deaths from this event.

There have been numerous additional earthquakes at Tabriz over the years that included loss of life, but none has been as devastating as those described.

Tai-Ching *earthquake, Taiwan*   On September 20, 1999, a devastating earthquake of MAGNITUDE 7.6 occurred. More than 2,400 people were killed, and 8,700 were injured; some 82,000 housing units were damaged or destroyed, and 600,000 people were left homeless; damage was estimated at more than $14 billion. The earthquake was generated by movement along 47 miles (75 km) of the Chelongpu Fault. Many strong AFTERSHOCKS with magnitudes up to 6.5 occurred for weeks after the main shock.

talus   When a ROCKFALL or ROCKSLIDE occurs, a mass of rock breaks free from the BEDROCK and slides or falls to the Earth either as a coherent mass or in pieces. When it strikes the ground, in most cases it breaks into a pile of rubble. This pile of rubble at the foot of a hill or cliff is called talus. With repeated events, this talus will build up into a talus slope, a cone-shaped deposit with a rough, barren surface. Talus slopes are common in areas of immature topography or repeated earthquake activity.

Tambora *volcano, near Java, Indonesia*   The eruption in 1815 of the 13,000-foot (3,960-m) volcano Tambora (also known as Tomboro) near Java was remarkable in three respects. It emitted an unusually large volume of ASH and other solid material, the eruption was recorded in detail by European observers in the vicinity, and Tambora's ash cloud was invoked later to account for a history-making change in global climate. Tambora was believed to be extinct during the early years of European settlement, but the volcano began to emit small showers of ash in 1814. A strong earthquake on the night of April 5, 1815, was followed by a series of explosions and an eruption of ash and smoke that darkened the sky for some 300 miles (483 km) distance. Ash fell as far away as 800 miles (1,287 km) from the volcano. The VEI of this eruption is 7.

Sir Stamford Raffles, founder of the British Colony of Singapore, served as military governor of Java when Tambora erupted on this occasion and reported that on Java the sky was darkened at noonday with ash clouds. An ASHFALL covered the land, while explosions like the sound of artillery or faraway thunder could be heard from the eruption. The sound of some explosions was heard in Sumatra, more than 900 miles (1,448 km) away. Tambora's eruption was most violent on April 11–12 but did not end until July, Raffles wrote. He reported that glowing LAVA appeared to cover the volcano, and stones the size of a person's head fell in the vicinity of Tambora. Twelve thousand natives perished in this eruption, Raffles wrote.

In all, more than 92,000 people died as the result of the eruption of Tambora. Some 80,000 of these were the result of starvation and disease caused by the destruction of the area.

Considerable subsidence was reported; for example, the site of the village of Tomboro was said to be covered by 18 feet (5.5 m) of water following the eruption. This eruption produced a deep CALDERA at the summit of the volcano. The collapse of this caldera may have been at least partly responsible for earthquakes that were felt up to approximately 300 miles (483 km) away. Between 1847 and 1913, a small cone and LAVA FLOW developed in the caldera, possibly in coincidence with a powerful earthquake, centered near the volcano, that occurred on January 3, 1909.

Tambora after the eruption of 1815 stood almost a mile shorter than before. Much of the estimated 36 cubic miles (150 km³) of solid material cast out from the volcano remained airborne in the form of very fine ash that formed a cloud in the upper atmosphere. This high-altitude cloud intercepted incoming sunlight. The resulting drop in insolation, or solar radiation reaching Earth's surface, was implicated in a dramatic change in climate and weather patterns in the Northern Hemisphere during the following year. The year 1816 is known as the YEAR WITHOUT A SUMMER because there was no warm season over much of the Northern Hemisphere. Although the winter of 1815–16 apparently was not unusually bitter, the cold season lasted well into the summer of 1816, with subfreezing temperatures and several inches of snowfall recorded in New England in June. Nonetheless, New England's harvest appears to have been adequate for that summer, partly because much of the harvest consisted of crops that can endure, and even thrive in, cool and moist weather.

Other parts of the Northern Hemisphere, however, experienced such poor harvests that starvation became a major cause of death among poor families in CANADA, and in some corners of Europe, the human population was reduced to eating rats. The economic effects of poor harvests were devastating. Grain prices reportedly rose fourfold in Switzerland, and when hungry Europeans and Canadians turned to the relatively well-off UNITED STATES for help, the foreigners bought up so much American grain that the price of everything connected with grain rose sharply and increased inflation in the United States. The year without a summer was accompanied by political turmoil in France, where an incipient famine, coming just after the devastation of the Napoleonic Wars, strained the social fabric to the point of rupture. Many farmers were afraid to take their produce to market for fear of being robbed and murdered by famished mobs along the way.

Farmers who dared take their crops to market required government troops in some instances to protect them from half-starved hordes who fought to reach the food.

Tambora's eruption has not been identified conclusively as the cause of the year without a summer because the unusual cold spell of 1816 is believed to be within the range of normal fluctuations in weather and climate, meaning that a summerless year such as that one may happen even in the absence of a major volcanic eruption the year before. This potential fluctuation is because volcanoes are not the only known influence on weather and climate. Many other factors have been identified or at least implicated in weather and climatic change, from sunspots to irregularities in Earth's motion. Yet, a U.S. Weather Bureau historical study of temperature as a function of solar radiation reportedly indicated that lesser eruptions than that of Tambora in 1815 have preceded significant drops in insolation and surface temperature. For example, total measured heat received from sunlight fell to about 88% of the normal figure, a reduction of 12% from that value and more than 16% from the previous year, immediately after the eruption of KRAKATOA in 1883. Insolation diminished 4% following the eruption of Alaska's BOGOSLOF volcano and several other volcanoes in 1889. Insolation fell about 13%, from 101% to 88% of normal, after the eruptions of Mount PELÉE and SOUFRIÈRE in 1902. A similar drop in incoming solar radiation, from 101% of normal to 84%, was recorded after the Mount KATMAI eruption in ALASKA in 1912. There is strong reason to suspect, then, that the eruption of Tambora in 1815 was a factor in bringing about a memorable year in both world climate and human history, long after the eruption itself was over.

Tanaga *volcano, Aleutian Islands, Alaska, United States* Tanaga volcano is located at the northern tip of Tanaga Island in a structure that has been interpreted as a CALDERA. Activity was reported on the island in 1763–70, possibly in 1791, and again in 1829 and 1914. In the 1914 activity, a LAVA FLOW reportedly formed.

Tancheng *earthquake, China* A massive earthquake occurred in the Shandong region of northeastern CHINA on July 25, 1668. It destroyed a 19.3-square-mile (50-km²) area, including the city of Tancheng, and damaged a 310-square-mile (800-km²) area. The MAGNITUDE was estimated at 8.5 on the RICHTER scale. Significant AFTERSHOCKS were said to have lasted 100 days. The earthquake triggered numerous LANDSLIDES, waterspouts, MUD VOLCANOes, and FISSURES. More than 50,000 people were estimated to have been killed in this event. There were so many bodies that relief workers buried them in mass graves, but even then, the pace was slow. Rain followed by scorching sun in subsequent days caused the bodies to decay, and the area was soon overwhelmed with plague.

Tangkubanparahu *See* SUNDA.

Tango *earthquake, Japan* On March 7, 1927, the town of Tango, JAPAN (now North Kyoto), was struck by a major earthquake. This event registered 7.3 on the RICHTER scale and caused great damage. The EPICENTER was estimated to have been in the shallow area of the Sea of Japan on the Yamada and Gomura Faults. The surface rupture on the Gomura Fault and was 11 miles (17.5 km) long, with a STRIKE-SLIP offset of up to nine feet (3 m) and a maximum DIP-SLIP offset of 27 inches (68.5 cm). A TSUNAMI was produced during this event with a maximum RUN-UP height of five feet (1.5 m). Abundant AFTERSHOCKS continued for about a month after the MAIN SHOCK, including the largest on April 1, with a magnitude of 6.5. In all, 2,900 people lost their lives in this event, and some 16,025 houses were destroyed.

Tangshan *China* The Tangshan earthquake of July 28, 1976, is perhaps the most destructive earthquake in history, whether destruction is measured in terms of lives lost or property damaged. Estimates put fatalities from the earthquake at about 655,000, with some 800,000 people injured, although official estimates list 250,000 fatalities. The city of Tangshan, approximately 85 miles (137 km) southeast of Beijing, was simply demolished. The physical damage at Tangshan has been compared to the destruction of Hiroshima. Two major earthquakes struck Tangshan on this occasion. The first, reported as MAGNITUDE 8.2 on the RICHTER scale, took place at 3:45 A.M. and lasted some two minutes. An AFTERSHOCK of magnitude 7.9 occurred several hours later. Each shock was roughly equivalent in magnitude to the earthquake that destroyed SAN FRANCISCO in 1906. Reports of the disaster told of widespread subsidence caused by the collapse of mine tunnels underground. The subsidence is said to have done particularly great damage to railway facilities. The Tangshan disaster occurred so close to Beijing that officials in the capital ordered the public to move outdoors in case an earthquake struck the city. Some 6 million people are believed to have slept outdoors in temporary shelters for more than two weeks. One intriguing aspect of this earthquake was the reported occurrence of EARTHQUAKE LIGHT. The red and white light is said to have been seen up to 200 miles (322 km) from the EPICENTER of the quake and, as viewed from Beijing,

This four-story concrete and brick building of the Tangshan People's Bank completely collapsed into a pile of rubble during the 1976 Tangshan, China, earthquake. *(Courtesy of the USGS)*

Complete destruction of all buildings (only tents for refugees are visible) by the great 1976 Tangshan earthquake. *(Courtesy of the USGS)*

allegedly lit up the sky as brightly as daylight in the direction of Tangshan.

Another thing that made the Tangshan earthquake significant is that it dispelled claims by the Chinese government that they could accurately predict earthquakes. In the previous year, Chinese scientists accurately predicted a major earthquake and evacuated the residents before it struck. The earthquake caused significant property damage but few fatalities. Ever since the Tangshan disaster, no such claims have been made.

Tao-Rusyr *caldera, Kuril Islands, Russia*   The Tao-Rusyr CALDERA is located on Onekotan Island in the KURIL chain and has an intracaldera cone, Krenitzyn Peak. Surrounding the cone is a CRATER lake that does not freeze in winter. The Tao-Rusyr caldera formed in 5550 B.C. in a very powerful explosive eruption (VEI = 6). In 1846 and 1879, SOLFATA-RAS were observed in the caldera, although it is not known whether this activity was merely ordinary or remarkable. An eruption occurred in 1952 (VEI = 3) at Krenitzyn Peak following and possibly the result of a powerful tectonic earthquake that took place several days earlier. Three days before the volcano erupted, there were reportedly disturbances in the

magnetic field in the vicinity of the caldera. A magnetic compass is said to have behaved in a strange fashion, decelerating slowly and unevenly in one direction.

Tarawera *volcano, North Island, New Zealand*   Tarawera is composed of 11 RHYOLITE domes that formed in three episodes at 12700 B.C., 9250 B.C., and A.D. 1070. The eruption of Tarawera along a FISSURE in 1886 was the first in historic times. It reportedly killed more than 150 people. An eruption on the night of June 9 of that year destroyed two native communities near the volcano. Early in the morning of the following day, strong earthquakes preceded a series of explosions from Wahanga, one of the peaks of Tarawera, that generated a huge cloud of ASH and vapor. Apparently INCANDESCENT material then burst from the CRATER, creating an appearance of flame leaping from the volcano's throat. In quick succession, other volcanoes nearby erupted, until a chain of mountains some nine miles (14.5 km) long was in eruption. Blasts of steam are also said to have occurred from Lake Rotomahana, which the eruptions caused to merge with adjacent Lake Rotomakiriri. These eruptions destroyed the beautiful lakeside geyserite formations known as Pink Terrace and White Terrace. A great FISSURE called Tarawera Chasm, more than a mile long and about 1,000 feet (305 m) deep, opened during Tarawera's 1886 eruption. The violent eruptions of June 10 lasted only about two hours, although lesser activity was observed for days afterward. During that time, some 0.5 cubic miles (2 km³) of LAVA were ejected. Vigorous GEYSER activity has been recorded along the fissure involved in that eruption.

Tarumai *volcano, Japan* The STRATOVOLCANO Tarumai has undergone explosive ERUPTIONS on more than 30 occasions since the mid-17th century. An especially large eruption in 1909 released great quantities of ASH and BOMBS.

Taupo *caldera, North Island, New Zealand* The Taupo CALDERA is located in the Taupo Volcanic Zone near Wairakei and is partly occupied by Lake Taupo. The boundaries of the caldera are poorly defined, but it is thought to be about 24 miles (39 km) wide. Roughly 2,000 years ago, during a series of violent eruptions that occurred here, a set of fissures opened and then filled with IGNIMBRITES. Slightly more than a mile outside the caldera is Tauhara volcano, a complex of domes of undetermined age. Tauhara is also the site of a GEOTHERMAL field. In 1895, a powerful earthquake occurred that caused LANDSLIDES and FISSURES in the vicinity of Wairakei. Earthquake activity was frequent in 1922 and involved displacement of approximately 10 feet (3 m) in places along FAULT lines. Subsidence of more than six feet (1.8 m) occurred along the northern shores of Lake Taupo. Several years later, further subsidence was reported, this time more than 10 feet (3 m). Fissures developed along the Kaiapo Fault, along the northeast rim of the caldera at the edge of the Wairakei field, and numerous fountains of water were reported seen along the fault.

A moderately strong earthquake occurred immediately west of Lake Taupo in 1953 and was followed by AFTER-SHOCKS. Earthquakes occurred again to the west and southwest of Lake Taupo in 1956–57.

Late in 1964, a swarm of more than 1,000 earthquakes greater than magnitude 2.5 started under Lake Taupo and continued through January 1965. This swarm is thought to have accompanied several inches of uplift along the east shore of the lake. In 1974 and 1981, HYDROTHERMAL explosions took place in the Tauhara geothermal field. An earthquake occurred several days before the 1981 explosion, but there is no proof that the earthquake was the cause of events leading to the explosion.

Very slight subsidence was noticed on the north shore of Lake Taupo in the early 1980s, and in 1983 EARTHQUAKE SWARMS occurred to the north of the lake, followed by several inches of uplift along the north shore. Earthquake activity intensified in mid-1983, followed by fast subsidence northwest of the Kaiapo Fault and uplift southeast of the fault. Slight subsidence preceded an earthquake swarm at the south side of the lake in 1984, but an equivalent amount of rebound occurred over the following months. The northeast shore was uplifted slightly in the latter part of 1984 and then subsided after a series of minor earthquakes in 1985.

In 1986, it appeared that an eruption might be imminent, because PUMICE was seen in the lake and gas bubbles were reported rising to the surface. Fears of eruption, however, proved to be groundless. The pumice was thought to have fallen into a stream that carried the material to the lake, and the gas bubbles evidently were no different from others that occur often in the lake. It is hard to predict eruptive activity at the Taupo caldera because of difficulty in separating magmatic from TECTONIC unrest.

Taupo Fault Zone *North Island, New Zealand* The Taupo Fault Zone on North Island has been the site of intensive volcanic activity both before and after European settlement of the islands in the late 18th century. Great quantities of volcanic material emerged from FISSURES in the Taupo Fault Zone in prehistoric times and were deposited as IGNIMBRITES. Similar conditions gave rise to the VALLEY OF TEN THOUSAND SMOKES in ALASKA, although the New Zealand deposits are far more voluminous than those in Alaska. There are some 50,000 square miles (129,499 km²) in area and thousands of feet deep (estimated at almost 200 cubic miles [820 km³] total) in the Taupo Fault Zone, compared to only about 50 square miles (130 km²) and several hundred feet deep (some seven cubic miles [28.7 km³] total) at the Valley of Ten Thousand Smokes. Eventually, eruptive activity was confined to a few locations but became more explosive. The Taupo Fault Zone was the site of the famous eruption of TARAWERA in 1886.

tectonic escape *See* EXTRUSION TECTONICS.

tectonics The study of the large-scale belts and terranes of rock and their relations and processes of formation. Tectonics may involve structural, stratigraphic, and petrologic relations. Typically a geologist studies a sequence of rocks in terms of their origin as well as the processes that have taken place since then. With this information, a model for the PLATE TECTONIC history of those rocks is proposed. Tectonics differs from other subdisciplines of geology in that it deals with the large scale processes of Earth by example or model.

teleseismic The arrival of seismic waves or indication of an earthquake that occurred a long distance away, commonly across a large ocean basin.

teletsunami In contrast to a local TSUNAMI, a teletsunami can cross an entire ocean basin and still have enough energy to be recognized. Tsunamis are generated with a certain amount of energy. They travel at high velocity and only grow into large waves as they come into a coast. Most tsunamis ATTENUATE so fast that they are only recognizable near their source. It takes a tsunami of remarkable power to cross an entire ocean basin and still be a threat. The BANDA ACEH event was the most recent true teletsunami, but the 1964 GOOD FRIDAY EARTHQUAKE and CHILEAN 1960 EARTHQUAKE both produced teletsunamis, as have numerous others.

Tengger *caldera, Java, Indonesia* The Tengger CALDERA played an important part in debate during the early 20th century about how calderas originate. It was determined that the caldera formed after an eruption of vast amounts of TUFF and formed during two separate episodes of collapse. The first episode formed the Ngadisari depression immediately to the northeast of Tengger, and the second episode of collapse created the Tengger caldera itself. Tengger is occupied partly by a lake, and a small cone called Bromo is located in the center of the caldera. Bromo undergoes frequent ERUPTIONS of the STROMBOLIAN type and has erupted at least 53 times since 1804. After Bromo erupted in 1835, a small lake formed in the CRATER. The lake grew in size following another eruption in 1838 and then began to change color from blue to green between 1838 and 1841. Sometime in the early 1840s, the floor of the crater appears to have bulged upward and given off gases. Several weeks later, the swelling subsided, and the crater floor collapsed. This swelling and subsequent collapse evidently were confined to the crater at Bromo.

In March 1858, Bromo became active again. This eruption was preceded by loud underground noises. Such noises also occurred before another eruption in 1858; before this eruption, earthquakes were felt up to a distance of about a mile (1.6 km) from the volcano. Bromo was blamed for a very strong earthquake that shook the island of Java in May 1865. Although the volcano was active for part of that year, there is no proof that the eruption and the earthquake were related, but some connection between the seismic and volcanic activity is possible. In 1888, Bromo emitted steam and SULFUR gas. Several days before an eruption in 1980, FUMAROLES became more active. The latest eruption was in 1984.

Tennessee *United States* Tennessee has an extensive history of strong seismic activity, partly because of its proximity to the NEW MADRID FAULT ZONE, which was responsible for a series of extremely powerful earthquakes during the winter of 1811–12. Another severe earthquake occurred in western Tennessee on January 4, 1843, and affected a wide area including portions of KENTUCKY, MISSOURI, ALABAMA, INDIANA, IOWA, SOUTH CAROLINA, and GEORGIA. Western Tennessee could undergo severe damage in the event of any future earthquakes comparable in intensity to the 1811–12 events because the area is settled much more heavily now than it was at the time of the great earthquakes of the early 19th century.

tension Tension is a type of STRESS that pull objects apart. Forces are directed in opposite directions (away from each other) in tension.

tephra Geologists define *tephra* as all solid particles emitted by a volcano during an ERUPTION. Tephra particles range in size from extremely fine ASH with approximately the consistency of confectioners' sugar to large pieces several feet in diameter (BOMBS). Tephra forms when MAGMA containing dissolved gases nears the surface; the gases bubble out of solution, forming cavities in the magma or LAVA as it solidifies. The resulting fragmented solid material is cast out of the volcano's throat and, under the right conditions, may travel for thousands of miles before settling to the surface. Almost all volcanoes release tephra when erupting, although some eruptions produce far more tephra than others. Tephra comes in various forms, including bombs, masses of fluid magma that

solidify in flight, and LAPILLI, gravellike bits of lava. Tephra may mix with extremely hot gases and flow downslope from the volcano as a *NUÉE ARDENTE* (fiery cloud), a phenomenon that can destroy a landscape as effectively as a nuclear explosion. A *nuée ardente* caused much of the destruction at Mount PELÉE in 1902. A high-altitude cloud of fine tephra from KRAKATOA appears to have altered global climate by intercepting incoming sunlight and lowering surface temperatures. A similar cloud from the 1815 eruption of TAMBORA has been implicated in the dramatic drop in temperatures the following year, the "YEAR WITHOUT A SUMMER."

Tephras constituents are known collectively as PYROCLASTICS. Pyroclastic deposits may occur either as pyroclastic flows (that is, as *nuée ardentes*) or as airfall deposits, which accumulate as tephra settles out of the atmosphere. Each kind of pyroclastic deposit shows a characteristic pattern when exposed vertically. In a pyroclastic flow, particles of many different sizes, large and small, are mixed together with little or no evidence of sorting or layering. Airfall deposits, by contrast, exhibit distinct layering and sorting into fine and coarse layers. MUDFLOW deposits consist of gravel of various sizes deposited amid fine silt. These deposits occur when large quantities of water flow down a volcano's flanks, picking up tephra along the way and finally depositing the mudflows in nearby stream valleys. Mudflows may occur when an eruption melts ice and snow on a volcano summit or casts water out of a lake in the CRATER. Heavy rains may also generate mudflows. A mudflow may overwhelm an entire community without any advance warning.

terrane A piece or belt of rock in Earth's CRUST that differs greatly in history and composition from those in adjacent areas. A terrane may represent a piece of land that was added to the continent through a collision. The terrane may represent a separate continent or ISLAND ARC and could have come from thousands of miles away. There are certain characteristics that the geologist searches out to prove the exotic nature of the terrane. Ophiolites, old pieces of OCEANIC CRUST, commonly lie along the boundaries between exotic terranes. There may also be highly deformed deep-sea sediments that formed in the trench of a SUBDUCTION ZONE but have been shoved onto land during collision. Exotic terranes are commonplace in the western United States because of the collision of several large tectonic plates and numerous smaller plates there. The phrase *collage tectonics* was coined for this area because geologic maps look like a collage of belts of unrelated sequences of rock. Terranes are interpreted by geologists who study TECTONICS.

See also PLATE TECTONICS.

Terror *See* EREBUS, MOUNT.

Texas *United States* Earthquake activity in Texas has provided interesting discoveries for geologists on occasion. For example, the July 30, 1925, earthquake in the panhandle region was a strong earthquake, felt 225 miles (362 km) away in Roswell, NEW MEXICO, and Leavenworth, KANSAS, 400 miles (644 km) distant. The area was not thought to be given to earthquakes (the only previous recorded earthquake in the Texas panhandle occurred in 1917), and for a while, oil drilling was suspected as a reason for the earthquake. The earthquake affected such a wide area, however, that drilling was eliminated as a possible cause. Another earthquake, at Mount Livermore in west Texas on August 16, 1931, caused heavy damage at Valentine and would have been extremely destructive if the event had occurred in a heavily settled area.

There are ancient volcanic deposits in west Texas related to volcanic activity on the RIO GRANDE RIFT. These deposits even spread into the sedimentary rocks of the Gulf of Mexico, where they decrease the oil and gas content of the strata.

Thera   *See* THIRA.

thermal gradient   The rate at which Earth's temperature increases with depth. Although an average figure of 30°C per kilometer is given for the thermal gradient, the rate of increase actually varies from one level of Earth's internal structure to another. The increase is believed to be rapid within the CRUST, where high concentrations of RADIONUCLIDES in granitic rock generate large amounts of heat by their decay. The thermal gradient is thought to diminish as the crust gives way to the outer layer of the MANTLE. In this layer, some about 36 to 60 miles (60 to 100 km) beneath the surface, temperatures are estimated at approximately 800 to 1200°C.

Thermal gradient also varies by location on Earth. In GEOTHERMAL areas such as ICELAND, YELLOWSTONE, or The Geysers, in CALIFORNIA, the thermal gradient can be very high, up to several hundred degrees per kilometer. Any area with significant igneous activity will have an elevated thermal gradient. In sedimentary basins that are high in radioactive elements and that insulate well, the thermal gradient can also be elevated but to a lesser extent. In basins with thick sequences of rocks that are low in radioactive elements, the thermal gradient might be depressed. In SUBDUCTION ZONES, the thermal gradient is also reduced because the OCEANIC CRUST is driven down into the mantle faster than it can heat up.

See also EARTH, INTERNAL STRUCTURE OF.

Thira   *volcano, Mediterranean basin, circa 1470 B.C.*   The island of Thira (also known as either SANTORINI or formerly Thera), between Greece and Turkey, is believed to have been almost totally destroyed in a volcanic explosion that wiped out the Minoan civilization. Thira became active after long dormancy around 1500 B.C. and appears to have erupted at intervals over some 30 years before a final, explosive eruption destroyed much of the island. Archaeological evidence indicates that powerful 200-foot (61-m)-high TSUNAMIs generated by the explosion caused widespread destruction on shores in the eastern MEDITERRANEAN. Apparently Minoan civilization never recovered from the natural disasters associated with this eruption of Thira. The destruction of Thira is believed to have given rise to the legend of ATLANTIS, the ancient civilization supposedly destroyed in a single day when the island of Atlantis sank in a great natural cataclysm. Initially, the legend placed the demolished island accurately in the vicinity of Crete, but the Greeks later imagined that Atlantis had been located farther west, in the great ocean beyond what we know as the Straits of Gibraltar. The Atlantic Ocean was named for the fictional island.

The explosion of Thira shifted the course of history in the Mediterranean. The dominant Minoan civilization was crippled by the tsunami. Previously, the Minoans had kept the "hill people" in Greece and Mycenae at bay with their technical superiority. Because the tsunami wiped out the coastal communities but left the hills intact, the Golden Age of Greece and empire of Alexander the Great were ushered in. The explosion also roughly corresponds to the flight of the Jews from Egypt. ASH in the atmosphere could have accounted for the bloodred skies. Could the parting of the Red Sea and drowning of the pharaoh's troops really been the tsunami from Thira?

Deep-sea coring in the Mediterranean has revealed a layer of thoroughly contorted and mixed-up sediment that researchers refer to as "homogenite." It is interpreted that this layer resulted from the scraping up and mixing (homogenizing) of existing sedimentary layers on the seafloor. This homogenite has been radiometrically dated at 1500 B.C. It occurs throughout the Mediterranean basin in thickness up to nine feet (3 m). Rough estimates indicate a VEI of 6 for this explosion.

Thira's eruption is sometimes compared to that of KRAKATOA. As in the case of Krakatoa, the explosion of the volcanic island left behind a CALDERA that filled with water. A group of small volcanic islands arose from the caldera and erupted intermittently from 1938 to 1941.

tholeiite   The most common BASALT type. Based on a characteristic bulk chemistry, tholeiitic trends and ranges are the common type to the ocean floor. They are produced in the middle stages of rifting and most HOT SPOTS as well as the MID-OCEAN RIDGE. FLOOD BASALTS are typically tholeiites as well.

Three Rivers   *earthquake, Quebec, Canada*   An earthquake on the order of that in NEW MADRID appears to have occurred in Quebec in 1663. The area is along a trend of earthquakes in the midcontinent that includes New Madrid. There is almost no information available except for the accounts given in *Jesuit Relations*:

> New lakes are seen where none was before. Some mountains were engulfed and disappeared. Many a waterfall is leveled and many a river is no more. The ground cracked in many places and opened crevices of unmeasured depth. And there has resulted such a disorder of fallen and splintered trees that one sees today fields of more than a thousand argents all leveled as if they had been recently tilled, in many places where there was nothing but forests.

Three Sisters   *volcano, Oregon, United States*   A cluster of volcanic peaks in western OREGON, the Three Sisters are part of the CASCADE MOUNTAINS located near Mount HOOD and CRATER LAKE. The most recent eruption was about 1,900 years ago.

thrust fault   A low-angle REVERSE FAULT.

tidal wave A misnomer for TSUNAMI.

Tien-Chi *caldera, China* Located on the border between China and Korea, the Tien-Chi CALDERA has a spectacular CRATER lake some three miles (4.8 km) wide and more than 1,000 feet (308 m) deep. The caldera is thought to have formed less than 1,000 years ago, during or after an eruption that deposited a huge volume of PUMICE over an area some 24 miles (39 km) wide around the caldera. Eruptions were reported at the caldera in 1597 and 1702, but these reports have not been confirmed. Tien-Chi is located near the volcano Paektusan, which is either a STRATOVOLCANO or a dome surmounting a vast lava shield.

See also BAITOUSHAN.

Toba *caldera, Sumatra, Indonesia* The Toba CALDERA is a volcano-tectonic depression. Although it has been argued that the depression is TECTONIC in origin, the vast quantities of TUFF expelled from eruptions here indicate to many geologists that volcanic activity was involved in the caldera's origin. The Young Toba Tuff was formed 74,000 years ago in likely the largest volcanic eruption in the last 2 million years (VEI = 8). The estimated volume of this eruption is 700 cubic miles (2,800 km^3), which is even larger than the YELLOWSTONE eruption of 2.2 million years ago. PYROCLASTIC FLOWS from this eruption covered some 7,722 square miles (20,000 km^2).

The Toba caldera marks the point where a chain of ANDESITIC volcanoes undergoes a change in relation to Sumatra's Semangko Fault Zone, a large STRIKE-SLIP FAULT. To the south of the Toba caldera, active volcanoes are located very close to the FAULT zone, but north of Toba the volcanoes are situated more to the east of the fault zone and are distributed over a broader area. Toba is thought to occupy a north-to-south boundary characterized by high seismicity, between two subducted plates of Earth's crust. Volcanism at Toba may be the result of increased production of MAGMA along this boundary zone.

The caldera has not experienced any eruptions within historical times, but earthquake activity is frequent in the vicinity of Toba. Powerful earthquakes have occurred at Toba on several occasions since the late 19th century. A very strong earthquake on May 17, 1892, centered about 75 miles (121 km) south of Toba, demolished numerous buildings and accompanied several feet of displacement along the DEXTRAL strike-slip Semangko Fault. A slump took place around the year 1914 at the upper end of the Asahan Valley on the southeast side of the caldera and made Lake Toba rise several feet, although no particular importance is attached to this event in terms of earthquake or volcanic activity. An earthquake occurred in November 1920 and was followed by EARTHQUAKE SWARMS in 1922. Damage occurred along the south shore of Lake Toba, where subsidence took place at several locations. FUMAROLES to the south and west of Lake Toba became more active after powerful earthquakes. A moderately strong earthquake in 1987, centered on the south shore of the lake near the community of Muara, caused significant damage.

Tofua *volcanic island, Tonga* Tofua Island is the site of Tofua CALDERA and has a history of eruptions and other unrest

dating back to the late 18th century. One ERUPTION in 1958–59 was severe enough to require the evacuation of the islands population for several months.

Tokachi *volcano, Hokkaidō, Japan* It is composed of overlapping STRATOVOLCANOes that have produced 17 historical ERUPTIONS. The eruptions are clustered into periods 1857, 1887, 1889, 1925–31, 1952–62, 1985, and 1989. Most eruptions are PHREATIC and small (VEI = 1–2). In the 1926 eruption, LAHARS swept down the volcano covering 12.4 miles (20 km) in 26 minutes. It destroyed more than 5,080 homes and killed at least 146 people. In the 1962 eruption, falling blocks killed five people.

See also DAISETSU-TOKACHI.

Tokai *See* KANTO.

Tokaido *earthquake and tsunami, Japan* On October 28, 1707, one of the greatest earthquakes in the history of JAPAN struck the Tokaido area. The earthquake was powerful enough to cause damage to houses and public buildings in 26 provinces. As is common in Japan, this submarine earthquake produce a massive TSUNAMI. This huge wave swept the coast from KWANTO to Kyūshū Island. In southern Shikoku near Koti, the wave reached a breathtaking height of 91 feet (27.5 m). The wave continued into the Inland Sea between the islands of Shikoku and Honshū. It rushed into Osaka Bay and up the two rivers that empty into it. More than 1,000 boats were swamped and sunk and several were carried upriver, where they damaged bridges. In all, it was reported that more than 4,900 people lost their lives, but the number is suspected as being much higher. It was also reported that over 29,000 houses were destroyed in the event.

Tokyo *Japan* The capital of JAPAN is known for its history of frequent and highly destructive earthquakes. The 1923 KANTO earthquake and subsequent fire destroyed much of Tokyo and nearby Yokohama. The earthquake was estimated at a magnitude of up to 8.3 on the RICHTER scale. A similar earthquake today would have a much more devastating effect than the 1923 earthquake for various reasons. The city is larger than in 1923, and therefore opportunities for damage are greater. Also, Tokyo now has numerous hazardous materials stored in it in large quantities, and those materials might pose serious dangers in the event of a major earthquake. The economic effects of a major earthquake in Tokyo would reach to numerous nations other than Japan, notably the UNITED STATES, and might cause serious hardship on an international scale because of Tokyo's importance as a financial center.

Tokyo is regularly struck by strong events. The last confirmed earthquake prior to Kanto was the November 11, 1855, event. The EPICENTER was within the city. Many charcoal braziers were upset and overturned by the shaking, and many fires broke out but were quickly contained. In all, some 50,000 houses were destroyed, and the DEATH TOLL was 6,757. There are reports of another earthquake on March 21, 1857, in Tokyo in which 107,000 were said to have perished largely as the result of fires, but this event could not be confirmed.

Tokyo is located near the volcano Mount FUJI, which has become a symbol of Japan. It is also near the volcano HAKONE. If either of these volcanoes reawakened, or if a new volcano formed in the area, the effects could be devastating on this densely populated area.

See also ECONOMIC EFFECTS OF EARTHQUAKES AND VOLCANOES.

Tolbachik *volcano, Kamchatka, Russia* Although the STRATOVOLCANO Tolbachick has erupted often since the mid-18th century, it is most famous for a major eruption in 1975 that was predicted so accurately that television crews were able to film its onset. Earthquake activity just before the ERUPTION made the prediction possible. The eruption resulted in the formation of four new cinder cones and was accompanied by an eruption column of eight miles (13 km) height. The resulting LAVA sheets are 260 feet (79 m) thick and cover 15 square miles (39 km²). The previous eruption was from 1967 to 1970.

Toledo *caldera, New Mexico, United States* The Toledo CALDERA is thought to have produced an estimated 50 cubic miles (205 km³) of TEPHRA in an ERUPTION some 1.3 million years ago. The Toledo caldera lies buried under a heavy ASH-FALL from the nearby VALLES caldera.

Toliman *See* ATITLÁN.

Tomboro *See* TAMBORA.

tomography Seismic tomography is the equivalent of a CAT scan of the Earth. By studying small changes in the velocity of seismic waves received by SEISMOGRAPHS on the surface, detailed images of the internal structure of the Earth may be constructed. These images have led to new understanding of the PLATE TECTONIC interactions and resulting thermal structure.

Tompaluan *See* TONDANO.

Tondano *caldera, (Celebes) Sulawesi, Indonesia* The Tondano caldera occupies the northern portion of Sulawesi and has a record of activity within historical times dating back to the early 19th century. Eruptions have been explosive, with some PHREATIC ERUPTIONS. Several andesitic STRATOVOLCA-NOES have formed along the rim of the CALDERA. A new CRA-TER called Tompaluan appeared in 1829 and started to smoke in 1893. A FISSURE some 300 feet (91 m) long that emitted smoke is thought to have formed in the caldera around 1898. Earthquakes in 1901 were ascribed to increased activity at the Soputan volcano on the south rim of the caldera; two mud pots formed at the foot of the volcano, and there was a report of ASH ejection and steaming from Soputan at this time. About a month later, a new SOLFATARA had appeared near the warm spring at Rumerega and was casting up boiling mud to a height of more than 700 feet (213 m), but the crater of Soputan reportedly showed no increase in activity.

In 1906, the new crater Aeseput formed at Soputan. That same year, earthquake activity increased, accompanied by fumes from the volcano and a glow at night. Soputan erupted on June 17, 1906, the same day an earthquake shook Tomohon, some 18 miles north-northeast of the volcano. Soputan continued erupting at intervals through 1913. A powerful TECTONIC earthquake occurred in 1932 but apparently did not affect the volcanoes. In 1958 and 1959, minor eruptions of ash occurred, and another series of minor eruptions took place from 1961 to 1962. Small explosions were reported between 1963 and 1968, and LAVA FLOWS and PYROCLASTIC FLOWS occurred at Soputan between 1966 and 1968. Small explosions occurred again in 1969–71. Gray gas, possibly containing ASH, emerged from Soputan in 1970, and fumarolic activity increased through a lava flow that had been laid down in 1966.

There was an increase in seismic activity at tile caldera in 1976, and the temperature of the crater lake increased sharply in early 1978. A strong explosive eruption occurred in August 1982 without any premonitory signs, but earthquakes did precede another eruption in November of that year. In April 1984, minor eruptions of gas through the lake killed plants as far as two miles (3.2 km) downwind, possibly from the action of acid rain or acid gases. An eruption in May 1984 showed no premonitory activity, but earthquake activity increased before an eruption in August 1984. Tectonic earthquake activity appears to have increased dramatically at the caldera in early 1985. In May 1985, the steam plume that normally rose from the volcano increased in height, and by May 19 the plume, now containing ash as well as steam, rose to an altitude of about three miles (4.8 km). In March 1986, the Tompulan crater sent up puffs of steam to an altitude of more than 1,000 feet (305 m), and PHRE-ATIC EXPLOSIONS produced LAHARS. In the second half of 1986, dozens of explosions occurred daily. Explosions continued, though less frequently, through May 1987. Between April and May 1987, light-colored fumes emerged from the Magawu (Mahawu) crater, where the crater lake rose in temperature and increased in volume.

Tongariro *volcano, North Island, New Zealand* Tongariro is a STRATOVOLCANO within the Tongariro volcanic complex that also includes RUAPEHO, Pihanga, and Kakaramea. NGAURUHOE sits in the middle of the Tongariro massif and is the most active volcano in NEW ZEALAND.

topographic monitoring Detailed altitude monitoring of the surface of volcanoes. When MAGMA moves into the MAGMA CHAMBER underneath a volcano before an ERUP-TION, it causes the volcano to inflate. This inflation will cause changes in elevations on the volcano. The monitoring involves the use of lasers and satellite altimetry and is very accurate. Even small changes can be detected. The inflation is accompanied by EARTHQUAKE SWARMS. Positive results from the topographic monitoring and earthquake swarms lead scientists to issue alerts because they usually herald an eruption.

Torfajökull *volcano, Iceland* Located in southeastern ICELAND, the Torfajökull volcano is associated with a CAL-DERA and is in an area of frequent earthquake activity. The volcano erupted around the year A.D. 900 and emitted both

LAVA and TEPHRA. Another eruption occurred about 1480, again involving both tephra and lava. Both these eruptions occurred during episodes of rifting on the Veidivötn FISSURE swarm, which lies northeast to southwest through the middle of Iceland. Tephra and lava in these eruptions were RHYOLITIC. Earthquakes have occurred at the caldera in recent years, sometimes at a rate of several dozen per day, but there have been no further eruptions since the 15th century.

Toribio *earthquake, Colombia* On June 6, 1994, an earthquake of MAGNITUDE 6.7 occurred. At least 295 people were killed with more than 500 missing. Severe damage to houses, highways, and bridges by both the earthquake and LANDSLIDES it triggered left some 13,000 homeless. One AVALANCHE blocked the Paez River, causing severe flooding.

Towada *caldera, Japan* The Towada CALDERA is located in the northern portion of the island of Honshū. The caldera contains a lake and is thought to have formed from the collapse of several small STRATOVOLCANOES. Towada had three very large (VEI = 5) prehistorical eruptions in 7550 B.C., 6650 B.C., and 3440 B.C. It has been quiet for approximately the last 1,000 years, but an eruption in approximately A.D. 915 is thought to have deposited airfall TEPHRA and PYROCLASTIC FLOWS over the surrounding area.

Tower Peak *See* AMBRIM.

Toya *caldera, Japan* Located on the island of Hokkaidō, the Toya CALDERA lies immediately north of Usu volcano and contains a lake, Lake Toya, in the middle of which a group of LAVA DOMES has formed an island. The historical record of activity at the caldera dates back to 1626 since which Usu has erupted nine times. Eruptions were recorded in 1663, 1769, 1822, 1853, 1910, 1944–45, 1977–82, and 2000. The most destructive of Mount Usu's eruptions occurred in 1822 when 59 people were killed. The eruption in 1910 was preceded by several days of earthquake activity, which was accompanied by the appearance of cracks and FAULTS aligned mostly east to west along the northwest foot of Usu volcano. Small CRATERs formed at the north foot of Usu, and numerous small mud cones formed along the seashore nearby. The explosive phase of the eruption was largely finished by early August.

In May 1932, an area of subsidence some 200 feet (61 m) wide was noticed off the north shore of Lake Toya; although this area subsided by more than 30 feet (9 m) in places, no eruption followed, although lake temperature rose considerably. Earthquakes started late in December 1943. An explosive eruption began in June 1944 and lasted through late October. Earthquake activity again showed a sharp increase in August 1977, followed by an explosive eruption of PUMICE. Magmatic explosions continued through August 14 on an intermittent basis. PHREATIC ERUPTIONS occurred on occasion in 1977 and 1978. Activity at Usu volcano has produced new hot springs, which in turn have generated opportunities for tourism because hot springs are popular attractions in JAPAN. In late March 2000, Mount Usu began to erupt. This was the first eruption in 22 years. During that eruption, two

people were killed, and 200 homes were destroyed. In this eruption, 16,000 people were evacuated of the 51,000 that live in the area.

T-phase A type of TSUNAMI with a short period that travels through the ocean at the speed of sound. They are typically only identified traveling in very deep ocean basins.

trachyte A volcanic rock of INTERMEDIATE composition that is typically PORPHYRITIC and has K-FELDSPAR and minor MAFIC minerals (BIOTITE, HORNBLENDE, or PYROXENE) as the chief components. With more QUARTZ, the trachyte would be a RHYOLITE, and with less K-feldspar, it would be a LATITE.

trajectory The direct line path of seismic wave traveling from its FOCUS (source) to the SEISMOGRAPH, where it is recorded.

transcurrent fault *See* STRIKE-SLIP FAULT.

transform fault A TECTONIC plate margin characterized by STRIKE-SLIP movement. About 99% of transform faults offset MID-OCEAN RIDGES, which results in the steplike patterns observed there. The few good examples of transform faults on land are highly seismically active. They include the SAN ANDREAS FAULT of CALIFORNIA, the MOTAGUA FAULT of GUATEMALA, and the South Alpine Fault of NEW ZEALAND.

See also PLATE TECTONICS.

transition zone A region where the upper and lower parts of the MANTLE meet. It is at 255–621 miles (410–1,000 km) depth and is characterized by an increase in density of rock by about 20% and a corresponding increase in seismic wave velocity. It is also called the C layer. It is speculated that the pressure is so great that the molecules in minerals in the rocks crush down to form a tighter, denser structure just as graphite crushes down to DIAMOND with pressure. There is also a transition zone from the INNER to the OUTER CORE. This transition is also recorded by seismic waves and density, but it is the result of a phase change from solid (inner core) to liquid (outer core).

Transverse Ranges *California, United States* These mountains form an east-west ridge through southern CALIFORNIA and include the San Bernardino, San Gabriel, Santa Susana, Santa Monica, and Santa Ynez Mountains. The Channel Islands off Santa Barbara are also seen as an extension of the Transverse Ranges. The Transverse Ranges appear to mark the zone where the LOS ANGELES basin is being driven to its northwest. Compression is evident in the Transverse Ranges and is thought to originate in this case from the movement of the PACIFIC CRUSTAL PLATE to the northwest. Uplift along the Transverse Ranges is thought to have been responsible for the highly destructive SAN FERNANDO earthquake of 1971. Spectacular displays of exposed sedimentary rock may be seen along highway U.S. 101 in the Santa Ynez Mountains near Santa Barbara and Buellton.

See also PLATE TECTONICS.

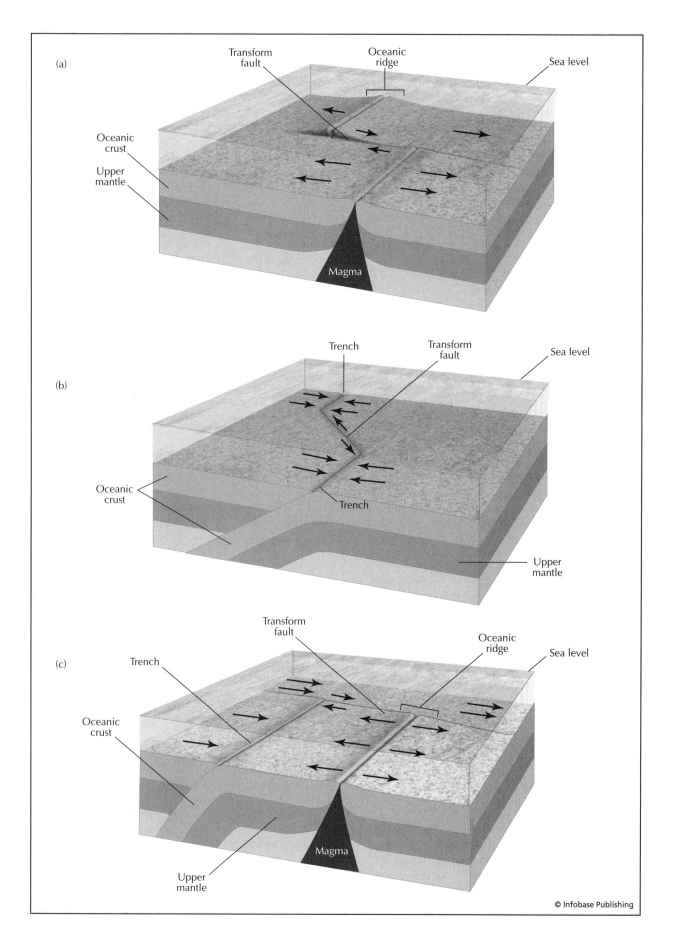

(*opposite page*) Transform margins are large strike-slip faults that separate plates. They must connect two other types of margins, convergent and/or divergent. There are three possible combinations: **(A)** divergent-divergent, **(B)** convergent-convergent, and **(C)** divergent-convergent. The most common are divergent-divergent and most offset the mid-ocean ridges.

tremor   Another less formal term for an earthquake.

Trident, Mount *Alaska, United States*   Mount Trident is a volcano located in the Aleutian Range near Mount KATMAI. It was originally named for the three cones near its summit, but a fourth was added in 1953. Trident has erupted 15 times since 1913, the latest in 1974. Movement of MAGMA from under Katmai toward Trident is believed to have played a part in the collapse of Katmai during its 1912 eruption.

triple junction   The location where three LITHOSPHERIC plates come together. Three divergent boundaries or arms emanate from the triple junction. These arms ideally meet at 120 degrees, so they are evenly spaced. The plates all move away from each other. The theory behind triple junctions is based on the physics of a sphere. If a sphere is broken at a point by a force from the inside, such as shooting a bullet through a glass light globe from the inside, three evenly spaced cracks will emanate from the hole. Pulses of MAGMA shooting from the MANTLE may from the triple junctions. Two arms of the triple junction will become active divergent margins with ocean basins, and the third arm will fail. The ideal example of a triple junction occurs at the southern tip of Saudi Arabia. The Red Sea forms one active arm, the Gulf of Aden forms the other active arm, and the GREAT RIFT VALLEY of East AFRICA forms the failed arm. This theory may explain how divergent boundaries develop, as well as why the continents tend to be triangular in shape.

Tristan de Cunha *volcanic island group, South Atlantic Ocean*   The islands of Tristan de Cunha, part of a MID-OCEAN RIDGE, made news worldwide in the early 1960s when earthquakes and an eruption of LAVA required the evacuation of the local population. The residents of Tristan de Cunha remained in exile for about a year and a half and then chose to return to their island home.

tsunami   A tsunami is a wave generated in a body of water by a physical disturbance. The wave may be produced by an earthquake, by a volcanic ERUPTION, or by several other processes. Tsunamis are known, erroneously, as "tidal waves," although tides are not a factor in their creation. The classic description of a tsunami goes roughly as follows. First, the waters withdraw from the shore, leaving the seabed exposed. Then the waters return as a breaker that may reach heights of 40 feet (12 m) or more. The wave carries away virtually everything in its path and may travel a mile (1.6 km) or even several miles inland before returning to the sea.

Actual tsunamis may not follow that exact pattern. Although some tsunamis are preceded by a withdrawal of water from the shore, other tsunamis may rise up from the sea with little or no warning. Nor does the wave always take the form of a curling breaker. A tsunami may manifest itself instead as a sudden upsurge of the sea. A train of tsunamis characteristically has such a great wavelength (the distance from crest to crest) that the waves may be imperceptible as they pass under ships in mid-ocean. As the waves approach land and "touch bottom" in shallow coastal waters, however, the tsunamis rise and take on a familiar wavelike form. In open oceans, the waves may travel at 400 to 700 miles (644 to 1,127 km) per hour, but near shore their velocity drops to only perhaps 50 or 60 miles (80 or 97 km) per hour so that the waves crowd together and may be spaced only a few thousand feet apart. This compressed spacing means that several towering waves may hit the shore in quick succession, with devastating effect. The exact characteristics of a

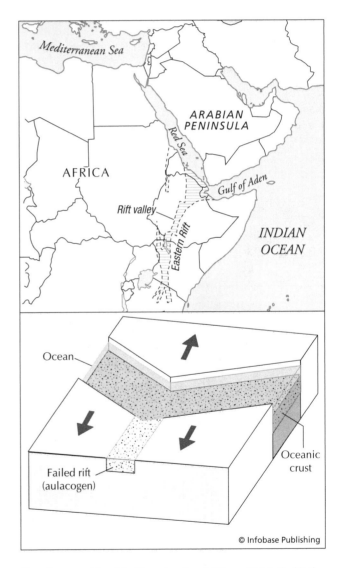

Above is a map of Saudi Arabia and northeast Africa, which is the ideal example of a triple junction. The Red Sea and Gulf of Aden are active arms floored by ocean crust. The East African Rift System is the failed arm. Below is a schematic diagram of a triple junction showing the two active arms that form an ocean basin and the failed arm that forms an aulacogen.

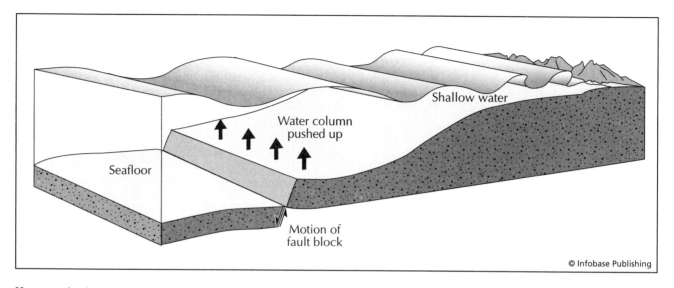

Movement of rock on a submarine fault during an earthquake (seaquake) that could generate a tsunami. There must be a vertical displacement of the seafloor to generate low-amplitude, long-wavelength waves that travel at hundreds of miles per hour. In the deep ocean, these waves are barely detectable, but when they come ashore, they grow into deadly monsters, as their long wavelengths make them grow and break far from shore.

tsunami depend on many factors, including the geometry of the seabed and the shoreline. A narrow inlet may "squeeze" a tsunami into a relatively small area and thus multiply its destructive power manifold. Tsunamis are associated closely with PACIFIC OCEAN shores because the rim of the Pacific coincides roughly with the "RING OF FIRE," a belt of intense earthquake and volcanic activity that exists where the PACIFIC CRUSTAL PLATE encounters adjacent plates. Along the

A tsunami destroys a bridge on the Wailuku River, near Hilo, Hawaii, on April 1, 1946. An earthquake of magnitude 7.4 in the Aleutian Islands, Alaska, produced catastrophic tsunamis all over the Pacific basin. Waves of 55 feet (18 m) in height killed 159 people and caused $26 million in property damage. Here the tsunami ripped apart a steel railroad bridge and carried it 850 feet (284 m) upstream. *(Courtesy of NOAA)*

The small, white arrow points to one of the victims of this devastating tsunami on April 1, 1946, in Hawaii. This photo shows the tsunami breaking over a pier in Hilo Harbor. There were 159 fatalities as the result of this event, including the man in the picture. *(Courtesy of NOAA)*

Ring of Fire, earthquakes and volcanic eruptions have generated numerous destructive tsunamis.

The National Oceanic and Atmospheric Administration (NOAA) Coast and Geodetic Survey's PACIFIC TSUNAMI WARNING SYSTEM was instituted following the tsunami that devastated HILO, Hawaii, in 1946. At the system's headquarters in Honolulu, Hawaii, seismic data and information on tides are monitored to provide early warning of tsunamis and potentially tsunamigenic events. If and when evidence indicates a tsunami poses a danger to shores around the Pacific Ocean, a warning goes out from Honolulu. The Coast and Geodetic Survey also maintains the Regional Tsunami Warning System in ALASKA. The survey's seismological observatory at Palmer, Alaska, maintains a watch on seismic data and issues a tsunami warning if a sufficiently powerful earthquake occurs in the ALEUTIAN ISLANDS or along the southeast shores of Alaska.

tsunamigenic Any process that can generate TSUNAMIS. Typically, these processes involve volcanic explosions, submarine earthquakes in which there is surface faulting, or mass movements either submarine or from land and into the ocean. By far, the most common are from earthquakes, but large and devastating tsunamis can result from any of these processes.

tuff A rock, made of PYROCLASTIC fragments fused together by the heat of an eruption.
See also ASHFALL; ASH-FLOW TUFF.

Tulik *See* OKMOK.

Tungurahua *volcano, Ecuador* An active STRATOVOLCANO that is also known as the Black Giant. The lava is mainly BASALT TO ANDESITE in composition. Tungurahua has had at least 17 ERUPTIONS in historical times. It is currently erupting. It began by ejecting 6,000 to 10,000 metric tons of material per day. Mushroom shaped ASH clouds rose to 35,000 feet (10,668 m). The ICE CAP was mostly turned into a LAHAR. At last report, it was still emitting EJECTA and LAVA. The previous eruption was in 1944.

turbidity current Loosely defined, a turbidity current is a submarine LANDSLIDE. When some disturbance, such as an earthquake, sets a turbidity current in motion, the current flows downslope as a dense suspension of mud, silt, and sand (though the composition of the sediment in it varies with location). The deposit of sediment laid down by a turbidity current is called a turbidite. As a rule, turbidity currents have little effect on human activities, but in at least one celebrated case they knocked out a deep-sea cable system. An earthquake beneath the GRAND BANKS off Newfoundland in 1929 set off a massive turbidity current that snapped cables on the seabed and interrupted telegraph communications. The maximum velocity of the current was estimated, by timing the breaks in various cables, at about 40 miles (64 km) per hour. At the EPICENTER, the earthquake caused a mass of sediment some 60 miles (97 km) long and 1,200 feet (366 m) thick to slide away.

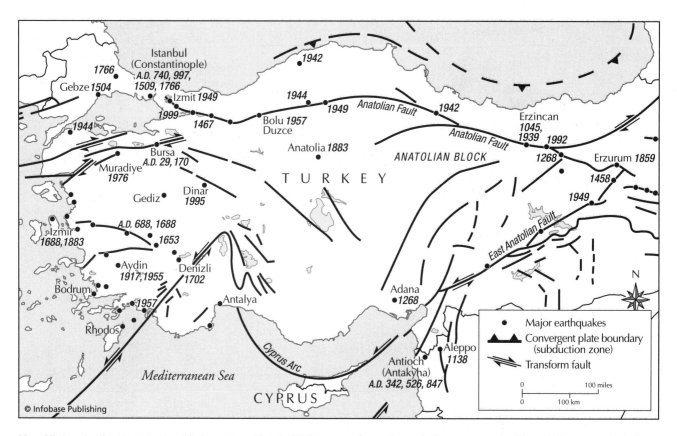

Map of Turkey showing the squeezing of Turkey eastward into the Mediterranean Sea as the result of escape tectonics. The solid lines are major faults, and the dots show historical epicenters with the year of the event. Major epicentral cities are also shown.

Altogether, the earthquake displaced perhaps 20 cubic miles (82 km³) of sediment spread over an area of roughly 36,000 square miles (93,240 km²).

Turkey The country in the eastern MEDITERRANEAN and Asia Minor is the site of regular devastating seismic activ-

The total collapse of a house in Lice, Turkey, as a result of an earthquake in 1975 *(Courtesy of the USGS)*

ity. The collision of the Arabian plate with the EURASIAN CRUSTAL PLATE is squeezing Turkey eastward like a watermelon seed would squirt out between thumb and forefinger. Most of the movement is taken up along the North Anatolian and related faults. They are seismically among the most active faults on Earth. They are right-lateral STRIKE-SLIP FAULTS. As recently as 1999, the IZMIT earthquake near ISTANBUL killed some 30,000 people; it was not an uncommon occurrence. Turkey has a rich history of devastating earthquakes; among them are the Istanbul earthquake of 1509 that reportedly killed 13,000 people; the Izmir quake of 1688 that killed over 20,000 people; DENIZLI earthquake of 1702, with a DEATH TOLL of more than 12,000; and the ERZINCAN earthquake of 1939 that killed some 33,000 people. Turkey is one of the more seismically active countries in the world.

Tuvio *See* AMBRIM.

tuya A tuya is a volcano formed under a glacier. Tuyas have characteristic flat tops as a result of not being able to penetrate the overlying ice. As the eruption continues, large quantities of water may be ponded above the volcano. PILLOW LAVAS are commonly formed in this water. There can also be deep erosion of the volcano where the melt water escapes.

U

ultramafic Igneous compositional term meaning exceptionally enriched in Iron (Fe) and magnesium (Mg). PLUTONIC rocks include PERIDOTITES and LHERZOLITES among others. The MANTLE is composed of ultramafic rocks. They are composed of various amounts of the minerals OLIVINE, various types of PYROXENE, and HORNBLENDE in some varieties. There are no recent discoveries of volcanic ultramafic rocks; the oldest go back about 3 billion years ago called KOMATIITES.

United States The United States of America has a long record of earthquake and volcanic activity. Seismic activity is most pronounced on the Pacific coast of the United States, notably in CALIFORNIA. There, ongoing movements among plates of Earth's CRUST have resulted in numerous strong earthquakes during the 20th century alone, most notably the SAN FRANCISCO earthquake of 1906, which destroyed much of the city through ground motion and a subsequent fire. Southern California, including the LOS ANGELES area, is extremely susceptible to earthquakes and is considered perhaps the most likely candidate for the next major earthquake (comparable in destructive power to the 1906 San Francisco earthquake). In the 20th century alone, the Los Angeles area has undergone several strong and destructive earthquakes, such as the LONG BEACH earthquake of 1933 and the SAN FERNANDO earthquake of 1971. The SAN ANDREAS FAULT, which extends several hundred miles along the California coast, has been responsible for some of the most powerful earthquakes to strike California in the last two centuries, but many other ACTIVE FAULTS in the state have generated, and remain capable of generating, strong earthquakes. (An example is the HAYWARD FAULT underlying the densely populated East Bay area of San Francisco.)

Other areas of strong earthquake activity in the United States include: New England, where the area around Boston, MASSACHUSETTS, has been struck by major earthquakes on several occasions; upstate NEW YORK, which includes part of the earthquake-prone SAINT LAWRENCE VALLEY; SOUTH CAR-OLINA, where one of the most powerful U.S. earthquakes of the 19th century hit CHARLESTON; the Mississippi Valley, site of the NEW MADRID, MISSOURI, earthquakes of 1811–12, thought to be the most powerful earthquakes in the history of the nation; ALASKA (site of the GOOD FRIDAY EARTHQUAKE), also subject to frequent volcanism; the HAWAIIAN ISLANDS, where earthquakes have accompanied the volcanic activity that created the islands; and OREGON and WASHINGTON, where strong earthquakes have been rare but still have the potential to occur and cause great destruction nonetheless. While few portions of the United States are completely free from earthquakes, some states, such as FLORIDA, are generally quiet from a seismic standpoint, compared to such active areas as California and Alaska.

Although California is the most seismically active state of the United States, the potential for highly destructive earthquakes actually may be greater in other portions of the nation, where major earthquakes are few and preparations for them are lacking. The danger from major earthquakes is especially great in portions of the eastern United States where active faults may underlie thick layers of sediment so that the faults give no sign of their activity until a powerful earthquake occurs. Adding to this threat is the fact that many large urban areas in the eastern United States are built on sediments in which ground water rises close to the surface. Under these conditions, LIQUEFACTION may cause soil to behave as a liquid during the passage of earthquake waves and cause massive destruction of buildings atop the liquefied soil. Liquefaction could cause tremendous property damage and loss of life in future earthquakes in such places as Boston, where much of the city is built on geologically unstable landfill, and cities in the Mississippi Valley around Memphis, TENNESSEE, and St. Louis, Missouri, which are built on the floodplain of the Mississippi River and are situated near the New Madrid Fault Zone, thought to be responsible for the tremendous earthquakes of 1811–12. The New Madrid Fault Zone is believed to have its origins in a great RIFT VALLEY that developed in the middle of what is now the United States and was buried under thick layers of sediment that today fill the Mississippi Valley. From an economic standpoint, earthquakes are a subject of some concern to the United States. It may be only a matter of time before one or more major U.S. cities will be destroyed, either in part or in whole, by an earthquake. Property damage alone in such an event would reach tremendous values, and the economic impact of a major earthquake would be amplified by America's growing dependence on services and products furnished by such cities, from financial transfers to electronic components. Total damage in the hundreds of billions of dollars, or even into the trillions, is conceivable as an outcome of a "superquake" in the densely settled areas of the United States, particularly the California coast.

In modern times, the danger from volcanism is largely confined to the Pacific Northwest. Here an encounter between the NORTH AMERICAN CRUSTAL PLATE and the JUAN DE FUCA CRUSTAL PLATE beneath the PACIFIC OCEAN has generated an active volcanic mountain range, the CASCADE MOUNTAINS, reaching from northern California through Oregon and Washington state into the Canadian province of British Columbia. Among the famous and spectacular volcanoes of the Cascades are Mount BAKER; Mount HOOD; Mount RAINIER; and Mount SAINT HELENS, which underwent an explosive and highly destructive eruption in 1980. In recent geologic time, the eruption and subsequent collapse of MAZAMA, a huge volcano in what is now Oregon, generated a CALDERA that filled with water and became CRATER LAKE, a popular tourist attraction. Volcanic activity has continued at Crater Lake since the destruction of Mazama and has created a small new volcano, Wizard Island, within the lake. The Cascade volcanoes are associated with an offshore trench belonging to the CASCADIA SUBDUCTION ZONE, where the Juan de Fuca plate is believed to have descended into Earth's MANTLE, melting along the way and producing MAGMA that rises back to the surface through eruptions of the volcanoes. In California, Mount SHASTA and LASSEN PEAK represent very recent volcanism on the geologic time scale (Lassen Peak erupted early in the 20th century), and renewed volcanic activity is a possibility for the near future in the vicinity of MONO LAKE, in east-central California near the NEVADA border.

Although modern volcanism is largely restricted to the far west United States, evidence of ancient volcanic activity is abundant in many parts of the nation. In NEW HAMPSHIRE, for example, ancient volcanism produced peculiar circular ridges visible today in the WHITE MOUNTAINS. Volcanic activity has produced spectacular calderas and other formations at LONG VALLEY, California, site of one of the most powerful eruptions in the history of North America; YELLOWSTONE NATIONAL PARK in Wyoming, where HYDROTHERMAL ACTIVITY continues to this day, producing (among other phenomena) the famous GEYSER "Old Faithful"; and the VALLES caldera in New Mexico. In the Northwest, vast flows of LAVA cover the COLUMBIA PLATEAU and SNAKE RIVER PLAIN. Volcanism in Alaska is frequent, widespread, and spectacular. Alaskan eruptions tend to be highly explosive, in contrast to the relatively tranquil eruptions of Hawaii's volcanoes. In this century, the eruption of Mount KATMAI in Alaska resulted in the collapse of the volcano's peak, the formation of a huge caldera and the generation of a plain of FUMAROLES, the VALLEY OF TEN THOUSAND SMOKES.

Volcanism and its associated phenomena have had considerable economic and historical importance for the United States. The deposits of GOLD that set off the gold rush of the mid-19th century, thus accelerating the settlement of the western United States, are thought to have their origins in chemical processes associated with hydrothermal activity. Also tied in one way or another to volcanism in the western United States are numerous deposits of rich metal ores that have yielded great quantities of silver, tin, copper, and other economically important metals. In some cases, the mines dug to exploit these minerals have encompassed a cubic mile (4 km^3) of rock with their tunnels.

The west coast of the United States is vulnerable to TSUNAMIS, or seismic sea waves, generated by earthquakes along U.S. shores or elsewhere around the Pacific Ocean basin. Perhaps the most famous tsunami in U.S. history accompanied the Alaska Good Friday Earthquake of 1964. This tsunami devastated shorelines in south Alaska, wiped out much of the state's commercial fishing fleet and retained enough power to cause widespread damage at CRESCENT CITY, California. Numerous other (though less destructive) tsunamis have occurred along the California coast, many of them apparently in connection with earthquakes there.

See also IMPACT STRUCTURE; PLATE TECTONICS.

Unzen, Mount *volcano, Japan* An andesitic STRATOVOLCANO that had relatively minor activity for nearly 200 years until 1990. In 1792, the collapse of the Mayuyama LAVA DOME created an AVALANCHE and TSUNAMI that killed an estimated 14,524 people. In the renewed activity, the lava dome began a sudden growth spurt and received world attention. As the dome quickly grew 300 feet (91 m) above the base of the CRATER, thousands of residents of the slopes and surrounding areas were evacuated. Many reporters and VOLCANOLOGISTS, on the other hand, flocked to the volcano to witness and record the collapse of the unstable dome and resulting PYROCLASTIC FLOWS. On June 3, 1991, a huge mass unexpectedly broke away and rolled down the slope at 60 miles (97 km) per hour. It killed 42 of the observers. Since then Mount Unzen has become the worldwide leader in pyroclastic flows, producing more than 7,000 since 1991.

upper mantle The top 621 miles (1,000 km) of the mantle beneath the CRUST. It contains the rigid mantle that is adhered to the crust to comprise the LITHOSPHERIC plates. It also includes the ASTHENOSPHERE upon which the plates float and are driven.

Ushiorashi-dake *See* DAISETSU-TOKACHI.

Usu, Mount *See* TOYA.

Utah *United States* Although Utah is not commonly associated with strong earthquakes, like nearby CALIFORNIA, Utah is extremely susceptible to strong earthquakes. Earthquakes are common in and near the Wasatch Mountains, and the potential earthquake hazard there puts much of the state population at risk. A U.S. Geological Survey study revealed that a powerful earthquake in Utah might kill many people by drowning if dams failed. Dam failure, the study suggested, might kill about 10 times more people than the earthquake itself.

A notable pair of earthquakes in Utah history occurred at Elsinore on September 29 and October 1, 1921. The earthquakes affected a wide area and followed several weeks of preliminary activity. A schoolhouse in Elsinore was dam-aged severely, and many people could have been killed if the damage had occurred during school hours. Another powerful earthquake took place on March 12, 1934, near Kosmo at the north end of the Great Salt Lake. This earthquake occurred in a lightly settled area but could have caused extensive destruction in more heavily settled territory. Large quantities of water emanated from FISSURES, and in some areas, the ground surface was altered extensively. Another strong shock followed several hours after this one.

V

Vaiont Dam *Italy* A huge construction project to dam the Piove River, ITALY, was completed in 1963. The Vaiont Dam was designed to produce a reservoir for water storage and recreational activities in the mountains. The dam was well designed and well constructed. The problem was that the river had deeply incised the moderately to steeply inclined limestone beds, thus removing lateral support. Limestone is interlayered with thin shale units that are rich in expanding clays. These clays expand upon exposure to water, which makes them very slick. The reservoir was filled with water, submerging the rocks. There was also a period of particularly heavy rain for two weeks. Some reports even claim that there was a minor earthquake. Whatever the cause or causes, a huge mass of rock 1.2 miles (2 km) long × 1.0 mile (1.6 km) wide × 500 feet (150 m) thick broke loose and produced a massive ROCKSLIDE into the reservoir. In less than one minute, the rockslide filled the 900-foot (270-m)-deep reservoir with TALUS to levels 600 feet (180 m) above the water line for 1.2 miles (2 km) upriver. The impact of the mass formed a SEICHE that splashed waves 850 feet (260 m) up the opposite side of the valley, stripping trees, rocks, and SOIL from the face. The huge mass of rock also displaced the water in the reservoir to the point where it topped the still intact dam and formed a huge wave that rushed into the valley below. The wave began its journey with an estimated height of 650 feet (195 m) and destroyed every town in its path for many tens of miles downstream. It was estimated that 3,000 people lost their lives in this disaster, but with the nature of the damage, the number is a best estimate. The rumor is that when the dam chief engineer learned of the disaster, he committed suicide.

Valles *New Mexico, United States* A huge basin in the Jemez Mountains in northern NEW MEXICO near LOS ALAMOS, the Valles CALDERA is one of the world's largest calderas and has an area of some 180 square miles (466 km²). The Valles caldera formed 1.12 million years ago and formed the extensive BANDELIER TUFF that is famous for the Native American (pueblo) cliff dwellings. The Valles caldera is inactive, although a body of partly molten rock several miles in diameter is thought to lie beneath the caldera. Experiments in tapping GEOTHERMAL ENERGY have been conducted around the edges of the caldera.

Valley of Ten Thousand Smokes *Alaska, United States* This area of steaming ground owes its existence to the simultaneous eruptions of Mount KATMAI and Novarupta in ALASKA in 1912. Siphoning MAGMA away from Katmai, Novarupta emitted a gigantic ASH FLOW more than 40 square miles (104 km²) in area and hundreds of feet thick. Vapors rising to the surface from still-hot TEPHRA deep below generated a plain of numerous FUMAROLES. A National Geographic Society expedition in 1916 discovered the fumarole-filled valley, and President Woodrow Wilson soon afterward proclaimed the Valley of Ten Thousand Smokes to be a national monument. Members of one expedition to the valley tried to cook food over one fumarole and discovered that the escaping gas (mostly water vapor, with traces of hydrogen sulfide, hydrochloric acid, and other compounds) would blast a frying pan out of one's hands unless one held it tightly.

Valparaiso *earthquake, Chile* The coastal city of Valparaiso, CHILE, has been struck directly by earthquakes or affected indirectly on numerous occasions. It was struck by a strong tremor on November 19, 1822, at 3 P.M. The estimated MAGNITUDE of this earthquake was 8.0 on the RICHTER scale. This earthquake was accompanied by uplift of the city on the order of four feet (1.3 m). It was said that the seafloor was also raised, producing a large area of dry land that was formerly seabed. There were reports tha approximately 10,000 people died in this event, but they cannot be confirmed.

The next major earthquake struck Valparaiso on August 16, 1906, at 12:40 P.M. The estimated Richter magnitude of this event was 8.4, and the damage in the city was estimated at XI on the modified MERCALLI scale. In addition to the widespread collapse of buildings, fire destroyed much of what was left. The DEATH TOLL from this event is a subject of debate. The three numbers of consideration are 1,500, 4,000, or 20,000. More than 100,000 people were said to have been made homeless, and the total cost of the damage was estimated at a staggering $256 million. Huge tent cities were set up all over town. The local government called in the military, who in turn established a strict martial law to prevent looting and general lawlessness.

velocity structure   SEISMIC WAVES travel at different velocities within the various rock layers, rock bodies, and groups of rock layers. The architecture of the different velocities for a given area is the velocity structure. Defining this velocity structure can be done for any size area, including the velocity structure of the Earth. The velocity structure will determine the speed at which waves will reach SEISMOGRAPHS and even damage to some degree, but it is also useful for exploration applications.

Venezuela   Located in northern SOUTH AMERICA, Venezuela has a history of highly destructive earthquakes. An earthquake on March 16, 1812, for example, destroyed CARACAS and is believed to have killed some 20,000 people. Another earthquake at Cua on May 14, 1878, wrecked the city, caused some 600 deaths, and was felt in Caracas.

Veniaminof *caldera, Alaska, United States*   Located on the Alaska Peninsula, Veniaminof is thought to have formed from the summit of a large STRATOVOLCANO between 3,500 and 4,000 years ago. A noisy, explosive eruption occurred in 1938, and ASH fell as far away as Mount KATMAI. Within historical times, the greatest eruption of Veniaminof occurred in 1892. Skies over southwest ALASKA turned dark during this eruption. Veniaminof is of particular interest to scientists because it is located next to the Shumagin Gap, a seismic gap that is under study as a possible site of strong earthquakes in the future. It has been suggested that Veniaminof was active for several years before and after an earthquake in 1938 but comparatively quiet for several decades before and after that span of years. This suggestion helped focus attention on Veniaminof's behavior when eruptive activity resumed there in 1983. There is reason to question whether activity at Veniaminof is a reliable indicator of seismic activity at the Shumagin Gap, however, because two eruptions occurred during the "quiet" period, in 1944 and 1956. During the eruption that began in 1983, Veniaminof produced explosions, FIRE FOUNTAINS, LAVA FLOWS and the growth of a VOLCANIC CONE in the CALDERA. Veniaminof emitted a dark ash cloud during an eruption in 1984, and a small amount of ash on another occasion in 1987. The most recent eruption was in 1993–94. It was a relatively minor eruption but produced LAVA as well as steam and ash clouds. Because Veniaminof is ice covered, JÖKULHLAUP is a danger during eruptions.

vent   An opening through which volcanic material escapes to Earth's surface. The material may be ASH, LAVA, and/or gas. This feature is the actual location where MAGMA becomes lava. There may be several vents from a single volcano that can erupt simultaneously, but more typically, they erupt one at a time. The vent may become blocked with time, and the entire pipe clogs with magma that eventually crystallizes. When the rest of the volcano erodes away, a plug like SHIP ROCK is left behind.

Venus   The second planet from the Sun is known to have volcanoes whose activity appears to have reworked the planet's surface extensively. Images returned in 1991 from the U.S. radar-mapping *Magellan* probe, in orbit around Venus, revealed evidence of fresh LAVA FLOWS from the volcano Maat Mons. The summit of Maat Mons was much darker, that is, less radar reflective, than the summits of other peaks on Venus. This relative darkness gave rise to speculation that Maat Mons might be crowned by recent lava flows. Scientists observing the radar images from Venus believed that the extreme heat and acidic atmosphere on Venus for long periods may have produced a layer of highly radar-reflective iron sulfide on mountains other than Maat Mons. Had volcanic activity occurred recently at Maat Mons, this sulfide coating would not yet have had time to form, and the volcano's summit would appear dark in radar images. At this writing, astronomers await additional images of Maat Mons to see if any changes appear that would indicate current volcanic activity. Maat Mons is about five miles (8 km) high, approximately the same height as MAUNA KEA in Hawaii, the tallest volcano on Earth. Formations resembling LAVA DOMES have also been observed on the surface of Venus. By the density of IMPACT CRATERS on the surface, it appears that the CRUST of Venus is less than 500 million years old. Therefore, by the density of volcanoes, Venus is clearly a volcanically active planet.

Venus also has a large number of FAULTS and FRACTURES visible on the surface. These faults offset features of all ages, showing that faulting is a continuing process on the planet. Although no "Venusquakes" have ever been detected, judging by the number of faults and the degree of their offset, they are expected.

A Magellan radar image of overlapping lava flows emanating from the Sapas Mons volcano. Sapas Mons is 240 miles (400 km) across and 0.9 mile (1.5 km) high. *(Courtesy of NASA)*

Vermont *United States* Located in New England and close to the SAINT LAWRENCE VALLEY, the state of Vermont experiences numerous minor earthquakes and an occasional strong one. Burlington experienced an earthquake on January 29, 1952, that affected an area of approximately 50 square miles (130 km²) and caused slight damage. Cracks in the ground some two miles (3.2 km) long were reported in the northern portion of the city. The earthquake of April 23, 1957, at St. Johnsbury, estimated at MERCALLI intensity V, caused much fright within an area about 30 miles (48 km) in diameter. On April 10, 1962, another earthquake estimated at Mercalli intensity V occurred in Vermont and affected much of the rest of New England as well as portions of NEW YORK; this earthquake caused considerable damage at the statehouse, as well as minor damage in Barre. Another moderate earthquake occurred in Vermont in the autumn of 1983.

vesicle A cavity within a volcanic rocks that is formed by the release and passage of gases within and through the molten rock immediately before it solidifies. The size of vesicles may vary from one kind and sample of volcanic rock to another. In PUMICE, a lightweight volcanic rock, vesicles are so numerous and tiny that they form, in effect, a "rock foam" that has had numerous applications over the centuries from abrasives to building material. The large numbers of vesicles make pumice so lightweight that it can float on water. Vesicles are less numerous but larger in SCORIA, another volcanic rock. Calculations of the volume of rock released in certain eruptions must allow for vesicles, which can expand the volume of volcanic rock considerably. Vesicles are abundant in rocks produced where gas-rich MAGMA rises to the surface from below. This situation occurs in many parts of the world where chains of volcanoes border oceans where SUBDUCTION ZONES carry great quantities of sediments from the seabed downward into the ASTHENOSPHERE. There, miles beneath the surface of Earth, water in the sediments is dissolved in the magma derived by melting of subducted CRUST, and mantle and carried back to the surface in the form of dissolved gas. This combination of processes appears to be active in the volcanoes of the CASCADE MOUNTAINS in the Pacific Northwest of the UNITED STATES.

If these vesicles are later filled with minerals precipitated out of solution in GROUNDWATER, they are called amygdules.

See also AMYGDALOIDAL.

Vestmann Islands *volcanic islands, Iceland* A short chain of volcanic islands off the south coast of ICELAND, the Vestmann Islands include the sites of two famous eruptions of recent years, those of SURTSEY in 1963 and HEIMAEY in 1973.

Vesuvius *volcano, Italy* The only active volcano on Europe's mainland, Vesuvius stands some 4,000 feet (1,219 m) high and is located near NAPLES. It sits within a large hooklike feature called the Somma Rim, which is believed to have formed in a volcanic collapse 17,000 years ago. In 5960 B.C. and 3580 B.C., Vesuvius erupted in some of the largest eruptions in Europe.

The most famous eruption of Vesuvius, in A.D. 79 is estimated to have had a VEI of 4. It destroyed the nearby cities of POMPEII AND HERCULANEUM and preserved them under ASH and mud until excavation began some 17 centuries later. Reports are conflicting but at least 3,360 and as many as 16,000 people are believed to have died in the two cities. The A.D. 79 eruption of Vesuvius altered the appearance of the volcano considerably. The eruption destroyed much of the cone, except for a semicircular remnant, now know as Mount Somma, on the northeast side of the mountain. The cone now known as Mount Vesuvius arose from the ruins of this older cone.

Records of activity at Vesuvius are unreliable for centuries after this eruption, and it seems likely that only major eruptions are mentioned. An eruption appears to have occurred around A.D. 203 and cast out ash and rocks. Another notable eruption occurred in A.D. 472, when ash emissions from the volcano are said to have fallen over much of Europe and the eruption reportedly was detected at Constantinople. Vesuvius erupted again in A.D. 512, 685, and 993. An eruption in 1036 was notable for its LAVA FLOWS, thought to have been the first to occur at Vesuvius within historical times. Eruptions are also recorded in 1049 and 1139. An ash eruption appears to have taken place in 1500.

An eruption in 1631 signified a change in the behavior of Vesuvius. Previously, it had been active only on an intermittent basis, with long periods of repose between eruptions. Now the mountain entered a time of virtually steady activity. The eruption of 1631 commenced in December with vigorous explosions and a cloud that darkened the adjacent land. Ash and cinders fell, and communities immediately around the volcano were evacuated. On December 17, FISSURES appeared on the southwest flank of the volcano and discharged large quantities of LAVA. The lava advanced down the side of the volcano rapidly and flowed through several communities, including La Scala, Portici, Pugliano, San Giorgio a Cremano, and Torre del Greco. The lava flow divided into many tongues before entering the sea, about two hours after emerging from the fissures. That evening, a mud rain began to fall on Naples, and large MUDFLOWS rolled through villages on the sides of the volcano. Lava flowed again from Vesuvius, this time from a fissure on the south flank. This eruption continued intermittently for several weeks before finally subsiding in January 1632 and caused widespread devastation. Lava and mudflows destroyed more than a dozen towns. The eruption is thought to have killed some 3,500 to 4,000 people. Many deaths are said to have resulted from poor judgment on the part of the governor of Torre del Greco, who waited until approaching lava reached the walls of the community before giving orders to evacuate. Before the residents could flee, lava breached the walls and flowed through the streets. The mountain lost more than 500 feet (152 m) of its summit, and the CRATER after the eruption was reportedly more than twice as wide as before.

The great eruption of 1767 was recorded in detail. It began in October with powerful explosions that lasted for two days and were followed by the opening of fissures on the northwest and south sides of the volcano. Lava flowed from these fissures toward nearby towns. An observer on the side of the volcano, within sight of one of the fissures when the lava emerged, described a FIRE FOUNTAIN that shot up high into the air and was accompanied by ashes and dark smoke. The people of Naples were alarmed and flocked to churches to pray for deliverance from the eruption. One account says that a mob set fire to the gates of the archbishop's residence after he refused to produce relics of Saint Januarius. (Near the end of the eruption, the cardinal gave in to public pressure and brought out the saint's head.) Three years later, Vesuvius resumed erupting and continued on and off, for several years.

In May 1779, another violent eruption occurred. A fissure opened on the north side of the volcano and emitted lava for weeks afterward. On August 8, strong explosions began, and a column of fire reportedly rose from Vesuvius to an altitude of approximately two miles (3.2 km). This eruption devastated the community of Ottaiano, which lay in the path of the advancing fiery column and was showered with glowing SCORIA. Stones as heavy as a hundred pounds reportedly fell on Ottaiano, and the population took shelter as best it could, in cellars and in archways, for about 25 minutes while the cloud was passing. The tempest of rock, ash, and SULFUR stopped abruptly, and the townspeople fled.

After this eruption, Vesuvius remained quiet for about five years. Then eruptive activity began again, culminating in the great eruption of 1793. A fissure between Torre del Greco and Resina opened and emitted lava from several locations. These individual flows joined and became a single flow of lava that overwhelmed Torre del Greco. The advancing front of lava is said to have been more than 1,000 feet (305 m) wide and extended more than 300 feet (91 m) into the sea before it stopped. This eruption was remarkable for its report of boiling water about 100 yards (92.5 m) offshore; one observer wrote that boats could not remain in the vicinity of this disturbance because the heat melted the pitch in their hulls.

Strong eruptions occurred again in 1822, 1838, and 1850. One of the volcano's most powerful eruptions within historical times took place in 1872. The volcano cracked open from summit to base on its north flank. Lava flowed from this great fissure and rolled over the towns of San Sebastiano and Massa, while ash from the crater fell across the surrounding land. This same year marked the start of an eruptive cycle that lasted until the early 20th century. Lava flowed from the crater in 1881, but the lava was so viscous that it tended to build up and form a domelike formation rather than flow rapidly downhill. A lava dome formed from 1891 to 1894. Another lava formation, the Colle Umberto, arose near the Vesuvian Observatory in the late 1890s. In the late 1800s, a cone was forming inside the crater, and by 1906 the cone had filled the crater entirely. That year, a fissure appeared on the south flank of the volcano. Lava flowed from the fissure, and great quantities of solid material were thrown up from the crater, falling back onto the volcano. Thick eruption clouds laced with lightning emerged from the volcano. Ottaiano again was devastated; the weight of the ASHFALL there crushed many buildings. More than 100 people were reported killed in San Giuseppe when the weight of ash caused the roof of a church to collapse on them. Toward the end of this eruption, a powerful outburst of gas blew away the top portion of the cone and reduced its height by more than 300 feet (91 m).

Except for minor eruptions of ash, Vesuvius remained quiet after this eruption until 1913, when eruptive activity began to rebuild the cone. The next powerful eruption occurred in 1944. Lava started to flow out of the crater on March 18, 1944. The lava flow proceeded in the direction of Massa and San Sebastiano and overwhelmed the towns. Huge amounts of scoria and ash were cast out from the crater. Three days later, the volcano began to send up tremendous fountains of lava. An outburst of gas similar to that of the 1906 eruption took place in 1944, again blasting away the upper part of the cone. The 1944 eruption caused extensive damage to a U.S. air base near Naples and interfered with allied operations in the region.

The character of Vesuvius's eruption has changed since the destruction of Pompeii. The 1036 eruption released lava as well as PYROCLASTIC material, and with the 1631 eruption, a new pattern of activity was observed in which the volcano emitted basaltic lava in cyclic eruptions. Several hypotheses have been put forward to explain the change in Vesuvius's activity. One hypothesis is that a fresh supply of BASALTIC magma became available underneath the mountain. Another hypothesis is that the eruption of A.D. 79 used up magma rich in dissolved gases that caused explosive eruptions so that basaltic magma could flow in to replace it. A third hypothesis is that contact with sedimentary rock near the surface changed the chemistry of the magma. It has also been suggested that some combination of these factors was responsible for the change in Vesuvius's eruptions.

Vesuvius remains one of the most dangerous volcanoes in the world. The population density is so great in the area that some 3 million people could be severely affected by a strong eruption. About 1 million people live and work within a four-mile (6.4 km) radius of the base of the volcano. This area would be devastated in the first 15 minutes of a major eruption. For these reasons, Vesuvius is closely monitored, and public education and evacuation procedures are serious issues in the area. Another eruption like that in A.D. 79 could be the most destructive ever in the world in terms of human life and dwellings.

Vesuvius is located near the PHLEGRAEAN FIELDS, an area noted for strong volcanic activity and for dramatic uplift and subsidence along the shore.

Villarrica *volcano, Chile* The STRATOVOLCANO Villarrica is located in central CHILE and has erupted explosively on some 61 occasions since 1558. A prehistoric eruption in 1810 B.C. is estimated to have had a VEI = 5. Villarrica is located where a north-northeast to south-southwest FAULT zone intersects with a chain of volcanoes oriented toward the northwest. The beautiful symmetry of the volcano is marred slightly by the rim of a CALDERA at slightly more than 7,000 feet (2,134 m). Eruptions of Villarrica commonly exhibit a variety of spectacular effects, including underground noises, loud explosive sounds, lofty ASH columns, LAVA FLOWS, and fountains of glowing debris. The eruptive history of Villarrica extends back to 1558 when an eruption destroyed the community of Villarrica. The town was destroyed again by another eruption in 1575 that killed more than 300 people. An eruption in 1640 released a flow of LAVA that extended to

Lake Villarrica and the Rio Tolten valley. Thereafter, unrest was reported at intervals of every several years through 1898. Loud noises from underground were noted in 1906, but there appears to have been only a slight emission of lava. An earthquake was reported in 1907. The volcano erupted in spectacular fashion in 1908, releasing an outburst of fire that set off an AVALANCHE of snow, ice, and rock from a glacier on the east side of the mountain. A small eruption occurred in 1908, and eruptions were recorded in 1909, 1913, 1915–18, and 1919.

A powerful earthquake on December 9, 1920, damaged a house about four miles from the CRATER and displaced furniture; three strong shocks were followed by numerous minor earthquakes that occurred all night at intervals of several minutes, with a stronger shock every 30 minutes or so. The earthquakes reportedly demolished several structures, including houses and stables. A violent, explosive eruption began on the night of December 10 with a fountain of fiery material. It lasted for a day and a half. In general, this eruption was characterized by explosions, at least one fountain of lava and minor emissions of TEPHRA.

Eruptions in late 1948 and early 1949 caused considerable loss of life. The explosions of October 18, 1948, and January 1 and 31, 1949, produced great avalanches (possibly PYROCLASTIC FLOWS) that are thought to have killed 60 people. Lava flows in January 1949 reportedly burned large areas of forest around the volcano. Eruptive activity between 1959 and 1961 was relatively minor; small explosions and emissions of tephra were reported. Destructive activity resumed in 1963 when avalanches cut several bridges in the vicinity of Lake Calafquen. In March 1964, MUDFLOWS rolled over part of a village and killed 22 people there. Mudflows and emissions of tephra occurred during eruptions that began in October 1971 and continued through December, killing 17 people and releasing lava that flowed some 10 miles (16 km) from a vent. Activity at FUMAROLES intensified in September 1979, and minor eruptions of tephra occurred in 1980. A red glow was observed at the summit in 1983, and a small emission of tephra was reported, along with minor explosions.

A short eruption of ash, together with an increase in earthquake activity, occurred in August 1984. On October 30, minor explosions and emission of lava began, and strong earthquake activity was recorded near the surface on November 4–5. Although instruments recorded no more earthquakes, a strong subterranean rumbling and a tremor were reported. Explosions and emissions of lava occurred between November 10 and 13, and mild unrest continued for several days afterward. A prominent bulge was seen on the southwest flank of the volcano on November 18, followed by a glow and emissions of ash through April 18. An increase in earthquake activity was reported again between June and November 1985.

A new PYROCLASTIC cone was built in a relatively active period from 1994 to 1995. The most recent activity was in 1999 when STROMBOLIAN ERUPTIONS, small explosions, and increased seismic activity forced local evacuations.

Virginia *United States* The state of Virginia is located in a region of moderate seismic activity in the southeast

UNITED STATES but has experienced strong earthquakes on occasion since colonial times. Virtually no portion of the state, from the coastal plain in the east to the mountains in the west, is completely free of seismic activity. For example, central Virginia experienced an earthquake on August 27, 1833, that was felt from Norfolk in the eastern part of the state to Lexington in the west, and from Baltimore, MARYLAND, south to Raleigh, NORTH CAROLINA. The shock was perceptible in Washington, D.C., and rattled windows energetically in Lynchburg. Public fright at Fredericksburg was considerable, and the shock at Charlottesville is described as "severe." There is also the Central Virginia Seismic Zone near the James River that regularly produces earthquakes, causing great concern for nuclear power plants there. The Southwestern Virginia Seismic Zone lies along the New River and has also experienced moderate seismic activity. There is a LANDSLIDE scar along the mountains of southwest Virginia. It appears to be about 25 million years old and is the largest identified scar. If this landslide is the result of seismic activity, there is a potential for much larger earthquakes than have been observed historically.

Like the other states along the Atlantic seaboard of the United States, Virginia has a broad coastal plain of unconsolidated sediment in the east. This area could undergo LIQUEFACTION, with a resultant potential for serious damage to property, in any future major earthquake, especially in areas where groundwater rises close to the surface.

vog Short for *volcanic smog,* this expression refers to a troublesome atmospheric phenomenon—an acidic haze—sometimes observed during eruptions at KILAUEA in Hawaii. Vog is thought to form partly from acid droplets released when LAVA reaches the sea and generates steam.

volatile Any substance that will be released as a gas during a volcanic eruption. When volcanoes erupt much of the propellant that shoots the ASH into the air are volatile components. The dominant volatile is water, but carbon dioxide, SULFUR dioxide, and nitrous oxides among others are also present in small quantities. The more volatile component, the more explosive the volcano. The reason that volatiles are so important in determining if a volcano will explode is because they change phase from liquid to gas during an eruption. Water, for example, is part of the liquid MAGMA underground. When the pressure is released as it erupts from the volcano, it changes to steam. One gallon of water vaporizes to 22.4 gallons of steam at room temperature and pressure. At higher temperatures, the expansion is much greater. This increase in volume happens instantaneously, forming an explosion. Magmas with higher volatile components produce more explosive volcanoes. The situation is analogous to a bottle of soda. If shaken, nothing happens to the soda until the cap is removed (pressure released). Then the carbon dioxide comes out of the liquid as a gas (volatile) all at once, and the soda foams over.

volcanic arc *See* ISLAND ARC.

volcanic cone A conical formation resulting from expulsion of LAVA and/or PYROCLASTIC material from a vent. Most

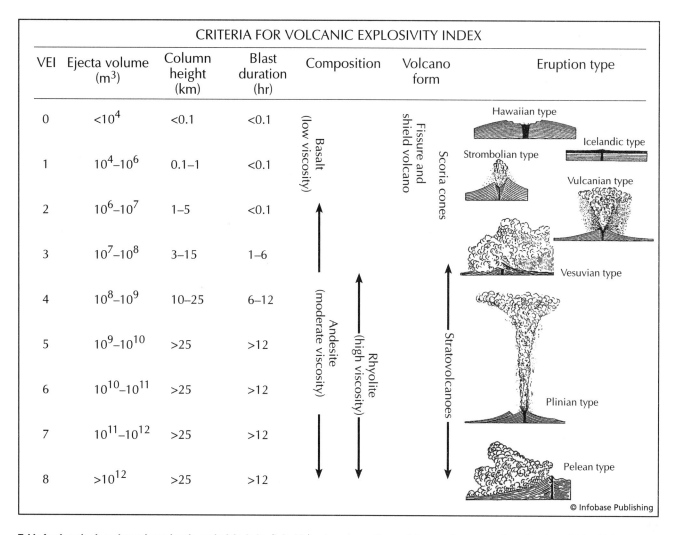

VEI	Ejecta volume (m^3)	Column height (km)	Blast duration (hr)	Composition	Volcano form	Eruption type
0	$<10^4$	<0.1	<0.1			
1	10^4–10^6	0.1–1	<0.1			
2	10^6–10^7	1–5	<0.1			
3	10^7–10^8	3–15	1–6			
4	10^8–10^9	10–25	6–12			
5	10^9–10^{10}	>25	>12			
6	10^{10}–10^{11}	>25	>12			
7	10^{11}–10^{12}	>25	>12			
8	$>10^{12}$	>25	>12			

CRITERIA FOR VOLCANIC EXPLOSIVITY INDEX

Basalt (low viscosity); Andesite (moderate viscosity); Rhyolite (high viscosity)

Fissure and shield volcano; Scoria cones; Stratovolcanoes

Hawaiian type; Icelandic type; Strombolian type; Vulcanian type; Vesuvian type; Plinian type; Pelean type

© Infobase Publishing

Table for the criteria to determine volcanic explosivity index (left side) and corresponding rock types, volcanoes, and eruption types (right side)

cones are composed primarily of pyroclastic material that is deposited layer upon layer from ASH FLOW and ASHFALL deposits. Wizard Island in CRATER LAKE is a familiar example of a volcanic cone.

See also VOLCANO.

volcanic dome A mound of thick, slowly flowing LAVA that formed over the VENT of a volcano. A volcanic dome has a characteristic mounded shape and has a surface covered with large chunks of IGNEOUS ROCK produced by the cooling of the lava. A whole series of domes may form, be destroyed, and then form again during explosive eruptive activity. The formation of a dome indicates that gas-rich lava has ceased (at least temporarily) to flow, and more viscous lava with lower dissolved-gas content has replaced it.

volcanic dust EJECTA that are even smaller than ASH. Dust is defined as less than 0.06 millimeters. It is shot high into the atmosphere where it can have a very high residence time. This dust reflects away the incoming solar radiation and can cool

Earth. Eruptions with a high dust content and a high eruption column can cause more of a climate change than much larger eruptions.

Volcanic Explosivity Index (VEI) A quantitative index to rate the intensity of volcanic eruptions that is based upon a set group of factors. The index ranges from 1 to 8; the higher the number, the more intense the eruption. The factors include volume of EJECTA, height of the ERUPTIVE COLUMN, and the duration of the explosion. Low VEIs are typical of FISSURE ERUPTIONS and SHIELD VOLCANOES (ICELAND and HAWAII). They are almost exclusive of BASALT eruptions. Moderate VEIs are typical of STROMBOLIAN and VULCANIAN eruptions from small volcanic cones. They are basaltic or ANDESITIC. High VEIs are typical of Vesuvian, PLINIAN, and Peléean eruptions with ANDESITES, DACITES, and RHYOLITES. Eruptions of low VEI are very common and happen almost every day of the year. Eruptions of high VEI are few and far between. The massive eruptions of Mount PINATUBO (1991) and El CHICHÓN (1982) had VEIs of about 5. To find eruptions with a VEI of 6 to 7, we must reach back to Mount PELÉE (1902), KRAKATOA (1883), and TAMBORA (1815). VEIs of 8 have not occurred in recent history, but TOBA, YELLOWSTONE (Huckleberry Ridge), and Fisher Creek in the San Juan Mountains of COLORADO produced such massive eruptions in prehistoric times. Considering the impact of Krakatoa and Tambora, imagine the impact on modern life if one of these giants erupted today.

volcanic neck A towerlike landform consisting of the igneous core of an extinct volcano whose outer layers have eroded away. The neck is composed of MAGMA that has crystallized in the conduit of the volcano. The layers that make up the cone are largely PYROCLASTIC and much more easily eroded than that in the neck. DEVILS TOWER, WYOMING, is a familiar example of a volcanic neck.

volcanic rock Volcanic rocks cooled from LAVA or deposited as EJECTA. They are classified on the basis of composition and texture. By composition, there are three main categories: FELSIC, INTERMEDIATE, and MAFIC. Felsic rocks include RHYOLITE and TRACHYTE, which can have visible K-FELDSPAR, but only rhyolite has QUARTZ. On the border between felsic and intermediate are LATITES and quartz latites, which can have PLAGIOCLASE and K-feldspar. Intermediate volcanic rocks include DACITES and ANDESITES; dacites can contain quartz. Mafic volcanic rocks include BASALTS. Textural terms can modify the name of the rock or completely name it. If there are grains in a volcanic rock that are large enough to see, the term *PORPHYRITIC* or *porphyry* is added to the name. The large grains are called phenocrysts. If there are gas holes in the rock (VESICLES), the term *vesicular* is added to the name. If there are a lot of holes, the term *SCORIA* is added. If the vesicles are filled with minerals that were later precipitated from groundwater, they are called amygdules and the term *AMYGDALOIDAL* is added. If the lava is cooled so fast that it forms a glass, the rock is called OBSIDIAN regardless of composition. Frothy glass is called PUMICE, regardless of composition. Ejecta including ASH, LAPILLI, and BOMBS that have been lithified into a rock is called TUFF. There are several types of tuff, depending on whether they are the result of an ASHFALL or an ASH FLOW.

volcanic tremor *See* HARMONIC TREMOR.

volcanism Any and all volcanic activity. Volcanism may be spectacular, involving great fountains of molten rock spouting high into the air, or tremendous explosions that are caused by the buildup of gases within a volcano. Alternatively, volcanic activity may be relatively quiet, releasing only small amounts of LAVA, or even restricted to minor emissions of vapor. Volcanism at a given site may be almost continuous over a period of centuries, or it may be comparatively infrequent with eruptions separated by intervals of centuries. Some volcanic events are highly destructive of human life and property; other incidents of volcanism may be dramatic but do little harm. Volcanism may take place in a wide variety of environments, from dry land to the bottom of the ocean. Eruptions have even occurred under glacial ice. Depending upon the circumstances of an eruption, products of volcanism

may include fine ASH, blobs of molten rock solidified in midair (called BOMBS), and floods of lava that can cover thousands of square miles. Not always an isolated phenomenon, volcanism may occur in combination with earthquakes, some of them capable of causing serious damage.

Volcanism also is thought to have a significant influence on global climate. Because ash is released into the upper atmosphere during a major eruption, it can intercept enough incoming sunlight to reduce the energy that Earth receives from the sun and thus reduce surface temperatures. In some cases, volcanic eruptions have been correlated with short-term changes in climate that resulted in major hardship for humans over large portions of Earth. Long-term changes are suspected as well but not as easily documented.

TSUNAMIS are also associated with certain explosive ERUPTIONS along the seacoasts of the world. A famous case in point is the 1883 eruption of the volcanic island KRAKATOA between Java and Sumatra; the explosion of that volcano generated a tsunami that circled the world several times before finally dissipating, and it was responsible for 36,000 deaths along shorelines near the eruption. LAHARS may cause extensive destruction in the vicinity of a volcano when eruptions melt ice and snow on the summit and send large quantities of mud and rock flowing down the flanks of the mountain. The flows may travel miles from their point of origin and bury the underlying land to a depth of many feet. Other phenomena capable of causing great destruction are the *nuée ardentes,* a fiery mixture of gas and glowing rock that is capable of incinerating a landscape almost as effectively as a nuclear explosion, and emissions of 800°C toxic gases, which may kill by incineration and asphyxia anyone caught in them.

The economic effects of volcanism are considerable. Negative effects include the destruction of homes, industrial facilities, communications equipment, vehicles, airports, highways, and other property, as well as damage to crops and to health. Pollution of water supplies is also a possibility. The economic impact of even a reasonably small eruption may run into the billions of dollars if the event occurs in or near a densely settled or intensively cultivated area. Cleanup alone may be so expensive that the most practical policy is to abandon the affected property, as the UNITED STATES military did when a 1991 eruption of Mount PINATUBO in the PHILIPPINE ISLANDS buried a nearby air base under ash and made the facility unusable. Although economic damage from many eruptions is largely confined to the area immediately around the volcano, under certain circumstances an eruption may cause a change in the global environment that spreads economic hardship over a wide area. An eruption of the Indonesian volcano TAMBORA, for example, is thought to have been at least partly responsible for a global cooling that had disastrous effects on agriculture in the Northern Hemisphere in the early 19th century.

Less directly, but no less significantly, ancient volcanism has influenced economics—and thus the rise and fall of nations and whole empires—by concentrating rich deposits of precious minerals, such as gold and silver, at certain locations. Volcanism and associated phenomena along the ANDES MOUNTAINS of SOUTH AMERICA gave them such a heavy concentration of gold-bearing rocks, for example, that the wealth of the Inca empire became an attraction for invaders from Spain. They conquered the Inca and plundered the mineral wealth of their empire. In similar fashion, plutonic formations in South Africa have yielded large quantities of diamonds, with consequent effects on the South African economy. Not all rich mineral deposits associated with volcanic activity are so exotic; copper and other minerals have played an equally important, if not greater, role in the wealth of the countries in which they have been mined. Volcanism may create new land that may be settled profitably, and the TEPHRA (ash and other ejecta) from an eruption may itself have considerable economic value, in applications ranging from abrasives to building material. A minor but colorful economic benefit of volcanism in some places is tourism. Certain volcanoes in the United States benefit especially from tourism: notably OREGON's CRATER LAKE; the spectacular water-filled CALDERA left by an ancient eruption; Mount SAINT HELENS, whose explosive eruption in 1980 made the then little-known volcano famous throughout the world; and the volcanoes of Hawaii, whose relatively safe but dramatic eruptions draw many visitors.

Although volcanism has been studied intensively for thousands of years, the science of volcanology operates under certain limitations. Except in a few locations such as Hawaii, eruptions are only intermittent and cannot be predicted far in advance. Many eruptions occur in isolated areas that are difficult to reach with the personnel and equipment required for on-site studies. Despite these inconveniences, however, the science of volcanology has advanced rapidly in the 20th century, thanks to a variety of new sensing techniques and computer analysis, which together have made possible the collection and processing of great quantities of data on eruptions. The 1980 eruption of Mount Saint Helens alone expanded the scientific knowledge of eruptions tremendously by giving volcanologists an active volcano to study within easy reach of numerous well-equipped laboratories and other research facilities.

See also "YEAR WITHOUT A SUMMER."

Volcán Nuevo *See* NEVADOS DE CHILIAN.

volcano, active *See* ACTIVE VOLCANO.

volcano A volcano is defined, in a broad sense, as an opening in Earth's crust through which MAGMA escapes to the surface where it is transformed into LAVA. More specifically, the word *volcano* refers to mountains produced by volcanic activity, known as volcanism or vulcanism. (The words *volcano*, VOLCANISM, and *vulcanism* all are derived from the Latin *Volasnus*, or Vulcan, the god of fire in Roman mythology.) In a volcanic ERUPTION, magma or just the hot gases from magma escape from an underground reservoir to the surface through a relatively narrow VENT, or conduit. Eruptions differ greatly in character from one volcano to another and sometimes within the history of the same volcano. Some eruptions are extremely violent and involve great outbursts of ASH, gas, and lava. These eruptions produce cinder cones and COMPOSITE VOLCANO.

A cinder cone is made up of fragments of rock ejected from the vent. PARICUTÍN in MEXICO is one example of a cinder cone. These fragments tend to be low-density rock that are formed when dissolved gases in the magma bubble out of solution as the rock solidifies, producing PUMICE that appears to be "rock foam," full of small cavities that give the rock a "frothy" texture. Rock fragments between about a half-inch (1 cm) and two inches (5 cm) in diameter are called cinders and constitute most of the cinder cone. Cinders are distinct from ash, finer material that winds may carry for great distances away from the volcano. Though there is a significant component of ash to every cinder cone. Mixed in with the cinders, in most cases, are volcanic BOMBS that are formed as large masses of lava are ejected intact. Cinder cones may grow rapidly, rising hundreds of feet in the first few days or weeks of their existence, but seldom reach heights of more than 1,000 feet (305 m). Cinder cones are typically steep-sided; the slope of a newly formed cinder cone can be 28°. The CRATER tends to be large and to have a rim higher on one side than the other because of prevailing winds that carry the volcano's output in a given direction. Cinder cones may occur virtually anywhere the appropriate kind of magma rich in dissolved gases can reach the surface. Clusters of cinder cones are commonplace. Eruptions of cinder cones may include LAVA FLOWS.

A composite volcano is more complex than a single cinder cone. Composite volcanoes are made up of layers of cinder and ash alternating with lava. Because of these alternating strata, composite volcanoes are known as STRATOVOLCANOes. A stratovolcano has steeply sloping sides, as cinder cones have, but has greater structural strength due to the rigid lava layers inside it. Stratovolcanoes may reach thousands of feet in height. Examples of stratovolcanoes are VESUVIUS in Italy, Mount FUJI in Japan, and Mount SAINT HELENS in the United States. Most stratovolcanoes are concentrated in two parts of the world: in the "RING OF FIRE," a belt of intense volcanic and earthquake activity encircling the PACIFIC OCEAN basin; and in the MEDITERRANEAN SEA. Eruptions of stratovolcanoes involve release of hot gases, ash, cinders, bombs, and lava.

Eruptions may occur with such violence that they demolish part of the mountain, as happened in the 1980 eruption of Mount Saint Helens. An even more violent eruption, that of the volcanic island KRAKATOA in 1883, destroyed most of the island and generated a TSUNAMI that inundated shorelines in the vicinity and killed 36,000 people. Eruptions of stratovolcanoes sometimes produce CALDERAS, when the entire central portion of the volcano is blasted away or, alternatively, collapses as magma within it is drained away. Young stratovolcanoes are characterized by their conical shape and symmetry. Older stratovolcanoes are less symmetrical because water and ice have eroded their flanks. Erosion produces a characteristic radial drainage pattern down the sides of the volcanoes. Another, separate drainage pattern occurs in some cases within the crater, as water flows downward from the rim to the floor of the crater. As erosion continues, the mountain is worn down gradually until only the VOLCANIC NECK, the solidified mass of magma where the vent once extended upward to the summit, remains. SHIP ROCK in NEW MEXICO is a famous example of a volcanic neck, from which DIKES radiate. Because of their great resistance to erosion, lava flows from stratovolcanoes may endure as mesas rising above the surrounding terrain.

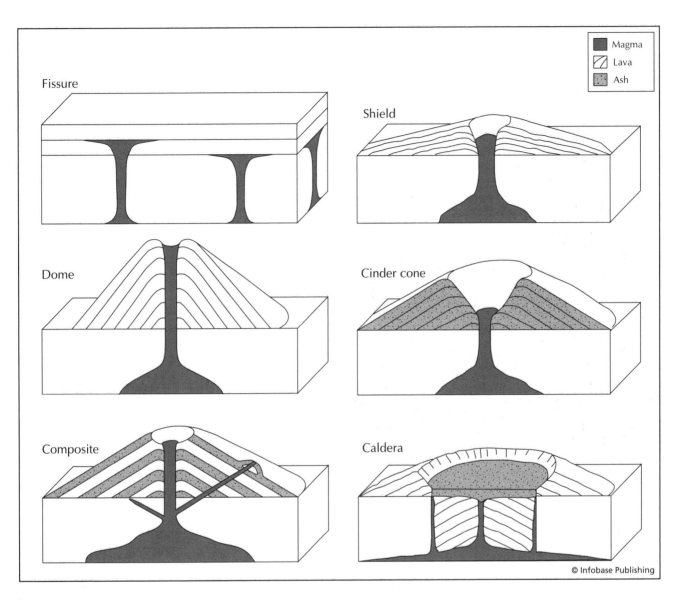

Magma
Lava
Ash

Fissure

Shield

Dome

Cinder cone

Composite

Caldera

© Infobase Publishing

Types of volcanoes

When eruptions are comparatively mild and involve only small, nonexplosive eruptions of lava, the product is a SHIELD VOLCANO, a gently sloping volcano (perhaps 5°) with a wide, flat top. The HAWAIIAN ISLANDS are shield volcanoes, as is OLYMPUS MONS on Mars. Shield volcanoes build up slowly through repeated outpourings of lava. Their lava has less dissolved gas in it than the lava of stratovolcanoes and does not form great quantities of ash and cinders. Lava from shield volcanoes may solidify as volcanic glass or BASALT. Shield volcanoes are also characterized by their large central depressions, which may be miles in diameter. On the floor of these central depressions, one may find FIRE FOUNTAINS of lava. Shield volcanoes produce many features similar to FISSURE ERUPTIONS, but there are no elevated areas at all in the latter. Fissure eruptions commonly occur in depressions.

See also SCORIA.

volcanologist A geologist who studies volcanoes. Their field of specialty can include any number of aspects, including the chemistry, landforms, hazards, history, PLATE TECTONICS, processes of formation, relation to FAULTS and fractures, and erosion.

Vulcanian-type eruption It is characterized by incandescent ejection of fragments of new LAVA as BOMBS in addition to ash. Small explosions produce fireworks-type displays in these volcanoes.

vulcanism *See* VOLCANISM.

Vulcano *volcano, Italy* The volcano that gave its name to volcanoes as a phenomenon and the VULCANIAN-TYPE ERUPTION, Vulcano is a STRATOVOLCANO located west of ITALY in the LIPARI ISLANDS. The eruptive history of Vulcano is difficult

A geologist (volcanologist) sampling lava from the interior of a lava tube, Kilauea volcano, Hawaii, 1973. *(Courtesy of USGS)*

to reconstruct for some periods because mythology has influenced accounts. It appears safe to say, however, that Vulcano has been active on a frequent basis for thousands of years. After the formation of the Lentia LAVA DOME, it collapsed forming the Lentia Caldera. Some 8,500 to 10,000 years ago, the Fossa cone began to grow within the CALDERA. At least four volcanic cycles built the cone.

Greek authors Aristotle and Thucydides both mention eruptions of Vulcano in the fourth and fifth centuries B.C. respectively. Aristotle apparently wrote about an eruption of Vulcano that was occurring at the time of writing. He reports that ASH from this eruption reached the mainland of Italy. A small island and youngest cone called Vulcanello (now connected to the north shore of the island of Vulcano) is thought to have emerged in the second century B.C., beginning in 183 B.C. An eruption in 126 B.C. is believed to have enlarged Vulcanello greatly. Vulcano is mentioned in the biography of Willebald, the Anglo-Saxon bishop of Eichstatt. Willebald lived in the eighth century A.D. and was reportedly one of the most widely traveled men of his day. On a journey through the MEDITERRANEAN, Willebald is said to have

stopped at Vulcano and desired to look into the crater during an eruption, but it proved impossible to climb the flanks of the volcano, covered as they were with SCORIA and ash. So Willebald and his companions had to stand at a distance and watch the flames of the eruption, which also involved a huge column of smoke and loud roaring sounds.

Recorded observations of eruptions at Vulcano are few in the centuries between the start of the Christian era and the early 1400s. This lack of information on Vulcano may be due in part to the sinister reputation that the Lipari Islands once had as a haven for pirates. An eruption in 1444 was documented by a Sicilian who reportedly described underwater eruptions and said that great rocks were cast out of the mountain. Later reports indicated that Vulcano and Vulcanello were separated by water after this eruption and that many new rocks had appeared in the waters around the island. A visitor to Vulcano in 1727 mentioned that the noise from a virtually nonstop eruption of the volcano was so tremendous that it prevented him from sleeping at night. At this time, the volcano reportedly had two CRATERS. If that was indeed the case, then some subsequent event apparently destroyed those craters and replaced them with a single crater because a visitor in the late 18th century mentioned only one crater existing in the volcano.

Vulcano is said to have erupted on a nearly continuous basis during the 1730s, but a period of relative quiet appears to have followed because observers of the volcano around 1768 to 1770 reported no dramatic activity. Eruptions occurred in 1771 and 1775. The 1775 eruption is said to have generated a flow of OBSIDIAN. A powerful eruption in 1786 lasted more than two weeks and cast out vast amounts of solid material. Underground noises from this eruption are said to have been heard throughout the surrounding islands. The island was evacuated as a result of this eruption and remained uninhabited for decades afterward. Volcanic activity resumed at Vulcano in 1873 when an eruption began that lasted more than a year and disrupted the mining of SULFUR from the crater. Another eruption commenced in 1888 and lasted through May 1890. This eruption involved powerful explosions and a thick cloud that spread white ash over a broad area. Some of the lava masses, or BOMBS, cast out by this eruption are said to have been gigantic. One had a reported volume of more than 300 cubic feet (8.5 m³). There have been no major eruptions of Vulcano since the late 19th century.

Vulsini *volcano, Italy* The volcano Vulsini is located near the west coast of central ITALY and is part of the Roman volcanic province. The volcano is thought to have grown in three phases of activity: PYROCLASTIC activity and extrusion of lava, producing the main portion of the Vulsini complex; large eruptions of IGNIMBRITE and the formation of a CALDERA by collapsing; and formation of parasitic vents after the caldera originated. Tertullian mentions a mysterious incident of "lightning" destroying Volsini, a nearby town, around 500 B.C., and eruption of flame was reported in 104 B.C. Immediately to the east of the caldera is the Lago di Bolsena, a lake that occupies a depression believed to have originated as a GRABEN.

Wasatch Mountains *Utah, United States* The Wasatch Mountains occupy part of the BASIN AND RANGE PROVINCE of the western UNITED STATES and are noted for earthquake activity. Exposures of unweathered rock along the base of the mountains indicate that large normal FAULTS dropped large areas of land down forming GRABENS that make the basins between the HORSTS that make up ranges. This fault movement was accompanied by powerful earthquakes in recent geologic time before UTAH was settled in the 19th century. The Wasatch Fault Zone has not generated a highly destructive earthquake since the mid-19th century, but the potential for such earthquakes exists and is cause for concern in Utah because most of the state's population lives close to the fault zones.

Washington *United States* The state of WASHINGTON in the Pacific Northwest at the Canadian border, is among the most seismically and volcanically active parts of the UNITED STATES because of the ongoing collision between the westward-moving continental landmass and the JUAN DE FUCA CRUSTAL PLATE beneath the PACIFIC OCEAN. This region has the pattern of volcanism and deep-FOCUS earthquake activity commonly associated with SUBDUCTION ZONES, in this case where a plate of OCEANIC CRUST is overridden by an advancing plate of CONTINENTAL CRUST. Washington has two of the world's most famous volcanoes, Mount RAINIER near Seattle and Mount SAINT HELENS near Portland, OREGON. Volcanism has been responsible for shaping much of the landscape of Washington through such mechanisms as ASHFALLS and LAVA FLOWS.

Earthquake activity in the state is cause for concern on the part of agencies responsible for disaster planning because the potential exists for highly destructive earthquakes in Washington, especially in the densely settled areas surrounding PUGET SOUND. Although the crust beneath the ocean immediately to the west of Washington is relatively quiet from a seismic standpoint, earthquake activity increases eastward across the Olympic Peninsula that separates Puget

Sound from the Pacific Ocean. Puget Sound overlies an area of intense earthquake activity, where shocks in the range of 4.0 on the RICHTER scale occur every few months on the average and earthquakes of approximately MAGNITUDES 2.0 to 3.0 are more commonplace. In a 1983 report to Congress, the U.S. Geological Survey (USGS) declared, "A growing body of data suggests that a great earthquake . . . could occur in the Pacific Northwest," meaning an earthquake of perhaps Richter magnitude 8.0 or above, roughly comparable to the earthquake that destroyed much of SAN FRANCISCO in 1906. "In fiscal year 1983," the report concluded, "two [USGS]-university seismological teams completed a study of earthquake potential in the Pacific Northwest. They concluded that the northwestern United States may [represent] a major seismic gap which is locked and presently seismically quiet, but which may fail in great earthquakes in the future." Special conditions in the Pacific Northwest make Washington highly susceptible to damage from powerful earthquakes. Much of the Puget Sound area, for example, is settled in the form of communities built largely on hillsides, where the potential for LANDSLIDES is great. Also, river deposits and other poorly consolidated materials, vulnerable to LIQUEFACTION in a major earthquake, underlie much of the developed areas in this region.

Although the state of Washington has been occupied by settlers of European descent for only about two centuries, and although the record of notable earthquakes during the early years of their occupation is not always reliable, Washington already has an impressive history of major and moderate earthquakes. Considerable damage occurred in Seattle as a result of a strong earthquake on August 6, 1932. One of the most powerful and destructive earthquakes in the history of Washington occurred on April 29, 1965, near Seattle. This earthquake estimated at Richter magnitude 6.5, caused property damage estimated at more than $12 million, and affected an area of more than 100,000 square miles (258,999 km²). The earthquake generated effects of MERCALLI intensity VII over a wide area and intensity VIII in some locations in the

Seattle area. An earthquake on May 18, 1980, preceded the violent eruption of Mount Saint Helens by only a few moments, was part of a sequence that started in late March, and led up to the catastrophic ERUPTION. The earthquake occurred so soon before the explosive eruption that it is difficult to tell the effects of the earthquake from the ACOUSTICAL effects of the explosion.

See also PLATE TECTONICS.

Watchung Mountains *New Jersey, United States* The Watchung Mountains represent a set of LAVA FLOWS that were produced by FISSURE ERUPTIONS during the separation of AFRICA from NORTH AMERICA some 200 million years ago. The Watchungs were produced as FLOOD BASALTS within the Newark Basin, but equivalent volcanic units in the Hartford Basin extend well into New England. Equivalent PLUTONIC units include the Palisades Sill of NEW JERSEY and NEW YORK. The Watchung lavas were erupted in three episodes of volcanic activity. LAVAS from these episodes make up First Watchung (Orange Mountain Formation), Second Watchung (Preakness Formation), and the Hook Mountain Formation. The equivalent volcanic flows in the Hartford Basin include the Talcott Formation, the Holyoke Formation, and the Hampden Formation.

water Water plays an important part in earthquake and volcanic activity. Water in SOIL near the surface, for example, may make the soil more vulnerable to LIQUEFACTION, the condition in which earthquake waves passing through the soil increase pore-water pressure enough to make the soil lose COHESION and behave as a fluid. As a result, structures built atop the soil may undergo severe damage. This effect of water near Earth's surface has been responsible for much of the damage in some of the most destructive earthquakes in history. The SAN FRANCISCO earthquake of 1906 was notable for the failure of buildings constructed on landfill, and the MEXICO CITY earthquake of 1985, where the destructive power of earthquake waves passing under the city, was augmented by the unconsolidated nature of the underlying sediments that were laid down as part of the bed of an ancient lake. The LOMA PRIETA, CALIFORNIA, and NIIGATA, JAPAN, earthquakes similarly showed much more damage in the areas of fill. Generally speaking, the soggier the soil, the greater its susceptibility to liquefaction.

GROUNDWATER has been known to behave in strange and sometimes violent ways during earthquakes, gushing out of the earth in torrents and even spurting high into the air in the form of fountains. Sand blows—"mud volcanoes," curious craterlike formations created when an earthquake forces wet sand or mud out at the surface—are another manifestation of water's behavior in seismic events. Sand blows formed prominently, for example, during the great CHARLESTON, SOUTH CAROLINA, earthquake and the NEW MADRID earthquake.

On even a more basic level, water can cause earthquakes. In Denver, COLORADO, the pumping of liquid waste under high pressure into deep wells caused increased seismic activity. There was a direct correlation between pumping and earthquakes during a period of two years. The reason that the activity increased is that the liquid pressure enhances the regional stress allowing it to exceed the strength of the rock. This process is called HYDROFRACTURING. The addition of water can also "lubricate" fault planes, reducing the strength of the fault rocks. There was even a James Bond movie where supercriminals threatened to drain the Salton Sea of southern California into the SAN ANDREAS FAULT to cause a silicon valley-destroying earthquake. At least the theory was sound in this story.

The levels of lakes and rivers have been seen to change markedly at the time of earthquakes, and disturbances on the surface of bodies of water during earthquakes can have devastating effects onshore. The most dramatic examples of such disturbances are, of course, TSUNAMIS, the giant seismic sea waves that ripple outward from the source of an earthquake and can carry destructive energy across an entire ocean in a matter of hours. The tsunami from the explosion of the volcano KRAKATOA in 1883, for example, was responsible for most of the deaths from that event and circled the globe several times before finally dissipating. Almost as powerful was the tsunami that followed the Alaskan GOOD FRIDAY EARTHQUAKE, devastating the shoreline of southern ALASKA and carrying destruction hundreds of miles south to CRESCENT CITY, California. On occasion, tsunamis have been known to reach miles inland from the shore before receding and have left oceangoing vessels stranded on dry land hundreds of yards from the water. SEICHES can also be created during earthquakes and volcanoes. They occur as a sloshing of water in enclosed bodies of water. Some excellent examples occurred in the harbors of Alaska during the 1964 Good Friday Earthquake but also in HEBGEN LAKE during the earthquake of 1959.

In volcanic eruptions, water dissolved in MAGMA or molten rock rising from deep underground affects the character and intensity of the eruptions. This effect is commonly observed along ocean margins, where SUBDUCTION ZONES are carrying water-bearing sediments deep into Earth's interior along with the descending slab of OCEANIC CRUST on which the sediments lie. As the descending piece of solid rock melts, the water in the sediments is dissolved in the molten rock under great pressure. When the magma rises through the MANTLE and the CRUST toward the surface, pressure on the magma is relieved, and the dissolved water in the molten rock has an opportunity to come out of solution as gas.

This process is comparable to what happens when an ordinary can of soda is opened: As pressure on the soda (which is filled with dissolved carbon dioxide) is relieved, the gas bubbles out of solution and creates a miniature "eruption" of gas and liquid from the can. Because gas takes up much more space than liquid, a tremendous expansion takes place in the release of pressure. If it is quick enough and there is enough gas, an explosion occurs. In a volcanic eruption, much the same thing may occur; only the molten rock solidifies as the gas within it bubbles out of solution, and the result may be one of several kinds of volcanic rock. Such rocks include PUMICE, a finely textured, "foamy" volcanic rock that is light enough to float on water, and SCORIA, in which the holes from gas bubbles, or VESICLES, are larger than in pumice. As this "foamy," solidified rock breaks up and is ground to bits in the explosion of a volcano, it becomes fine ash. Under certain circumstances, gases released from solution in molten rock during an eruption may build up inside the volcano and eventually produce a colossal explosion. This is what happened at Krakatoa, where the effect of clogging up the VENT of the volcano was to stop it up and allow gases released from the magma to attain tremendous pressures. Ultimately, this situation caused the island to explode with devastating effects. At Mount SAINT HELENS in WASHINGTON State in 1980, gases built up under the north flank of the volcano until that flank burst open and sent a powerful blast sweeping over the landscape, leveling forests. The potential for such events is relatively high where SUBDUCTION ZONES carry water-saturated sediments into Earth's interior near chains of volcanoes, such as the CASCADE MOUNTAINS, of which Mount Saint Helens is part.

MUDFLOWS or LAHARS are another destructive manifestation of water in volcanic eruptions. Heat released from an eruption on the summit of a volcano may melt ice or expel the contents of a crater lake mixed with vast quantities of ash, sending great flows of mud down the flanks of a mountain. These flows may accumulate to great depths and bury entire communities. Many volcanoes exhibit the phenomenon of PHREATIC ERUPTIONS, in which heated water burst out explosively in the form of steam. When phreatic eruptions occur in combination with eruptions of magma, the result is called a phreato-magmatic eruption.

HYDROTHERMAL ACTIVITY, the result of water heated underground by bodies of molten rock near Earth's surface, is a highly complex phenomenon that has had a considerable impact on human history. At some locations, the heated underground water may be tapped for generating electrical power or for heating. Elsewhere, minerals dissolved in the water may be deposited in the form of rich ores of GOLD, silver, copper, and other metals of economic importance. Such activity is thought to have formed many of the valuable ore deposits of the western UNITED STATES and western SOUTH AMERICA, for example. One especially violent and photogenic kind of hydrothermal activity is the GEYSER, a fountain of water and vapor that may reach heights of more than 100 feet (30 m). Some of the world's most famous geysers are located in YELLOWSTONE NATIONAL PARK in the United States and in ICELAND.

Sometimes the interaction of water and volcanic heat creates formations of rare beauty and complexity. At Mount RAINIER in the Cascade range of the United States, for example, volcanic heat at the summit of the mountain has reached an equilibrium with the accumulation of ice and snow to produce an intricate and fragile network of caverns within the ice. These caverns are large enough to accommodate a person.

See also GEOTHERMAL ENERGY.

wavelength The distance between corresponding points on two adjacent sinusoidal waves; for example the measured distance between peaks on two successive waves in a line. Wavelength is measured for seismic waves.

Wegener, Alfred (1880–1930) A German meteorologist, Wegener is widely credited with originating the concept of *continental drift,* which gave rise to the theory of PLATE TECTONICS, the basis for much of modern geology. Although Wegener appears to have taken the idea of continental displacements from other authors, notably the U.S. geologist Frank Bursley Taylor, Wegener presented a comprehensive model of continental motion and the formation of Earth's landmasses in his book *The Origins of Continents and Oceans.* Wegener suggested that the continents are not fixed in position but instead float atop denser rock in Earth's MANTLE, just below the CRUST. Wegener believed the continents as arranged today were once united as a single supercontinent, which Wegener called PANGAEA, meaning "all land." This supercontinent was surrounded by a world ocean he called Panthalassa. The ATLANTIC OCEAN basin, Wegener wrote, was created during the breakup of Pangaea, as the continents moved apart from one another. The energy for the continents' movements, Wegener suggested, came from a tidal motion that supposedly produced a net westward motion of landmasses. Additional energy for continental motion was believed to come from a *Polftucht,* or "flight from the poles," resulting from Earth's rotation.

Wegener's formulation had numerous flaws. He supported his account of the formation of the Atlantic Ocean basin, for example, with inaccurate data that indicated Greenland was moving westward several kilometers per year, when the actual figure is nearer several centimeters per year. For such shortcomings, Wegener's work was criticized.

Wegener graduated from Berlin University and served later on the faculty of Marburg University. He also served as an officer in the German army during World War I. Wegener himself appears to have been a colorful figure. Athletic and adventurous, he took part in the exploration of the atmosphere, setting a world record in 1902 for long-distance balloon flight (56 consecutive hours aloft) in the company of his brother Kurt. Wegener was fascinated with Greenland and joined several expeditions to the island. On his final visit in 1930, he died, presumably of a heart attack, and was buried on the ice cap.

welded tuff After one ASHFALL or ASH-FLOW TUFF is deposited, another ashflow tuff is deposited on top. The heat of the later flow weld underlying ash together into a very dense, glassy, and layered rock. Successive flows continue to weld the underlying layers together. Welded tuffs are common in RHYOLITE and DACITE volcanoes.

Westdahl *volcano, Aleutian Islands, Alaska, United States* Westdahl, in the ALEUTIAN ISLANDS on Unimak Island, stands about 6,000 feet (1,830 m) high and is located approximately 700 miles (1,127 km) southwest of Anchorage. It is a large SHIELD VOLCANO covered by an icecap. It has erupted eight times in historical times, including 1795, 1796, 1820, 1827–30, 1964–65, 1978, 1979, and 1991–92. The ASH cloud from 1991 FISSURE ERUPTION reached an altitude greater than four miles (6.4 km) and LAVA FOUNTAINS yielded flows that are 4.4 miles (7 km) long.

West Virginia *United States* The state of West Virginia lies mostly within a region of minor seismic risk, although the eastern border of the state lies at the edge of a region of higher seismic activity that extends through the western portion of VIRGINIA. An earthquake on May 2, 1853, affected some 72,000 square miles (186,479 km²) in West Virginia (which had not yet become a separate state) and Virginia and was felt very strongly in Lynchburg, Virginia. This earthquake also alarmed residents of Wheeling and Parkersburg in West Virginia.

Whakatane Fault *North Island, New Zealand* A zone of intense volcanic activity, the Whakatane Fault is the location of several active volcanoes, including NGAURUHOE, RUAPEHU, TARAWEA, and TONGARIRO. Numerous other volcanic mountains, not presently active, are also found here. It is likely an extension of the South Alpine fault of the South Island. The South Alpine fault is a TRANSFORM FAULT that has been compared to the SAN ANDREAS FAULT for its style and amount of seismic activity.

White Island *volcano, New Zealand* The summit of two overlapping STRATOVOLCANOes forms one of New Zealand's most active volcanoes. It has produced some 35 small to moderate steam and TEPHRA eruptions (VEI = 2–3) since 1826. The only reported casualties from this volcano resulted from a LANDSLIDE in 1914 that killed 11 workers. However, the volcano was not erupting at the time.

White Mountains *New Hampshire, United States* The White Mountains are remarkable for the role of PLUTONs in their formation. The mountains exhibit interesting circular or crescent-shaped ranges such as the Ossipee Mountains, the Pliny Range, and the Crescent Range. These ranges are thought to have formed when a MAGMA CHAMBER several miles below the surface moved a cylindrical mass of rock directly above the magma, possibly pushing the cylinder upward and then allowing it to fall as pressure in the magma chamber diminished. In subsequent eruptions, according to this scenario, MAGMA flowed upward around the edges of the fallen cylinder and formed nested ring-shaped plutons called RING DIKES. (Similar structures are found in AFRICA, AUSTRALIA, Scotland, and Scandinavia.) The Rattlesnakes, a pair of mountains north of Bristol and Laconia, are thought to have originated in this manner. Ring dikes north of the Presidential Mountains indicate that a large volcano once must have stood in that location.

On a larger scale, the White Mountains are believed to have originated from a rising plume, HOT SPOT, of molten rock inside the MANTLE. It is thought to have given rise to the hills near Montreal, Quebec, as well as the New England SEAMOUNTS, a string of SUBMARINE VOLCANOes extending from southern New England out into the ATLANTIC OCEAN for approximately a thousand miles (1,600 km). This scenario for the formation of the mountains and seamount chain has been questioned, however, because it would require a progressive increase in the age of formations westward, and estimates of age do not support this scenario. A structure similar to the ring-shaped formations of the White Mountains has been detected off the east coast of Massachusetts and is suspected of involvement with the two powerful earthquakes that

occurred in the Boston area in 1727 and 1755. Earthquake activity in the White Mountains has resulted in damage on occasion, notably in December 1940, when two earthquakes knocked down chimneys in the vicinity of the Ossipee formation and generated cracks in the crust on snow there.

White Terrace *North Island, New Zealand* A terraced formation produced by HYDROTHERMAL ACTIVITY, the geyserite White Terrace was located at Lake Rotomahana. Eruptions of a GEYSER supplied the mineralized waters that gave rise to the White Terrace, which covered more than seven acres. A nearby, similar formation was the Pink Terrace. The two formations were considered among the great natural treasures of New Zealand. They were destroyed during an eruption of TARAWERA in 1886.

Wisconsin *United States* Wisconsin is not noted for strong earthquake activity but is affected on occasion by earthquakes in nearby ILLINOIS, northern MICHIGAN, and OHIO.

Witori *volcano, Papua New Guinea* The Witori volcano is located on New Britain Island and stands approximately 2,000 feet (610 m) high. The active cone of the volcano is called Pago and erupted in two intervals from 1911 to 1918 and 1920 to 1933, releasing a flow of LAVA that formed a pool on the floor of the CALDERA. Another eruption may have occurred around the year 1900, but this is not certain. The caldera is believed to have formed some 2,350 years ago in a huge eruption of estimated VEI = 6.

Wizard Island *Oregon, United States* A cinder cone, Wizard Island rises from the waters of CRATER LAKE on Mount MAZAMA.

Wolf *volcano, Galápagos Islands, Ecuador* A SHIELD VOLCANO on Isabela (Albemarle) Island, Wolf has a history of unrest dating back to 1800 and possibly earlier. An explosive eruption was recorded in 1800, and LAVA FLOWS were reported in 1925–26 and 1933. Other eruptions may have occurred in 1935 and 1938. A 1948 eruption involved explosive activity and lava flows, and lava flowed again from the volcano in 1963. In March 1973, an EARTHQUAKE SWARM occurred under the southeast flank of the volcano. Strong and steady rumbling sounds from the CALDERA were heard late in 1973, but no evidence of an actual eruption was found. An eruption in 1982 was remarkable in that it occurred in two hemispheres at the same time. One eruption of lava, in the caldera, took place in the Northern Hemisphere, while another lava flow only six miles (9.6 km) away occurred in the Southern Hemisphere.

Wonchi *See* ASAWA.

Wrangell Mountains *Alaska, United States* The Wrangell Mountains include a large group of volcanoes, including Mount Drum, Mount Gordon, Mount Sanford, Mount Churchill Bona, and Mount Wrangell. These volcanoes are thought to have grown with unusual rapidity. Two CALDERAS, one approximately 10 miles (16 km) wide and the other some three to four miles (4.8 to 6.4 km) wide on average, have been described on Mount Wrangell.

The summit caldera includes three CRATERS: North Crater, East Crater, and West Crater. Eruptions may have occurred at Wrangell in 1819 and 1884, although this is not certain. In 1899 Wrangell reportedly started to emit steam in great quantities following a powerful earthquake and continued smoking afterward. Steaming intensified in 1907. Melting of ice accelerated at Wrangell around 1965, and since that time much of the ice that existed in North Crater before 1965 has disappeared. Temperatures increased markedly at FUMAROLES in the early 1980s, and plumes were reported rising more than a half-mile (0.8 km) above the volcano in April 1986. It has been suggested that the recent increase in heating at Wrangell had something to do with the powerful GOOD FRIDAY EARTHQUAKE in Alaska in 1964.

wrench fault An old term for a STRIKE-SLIP FAULT. Movement between the blocks resembles a monkey wrench.

Wyoming *United States* The state of Wyoming shows most of its seismic activity along its western border, where it borders on a zone of high seismic risk extending from southwest MONTANA southward into southwest UTAH. Wyoming has a history of abundant volcanic activity, especially in the vicinity of Yellowstone. HYDROTHERMAL ACTIVITY is found at YELLOWSTONE NATIONAL PARK in the form of numerous GEYSERS. Volcanic and related activity at Yellowstone is believed to be the work of a giant reservoir of MAGMA beneath the Yellowstone CALDERA.

xenolith When a PLUTON intrudes the crust, pieces of the preexisting surrounding COUNTRY ROCK are broken off into the MAGMA. They float in the magma until it crystallizes. Then the piece of rock, called a xenolith or inclusion, is frozen into the IGNEOUS ROCK. If from deep in Earth, the xenoliths tend to be rounded and deformed. If shallow, they tend to be more angular. Xenoliths can also occur in LAVA FLOWS. They are ripped out of the sides of the feeder VENT to the volcano and carried along in the flow. They can also be picked up from the ground as the lava flows away from the volcano.

Xichang *earthquake, China* A major earthquake struck the Xichang area of eastern China on September 12, 1850. The EPICENTRAL area experienced damage equivalent to X on the modified MERCALLI scale, and the estimated RICHTER magnitude was 7.5. Ground effects in the area included FISSURES, MUD VOLCANOes, waterspouts, and SUBSIDENCE. Collapsed buildings from SURFACE WAVES and LIQUEFACTION caused most casualties. The total loss of life numbered 20,650 victims, making it one of the larger disasters in the area.

Xining *earthquake, China* One of the most devastating earthquakes in history was reported to have struck Xining

and Qinghai (Kansu Province and Wuwei), China, at 10:30 A.M. on May 23, 1927. The RICHTER magnitude of the event was 8.3, and the INTENSITY reached a maximum of XI on the modified MERCALLI scale. Vibrations from the event were recorded on SEISMOGRAPHs halfway around the world for 2.5 minutes with the needle jumping off of the seismograph in many cases. FISSURES appeared all over the area, some reportedly up to 43 feet (13 m) wide. SURFACE RUPTURES were traced for over 8.5 miles (14 km). Huge LANDSLIDES up to one mile (1.5 km) wide buried villages and, in one case, blocked a river, forming an earthquake lake. There were also reports of black water and SAND BOILS. There were reports of subsidence of 10 to 16.5 feet (3 to 5 m) in some areas. Damage to towns was severe. Over 90% of houses simply collapsed in many areas, and many fortifications fell. The U.S. Geological Survey and other sources list the DEATH TOLL at over 200,000, making this one of the deadliest earthquake disasters ever. A recent high-quality translation of official records reported the loss of over 200,000 livestock but a total of 39,000 people. It is still a great disaster but not one of the deadliest even for China. Reports in foreign languages, if not translated properly, can lead to errors that are propagated by numerous sources.

Y

Yake-dake *Honshū, Japan* An andesitic STRATOVOLCANO with at least 26 ERUPTIONS since its first documented eruption in A.D. 686. Since then it erupted in 1585, 23 times between 1907 and 1939, and in 1962–63. The 1585 eruption was large (VEI = 3+) and produced a major LAHAR. Other lahars were produced in 1915, 1931, and 1962–63. Most eruptions are PHREATIC. Yake-dake is near the ON-TAKE caldera.

Yamakawa *volcano, Japan* See ATA.

"year without a summer" The year 1816 is commonly called the year without a summer because of its extremely cool temperatures that have been attributed at least in part to the eruption of the volcano TAMBORA the previous year. Finely divided solid material ejected by Tambora is thought to have entered and lingered in the upper atmosphere, intercepting sunlight and reducing temperatures on the ground. The winter weather of 1815–16 lingered well into the spring, and summer temperatures in 1816 were so far below normal that snow reportedly fell in New England into July. The unusually cold winter disrupted agriculture in Europe and North America, and near-famine conditions are thought to have developed in portions of CANADA and other areas of the world. Apparently, the UNITED STATES produced enough food to sustain itself, but demand from other areas drove prices up everywhere.

See also CLIMATE, VOLCANOES AND.

Yellowstone National Park *Wyoming, United States* Best known for its GEYSERS, such as the famous "Old Faithful," Yellowstone National Park occupies part of the Yellowstone Plateau, which is contiguous with the SNAKE RIVER PLAIN and is seen as an extension of it. The Yellowstone Plateau stands approximately 8,000 feet (2,438 m) above sea level and is thought to have been formed by a tremendous set of volcanic eruptions as the NORTH AMERICAN CRUSTAL PLATE moved westward over a HOT SPOT. According to this scenario, crustal rock melted over the hot spot and produced large amounts of MAGMA that made their way to the surface in eruptions. Some of the magma involved in the Yellowstone eruptions originated deep in the MANTLE producing BASALTS, as well as by melting of CONTINENTAL CRUST near the surface to produce the RHYOLITES.

Although early eruptions are believed to have been small, the Yellowstone area later was the site of three great eruptions. As huge amounts of magma were expelled, the roof of the MAGMA CHAMBER underlying Yellowstone collapsed and created a caldera occupying an area of about 1,000 square miles (2,560 km²). This eruption occurred about 2.2 million years ago and produced about 610 cubic miles (2,500 km³) of EJECTA, now represented by the Huckleberry Ridge tuff. Another cycle of eruptions released some 70 cubic miles (287 km³) of magma and produced the Island Park caldera, some 17 miles (27.2 km) wide. A third and more powerful cycle of eruptions, though not as great at the initial cycle, is thought to have deposited ASH over much of North America. Deposits linked to this eruption have been found in KANSAS. Again, a caldera formed, this one known today as the Yellowstone caldera and measuring about 30 by 50 miles (48 by 80 km). This last great eruption occurred about 600,000 years ago and produced the Lava Creek Tuff with an estimated volume of 244 cubic miles (1,000 km³). Compare these eruptions to that at Mount SAINT HELENS, which produced 0.25 cubic miles (1 km³) of EJECTA to appreciate their immensity. Further eruptions at Yellowstone are possible but a strong shift in the TECTONICS of the area would be necessary.

Besides its numerous geysers and other evidence of HYDROTHERMAL ACTIVITY, Yellowstone is noted for its petrified forest, where eruptive activity buried and preserved more than 25 layers of forest growth believed to represent some 20,000 years of geologic time. The petrified forest has been exposed in a section more than 1,000 feet (305 m) thick along the Lamar River in Yellowstone Park. The petrified forest of Yellowstone should not be confused with the petrified forest of ARIZONA, which appears to have formed under different conditions than that of Yellowstone. Whereas the Yellowstone forests stand where they were buried by volcanic eruptions, the trees of the Arizona petrified forest were apparently deposited and buried by floodwaters.

APPENDIX A

CHRONOLOGY OF EARTHQUAKES AND VOLCANIC ERUPTIONS

The following list represents a selection, though not a complete list, of major earthquakes and volcanic eruptions in history. In most cases, casualty figures are estimates. (There can be considerable disagreement among various sources where casualty figures are concerned, and readers should bear in mind that many of the statistics cited here on numbers of people killed in earthquakes, tsunamis, and eruptions represent only informed guesses.) For earthquakes that occurred before the development of modern seismological instruments and the Richter scale of earthquake magnitude, magnitudes also represent estimates. Exact magnitude figures were not available for some earthquakes of the 20th century.

c. 1470 B.C. *Santorini, Mediterranean, volcanic eruption.* Eruption is thought to have killed thousands and destroyed the Minoan civilization.

1226 B.C. *Etna, Sicily, volcanic eruption.* Other eruptions occurred in 1170 B.C., 1149 B.C., 525 B.C., 477 B.C., 396 B.C., 140 B.C., 126 B.C., 122 B.C., A.D. 1169, 1329, 1536, 1669, 1693, 1755, 1843, 1928.

464 B.C., **date uncertain.** *Sparta, Greece, earthquake.* Mercalli intensity X. 20,000 killed.

373 B.C., **date uncertain.** *Helike, Greece, earthquake.* Estimated magnitude 6.8. Tsunami destroys city, and it is abandoned. No record of survivors.

217 B.C., **June.** *North Africa, earthquake.* More than 100 cities were destroyed, and more than 50,000 people are estimated to have been killed.

A.D. 19. *Syria, earthquake.* More than 100,000 people believed killed.

79, August 24. *Vesuvius, Italy, volcanic eruption.* This eruption destroyed—and simultaneously preserved—the cities of Pompeii and Herculaneum. Other notable eruptions occurred in 203, 472, 1036, 1049, 1198, 1302, 1538, 1631, 1707, 1779, 1794, 1872, 1905, 1929, and 1944.

365, July 21. *Alexandria, Egypt, earthquake.* This earthquake destroyed Alexandria and the giant lighthouse for which the city was famous. Some 50,000 people are thought to have been killed.

526, May 26. *Antioch, Syria, earthquake.* Approximately 250,000 people are believed to have died.

536. *Indonesia area, volcanic eruption.* Evidence for perhaps the largest historical eruption but no volcano identified. Two-year climate change.

688, date uncertain. *Smyrna, Turkey, earthquake.* Some 20,000 people killed.

856, December. *Corinth, Greece, earthquake.* More than 45,000 people killed.

856, December 22. *Damghan, Iran, earthquake.* Up to 200,000 people killed.

893, date uncertain. *Ardabil, Iran, earthquake.* 100,000–180,000 people killed.

893, date uncertain. *Debal, Pakistan, earthquake.* 150,000 people killed. Could be same event as Ardabil.

1038, January 9. *Shensi, China, earthquake.* More than 23,000 killed.

1042, November 11. *Tabriz, Iran, earthquake.* 40,000–50,000 people killed.

1057, date unknown. *Chihli, China, earthquake.* 25,000 killed.

1101, date uncertain. *Khoresan, Iran, earthquake.* Mercalli intensity X. Death toll 60,000.

1138, September 8. *Aleppo, Syria, earthquake.* 100,000–250,000 killed.

1170, date unknown. *Sicily, Italy, earthquake.* 15,000 killed.

1268, date unknown. *Cilicia, Asia Minor, earthquake.* 60,000 killed.

1183, date uncertain. *Tripoli, Libya, earthquake.* More than 20,000 people killed.

1290, September 27. *Chihli, China, earthquake.* 100,000 killed.

1293, May 20. *Kamakura, Japan, earthquake.* 30,000 killed.

1303, August 8. *Rhodes, Greece, earthquake.* More than 25,000 killed, including 10,000 in Alexandria, Egypt.

1362, date uncertain. *Öraefajökull, Iceland, volcanic eruption.* Several dozen farms destroyed.

1456, date unknown. *Naples, Italy, earthquake.* 60,000 killed.

1493, October 3. *Cos, Greece, earthquake.* Estimated magnitude 6.8. Death toll 5,000.

1509, September 10. *Constantinople (Istanbul), Turkey, earthquake.* Mercalli intensity X. More than 13,000 killed.

1522, date uncertain. *Masaya, Nicaragua, volcanic eruption.* Several towns destroyed.

1531, January 26. *Lisbon, Portugal, earthquake.* 30,000 killed.

1549, date uncertain. *Nicaragua, eruption of Volcan de Agua.* Thousands of people killed.

1555, September. *Kashmir, India, earthquake.* Mercalli intensity XII. Death toll up to 60,000.

1556, January 3. *Shensi, China, earthquake.* 830,000 killed.

1591, date uncertain. *Taal, Luzon, Philippine Islands, volcanic eruption.* Thousands of people reportedly died in this eruption, the first recorded eruption of Taal. Other major eruptions occurred in 1749 and 1754, on January 28 and 30, 1911, and on September 28, 1965.

1614, November 26. *Echigo, Takata, Japan, earthquake.* Thousands of people are thought to have been killed by an earthquake and subsequent tsunami.

1616, date uncertain. *Mayon, Philippine Islands, volcanic eruption.* This was the first recorded eruption of Mayon. Numerous villages were destroyed, and thousands of people were killed. Another major eruption occurred on October 23–30, 1766, killing some 2,000 people. Mayon erupted again on February 1, 1814, killing more than 2,000. Several dozen were killed in another eruption in 1853. The eruption of July 9, 1888, destroyed portions of two towns. An eruption on June 23, 1897, killed several hundred people and caused widespread destruction. Some 200 people reportedly died in an eruption on July 1, 1928.

1623, date uncertain. *Irazu, Costa Rica, volcanic eruption.* The death toll from this eruption is thought to be in the hundreds.

1626, July 30. *Naples, Italy, earthquake.* This earthquake wrecked the city, destroyed numerous villages, and killed some 70,000 people. Another major earthquake in 1693 killed more than 90,000.

1650, date uncertain. *Asamayama, Japan, volcanic eruption.* This eruption is believed to have killed several hundred people. Another eruption in 1783 killed about 5,000 people.

1653, June 16. *Khoresan, Iran, earthquake.* Mercalli intensity IX. More than 5,600 killed.

1663, February 5. *Saint Lawrence River Valley, Canada, earthquake.* Maximum Mercalli intensity X. Casualty figures unavailable. This earthquake was so powerful that it caused damage to buildings in southern New England.

1667, date unknown. *Shemaka, Caucasia, earthquake.* More than 80,000 killed.

1667, April 6. *Monte Negro, Croatia, earthquake.* Mercalli intensity IX. More than 5,000 killed.

1668, July 25. *Tancheng, China, earthquake.* More than 50,000 people killed.

1688, July 10. *Izmir (Smyrna), Turkey, earthquake.* Mercalli intensity X. More than 20,000 killed.

1692, June 7. *Port Royal, Jamaica, earthquake.* More than 1,000 people died in this earthquake, which submerged a large portion of the city.

1693, January 9–11. *Catania, Italy, earthquake.* 60,000 killed, although another estimate puts the death toll at 100,000. In the same year, in an earthquake in Naples, Italy, some 93,000 were killed by another earthquake.

1698, date uncertain. *Cotopaxi, Ecuador, volcanic eruption.* Hundreds of people were reported killed in this eruption. Other major eruptions occurred in 1741 and 1744 and on June 26, 1877.

1702, February 25. *Denizili, Turkey, earthquake.* Mercalli intensity X. More than 12,000 killed.

1703. *Tokyo (Edo), Japan, earthquake.* This earthquake destroyed the city and reportedly took 200,000 lives.

1707, October 28. *Tokaido, Japan, earthquake. Tsunami kills 4,900 people.*

1717, November 19. *Denizili, Turkey, earthquake.* Mercalli intensity IX. More than 6,000 killed.

1721, April 26. *Tabriz, Iran, earthquake.* Likely 80,000 killed but may be up to 250,000.

1727, November 18. *Tabriz, Iran earthquake.* 77,000 killed but could be confused with the 1721 event.

1731, November 30. *Peking (Beijing), China, earthquake.* Some 100,000 people were reportedly killed.

1737, October 11. *Calcutta, India, earthquake.* 300,000 killed.

1754, September 10. *Cairo, Egypt, earthquake.* Magnitude 5.4. Death toll 40,000.

1755, June 7. *Persia (in the vicinity of present-day Iran), earthquake.* 40,000 killed.

1755, November 1. *Lisbon, Portugal, earthquake associated with a destructive tsunami.* Up to 70,000 killed.

1766, May 22. *Constantinople (Istanbul), Turkey, earthquake.* Mercalli intensity IX with tsunami. Death toll 4,500.

1772, date uncertain. *Papandajan, Java, Indonesia, volcanic eruption.* Some 3,000 people were reported killed in this eruption, which destroyed numerous villages as well as plantations. The volcano lost more than 3,000 feet (923 m) of its height in this eruption.

1779–1780, December through January. *Sakurajima, Japan, volcanic eruption.* This eruption destroyed several villages. Another eruption and accompanying earthquakes on January 12, 1914, destroyed many villages.

1780, February 28. *Tabriz, Iran, earthquake.* Death toll 50,000–60,000.

1783, February 4. *Calabria, Italy, earthquake.* Perhaps 40,000 killed. Another earthquake in 1797 is thought to have killed an additional 50,000 people.

1783. *Iceland and Japan, volcanic eruptions.* A 1783 eruption of the volcano Asamayama in Japan killed several thousand people. Between mid-June and early August, an eruption of Skaptarjöokull in Iceland caused widespread destruction of property and loss of life, killing several thousand people. Also in June, a great flow of lava occurred at Laki in Iceland, covering more than 100,000 acres of land with molten rock and killing some 10,000 people.

1790, October 9. *Oran, Algeria, earthquake.* Mercalli intensity X. Death toll 3,000.

1792, February 10. *Unzen, Japan, volcanic eruption.* This eruption killed some 15,000 people and destroyed two cities.

1793, date uncertain. *Miyi-Yama, Java, Indonesia, volcanic eruption.* This eruption is believed to have killed more than 50,000 people.

1797, February 4. *Quito, Ecuador, earthquake.* 40,000 killed.

1800, date uncertain. *Mount Guntur, Java, Indonesia, volcanic eruption.* This eruption killed several hundred people, and a lava flow reportedly filled a valley.

1811, December 16. *New Madrid, Missouri, United States, earthquake.* Estimated at Mercalli intensity XI, this earthquake was one of a series of extremely powerful temblors that struck the New Madrid area (near St. Louis, Missouri, and Memphis, Tennessee) in the winter of 1811–12. Other great earthquakes in this series took place on January 23 and February 7, 1812. Casualties occurred but are thought to have been few in number because the area affected was sparsely settled at the time.

1812, March 26. *Caracas, Venezuela, earthquake.* More than 10,000 killed.

1812, December 21. *Near Santa Barbara, California, United States, earthquake, with possible associated tsunami.* Mercalli intensity X. Several persons are thought to have been injured.

1815, April 5. *Tambora, near Java, Indonesia, volcanic eruption.* This volcano's 1815 eruption was one of the most powerful in history and killed tens of thousands of people. The exact final death toll from Tambora's 1815 eruption is difficult to estimate because airborne particulate from the explosion are believed to have played a part in cooling global climate over the following year, disrupting agriculture and causing the famous "year without a summer."

1819, June 16. *Kutch, India, earthquake.* More than 1,500 killed.

1822, October. *Galung Gung, Java, Indonesia, volcanic eruption.* Two eruptions of Galung Gung killed several thousand people altogether and destroyed hundreds of villages.

1822, November 19. *Valparaiso, Chile, earthquake.* Death toll 10,000.

1822, date unknown. *Aleppo, Asia Minor, earthquake.* Some 22,000 killed.

1828, December 18. *Echigo, Japan, earthquake.* 30,000 killed.

1842, May 7. *Cap-Hatien, Haiti, earthquake.* Tsunami killed 5,000–10,000.

1847, date unknown. *Zenkoji, Japan, earthquake.* Some 34,000 killed.

1850, September 12. *Xichang, China, earthquake.* Mercalli intensity X. More than 20,650 killed.

1853, May 4. *Shiraz, Iran, earthquake.* Mercalli intensity VIII. Death toll 12,000.

1854, December 23. *Ansei Tokai, Japan, earthquake.* Between 5,000 and 31,000 people killed.

1855, November 11. *Tokyo, Japan, earthquake.* Death toll 6,757.

1857, January 9. *Fort Tejon, California, United States, earthquake.* One of the most powerful earthquakes in American history, the Fort Tejon earthquake is ranked at Mercalli intensity X or above and involved a rupture of the San Andreas Fault.

1857, March 21. *Tokyo, Japan, earthquake.* More than 100,000 people are thought to have died in this earthquake.

1859, March 22. *Quito, Ecuador, earthquake.* Death toll 5,000.

1863, August 13. *Bolivia and Peru, earthquake.* 25,000 killed.

1868, August 16. *Colombia and Ecuador, earthquake.* 70,000 killed.

1872, March 26. *Owens Valley, California, United States, earthquake.* The Owens Valley earthquake is thought to have killed some 50 to 60 people.

1875, May 18. *Cucuta, Colombia, earthquake.* Estimated magnitude 7.3. Death toll 10,000–16,000 people.

1881, April 3. *Chios, Greece, earthquake.* Estimated magnitude 6.5. Tsunami kills 7,000 people and injures 20,000.

1883, August 27. *Krakatoa, Sunda Strait, Indonesia, volcanic eruption.* The 1883 eruption of Krakatoa killed perhaps 50,000 people and generated a tsunami, or seismic sea wave, that circled the globe several times before dissipating. The wave caused tremendous destruction along shorelines near the volcano.

1884. *Ramapo Fault, New Jersey, earthquake.* Magnitude 5.4. Rang church bells in Connecticut.

1885, May 30. *Sringar, India, earthquake.* Death toll 3,200.

1886, August 31. *Charleston, South Carolina, United States, earthquake.* The Charleston earthquake caused tremendous property damage and killed some 60 people. Exactly what caused this earthquake is still uncertain.

1887, February 23. *Riviera, France-Italy, earthquake.* Mercalli intensity X. Tsunami kills more than 2,000 people.

1891, October 28. *Mino-Owari, Japan, earthquake.* 7,000 killed.

1895, January 17. *Quchan, Iran, earthquake.* Up to 11,000 killed.

1896, June 15. *Riku-Ugo, Japan, earthquake.* More than 20,000 killed. A tsunami was associated with this earthquake.

1897, June 12. *Assam, India, earthquake.* Magnitude 8.7. More than 1,000 killed.

1898, November 17. *Quchan, Iran, earthquake.* Up to 18,000 killed. Possible confusion with 1895 event.

1902, May 6–13. *Soufrière, Saint Vincent, Caribbean, volcanic eruption.* The death toll from this eruption of Soufrière is estimated at 3,000.

1902, May 8. *Mount Pelée, Martinique, Caribbean, volcanic eruption.* The 1902 eruption of Pelée is among the most spectacular and thoroughly documented eruptions. More than 30,000 people are thought to have died in this eruption, which destroyed the city of Saint Pierre. Another eruption of Pelée on August 30 added to the destruction.

1902, October 24. *Santa María, Guatemala, volcanic eruption.* Some 6,000 people were killed.

1902, December 16. *Andijan, Turkistan, earthquake.* Magnitude 6.4. Between 4,500 and 10,000 killed.

1905, April 4. *Kangra, India, earthquake.* Magnitude 7.8. About 20,000 killed.

1906, April 18. *San Francisco, California, United States, earthquake.* Magnitude approximately 8.2. The San Francisco earthquake and fire that followed destroyed the city. An estimated 700 to 2,000 people were killed.

1906, August 16. *Valparaiso, Chile, earthquake.* Magnitude 8.4. Between 1,500 and 20,000 deaths.

1908, December 28. *Messina, Sicily, Italy, earthquake.* Magnitude 7.5. 120,000 killed, although another estimate says 160,000.

1915, January 13. *Avezzano, Italy, earthquake.* Magnitude 7.0. 30,000 killed.

1918, February 15. *Kwangtung, China, earthquake.* Magnitude 7.3. More than 10,000 deaths.

1920, December 16. *Kansu, China, earthquake.* An earthquake caused extensive destruction and took an estimated 180,000 lives.

1923, September 1. *Tokyo, Japan, Kanto earthquake.* Magnitude 7.9 to 8.3. The earthquake was accompanied by fire and destroyed much of Tokyo. More than 140,000 were killed.

1927, March 7. *Tango, Japan, earthquake.* Tsunami kills more than 2,900 people.

1927, May 22. *Kansu, China, earthquake.* Estimated 100,000 people killed.

1931, December 13–28. *Merapi, Java, Indonesia, volcanic eruption.* More than 1,000 people are thought to have died in this eruption.

1932, December 25. *Kansu, China, earthquake.* Estimated 70,000 people killed.

1934, January 16. *Nepal, earthquake.* More than 9,000 are believed to have been killed.

1935, May 31. *Quetta, India, earthquake.* Magnitude 7.5. 60,000 killed.

1939, January 24. *Chillan, Chile, earthquake.* Magnitude approximately 7.7. Perhaps 50,000 killed.

1939, December 26. *Erzincan, Turkey, earthquake.* Magnitude 7.9. Tsunami on the Black Sea. More than 33,000 killed.

1941, June 26. *Andaman Islands, India, earthquake.* Magnitude 7.7 Earthquake and tsunami kills more than 5,000 people.

1944, June 10. *Paricutín, Mexico, volcanic eruption.* Paricutín, a volcano that arose rapidly in a field beginning in 1943 and grew to be more than 1,200 feet (369 m) tall, erupted violently, and destroyed nearby towns.

1945, November 27. *Makran, Pakistan, earthquake.* Magnitude 8.0. Tsunami kills more than 4,000 people.

1946, November 10. *Ancash, Peru, earthquake.* Magnitude 7.4. Death toll 1,400.

1948, October 6. *Ashgabat, Turkmenistan, earthquake.* Magnitude 7.3. Death toll 60,000–70,000.

1949, August 5. *Pelilo, Ecuador, earthquake.* Magnitude 7.5. Death toll 6,000, with 20,000 injured and 100,000 homeless.

1950, August 15. *Arunachal Pradesh, India, earthquake.* Magnitude 8.6. Death toll 1,526.

1951, January 21. *Mount Lamington, Papua New Guinea, volcanic eruption.* One of the most destructive of modern times, this eruption killed some 6,000 people.

1960, February 29. *Agadir, Morocco, earthquake.* Magnitude 5.9. Estimated 12,000 killed.

1962, September 1. *Northwestern Iran, earthquake.* Magnitude 7.3. 14,000 killed.

1963, July 26. *Skopje, Yugoslavia, earthquake.* 1,500 to 2,000 killed; most of city destroyed.

1964, March 27. *Alaska, United States, earthquake.* Magnitude 9.1. The Good Friday earthquake, as this event was known, was accompanied by a highly destructive and far-reaching tsunami. More than 100 people were killed.

1968, August 31. *Iran, earthquake.* Magnitude 7.4. More than 15,000 killed.

1970, May 30. *Peru, earthquake.* Magnitude 7.8. More than 60,000 killed.

1971, February 9. *San Fernando, California, United States, earthquake.* Magnitude 6.5. The San Fernando earthquake killed fewer than 100 people, but that figure could have been much higher if the earthquake had lasted slightly longer.

1972, December 23. *Managua, Nicaragua, earthquake.* Magnitude 6.2. Estimated 11,000 killed.

1974, December 28. *Patan, Pakistan, earthquake.* Magnitude 6.3. Death toll 5,300, with 15,000 injured.

1976, February 4. *Guatemala, earthquake.* Magnitude 7.9. More than 23,000 killed.

1976, July 28. *Tangshan, China, earthquake.* Magnitude 7.6. Approximately 700,000 killed.

1976, August 17. *Moro Gulf, Philippines, earthquake.* Magnitude 7.9. Tsunami kills about 7,000 people, with 10,000 injured.

1976, November 24. *Muradive, Turkey, earthquake.* Magnitude 7.3–7.9. More than 5,000 killed.

1977, March 4. *Bucharest, Romania, earthquake.* Magnitude 7.4. Between 1,570 and 2,000 killed.

1978, September 16. *Iran, earthquake.* 25,000 killed.

1980, May 18. *Mount Saint Helens, Washington, United States, volcanic eruption.* One of the most intensively documented and studied eruptions of all time, the eruption destroyed some 150 square miles (385 km²) of timber and caused $1.5 billion or more in total damage. More than 100 people were reported dead or missing.

1980, October 10. *El Asnam, Algeria, earthquake.* Magnitude 7.7. More than 3,000 killed.

1982. *El Chichón, southern Mexico, volcanic eruption.* The eruption injected some 20,000,000 metric tons of sulfuric acid droplets into the atmosphere. The aerosols lowered the temperature in the Northern Hemisphere by 0.5°C.

1982, December 13. *Northern Yemen, earthquake.* 2,800 killed.

1983, May 2. *Coalinga, California, United States, earthquake.* $30 million in damage reported.

1983, May 26. *Honshū, Japan, tsunami.* 81 killed.

1985, September 19. *Mexico City, Mexico, earthquake.* Magnitude 8.1. Some 8,500 killed. Although this earthquake originated along the Pacific coast of Mexico, it caused great destruction in Mexico City because portions of the city were built on the unstable sediments of a former lake bed.

1986, August 21. *Cameroon, volcanic event.* An escape of volcanic gas from a lake bed killed some 1,700 people.

1987, October 1. *Southern California, United States, earthquake.* $200 million damage.

1988, December 7. *Armenia, earthquake.* Magnitude 6.8. 25,000 killed.

1989, October 17. *San Francisco, California, United States, earthquake.* 67 killed; $5 billion in damage.

1990, June 21. *Gilan, Iran, earthquake.* Magnitude 7.7. More than 50,000 people killed, 200,000 injured, and 500,000 homeless.

1990, July 16. *Luzon, Philippine Islands, earthquake.* 700 killed.

1991, June 10–17. *Mount Pinatubo, Philippine Islands, volcanic eruption.* 900 killed.

1992, December 10. *Island of Flores, Indonesia, earthquake.* More than 1,000 killed.

1993, September 30. *Maharashta, India, earthquake.* 16 villages destroyed; many others damaged. Tens of thousands killed.

1994, January 17. *Northridge, California, earthquake.* Magnitude 6.7. Some $20 billion in damage, the Los Angeles area was devastated. Aftershocks continued for more than a year causing widespread psychological problems.

1995, January 22. *Kobe, Japan, earthquake.* Magnitude 6.8. More than 5,500 people killed. Extensive damage.

1995, May 28. *Sakhalin, Russia, earthquake.* Magnitude 7.6. About 2,000 killed.

1995–1997. *Soufrière Hills, Montserrat, volcanic eruption.* The island resort is completely destroyed.

1998, May 30. *Mazar-e Sharif, Afghanistan, earthquake.* Magnitude 6.5. More than 4,000 people killed.

1998, July 17. *Sissano, New Guinea, earthquake.* Magnitude 7.1. Some 2,700 people killed, including those from a resulting tsunami.

1999, January 25. *Armenia, Colombia, earthquake.* Magnitude 6.2. Some 1,900 people killed.

1999, August 7. *Izmit, Turkey, earthquake.* Magnitude 7.6. More than 17,000 killed.

1999, September 20. *Tai-Chung, Taiwan, earthquake. Magnitude 7.6. More than 2,400 people killed. More than $14 billion in damage.*

2000, May 4. *Sulawesi, Indonesia, earthquake.* Magnitude 7.6. More than 46 people killed; 264 injured. Extensive damage and power outages.

2000, June 4. *Southern Sumatra, Indonesia, earthquake.* Magnitude 8.0. More than 103 people killed; 2,174 injured.

2000, July 14. *Kodiak Island region, Alaska, earthquake.* Magnitude 6.8.

2000, September 3. *Northern California, United States, earthquake.* Magnitude 5.2. More than 41 people injured in the Napa area. Felt in much of northern California from Sacramento to Santa Rosa and south to the San Francisco Bay.

2001, January 1. *Mindanao, Philippines, earthquake.* Magnitude 7.3.

2001, January 13. *Off the coast of Central America, earthquake.* Magnitude 7.7. More than 844 people killed; 4,723 people injured. 108,226 houses destroyed and 150,000 buildings damaged in El Salvador. Felt from Mexico City to Colombia.

2001, January 26. *Gujarat, India, earthquake.* Magnitude 7.9. Confirmed at least 20,000 people killed; another 166,812 injured. 600,000 homeless.

2001, February 13. *El Salvador, earthquake.* Magnitude 6.6. At least 315 people killed; 3,000 injured. Extensive damage. Landslides occurred in various areas of the country.

2001, February 28. *Seattle, Washington, United States, earthquake.* Magnitude 6.8. More than 215 people injured. Damage is estimated to be in the billions.

2001, March 24. *Hiroshima, Japan, earthquake.* Magnitude 6.4. Approximately 150 injured; 3 people reported killed. 500 homes destroyed.

2001, March 28. *Minahasa, North Sulawesi, Indonesia, volcanic eruption and earthquake.* Magnitude 5.4. Mount Lokon erupted at practically the same time as an earthquake hit Bengkulu town. No casualties reported to date.

2003, May 21. *Algiers, Algeria, earthquake.* Magnitude 6.8. Death toll 2,266, with 150,000 homeless.

2003, December 26. *Bam, Iran, earthquake.* Magnitude 6.7. More than 43,000 killed and 90,000 left homeless.

2004, March 28. *Northern Indonesia, aftershock earthquake.* Magnitude 8.7. More than 1,000 people killed.

2004, December 26. *Banda Aceh, Indonesia, earthquake.* Magnitude 9.0. Huge tsunami devastates entire Indian Ocean basin. More than 283,100 people killed and 1,126,900 homeless.

2005, October 8. *Muzaffarabad, Pakistan, earthquake.* Magnitude 7.6. Some 86,000 people killed and at least 2.5 million homeless.

2006, May 27. *Bantul, Indonesia, earthquake.* Magnitude 6.3 earthquake struck the Bantul-Yogyakarta area during a nearby eruption of Mount Merapi. Approximately 6,300 people killed.

APPENDIX B

EYEWITNESS ACCOUNTS OF MAJOR ERUPTIONS AND QUAKES

Eruption of Vesuvius, A.D. 79

Roman naturalist and naval officer Pliny the Elder and his nephew Pliny the Younger are remembered in connection with the catastrophic eruption of Vesuvius in A.D. 79, which destroyed the nearby cities of Pompeii and Herculaneum. Pliny the Elder was killed in the eruption, and Pliny the Younger described it, together with his uncle's death, in a pair of letters written to the historian Tacitus. Although thought to be inaccurate in some respects, Pliny's letters are worth quoting at length:

Your request that I should send an account of my uncle's death, in order to transmit a more exact relation of it to posterity, deserves my acknowledgments. . . .

He was at that time with the fleet under his command at Misenum. On the 24th of August, about one in the afternoon, my mother desired him to observe a cloud which appeared of a very unusual size and shape. He had just returned from taking the benefit of the sun, and, after bathing himself in cold water, and taking a slight repast, had retired to his study. He immediately arose and went out upon an eminence, from whence he might distinctly view this very uncommon appearance. It was not at that distance discernible from what mountain the cloud issued, but it was found afterward to ascend from Mount Vesuvius. I cannot give a more exact description of its figure than by comparing it to that of a pine tree, for it shot up to a great height in the form of a trunk, which extended itself at the top into a sort of branches . . . it appeared sometimes bright, and sometimes dark and spotted, as it was more or less impregnated with earth and cinders.

This extraordinary phenomenon excited my uncle's philosophical curiosity to take a nearer view of it. He ordered a light vessel to be got ready. . . . As he was passing out of the house he received dispatches: the marines at Retina, terrified by the imminent peril (for the place lay beneath the mountain, and there was no retreat but by ships), entreated his aid in this extremity. He accordingly changed his first design, and what he began with a philosophical he now pursued with an heroical turn of mind.

He ordered the galleys to put to sea, and went himself on board with an intention of assisting not only Retina but many other places, for the population is thick on that beautiful coast. When hastening to the place from which others fled with the utmost terror, he steered a direct course to the point of danger, and with so much calmness and presence of mind, as to be able to make and dictate his observations upon the motion and figure of that dreadful scene. He was now so nigh the mountain that the cinders, which grew thicker and hotter the nearer he approached, fell into the ships, together with pumice-stones, and black pieces of burning rock; they were in danger of not only being left around by the sudden retreat of the sea, but also from the vast fragments which rolled down from the mountain, and obstructed all the shore.

Here he stopped to consider whether he should return back again; to which the pilot advised him. "Fortune," said he, "favors the brave; carry me to Pomponianus." Pomponianus was then at Stabiae, separated by a gulf, which the sea, after several insensible windings, forms upon the shore. He (Pomponianus) had already sent his baggage on board; for though he was not at that time in actual danger, yet being within view of it, and indeed extremely near, if it should in the least increase, he was determined to put to sea as soon as the wind should change. It was favorable, however, for carrying my uncle to Pomponianus, whom he found in the greatest consternation. He embraced him with tenderness, encouraging and exhorting him to keep up his spirits; and the more to dissipate his fears he ordered, with an air of unconcern, the baths to be got ready; when, after having bathed, he sat down to supper with great cheerfulness, or at least (what is equally heroic) with all the appearance of it.

In the meantime, the eruption from Mount Vesuvius flamed out in several places with much violence, which the darkness of the night contributed to render still more visible and dreadful. But my uncle, in order to soothe the apprehensions of his friend, assured him it was only the burning of the villages, which the country people had abandoned to the flames; after this he retired to rest, and it was most certain he was so little discomposed as to fall into a deep sleep; for, [as he was] pretty fat, and breathing hard, those who attended him actually heard him snore. The court which led to his apartment being now almost filled with stones and ashes, if he had continued there any longer it would have been impossible for him to have made his way out; it was thought proper, therefore, to awaken him. He got up and went to Pomponianus and the rest of his company.... They consulted together whether it would be most prudent to trust to the houses, which now shook from side to side with frequent and violent concussions; or to fly to the open fields, where the calcined stone and cinders, though light indeed, yet fell in large showers and threatened destruction. In this distress they resolved for the fields as the less dangerous situation of the two—a resolution which, while the rest of the company were hurried into it by their fears, my uncle embraced upon cool and deliberate consideration.

They went out, then, having pillows tied upon their heads with napkins; and this was their whole defense against the storm of stones that fell around them. It was now day everywhere else, but there a deeper darkness prevailed than in the most obscure night; which, however, was in some degree dissipated by torches and other lights of various kinds. They thought proper to go down further upon the shore, to observe if they might safely put out to sea; but they found that the waves still ran extremely high and boisterous. There my uncle, having drunk a draught or two of cold water, threw himself down upon a cloth which was spread for him, when immediately the flames, and a strong smell of sulfur which was the forerunner of them, dispersed the rest of the company, and obliged him to rise. He raised himself up with the assistance of two of his servants, and instantly fell down dead, suffocated, as I conjecture, by some gross and noxious vapor, having always had weak lungs, and being frequently subject to a difficulty of breathing.

As soon as it was light again, which was not until the third day after this melancholy accident, his body was found entire, and without any marks of violence upon it, exactly in the same posture as that in which he fell, and looking more like a man asleep than dead . . .

My uncle having left us [Pliny the Younger writes], I pursued [my] studies till it was time to bathe. After which I went to suppose, and from thence to bed, where my sleep was greatly broken and disturbed. There had been, for many days before, some shocks of an earthquake, which the less surprised us as they are extremely frequent in Campania; but they were so particularly violent that night, that they not only shook everything around us, but seemed, indeed, to threaten total destruction. My mother flew to my chamber, where she found me rising in order to awaken her. We went out into a small court belonging to the house, which separated the sea from the buildings. As I was at that time but eighteen years of age, I know not whether I should call my behavior, in this dangerous juncture, courage or rashness; but I took up [the historian] Livy, and amused myself with turning over that author, and even making extracts from him, as if all about me had been in full security. While we were in this posture, a friend of my uncle's, who was just come from Spain to pay us a visit, joined us; and observing me sitting with my mother with a book in my hand, greatly condemned her calmness at the same time that he reproved me for my careless [air of] security. Nevertheless, I still went on with my author.

Though it was now morning, the light was exceedingly faint and languid; the buildings all around us tottered; and, though we stood upon open ground, yet as the place was narrow and confined, we therefore resolved to quit the town. The people followed us in the utmost consternation, and, as to a mind distracted with terror every suggestion seems more prudent than its own, pressed in great crowds about us in our way out.

Being got to a convenient distance from the houses, we stood still, in the midst of a most dangerous and dreadful scene. The chariots which we had ordered to be drawn out were so agitated backwards and forwards, though upon the most level ground, that we could not keep them steady, even by supporting them with large stones. The sea seemed to roll back upon itself, and to be driven from its banks by the convulsive motion of the earth; it is certain at least that the shore was considerably enlarged, and many sea animals were left upon it. On the other side a black and dreadful cloud, bursting with an igneous serpentine vapor, darted out a long train of fire, resembling flashes of lightning, but much larger.

Upon this the Spanish friend whom I have mentioned, addressed himself to my mother and me with great warmth and earnestness; "If your brother and your uncle," said he, "is safe, he certainly wishes you to be so too; but if he has perished, it was his desire, no doubt, that you might both survive him: why therefore do you delay your escape for a moment?" We could never think of our own safety, we said, while we were uncertain of his. Hereupon our friend left us, and withdrew with the utmost precipitation. Soon afterward, the cloud seemed to descend, and cover the whole ocean; as it certainly did the island of Capreae, and the promontory of Misenum. My mother strongly [urged] me to make my escape . . . as for herself, she said, her age and corpulency rendered all attempts of that sort impossible. . . . But I absolutely refused to leave her, and taking her by the hand, I led her on; she complied with great reluctance, and not without many reproaches to herself for retarding my flight.

The ashes now began to fall upon us, though in no great quantity. I turned my head and observed behind us a thick smoke, which came rolling after us like a torrent. I proposed, while we yet had any light, to turn out of the high road lest she should be pressed to death in the dark by the crowd that followed us. We had scarce stepped out of the path when darkness overspread us, not like that of a cloudy night, or when there is no moon, but of a room when it is all shut up and all the lights are extinct. Nothing then was to be heard but the shrieks of women, the screams of children and the cries of men; some calling for their children, others for their parents, others for their husbands, and only distinguishing each other by their voices; one lamenting his own fate, another that of his family; some wishing to die from the very fear of dying. . . . Among them were some who augmented the real terrors by imaginary ones, and made the freighted multitude believe that Misenum was actually in flames.

At length a glimmering light appeared, which we imagined to be rather the forerunner of an approaching burst of flames, as in truth it was, than the return of day. However, the fire fell at a distance from us; then again we were immersed in thick darkness, and a heavy shower of ashes rained upon us, which we were obliged every now and then to shake off, otherwise we should have been crushed and buried in the heap. . . . At last this dreadful darkness was dissipated by degrees, like a cloud of smoke; the real day returned, and soon the sun appeared, though very faintly, and as when an eclipse is coming on. Every object that presented itself to our eyes (which were extremely weakened) seemed changed, being covered over with white ashes, as with a deep snow. We returned to Misenum, where we refreshed ourselves as well as we could, and passed an anxious night between hope and fear, for the earthquake still continued, while several greatly excited people ran up and down, heightening their own and their friends' calamities by terrible predictions. . . .

Earthquake at Concepción, Chile, February 20, 1835

In his memoir *The Voyage of the* Beagle, Charles Darwin described the February 20, 1835, earthquake at Concepción, Chile and its results:

This day has been memorable in the annals of Valdavia, for the most severe earthquake experienced by the oldest inhabitant. I happened to be on shore, and was lying down in the wood to rest myself. It came on suddenly, and lasted two minutes, but the time appeared much longer. The rocking of the ground was very sensible. The undulations appeared to my companion and myself to come from due east, whilst others thought they proceeded from southwest: this shows how difficult it sometimes is to perceive the directions of the vibrations. There was no difficulty in standing upright, but the motion made me almost giddy: it was something like the movement of a vessel in a little cross-ripple, or still more like that felt by a person skating over thin ice, which bends under the weight of his body.

A bad earthquake at once destroys our oldest associations: the earth, the very emblem of solidity, has moved beneath our feet like a thin crust over a fluid;—one second of time has created in the mind a strange idea of insecurity, which hours of reflection would not have produced. In the forest, as a breeze moved the trees, I felt only the earth tremble, but saw no other effect. Captain FitzRoy and some officers were in town during the shock, and

there the scene was more striking; for although the houses, from being built of wood, did not fall, they were violently shaken, and the boards creaked and rattled together. The people rushed out of doors in the greatest alarm. It is these accompaniments that create the perfect horror of earthquakes, experienced by all who have thus seen, as well as felt, their effects. Within the forest it was a deeply interesting, but by no means an awe-exciting phenomenon. The tides were very curiously affected. The great shock took place at the time of low water; and an old woman who was on the beach told me that the water flowed very quickly, but not in great waves, to high-water mark, and then as quickly returned to its proper level; this was also evident by the line of wet sand. The same kind of quick but quiet movement in the tide happened a few years since at Chiloe, during a slight earthquake, and created much causeless alarm. In the course of the evening there were many weaker shocks, which seemed to produce in the harbor the most complicated currents, and some of great strength.

March 4th.—We entered the harbor of Concepcion. While the ship was beating up to the anchorage, I landed on the island of Quiriquina. The mayor-domo of the estate quickly rode down to tell me the terrible news of the great earthquake of the 20th:—"That not a house in Concepcion or Talcahuano (the port) was standing; that seventy villages were destroyed; and that a great wave had almost washed away the ruins of Talcahuano." Of this latter statement I soon saw abundant proofs—the whole coast being strewn over with timber and furniture as if a thousand ships had been wrecked. Besides chairs, tables, bookshelves, etc., in great numbers, there were several roofs of cottages, which had been transported almost whole. The storehouses at Talcahuano had been burst open, and great bags of cotton, yerba, and other valuable merchandise were scattered on the shore. During my walk round the island, I observed that numerous fragments of rock, which, from the marine productions adhering to them, must have been lying recently in deep water, had been cast up high on the beach; one of these was six feet long, three broad, and two thick.

The island itself as plainly showed the overwhelming power of the earthquake, as the beach did that of the consequent great wave. The ground in many parts was fissured in north and south lines, perhaps caused by the yielding of the parallel and steep sides of this narrow island. Some of the fissures near the cliffs were a yard wide. Many enormous masses had already fallen on the beach; and the inhabitants thought that when the rains commenced far greater slips would happen. The effect of the vibration on the hard primary slate, which composes the foundation of the island, was still more curious: the superficial parts of some narrow ridges were as completely shivered as if they had been blasted by gunpowder. This effect, which was rendered conspicuous by the fresh fractures and the displaced soil, must be confined to near the surface, for otherwise there would not exist a block of solid rock throughout Chile; nor is this improbable, as it is known that the surface of a vibrating body is affected differently from the central part. It is, perhaps, owing to this same reason, that earthquakes do not cause quite such terrific havoc within deep mines as would be expected. I believe this convulsion has been more effectual in lessening the size of the island of Quiriquina, than the ordinary wear-and-tear of the sea and weather during the course of a whole century.

The next day I landed at Talcahuano, and afterwards rode to Concepcion. Both towns presented the most awful yet interesting spectacle I ever beheld. To a person who had formerly known them, it possibly might have been still more impressive; for the ruins were so mingled together, and the whole scene possessed so little the air of a hospitable place, that it was scarcely possible to imagine its former condition. The earthquake commenced at half-past eleven o'clock in the forenoon. If it had happened in the middle of the night, the greater number of the inhabitants (which in this one province must amount to many thousands) must have perished, instead of less than a hundred; as it was, the invariable practice of running out of doors at the first trembling of the ground, alone saved them. In Concepcion each house, or row of houses, stood by itself, a heap or line of ruins; but in Talcahuano, owing to the great wave, little more than one layer of bricks, tiles and timber, with here and there part of a wall left standing, could be distinguished. From this circumstance Concepcion, although not so completely desolated, was a more terrible, and if I may so call it, picturesque sight. The first shock was very sudden. The mayor-domo at Quiriquina told me, that the first notice he received of it, was finding both the horse he rode and himself, rolling together on the ground. Rising up, he was again thrown down. He also told me that some cows which were standing on the steep side of the island were rolled into the sea. The great wave caused the destruction of many cattle; on one low island, near the head of the bay, seventy animals were washed off and drowned. It is generally thought that this has been the worst earthquake ever recorded in Chile; but as the very severe ones occur only after long intervals, this cannot easily be known; nor indeed would a much worse shock have made any difference, for the ruin was now complete. Innumerable small tremblings followed the great earthquake, and within the first twelve days no less than three hundred were counted.

After viewing Concepcion, I cannot understand how the greater number of inhabitants escaped unhurt. The houses in many parts fell outwards; thus forming in the middle of the streets little hillocks of brickwork and rubbish. Mr. Rouse, the English consul, told us that he was at breakfast when the first movement warned him to run out. He had scarcely reached the middle of the courtyard, when one side of his house came thundering down. He retained presence of mind to remember, that if he once got on top of that part which had already fallen, he would be safe. Not being able from the motion of the ground to stand, he crawled up on his hands and knees; and no sooner had he ascended this little eminence, than the other side of the house fell in, the great beams sweeping close in front of his head. With his eyes blinded, and his mouth choked with the cloud of dust which darkened the sky, at last he gained the street. As shock succeeded shock, at the interval of a few minutes, no one dared approach the shattered ruins; and no one knew whether his dearest friends and relations were not perishing from the want of help. Those who had saved any property were obliged to keep a constant watch, for thieves prowled about, and at each little trembling of the ground, with one hand they beat their breast and cried "misericorda!" and then with the other filched what they could from the ruins. The thatched roofs fell over the fires, and flames burst forth in all parts. Hundreds knew themselves ruined, and few had the means of providing food for the day.

Earthquakes alone are sufficient to destroy the prosperity of any country. If beneath England the now inert subterranean forces should exert those powers, which most assuredly in former geological ages they have exerted, how completely would the entire condition of the country be changed! What would become of the lofty houses, thickly packed cities, great manufactories, the beautiful public and private edifices? If the new period of disturbance were first to commence by some great earthquake in the dead of the night, how terrific would be the carnage! England would at once be bankrupt; all papers, records, and accounts would from that moment be lost. Government being unable to collect the taxes, and failing to maintain its authority, the hand of violence and rapine would remain uncontrolled. In every large town famine would go forth, pestilence and death following in its train.

Shortly after the shock, a great wave was seen from the distance of three or four miles, approaching in the middle of the bay with a smooth outline; but along the shore it tore up cottages and trees, as it swept onwards with irresistible force. At the head of the bay it broke in a fearful line of white breakers, which rushed up to a height of 23 vertical feet above the highest spring-tides. Their force must have been prodigious; for at the Fort a cannon with its carriage, estimated at four tons in weight, was moved 15 feet inwards. A schooner was left in the midst of the ruins, 200 yards from the beach. The first wave was followed by two others, which in their retreat carried away a vast wreck of floating objects. In one part of the bay, a ship was pitched high and dry on shore, was carried off, again driven on shore, and again carried off. In another part, two large vessels anchored near together were whirled about, and their cables were thrice wound round each other; though anchored at a depth of 36 feet, they were for some minutes aground. The great wave must have traveled slowly, for the inhabitants of Talcahuano had time to run up the hills behind the town; and some sailors pulled out seaward, trusting successfully to their boat riding securely over the swell, if they could reach it before it broke. One old woman with a little boy, four or five years old, ran into a boat, but there was nobody to row it out: the boat was consequently dashed against an anchor and cut in twain; the old woman was drowned, but the child was picked up some hours afterwards clinging to the wreck. Pools of saltwater were still standing amidst the ruins of the houses, and children, making boats with old tables and chairs, appeared as happy as their parents were miserable. It was, however, exceedingly interesting to observe, how much more active and cheerful all appeared than could have been expected. It was remarked with much truth, that from the destruction being universal, no one individual was humbled more than another, or could suspect his friends of coldness—that most grievous result of the loss of wealth. Mr. Rouse, and a large party whom he kindly took under his protection, lived for the first week in a garden beneath some apple-trees. At first they were as merry as if it had been a picnic; but soon afterward heavy rain caused much discomfort, for they were absolutely without shelter.

In Captain Fitzroy's excellent account of the earthquake, it is said that two explosions, one like a column of smoke and another like the blowing of a great whale, were seen in the bay. The water also appeared everywhere to be boiling; and it "became black, and exhaled a most disagreeable sulphurous smell." These latter circumstances were observed in the Bay of Valparaiso during the earthquake of 1822; they may, I think, be accounted for, by the disturbance of the mud at the bottom of the bay containing organic matter in decay. In the Bay of Callao, during a calm day, I noticed, that as the ship dragged her cable over the bottom, its course was marked by a line of bubbles. The lower orders in Talcahuano thought that the earthquake was caused by some old Indian women, who two years ago, being offended, stopped the volcano of Antuco. This silly belief is curious, because it shows that experience has taught them to observe, that there exists a relation between the suppressed action of the volcanos, and the trembling of the ground. It was necessary to apply the witchcraft to the point where their perception of cause and effect failed; and this was the closing of the volcanic vent. This belief is the more singular in this particular instance, because, according to Captain FitzRoy, there is reason to believe that Antuco was noways affected.

The town of Concepcion was built in the usual Spanish fashion, with all the streets running at right angles to each other; one set ranging S.W. by W, and the other set N.W. by N. The walls in the former direction certainly stood better than those in the latter; the greater number of the masses of brickwork were thrown down towards the N.E. Both these circumstances perfectly agree with the general idea, of the undulations having come from the S.W., in which quarter subterranean noises were also heard; for it is evident that the walls running S.W. and N.E. which presented their ends to the point whence the undulations came, would be much less likely to fall than those walls which, running N.W. and S.E., must in their whole lengths have been at the same instant thrown out of the perpendicular; for the undulations, coming from the S.W., must have extended in N.W. and S.E. waves, as they passed under the foundations. This may be illustrated by putting books edgeways on a carpet, and then . . . imitating the undulations of an earthquake; it will be found that they fall with more or less readiness, according as their direction more or less coincides with the line of the waves. The fissures in the ground generally, though not uniformly, extended in a S.E. and N.W. direction, and therefore corresponded to the lines of undulation or of principal flexure. Bearing in mind all these circumstances, which so clearly point to the S.W. as the chief focus of disturbance, it is a very interesting fact that the island of S. Maria, situated in that quarter, was, during the general uplifting of the land, raised to nearly three times the height of any other part of the coast.

The different resistance offered by the walls, according to their direction, was well exemplified in the case of the Cathedral. The side which fronted the N.E. presented a grand pile of ruins, in the midst of which door-cases and masses of timber stood up, as if floating in a stream. Some of the angular blocks of brickwork were of great dimensions; and they were rolled to a distance on the level plaza, like fragments of rock at the base of some high mountain. The side walls (running S.W. and N.E.), though exceedingly fractured, yet remained standing; but the vast buttresses (at right angles to them, and therefore parallel to the walls that fell) were in many cases cut clean off, as if by a chisel, and hurled to the ground. Some square ornaments on the coping of these same walls, were moved by the earthquake into a diagonal position. A similar circumstance was observed after an earthquake at Valparaiso, Calabria, and other places, including some of the ancient Greek temples. This twisting displacement, at first appears to indicate a vorticose movement beneath each point thus affected; but this is highly improbable. May it not be caused by a tendency in each stone to arrange itself in some particular position, with respect to the lines of vibration,—in a manner somewhat similar to pins on a sheet of paper when shaken? Generally speaking, arched doorways or windows stood much better than any other part of the buildings. Nevertheless, a poor, lame old man, who had been in the habit, during trifling shocks, of crawling to a certain doorway, was this time crushed to pieces.

I have not attempted to give any detailed description of the appearance of Concepcion, for I feel that it is quite impossible to convey the mingled feelings which I experienced. Several of the officers visited it before me, but their strongest language failed to give a just idea of the scene of desolation. It is a bitter and humiliating thing to see works, which have cost man so much time and labour, overthrown in one minute; yet compassion for the inhabitants was almost instantly banished, by the surprise in seeing a state of things produced in a moment of time, which one was accustomed to attribute to a succession of ages. In my opinion, we have scarcely beheld, since leaving England, any sight so deeply interesting.

In almost every severe earthquake, the neighbouring waters of the sea are said to have been greatly agitated. The disturbance seems generally, as in the case of Concepcion, to have been of two kinds: first, at the instant of the shock, the water swells up high on the beach with a gentle motion, and then as quietly retreats; secondly, some time afterwards, the whole body of the sea retires from the coast, and then returns in waves of overwhelming force. The first movement seems to be an immediate consequence of the earthquake affecting differently a fluid and a solid, so that their respective levels are slightly deranged: but the second case is a far more important phenomenon. During most earthquakes, and especially during those on the west coast of America, it is certain that the first great movement of the waters has been a retirement. Some authors have attempted to explain this, by supposing that the water retains its level, whilst the land oscillates upward; but surely the waters close to the land, even on a rather steep coast, would partake of the motion of the bottom: moreover . . . similar movements of the sea have occurred at islands far distant from the chief line of disturbance, as was the case with Juan Fernandez during this earthquake, and with Madeira during the famous Lisbon shock. I suspect (but the subject is a very obscure one) that a wave, however produced, first draws the water from the shore, on which it is advancing to break: I have observed that this happens with the little waves from the paddles of a steam-boat. It is remarkable that whilst Talcahuano and Callao (near Lima), both situated at the head of large shallow bays, have suffered during every severe earthquake from great waves, Valparaiso, seated close to the edge of profoundly deep water, has never been overwhelmed, though so often shaken by the severest shocks. From the great wave not immediately following the earthquake, but sometimes after the interval of even half an hour, and from distant lands being affected similarly with the coasts near the focus of the disturbance, it appears that the first wave rises in the offing; and as this is of general occurrence, the cause must be general: I suspect we must look to the line, where the less disturbed waters of the deep ocean join the water nearer the coast, which has partaken of the movements of the land, as the place where the great wave is first generated; it would also appear that the wave is larger or smaller, according to the extent of shoal water which has been agitated together with the bottom on which it rested.

The most remarkable effect of this earthquake was the permanent elevation of the land; it would probably be far more correct to speak of it as the cause. There can be no doubt that the land round the Bay of Concepcion was upraised two or three feet; but it deserves notice, that owing to the wave having obliterated the old lines of tidal action on the sloping sandy shores, I could discover no evidence of this fact, except in the united testimony of the inhabitants, that one little rocky shoal, now exposed, was formerly covered with water. At the island of S. Maria (about thirty miles distant) the elevation was greater on one part, Captain FitzRoy found beds of putrid mussel-shells still adhering to the rocks, ten feet above highwater mark: the inhabitants had formerly dived at lower-water spring-tides for these shells. The elevation of this province is particularly interesting, from its having been the theater of several other violent earthquakes, and from the vast numbers of seashells scattered over the land, up to a height of certainly 600, and I believe, of 1,000 feet. At Valparaiso, as I have remarked, similar shells are found at the height of 1,300 feet: it is hardly possible to doubt that this great elevation has been effected by successive small uprisings, such as that which accompanied or caused the earthquake of this year, and likewise by an insensibly slow rise, which is certainly in progress on some parts of this coast.

The island of Juan Fernandez, 360 miles to the N.E., was, at the time of the great shock of the 20th, violently shaken, so that the trees beat against each other, and a volcano burst forth under water close to the shore: these facts are remarkable because this island, during the earth-

quake of 1751, was then also affected more violently than other places at an equal distance from Concepcion, and this seems to show some subterranean connection between these two points. Chiloe, about 340 miles southward of Concepcion, appears to have been shaken more strongly than the intermediate district of Valdivia, where the volcano of Villarica was noways affected, whilst in the Cordillera in front of Chiloe, two of the volcanos burst forth at the same instant into violent action. These two volcanos, and some neighboring ones, continued for a long time in eruption, and ten months afterwards were again influenced by an earthquake at Concepcion. Some men, cutting wood near the base of one of these volcanos, did not perceive the shock of the 20th, although the whole surrounding Province was then trembling; here we have an eruption relieving and taking the place of an earthquake. . . . Two years and three-quarters afterwards, Valdivia and Chiloe were again shaken, more violently than on the 20th, and an island in the Chonos Archipelago was permanently elevated more than eight feet. It will give a better idea of the scale of these phenomena, if . . . we suppose them to have taken place at corresponding distances in Europe:— then would the land from the North Sea to the Mediterranean have been violently shaken, and at the same instant of time a large tract of the eastern coast of England would have been permanently elevated, together with some outlying islands,—a train of volcanos on the coast of Holland would have burst forth in action, and an eruption taken place at the bottom of the sea, near the northern extremity of Ireland—and lastly, the ancient vents of Auvergne, Cantal, and Mont d'Or would each have sent up to the sky a dark column of smoke, and have long remained in fierce action. Two years and three-quarters afterwards, France, from its center to the English Channel, would have been again desolated by an earthquake, and an island permanently upraised in the Mediterranean.

The space, from under which volcanic matter on the 20th was actually erupted, is 720 miles in one line, and 400 miles in another line at right angles to the first: hence, in all probability, a subterranean lake of lava is here stretched out, of nearly double the area of the Black Sea. From the intimate and complicated manner in which the elevatory and eruptive forces were shown to be connected during this train of phenomena, we may confidently come to the conclusion, that the forces which slowly and by little starts uplift continents, and those which at successive periods pour forth volcanic matter from open orifices, are identical. From many reasons, I believe that the frequent quakings of the earth on this line of coast are caused by the rending of the strata, necessarily consequent on the tension of the land when upraised, and their injection by fluidified rock. This rending and injection would, if repeated often enough (and we know that earthquakes repeatedly affect the same areas in the same manner), form a chain of hills;—and the linear island of S. Mary, which was upraised thrice the height of the neighbouring country, seems to be undergoing this process. I believe that the solid axis of a mountain differs in its manner of formation from a volcanic hill only in the molten stone having been repeatedly injected, instead of having been repeatedly ejected. . . .

Eruption of Mount Pelée, Martinique, May 8, 1902

Many firsthand accounts of the eruption of Mount Pelée and the destruction of Saint Pierre and vicinity were recorded, despite the virtually complete loss of life on shore. One of the most comprehensive and vivid accounts of the catastrophe comes from Comte de Fitz-James, a French traveler, who with his companion Baron Fontenilliat witnessed the destruction of Saint Pierre from the relative safety of a boat in the harbor:

From the depths of the earth came rumblings, an awful music which cannot be described. I called my companion's name, and my voice echoed back at me from a score of angles. All the air was filled with the acrid vapors that had belched from the mouth of the volcano. . . .

From a boat in the roadstead. . . I witnessed the cataclysm that came upon the city. We saw the shipping destroyed by a breath of fire. We saw the cable ship *Grappler* keel over under the whirlwind, and sink as through drawn down into the waters of the harbor by some force from below. The *Roraima* was overcome and burned at anchor. The *Roddam*. . . was able to escape like a stricken moth which crawls from a flame that has burned its wings. . . .

Our own danger was great, and had it not been for the bravery and courage of the Baron I would have perished. . . . I was stunned, unable to lift a hand to assist myself. Baron de Fontenilliat dragged me from the boat into the water, where he supported me. . . .

[Before the eruption, it] was such a morning as . . . is almost impossible to describe. Low hanging clouds gave the scene a dismal appearance, and this was heightened by the fine volcanic dust which filled the atmosphere, making respiration difficult. This dust was next to impalpable. It could not be seen as it floated in the air, but it settled so rapidly that my hand, resting upon the edge of the boat, was covered completely in less than three minutes.

As we made our way across the water we more than half faced Mont Pelee, which was throwing off a heavy cloud of smoke, steam and ashes. No flames were to be seen. On shore the inhabitants were making their way about the waterfront. The city was to our right. Small craft plied about the harbor, some trading with the ships that were at anchor, while in some fishermen were going out to the fishing grounds, just off Carbet. . . .

[The] calm of that morning was almost abnormal. Not a ripple was to be seen upon the face of the sea. Not a breath of air was stirring, which made it more difficult for us to breathe. . . .

The rumblings from the bowels of the mountain were majestic in tone. I cannot tell you just how they sounded, but perhaps you can imagine a mighty hand playing upon the strings of a harp greater than all the world. The notes produced were deep and full of threatenings. There was a jarring sensation, and every now and then there was a commotion of the waters that caused a swell without making the surface break.

Out from the shore put a small launch carrying the pennant of Governor Mouttet. The Governor at the last moment had realized that the situation was filled with a terrible danger. He was attempting to escape with his family and a few friends. I had commented to Baron de Fontenilliat upon the appearance of the Governor's craft. . . . [The Governor, as evidence proved, was too late in his attempt at flight.]

While we were talking there came an explosion that . . . I can liken only to a shot from a mammoth cannon. The breath of fire swept down upon the city and waterfront with all of the force that could have been given to it by such a cannon. Of this comparison I shall have more to say later. For the present it will do to add that the explosion was without warning and that the effect was instantaneous. Cinders were shot into our [faces] with stinging effect.

The air was filled with flame. Involuntarily we raised our hands to protect our faces. I noted the same gesture when I saw the bodies of victims on shore; arms had been raised and the hands were extended with palms outward, a gesture that in a peculiar manner indicated dread and horror.

[When] the frightful explosion came, our two boatmen were either thrown from the boat or with a quick impulse they sprang overboard. It was the one thing [they could] do to save their lives; but . . . they lost their presence of mind and, instead of staying by the side of the boat, they swam away in the direction of Precheur, which we were approaching when the [explosion] came. It was impossible for them to land at Precheur, so they were compelled to put back. They then struck out across the bay, evidently hoping to reach Carbet. We saw neither of them again, and I have no doubt they were drowned.

My brave companion . . . sprang into the water, and when he saw that I did not move he reached up and catching me by the shoulder, dragged me from the boat. I was stunned at first, and, though it was not a physical injury, I could not move of my own volition until the cold water restored my senses. It was thus that we could see all that happened about us.

The *Grappler* rushed through the water as far as her anchor-cable would permit. Then she seemed to rise by the bow, and when she settled back she sank almost before the force of the explosion had spent itself.

The *Roraima* was all a mass of flames for several seconds. We could see the poor wretches aboard . . . her rushing about in a vain attempt to escape from the fire that enveloped them. Captain Muggah—or, at least, I suppose that it was he—made an attempt to give orders to the . . . crew. Then he staggered to the railing and fell overboard.

The *Roddam* was also overcome. Her gangway was over the side. Her upper works were wrecked, but by heroic effort those on board were able to let slip the anchor chain, and, after many attempts, the ship began to move. She . . . crawled away. It was a splendid display of courage.

At least three hours elapsed after the explosion before the *Roddam* cleared the harbor.

On shore all was aflame. The city burned with a terrible roar. We realized that the inhabitants had all died, as not one was to be seen making an attempt to escape. Not a cry was heard save from the ships that were in the harbor.

Our own condition was desperate in the extreme. The heat was intense. We were able to keep our faces above the surface of the water for a second at a time at the most. We would take a mouthful of air and then sink into the water to stay there until forced to come to the surface again. This lasted only about three minutes. After that we were able to float by the side of the boat, dipping only occasionally.

The water began to get so warm that I feared we had escaped roasting only to be boiled to death. In reality the water did not get so warm as to be uncomfortable. [The water] at the surface was many degrees warmer than that a foot below.

. . . St. Pierre was mantled by a dense black cloud. Our eyes could not penetrate it, but it lifted a few seconds, revealing below it a second cloud, absolutely distinct from it. The second cloud was yellow, apparently made up of sulphurous gases. It lifted as did the first, both rising like blankets, and in a similar manner they floated away. Then, as the yellow cloud lifted from the earth, we saw the flames devouring the city, from which all show of life had disappeared. . . .

When we could sustain the heat that filled the air we clambered into the boat and rowed back to Carbet. The *Roddam* had just gone out from the harbor, the *Roraima* was a smoking wreck, the *Grappler* had disappeared entirely, and little was to be seen of the other craft.

At Carbet we found the village absolutely deserted. Two portions of it had been ruined. [The portion by] the water's edge had been swept by the great wave which followed the explosion. . . . [The] wave . . . was of terrific force, and it added to the confusion all along the shore. Part of Carbet had been struck by the wave of fire from the volcano, but the greater portion of the village was left uninjured.

When we got ashore we called aloud, and only the echo of our voices answered us. Our fear was great, but we did not know which way to turn, and had it been our one thought to escape we would not have known how to do so. It was about one o'clock in the afternoon when we reached shore. Our weariness was beyond description. Sleep was the one thing that I wanted, but I overcame the desire and, with Baron de Fontenilliat, set off to make our way to St. Pierre, hoping that we might still render some assistance to the injured.

Not knowing the paths, we attempted to enter the city from the direction traveled by the blast of the volcano. That brought us to the flames and we were driven back. Then we went further into the country, and so happened to meet two soldiers who . . . had been in camp at Colson, far back from St. Pierre, but, on leave, had wandered in toward the city. They heard the explosion and rushed down from the hills to give aid where it was needed. When they went in through the streets it was at the risk of their lives. They were the only ones who ventured into St. Pierre that afternoon. They came upon a sailor so injured that he could not move. Picking him up, they carried him back out of the danger zone. . . .

Again entering the city, [the soldiers] found five women in a hut. They were much injured, but were not dead. The soldiers gave them drink and put food within their reach, and then left them, promising to return with assistance as soon as possible. . . .

Now, to show the folly of those upon whom responsibility fell in that hour of terrible disaster, I may say that when those two soldiers reached their camp they were sent to the guard-house for having remained away after hours. They told of the five suffering women, and their officer insisted that the tale had been arranged by them for the purpose of escaping punishment. They were kept under guard all Thursday night and all of the next day and the following night.

During those thirty-six hours the two soldiers made no complaint of their own treatment, but they continued to beg that assistance be sent to the women whom they had left so badly injured. Finally their plea prevailed, and on Saturday they were permitted to lead a rescue party to St. Pierre. Then their story was fully verified. One of the women was still alive. She told how . . . her four friends . . . had died late Friday night. She was taken to Fort de France, where she died a few days later. . . .

Our shoes were burned to a crisp, but we plodded about those hills as long as we were able to move. Then we returned to Carbet, and remained there that night. . . .

It is impossible to describe even in the most faint manner the horror of St. Pierre. There were some things that can be made clear, but many more that cannot be explained by anything known to human reason.

It happened that one of the first bodies found . . . [when] we entered St. Pierre on Monday was that of a pretty little girl about four years old. She sat in a lifelike position by the side of a box containing her toys. But how shall we explain the fact that the house in which she was found was in absolute ruins, and, instead of being under the debris, the body was on top of it all? It was as though the little girl and her box of toys had been lifted into the air, and, after the building had fallen into ruins, had been dropped back to earth.

So it was in the streets. The explosion happened just before eight o'clock. It was a feast day. Mass was called for eight o'clock, and many were on their way to the cathedral. All of these had been lifted into the air, and after the ruins had fallen the bodies dropped back. . . .

We saw great stones that seemed to be marvels of strength, but when touched with the toe of a boot they crumbled into impalpable dust. I picked up a bar of iron. It was about an inch and a half thick and three feet long. It had been manufactured square and then twisted so as to give it greater strength. The fire that came down from Mont Pelee had taken from the iron all of its strength and had left it so that when I twisted it, it fell into filaments, like so much broom straw.

Back of the cathedral was the savannah. Great trees had been torn up by the roots, leaving holes twenty feet deep and thirty to forty feet across. Then these holes had been filled by the ashes that poured down from the volcano. Trees were cut off as though by a mighty knife in the hands of a giant reaper. Everywhere were banks of cinders and ashes.

When the Baron and I first went into the ruined city we were too awe-struck to speak. Then . . . I called to him. His name echoed back to us from a score of standing walls. All about us were bodies. On few faces was to be seen the peace which I have seen mentioned by others. I believe that almost all had time to realize what was upon them, but they did not have time to suffer. Their arms were outstretched. . . The hands were open and the fingers were spread. It was a common gesture, and I believe that it was the act of men and women who threw up their arms to ward off a blow which they knew was descending upon them. . . .

I know that the explosion of Mont Pelee was not accompanied by anything like an earthquake, for . . . when we entered St. Pierre we found the fountains all flowing, just as though nothing had happened. They continued to flow, and are flowing still, unless destroyed by the later explosions.

There was no flow of lava. It was all ashes, dust, gas and mud. . . .

One vessel caught in the eruption was the steamship *Roraima,* of the Quebec Steamship Company. The ship arrived in St. Pierre around 7 A.M. on the morning of the eruption and had trouble making its way into the port because of darkness and a heavy ashfall. Assistant Purser Thompson's account of the explosion of St. Pierre and the destruction of St. Pierre follows:

I saw St. Pierre destroyed. It was blotted out by one great flash of fire. Nearly 40,000 persons were all killed at once. Out of 18 vessels lying in the roads [that is, the harbor] only one, the British steamer *Roddam,* escaped, and she, I hear, lost more than half on board. It was a dying crew that took her out.

Our boat . . . arrived at St. Pierre early Thursday morning. For hours before we entered the roadstead we could see flames and smoke rising from Mont Pelee. No one on board had any idea of danger. Captain G. T. Muggah was on the bridge, and all hands got on deck to see the show.

The spectacle was magnificent. As we approached St. Pierre we could distinguish the rolling and leaping of the red flames that belched from the mountain in huge volumes and gushed high into the sky. Enormous clouds of black smoke hung over the volcano.

When we anchored at St. Pierre I noticed the cable steamer *Grappler,* the *Roddam,* three or four American schooners and a number of Italian and Norwegian barks. The flames were then spurting straight up in the air, now and then waving to one side or the other for a moment and again leaping suddenly higher up.

There was a constant muffled roar. It was like the biggest oil refinery in the world burning up on the mountain top. There was a tremendous explosion about 7:45 o'clock, soon after we got in. The mountain was blown to pieces. There was no warning. The side of the mountain was ripped out, and there was hurled straight toward us a solid wall of flames. It sounded like thousands of cannon.

The wave of fire was on us and over us like a lightning flash. It was like a hurricane of fire. It saw it strike the . . . *Grappler* broadside on and capsize her. From end to end she burst into flames and then sank. The fire rolled in mass straight down upon St. Pierre and the shipping. The town vanished before our eyes and the air grew stifling hot, and we were in the thick of it.

Wherever the mass of fire struck the sea the water boiled and sent up vast clouds of steam. The sea was torn into huge whirlpools that careened toward the open sea. One of these horrible hot whirlpools swung under the *Roraima* and pulled her down on her beam ends with the suction. She careened way over to port, and then the fire hurricane from the volcano smashed her, and over she went on the opposite side. The fire wave swept off the mass and smokestack as if they were cut with a knife.

Captain Muggah was the only one on deck not killed outright. He was caught by the fire wave and terribly burned. He yelled to get up the anchor, but, before two fathoms were heaved in the *Roraima* was almost upset by the boiling whirlpool, and the fire wave had thrown her down on her beam ends to starboard. Captain Muggah was overcome by the flames. He fell unconscious from the bridge and toppled overboard.

The blast of fire from the volcano lasted only a few minutes. It shriveled and set fire to everything it touched. Thousands of casks of rum were stored in St. Pierre, and these were exploded by the terrific heat. The burning rum ran in streams down every street and out to the sea. This blazing rum set fire to the *Roraima* several times. Before the volcano burst the landings of St. Pierre were crowded with people. After the explosion not one living being was seen on land. Only 25 of those on the *Roraima* out of 68 were left after the first flash.

The French cruiser *Suchet* came in and took us off at 2 p.m. She remained nearby, helping all she could, until 5 o'clock, then went to Fort de France with all the people she had rescued. At that time it looked as if the entire north end of the island was on fire.

Another crew member of the *Roraima* witnessed a horrible spectacle on deck:

Hearing a tremendous report and seeing the ashes falling thicker, I dived into a room, dragging with me Samuel Thomas, a gangway man and . . . [shut] the door tightly. Shortly after I heard a voice, which I recognized as that of the chief mate, Mr. Scott. Opening the door with great caution, I drew him in. The nose of Thomas was burned by the intense heat.

We three and Thompson, the assistant purser . . . were the only persons who escaped practically uninjured. The heat being unbearable, I emerged in a few moments, and the scene that presented itself to my eyes baffles description. All around on the deck were the dead and dying covered with boiling mud. There they lay, men, women and little children, and the appeals of the latter for water were heart-rending. When water was given them they could not swallow it, owing to their throats being filled with ashes or burnt with the heated air.

The ship was burning aft, and I jumped overboard, the sea being intensely hot. I was at once swept seaward by a tidal wave, but, the sea receding a considerable distance, the return wave washed me against an upturned sloop to which I clung. I was joined by a man so dreadfully burned and disfigured as to be unrecognizable. Afterwards I found he was the captain of the *Roraima,* Captain Muggah. He was in dreadful agony, begging piteously to be put on board his ship.

Picking up some wreckage which contained bedding and a tool chest, I, with the help of five others who had joined me on the wreck, constructed a rude raft, on which we placed the captain.... Seeing the *Roddam,* which arrived in port shortly after we anchored, making for the *Roraima,* I said goodbye to the captain and swam back to the *Roraima.* [The captain's body was recovered later.]

The *Roddam,* however, burst into flames and put to sea. I reached the *Roraima* at about half-past 2, and was afterwards taken off by a boat from the French warship *Suchet.* Twenty-four others with myself were taken on to Fort de France. Three of these died before reaching port. A number of others have since died.

Ellery Scott, mate of the *Roraima,* recounted his view of the eruption and of the last moments of Captain Muggah:

All hands had had breakfast. I was standing on the fo'c's'l head trying to make out the marks on the pipes of a ship 'way out and heading for St. Lucia. I wasn't looking at the mountain at all. But I guess the captain was, for he was on the bridge, and the last time I heard him speak was when he shouted, "Heave up, Mr. Scott; heave up [raise the anchor]." I gave the order to the men, and I think some of them did jump to get the anchor up, but nobody knows what really happened for the next fifteen minutes. I turned around toward the captain and then I saw the mountain.

Did you ever see the tide come into the Bay of Fundy? It doesn't sneak in a little at a time as it does 'round here. It rolls in, in waves. That's the way the cloud of fire and mud and white-hot stones rolled down from that volcano over the town and over the ships. It was on us in almost no time, but I saw it, and in the same glance I saw our captain bracing to meet it on the bridge. He was facing the fire cloud with both hands gripped hard to the bridge rail, his legs apart and his knees braced back stiff. I've seen him brace himself that same way many a time in a rough sea with the spray going mast-head high and green water pouring along the decks.

I saw the captain ... at the same time I saw that ruin coming down on us. I don't know why, but that last glimpse of poor Muggah on his bridge will stay with me just as long as I remember St. Pierre, and that will be long enough.

In another instant it was all over for him. As I was looking at him he was all ablaze. He reeled and fell on the bridge with his face toward me. His mustache and eyebrows were gone in a jiffy. His hat had gone, and his hair was aflame, and so were his clothes from head to foot. I knew he was conscious when he fell, by the look in his eyes, but he didn't make a sound.

That all happened a long way inside of half a minute; then something new happened. When the wave of fire was going over us, a tidal wave of the sea came out from the shore and did the rest. That wall of rushing water was so high and so solid that it seemed to rise up and join the smoke and flame above. For an instant we could see nothing but the water and the flame.

That tidal wave picked the ship up like a canoe and then smashed her. After one list to starboard the ship righted, but the masts, the bridge, the funnel and all the upper works had gone overboard.

I had saved myself from fire by jamming a metal ventilator cover over my head and jumping from the fo'c's'l head. Two St. Kitts [natives] saved me from the water by grabbing my legs and pulling me down into the fo'c's'l after them. Before I could get up, three men tumbled in on top of me. Two of them were dead.

Captain Muggah went overboard, still clinging to the fragments of his wrecked bridge. Daniel Taylor, the ship's cooper, and a Kitts native jumped overboard to save him. Taylor managed to push the captain on to a hatch that had floated off from us, and then they swam back to the ship for more assistance, but nothing could be done for the captain. Taylor wasn't sure he was alive. The last we saw of him or his dead body, it was drifting shoreward on that hatch ...

[After] staying in the fo'c's'l about twenty minutes, I went out on deck. There were just four of us left aboard who could do anything. ... It was still raining fire and hot rocks, and you could hardly see a ship's length for dust and ashes, but we could stand that. There were burning men and some women and two or three children lying around the deck. Not just burned, but burning, then, when we got to them. More than half the ship's company had been killed in that first rush of flame. Some had rolled overboard when the tidal wave came, and we never saw so much as their bodies. The cook was burned to death in his galley. He had been paring potatoes for dinner, and what was left of his right hand held the shank of his potato knife. The wooden handle was in ashes. All that happened to [the] man in less than a minute. The donkey engineman was killed on deck sitting in front of his boiler. We found parts of some bodies—a hand, or an arm or a leg. Below decks there were some twenty alive.

The ship was on fire, of course, what was left of it. The stumps of both masts were blazing. Aft she was like a furnace, but forward the flames had not got below deck, so we four carried those who were still alive on deck into the fo'c's'l. All of them were burned and most of them were half strangled. ...

My own son's gone, too. It had been his trick at lookout ahead during the dog watch that morning, when we were making for St. Pierre, so I supposed at first when the fire struck us that he was asleep in his bunk and safe. But he wasn't. Nobody could tell me where he was. I don't know whether he was burned to death or rolled overboard and drowned. . . .

After getting all hands that had any life left in them below and [attending] to the best we could, the four of us that were left halfway ship-shape started in to fight the fire. . . . Thanks to the tidal wave that cleared our decks, there wasn't much left to burn, so we got the fire down so's we could live on board with it several hours more, and then [we] turned to knock a raft together out of what timber and truck we could find below. Our boats had gone overboard with the masts and funnel.

We made that raft for something over thirty that were alive. . . . But we did not have to risk the raft, for . . . the *Suchet* came along and took us all off. We thought for a minute just after we were wrecked that we were to get help from a ship that passed us . . . but she kept on. We learned afterward that she was the *Roddam*.

One of the strangest experiences associated with this eruption was reported by Captain Eric Lillienskjold of the Danish steamship *Nordby,* which was several hundred miles from the island:

We were plodding along slowly that day. About noon I took the bridge to make an observation. It seemed to be hotter than ordinary. I shed my coat and vest and got into what little shade there was. As I worked it grew hotter and hotter, I didn't know what to make of it. Along about 2 o'clock in the afternoon it was so hot that all hands got to talking about it. We reckoned that something queer was coming off, but none of us could explain what it was. You could almost see the pitch softening in the seams.

Then, as quick as you could toss a biscuit over its rail, the *Nordby* dropped—regularly dropped—three or four feet down into the sea. No sooner did it do this than big waves, that looked like they were coming from all directions at once, began to smash against our sides. This was queerer yet, because the water a minute before was as smooth as I ever saw it. I had all hands piped on deck, and we battened down everything loose to make ready for a storm. And we got it all right—the strangest storm you ever heard tell of.

There was something wrong with the sun that afternoon. It grew red and then dark red and then, about a quarter after 2, it went out of sight altogether. The day got so dark that you couldn't see half a ship's length ahead of you. We got our lamps going, and put on our oilskins, ready for a hurricane. All of a sudden there came a sheet of lightning that showed up the whole tumbling sea for miles and miles. We sort of ducked, expecting an awful crash of thunder, but it didn't come. There was no sound except the big waves pounding against our sides. There wasn't a breath of wind.

Well, sir, at that minute there began [the] most exciting time I've ever been through, and I've been on every sea on the map for twenty-five years. Every second there'd be waves 15 or 20 feet high, belting us head-on, stern-on and broadside, all at once. We could see them coming, for without any stop at all, flash after flash of lightning was blazing all about us.

Something else we could see, too. Sharks! There were hundreds of them on all sides, jumping up and down in the water. And sea birds! A flock of them, squawking and crying, made for our rigging and perched there. They seemed like they were scared to death. But the queerest part of it all was the water itself. It was hot—not so hot that our feet could not stand it when it washed over the deck, but hot enough to make us think that it had been heated by some kind of a fire.

Well, that sort of thing went on hour after hour. The waves, the lightning, the hot water and the sharks, and all the rest of the odd things happening, frightened the crew out of their wits . . . Mighty strange things happen on the sea, but this topped them all.

I kept to the bridge all night. When the first hour of morning came, the storm was still going on. We were all pretty much tired out by that time, but there was no such thing as trying to sleep. The waves were still us around, and we didn't know whether we were one mile or a thousand miles from shore. At two o'clock in the morning all the queer going on stopped just the way they began—all of a sudden. We lay to until daylight; then we took our reckonings and started off again. We were about 700 miles off Cape Henlopen. . . . None of us was hurt, and the old *Nordby* herself pulled through all right, but I'd sooner stay ashore than see waves without wind and lightning without thunder.

Captain Freeman of the *Roddam* recorded his experiences of the eruption and its aftermath:

I went to anchorage between 7 and 8 and had hardly moored when the side of the volcano opened out with a terrible explosion. A wall of fire swept over the town and the bay. The *Roddam* was struck boardside by the burning mass. The shock to the ship was terrible, nearly capsizing her.

Hearing the awful report of the explosion and seeing the great wall of flames approaching the steamer, those on deck sought shelter wherever it was possible, jumping into the cabin, the forecastle and even into the hold. I was in the chart room, but the burning embers were borne by so swift a movement of the air that they were swept in through the door and port holes, suffocating and scorching me badly. I was terribly burned by these embers about the face and hands, but managed to reach the deck. Then, as soon as it was possible, I mustered the few survivors who seemed able to move, ordered them to slip the anchor, leaped for the bridge and [rang] the engine for full speed astern. The second and the third engineer and a fireman were on watch below and so escaped injury . . . but the men on deck could not work the steering gear because it was jammed by the debris from the volcano. We accordingly went ahead and astern until the gear was free, but in this running backward and forward it was two hours after the first shock before we were clear of the bay.

One of the most terrifying conditions was that, the atmosphere being charged with ashes, it was totally dark. The sun was completely obscured, and the air was only illuminated by the flames from the volcano and those of the burning town and shipping. It seems small to say that the scene was terrifying in the extreme. As we backed out we passed close to the *Roraima,* which was one mass of blaze. The steam was rushing from the engine room, and the screams of those on board were terrible to hear. The cries for help were all in vain, for I could do nothing but save my own ship. When I last saw the *Roraima* she was settling down by the stern. . . .

When the *Roddam* was safely out of the harbor of St. Pierre, with its desolations and horrors, I made for St. Lucia. Arriving there, and when the ship was safe, I mustered the survivors as well as I was able and searched for the dead and injured. Some I found in the saloon where they had . . . sought for safety, but the cabins were full of burning embers that had blown in through the port holes. Through these the fire swept as through funnels and burned the victims where they lay or stood, leaving a circular imprint of scorched and burned flesh. I brought ten on deck who were thus burned; two of them were dead, the others survived, although in a dreadful state of torture from their burns. Their screams of agony were heartrending. . . . The ship was covered from stem to stem with powdered lava, which retained its heat for hours after it had fallen. In many cases it was practically incandescent, and to move about the deck in this burning mass was not only difficult but absolutely perilous. I am only now able to begin thoroughly to clear and search the ship for any damage done by this volcanic rain, and to see if there are any corpses in out-of-the-way places. For instance, this morning, I found one body in the peak of the forecastle. The body was horribly burned and the sailor had evidently crept in there in his agony to die.

On the arrival of the *Roddam* at St. Lucia the ship presented an appalling appearance. Dead and calcined bodies lay about the deck, which was also crowded with injured, helpless and suffering people. . . . The woodwork of the cabins and bridge and everything inflammable on deck were constantly igniting, and it was with great difficulty that we few survivors managed to keep the flames down. My ropes, awnings, tarpaulins were completely burned up.

I witnessed the entire destruction of St. Pierre. The flames enveloped the town in every quarter with such rapidity that it was impossible that any person could be saved. As I have said, the day was suddenly turned to night, but I could distinguish by the light of the burning town people distractedly running about on the beach. The burning buildings stood out from the surrounding darkness like black shadows. All this time the mountain was roaring and shaking, and in the intervals between these terrifying sounds I could hear the cries of despair and agony from the thousands who were perishing. These cries added to the terror of the scene, but it is impossible to describe its horror or the dreadful sensations it produced. It was like witnessing the end of the world.

Captain Cantell of the British steamship *Etona* visited the *Roddam* at Saint Lucia on May 11 and described the damage to the vessel afterward:

The *Roddam* was covered with a mass of fine bluish gray dust or ashes of cementlike appearance. In some places it lay two feet deep on the decks. This matter had fallen in a red-hot state all over the steamer, setting fire to everything it struck that was burnable, and, when it fell on the men on board, burning off limbs and large pieces of flesh. This was shown by finding portions of human flesh when the decks were cleared of the debris. The rigging, ropes, tarpaulins, sails, awnings, etc., were charred or burned, and most of the upper stanchions and spars were swept overboard or destroyed by fire. Skylights were smashed and cabins were filled with volcanic dust. The scene of ruin was deplorable.

The captain, though suffering the greatest agony, succeeded in navigating his vessel to the port of Castries, St. Lucia, with 18 dead bodies on the deck and human limbs scattered about. A sailor stood by constantly wiping the captain's injured eyes. . . .

Captain Cantell had witnessed the eruption from a relatively safe distance:

The weather was clear and we had a fine view, but the old outlines of St. Pierre were not recognizable. Everything was a mass of blue lava, and the formation of the land itself seemed to have changed. When we were about eight miles off the northern end of the island Mount Pelee began to belch a second time. Clouds of smoke and lava shot into the air and spread all over the sea, darkening the sun. Our decks in a few minutes were covered with a substance that looked like sand dyed a bluish tint, and which smelled like phosphorus. . . .

We were about four miles off the northern end of the island when suddenly there shot up in the air to a tremendous height a column of smoke. The sky darkened and the smoke seemed to swirl down upon us. In fact, it spread all around, darkening the atmosphere as far as we could see. I called Chief Engineer Farrish to the deck.

"Do you see that over there?" I asked, pointing to the eruption. . . .

"Well, Farrish, rush your engines as they have never been rushed before," I said to him. . . .

We began to cut through the water at almost twelve knots. Ordinarily we make ten knots. We could see no more of the land contour, but everything seemed to have been enveloped in a great cloud. There was no fire visible, but the lava dust rained down upon us steadily. In less than an hour there were two inches of it upon our deck.

The air smelled like phosphorus. No one dared look up to try to locate the sun, because one's eyes would fill with lava dust. Some of the blue lava dust is sticking to our mast yet, although we have swabbed decks and rigging again and again to be clear of it.

After little more than an hour's fast running we saw daylight ahead and began to breathe easier. . . .

Eruption of Soufrière, Saint Vincent, May 7, 1902

Following are several eyewitness accounts:

I was fishing at some distance from the shore when my boatman said to me, "Look at the Soufrière, sir. It is smoking!"

From the top of the cone, reaching far up into the heavens, a dark column of smoke arose, while the mouth of the crater itself glowed like a gigantic forge belching a huge jet of yellow flame. The mass of smoke spread out into branshes extending for miles, and clouds of sulphurous vapor, overflowing, as it were, the bowl of the crater, began to roll down the mountain slopes.

We reached shore and started to run for our lives. We were soon enveloped in impenetrable darkness, and I was unable to distinguish the white shirt of my boatman at a yard's distance. But as he knew every inch of the ground, I held on to a stick he had, and so we stumbled on until we reached a place of safety. The incessant roar of the volcano, the rumbling of the thunder, the flashes of the lightning added to the terrific grandeur of the scene. At last we emerged from the pall of death, half suffocated, and with our temples throbbing as if they were going to burst.

* * *

We [a group of several persons who were rescued from a house] heard the mountain roaring the whole morning, but we thought it would pass off, and we did not like to abandon our homes, so we chanced it. About half-past one it began to rain pebbles and stones, some of which were alight; but then, although we were afraid, we could not leave. The big explosion must have taken place at half-past two o'clock. There was fire all around me, and I could not breathe. My hands and feet got burned, but I managed to reach the house where the others were.

In two hours everything was over, although pebbles and dust fell for a long time after. My burns got so painful and stiff that I could not move. We remained until Sunday morning without food or water. Five persons died, and as none of us could throw the bodies out, or even move, we had to lie alongside the bodies until we were rescued.

* * *

From Sunday night, May 4, the heat had been oppressive. Never had I experienced such heat before. It was with the utmost difficulty one could breathe, and to sleep was impossible . . .

On Tuesday I learned . . . that the Chateau Belaire side of the mountain was showing signs of activity. On Wednesday morning, between nine and ten o'clock, the lightning and thunder began. Such lightning and such thunder! Oh, it is terrible to remember, and thrice terrible it was to behold! Blinding flashes that zigzagged with hissing fury and a lurid light ominous of destruction. . . .

In the meantime some fisher girls who came down from the mountain said they had observed the water in the mountain lake to be boiling rapidly and the grass in the vicinity to be torn up. Then, you will understand, I got anxious. The storm grew in fury. The thunder became louder and louder. . . .

Amid the crashing thunder peals and the dreadful lightning there began to fall a shower of small pebbles, and later on there were stones as big as your fist. Meanwhile dismal rumblings were heard, as though the moun-

tain groaned under the weight of accumulated fury, and the earth swayed in deep sympathy.

At half-past two the explosion occurred and darkness fell upon the land. . . . The sounds were weird and abysmal, and caused our hearts to quiver with fear.

The rain of big stones continued up to about eleven o'clock at night, when sand began to fall. From where we were, we could see the reflection of the fire in the sky, but could not see the blaze . . . But at last morning broke . . . a dull, dismal, dreary day came, not much distinguishable from the preceding night. . . . But it was day, and that fact afforded some measure of relief. We could see and hear others in the town. . . . Among those who came into [Georgetown] or were brought in were many who had been stricken by lightning and were paralyzed, or who had been scorched by the burning hot sand and were blistered and sore.

* * *

In company with several gentlemen [a clergyman wrote], on Wednesday at noon I left in small rowboat to go to Cheateau Belaire, where we hoped to get a better view of the eruption. As we passed Layou, the first town on the leeward coast, the smell of sulphuretted hydrogen was very perceptible. Before we got halfway on our journey a vast column of steam, smoke and ashes ascended to a prodigious elevation, falling apparently in the vicinity of Georgetown. . . . We were about eight miles from the crater, as the crow flies, and . . . I judge that the awful pillar [the eruption column] was fully eight miles in height.

We were proceeding rapidly to our point of observation, when an immense cloud, dark, dense and apparently thick with volcanic material, descended over our pathway. . . . This mighty bank of sulphurous vapor and smoke assumed at one time the shape of a gigantic promontory, then appeared as a collection of twirling, revolving cloud whorls, turning with rapid velocity; now assuming the shape of gigantic cauliflowers, then efflorescing into beautiful flower shapes, some dark, some effulgent, some bronze, others pearly white and all brilliantly illuminated by electric flashes.

Darkness, however, soon fell upon us. The sulphurous air was laden with fine dust that fell thickly upon and around us, discoloring the sea. A black rain began to fall, followed by another rain of [volcanic material]. The electric flashes were marvelously rapid in their motions, and numerous beyond all computation. These, with the thundering roar of the mountain, mingled with the dismal roar of the lava, the shocks of earthquakes, the falling stones, the enormous quantity of material ejected from the belching crater, producing a darkness as dense as a starless midnight . . . combined to make up a scene of horror.

An unnamed press correspondent who visited the island provided some idea of the destruction visited upon Saint Vincent by the volcano:

The entire northern portion of the island is covered with ashes to an average depth of eighteen inches, varying from a thin layer at Kingstown to two feet or more at George-town. The crops are ruined, nothing green can be seen, the streets of Georgetown are cumbered with snowdriftlike heaps of ashes, and ashes rest so heavily on the roofs that in several cases they have caused them to fall in. There will soon be 5,000 destitute persons in need of assistance from the government, which is already doing everything possible to relieve the sufferers. There are a hundred injured people in the hospital at Georgetown, gangs of men are searching for the dead or rapidly burying them in trenches, and all that can be done under the circumstances is being accomplished.

The arrival here of the first detachment of the Ambulance Corps, which brought sufferers from Georgetown, caused a sensation. This batch consisted of a hundred persons, whose charred bodies exhaled fetid odors, and whose loathsome faces made even the hospital attendants shudder. All these burned persons were suffering fearfully from thirst and uttering, when strong enough to do so, agonizing cries for water. It is doubtful whether any of the whole party will recover.

While the outbreak of the volcano on the island of Martinique killed more people outright, more territory has been ruined in St. Vincent, hence there is greater destitution here. The injured persons were horribly burned by the hot grit, which was driven along with tremendous velocity. Twenty-six persons who sought refuge in a room ten feet by twelve were all killed. One person was brained by a huge stone some nine miles from the crater.

Rough coffins are being made to receive the bodies of the victims. The hospital here is filled with dying people. Fifty injured persons are lying on the floor of that building, as there are no beds for their accommodation, though cots are being rapidly constructed of boards. . . .

Since midnight on Tuesday the subterranean detonations here have ceased, and the Soufriere on Wednesday relapsed, apparently, into perfect repose, no smoke rising from the crater, and the fissures emitting no vapor. The stunted vegetation that formerly adorned the slopes of the mountain has disappeared, having given place to gray-colored lava, which greets the eye on every side. The atmosphere is dry. Rain would be welcome, as there is a great deal of dust in the air, which is disagreeable and irritating to the throats and eyes, and is causing the merchants to put all their drygoods under cover. . . . [People] who have remained on the estates are half-starved, and the few Carib survivors are leaving their caves and pillaging abandoned dwelling houses and shops. . . .

The report that the volcanic lake which occupied the top of the mountain has disappeared, now appears to be confirmed. A sea of lava, emitting sulfurous fumes, now apparently occupies the place, and several new craters have been formed. The last time the volcano showed activity, on Tuesday last, the craters, old and new, and numer-ous fissures in the mountain sides discharged hot vapor, deep subterranean murmurings were heard, the ground trembled at times, from the center of the volcano huge volumes of steam rose like gigantic pine trees toward the sky, and a dense black smoke, mingling with the steam, issued from a new and active crater, forming an immense pall over the northern hills, lowering into the valleys and then rising and spreading until it enveloped the whole island in a peculiar gray mist. . . .

The sulphurous vapors, which still exhale all over the island, are increasing the sickness and mortality among the surviving inhabitants, and are causing suffering among the new arrivals. . . .

The stench in the afflicted districts is terrible beyond description. Nearly all the huts left standing are filled with dead bodies. In some cases disinfectants and the usual means of disposing of the dead are useless, and cremation has been resorted to. When it is possible the bodies are dragged with ropes to the trenches and are there hastily covered up, quicklime being used when available. Many of the dead bodies were so covered with dust that they were not discovered until walked upon by visitors, or by the relieving officers or their assistants. . . .

The volcano resumed activity on the night of May 18. Earthquakes were felt on the island, and smoke emanated from fissures and craters on Soufriere. As churchgoers returned from services around 8:30 P.M., one account of the eruption says, "an alarming luminous cloud suddenly ascended many miles high in the north of the island, and drifted sluggishly to the northeast. Incessant lightning fell on the mountain, and one severe flash seemed to strike about three miles from Kingstown. The thunderous rumblings in the craters lasted for two hours and then diminished until they became mere murmurings. During the remainder of the night the volcano was quiet, though ashes fell from 10 o'clock until midnight. The inhabitants were frenzied with fear at the time of the outbreak, dreading a repetition of the catastrophe which had caused such terrible loss of life on the island. They ran from the streets into the open country, crying and praying for preservation from another calamity. No one on the island of St. Vincent slept that night. . . .

The continuous agitation of the volcano and the absence of rain caused the vicinity of the afflicted villages to look like portions of the Desert of Sahara. A thick, smoky cloud overspread the island, all business was suspended, the streets were empty and everyone was terror-stricken. The feeling of suspense grew painful. People passed their time gazing at the northern sky, where the thunder clouds gathered and the mournful roaring of the volcano was heard. Ashes and pumice fell slowly in the [outlying] districts, and a new reign of terror existed in the island. But during the next day the volcanic disturbances moderated, and some degree of calm returned to the afflicted islanders."

Appendix C

FURTHER READING AND WEB SITES

Because the literature on earthquakes and volcanoes is so vast, the following list represents only a select bibliography of relevant materials on the subject.

Aaronson, S. "The Social Cost of Earthquake Prediction." *New Scientist* 3/17, 1977, 634–636.

Abbott, P. *Natural Disasters, 2nd Edition,* Boston: WCB/McGraw-Hill, 1999.

Adams, W., ed. *Tsunamis in the Pacific Ocean.* Honolulu: East-West Center Press, 1970.

American Geological Institute. *Glossary of Geology,* 1999.

Anderson, C. "Animals, Earthquakes and Eruptions." *Field Museum of Natural History Bulletin* 44, 5 (1973): 9–11.

Ballard, R., and J. Grassle. "Return to Oases of the Deep." *National Geographic,* November 1979, 686–705.

Barnea, J. "Geothermal Power." *Scientific American,* January 1972, 70–77.

Bird, J., and B. Isacks, eds. *Plate Tectonics: Selected Papers from the Journal of Geophysical Research.* Washington, D.C.: American Geophysical Union, 1972.

Blong, R. J. *Volcanic Hazards.* Orlando, Fla.: Academic Press, 1984.

Bolt, B. *Earthquakes: A Primer.* New York: Freeman, 1978.

Bolt, B., W. Horn, G. Macdonald, and R. Scott. *Geological Hazards.* Berlin: Springer Verlag, 1975.

Booth, B. "Predicting Eruptions." *New Scientist,* September 9, 1976, 526–528.

Boyd, F., and H. Meyer, eds. *Kimberlites, Diatremes and Diamonds: Their Geology, Petrology and Geochemistry.* Washington, D.C.: American Geophysical Union, 1979.

Bradford, M., and R. S. Carmichael. *Natural Disasters, Volumes I–III.* Pasadena, Calif.: Salem Press, 2001.

Briggs, P. *Will California Fall into the Sea?* New York: McKay, 1972.

Bullard, F. *Volcanoes in History, in Theory, in Eruption.* Austin: University of Texas Press, 1962.

———. *Volcanoes of the Earth.* Austin: University of Texas Press, 1984.

Burke, K., and J. Wilson. "Hot Spots on the Earth's Surface." *Scientific American,* August 1976, 46–57.

Clague, D., and C. Dalrymple. "The Hawaiian-Emperor Volcanic Chain." *Volcanism in Hawaii.* U.S. Geological Survey, Professional Paper 1350, 1987, 5–54.

Cox, A., ed. *Plate Tectonics and Geomagnetic Reversals.* New York: Freeman, 1973.

Crandell, D., and D. Mullineaux. *Potential Hazards from Future Eruptions of Mount St. Helens Volcano.* U.S. Geological Survey, Bulletin 1383-C. 1978.

Davis, L. *Natural Disasters.* Rev. ed. New York: Facts On File, 2002.

Decker, R., and B. Decker. *Volcanoes.* Rev. ed. New York: Freeman, 1989.

Decker, R., T. Wright, and P. Stauffer, eds. *Volcanism in Hawaii.* U.S. Geological Survey, Professional Paper 1350. 2 vols. 1987.

Easterbrook, D. J. *Surface Process and Landforms.* Englewood Cliffs, N.J.: Prentice Hall, 1993.

Eiby, G. *Earthquakes.* Auckland, New Zealand: Heineman, 1980.

Fisher, R., and H.-U. Schmincke. *Pyroclastic Rocks.* Berlin: Springer Verlag, 1984.

Francis, P. *Volcanoes.* Middlesex, England: Penguin Books, 1976.

———. *Volcanoes: A Planetary Perspective.* New York: Oxford University Press, 1994.

Fried, J. *Life along the San Andreas Fault.* New York: Saturday Review Press, 1973.

Gold, T., and S. Sofer. "The Deep-Earth-Gas Hypothesis." *Scientific American,* June 1980, 154–161.

Green, J., and N. Short, eds. *Volcanic Landforms and Surface Features.* New York: Springer Verlag, 1971.

Gutenberg, B., and C. Richter. *Seismicity of the Earth and Associated Phenomena.* Princeton, N.J.: Princeton University Press, 1954.

Haas, J., and D. Mileti. *Socioeconomic Impact of Earthquake Prediction on Government, Business and Industry*. Boulder: Institute of Behavioral Sciences, University of Colorado, 1976.

Halacy, D. *Earthquakes: A Natural History*. Indianapolis: Bobbs-Merrill, 1974.

Heck, N. H. "Japanese Earthquakes." *Bulletin of the Seismological Society of America* 34, 1994, 117–136.

Heezen, B., and C. Hollister. *The Face of the Deep*. New York: Oxford University Press, 1971.

Iacopi, R. *Earthquake Country*. Menlo Park, Calif.: Lane, 1973.

International Association of Volcanology and Chemistry of the Earth's Interior. *Catalog of Active Volcanoes of the World*. Parts 1–22. Rome: Int'l. Assn. of Volcanology, 1951–75.

Kerr, R. "Earthquakes: Prediction Proving Elusive." *Science*, April 28, 1978, 419–421.

Kruger, P., and C. Otte, eds. *Geothermal Energy*. Stanford, Calif.: Stanford University Press, 1973.

Latter, J. H. "Natural Disasters." *The Advancement of Science* 25, 1969, 363–380.

Lawson, A. *The California Earthquake of April 18, 1906*. Report of the State Earthquake Investigation Commission. Washington, D.C.: Carnegie Institution, 1908.

Leet, L., and F. Leet. *Earthquake: Discoveries in Seismology*. New York: Dell, 1964.

Lipman, P., and D. Mullineaux, eds. *The 1980 Eruptions of Mount St. Helens*. U.S. Geological Survey, Professional Paper 1250, 1981.

Luhr, J. F., and T. Simkin. *Paricutín: The Volcano Born in a Mexican Cornfield*. Phoenix: Geoscience Press, 1993.

Macdonald, G. *Volcanoes*. Englewood Cliffs, N.J.: Prentice Hall, 1972.

Macdonald, G., A. Abbott, and F. Peterson. *Volcanoes in the Sea*. Honolulu: University of Hawaii Press, 1983.

McClelland, L., T. Simkin, M. Summers, E. Nielsen, and T. C. Stein. *Global Volcanism 1975–1985*. Englewood Cliffs, N.J.: Prentice Hall, 1989.

Menard, H. *Marine Geology of the Pacific*. New York: McGraw-Hill, 1964.

Moores, E. M., ed. *Shaping the Earth: Tectonics of the Continents and Oceans*. New York: W. H. Freeman, 1990.

Morris, C. *The Volcano's Deadly Work*. New York: Scull, 1902.

Muffler, L., ed. *Assessment of Geothermal Resources of the United States—1978*. U.S. Geological Survey, Circular 790, 1979.

Nabavi, M. S. "Historical Earthquakes in Iran, c. 300 B.C.– A.D. 1900." *Journal of the Earth and Space Physics* 7, 1978, 70–117.

Newhall, C., and D. Dzurisin. *Historical Unrest at Large: Calderas of the World*. U.S. Geological Survey, Bulletin 1855, 1988.

Oakeshott, G. *California's Changing Landscapes*. New York: McGraw-Hill, 1978.

———. *Volcanoes and Earthquakes*. New York: McGraw-Hill, 1976.

Panel on Earthquake Prediction of the Committee on Seismology. *Predicting Earthquakes*. Washington, D.C.: National Academy of Sciences, 1976.

Papazachos, B., and Papazacou, C. *The Earthquakes of Greece*. Thessaloniki, Greece, 1997.

Park, C., and R. MacDiarmid. *Ore Deposits*. New York: Freeman, 1975.

Paterson, D. "Methane from the Bowels of the Earth." *New Scientist* 6/29, 1978, 896–897.

Press, F., and R. Siever. *Earth*. New York: Freeman, 1986.

Richter, C. *Elementary Seismology*. San Francisco: Freeman, 1958.

Rikitake, T. *Earthquake Prediction*. Amsterdam: Elsevier, 1976.

Ritchie, D. *The Ring of Fire*. New York: Atheneum, 1981.

———. *Superquake!* New York: Crown, 1988.

Rona, P. "Plate Tectonics and Mineral Resources." *Scientific American*, July 1973, 86–95.

Rybach, L., and L. Muffler. *Geothermal Systems: Principles and Case Histories*. New York: Wiley, 1981.

Sigurdsson, H., B. Houghton, S. R. McNutt, H. Rymer, and J. Stix, eds. *Encyclopedia of Volcanoes*. San Diego: Academic Press, 1999.

Simkin, T., and L. Siebert. *Volcanoes of the World*. Phoenix: Geoscience Press, 1994.

Simkin, T., D. J. Unger, R. I. Tilling, P. R. Vogt, and H. Spall, compilers. *This Dynamic Planet: World Map of Volcanoes, Earthquakes, Impact Craters and Plate Tectonics*. U.S. Geological Survey, 1994.

Smith, F. *The Seas in Motion*. New York: Crowell, 1973.

Smith, R. "Waiting for the Other Plate to Drop in California." *Science* 2/24, 1978, 797.

Smithsonian Institution's Global Volcanism Network. "Summary of Recent Activity." *Bulletin of Volcanology*, Continuing Volumes and Years.

Stearns, H. *Geology of the State of Hawaii*. 2d ed. Palo Alto, Calif.: Pacific Books, 1985.

Steiribrugge, K. *Earthquakes, Volcanoes and Tsunamis*. New York: Scandia America Group, 1982.

Stommel, H., and E. Stommel. "The Year without a Summer." *Scientific American*, June 1979, 176–180.

Sugimura, A., and S. Uyeda. *Island Arcs: Japan and Its Environs*. Amsterdam: Elsevier, 1973.

Sullivan, R., with D. Mustart and J. Galehouse. "Living in Earthquake Country: A Survey of Residents Living along the San Andreas Fault." *California Geology*, January 1977, 3–8.

Tarling, D., and M. Tarling. *Continental Drift*. New York: Pelican, 1974.

Tazieff, H. *When the Earth Trembles*. New York: Harcourt, Brace and World, 1964.

Thomas, G., and M. Witts. *The San Francisco Earthquake*. New York: Stein and Day, 1971.

Thorarinsson, S. *Surtsey*. New York: Viking, 1967.

Tilling, R. I. *Volcanoes*. Denver: U.S. Geological Survey Information Circular, 1996.

U.S. Geological Survey. *The San Fernando, California, Earthquake of February 9, 1971*. Professional Paper 733, 1971.

Uyeda, S. *The New View of the Earth*. New York: Freeman, 1978.

Verney, P. *The Earthquake Handbook*. London: Paddington, 1979.

Wegener, A. *The Origins of the Oceans and Continents.* New York: Methuen, 1967.

Wexier, H. "Volcanoes and World Climate." *Scientific American,* April 1952, 74–81.

Williams, H., and A. McBirney. *Volcanology.* San Francisco: Freeman, Cooper, 1979.

Wood, C. A., and J. Kienle. *Volcanoes of North America: United States and Canada:* New York: Cambridge University Press, 1990.

Wyllie, P. "The Earth's Mantle." *Scientific American,* March 1975, 50–57.

Yeats, R. S., K. Sieh, and C. R. Allen. *The Geology of Earthquakes.* New York: Oxford University Press, 1997.

Ziony, J., ed. *Evaluating Earthquake Hazards in the Los Angeles Region—An Earth-Science Perspective.* U.S. Geological Survey, Professional Paper 1360, 1985.

INTERNET RESOURCES

There are many Web sites available for both volcanoes and earthquakes. Search engines can lead you to them. Below are lists of some of the better resources and sites that can connect you to even more resources. These lists emphasize the major government sites, several exceptional sites, and sites with substantive links.

General

American Geological Institute. URL: http://www.agiweb.org. Accessed November 30, 2005. Contains extensive information on geology, including earthquakes and volcanoes at a technical and non-technical level.

American Geophysical Union. URL: http://www.agu.org. Accessed November 30, 2005. The premier academic geophysical organization in the United States. Contains technical information on earthquakes and volcanoes.

Geohazards, U.S. Geological Survey. URL: http://geohazards.cr.usgs.gov. Accessed November 30, 2005. The best repository of earthquake and volcano information available. Most of the specific earthquake and volcano sites can be accessed from here.

Geological Society of America. URL: http://www.geosociety.org. Accessed November 30, 2005. The premier academic geologic organization in the United States. Contains technical information on earthquakes and volcanoes.

National Geophysical Data Center, National Oceanographic and Atmospheric Agency. URL: http://www.ngdc.noaa.gov/seg/hazard/hazards.shtml. Accessed November 30, 2005. Outstanding repository of information and photos of natural disasters worldwide.

Earthquakes

ASC: Amateur Seismic Centre. URL: http://asc-india.org. Accessed November 1, 2005. Contains detailed information and accounts of significant earthquakes from India, Pakistan, and Nepal as well as images.

Earthquake Hazards Program. U.S. Geological Survey. URL: http://earthquake.usgs.gov. Accessed November 11, 2005. Information on current seismic activity, historical earthquakes, regions, educational materials, tables of strongest earthquakes, plate tectonics, and virtually all important links.

EQIIS: Earthquake Image Information System, University of California at Berkeley. URL: http://nisee.berkeley.edu/eqiis.html. Accessed November 11, 2005. Contains a catalog of primarily historical earthquake images from around the world.

Significant Earthquakes Database. National Geophysical Data Center of National Oceanographic and Atmospheric Agency. URL: http://www.ngdc.noaa.gov/seg/hazard/sig/hazard/sig_srch_idb.shtml. Accessed November 11, 2005. Contains an extensive, easily searchable database of all reported significant earthquakes, including all statistics, pertinent information, and primary references.

Volcanoes

Alaskan Volcano Observatory, U.S. Geological Survey. URL: http://www.avo.alaska.edu. Accessed November 30, 2005. Contains real-time monitoring (including Web cams) and historical data for volcanic activity in Alaska.

Cascades Volcano Observatory, U.S. Geological Survey. URL: http://vulcan.wr.usgs.gov. Accessed November 30, 2005. Contains real-time monitoring and historical data for volcanic activity in the Cascade Range of Washington, Oregon, and California.

Hawaiian Volcano Observatory, U.S. Geological Survey. URL: http://hvo.wr.usgs.gov. Accessed November 30, 2005. Contains real-time monitoring and historical data for volcanic activity in Hawaii.

Volcanoes of the World, Global Volcanism Program of the Smithsonian Institution. URL: http://www.volcano.si.edu/world. Accessed November 11, 2005. Information and images of most of the volcanoes of the world and the processes that control them.

Volcano World. URL: http://www.volcano.und.nodak.edu. Accessed November 11, 2005. Contains reports and images of many of the world's volcanoes in simple language and also has educational materials.

APPENDIX D

The Deadliest Earthquakes

Date	Location	Deaths	Magnitude
January 23, 1556	Shansi, China	830,000	
July 27, 1976	Tangshan, China	655,000	8.2
December 26, 2004	Sumatra	283,106	9
August 9, 1138	Aleppo, Syria	230,000	
May 22, 1927	Xining, China	200,000	7.9
December 22, 856+	Damghan, Iran	200,000	
December 16, 1920	Gansu, China	200,000	8.6
March 23, 893+	Ardabil, Iran	150,000	
September 1, 1923	Kanto, Japan	143,000	7.9
October 5, 1948	USSR (Turkmenistan)	110,000	7.3
December 28, 1908	Messina, Italy	70,000–100,000	7.2
September, 1290	Chihli, China	100,000	
October 8, 2005	Muzafarrabad, Pakistan	86,000	7.6
November, 1667	Shemakha, Caucasia	80,000	
November 18, 1727	Tabriz, Iran	77,000	
November 1, 1755	Lisbon, Portugal	70,000	8.7
December 25, 1932	Gansu, China	70,000	7.6
May 31, 1970	Peru	66,000	7.9
1268	Silicia, Asia Minor	60,000	
January 11, 1693	Sicily, Italy	60,000	
May 30, 1935	Quetta, Pakistan	30,000–60,000	7.5
February 4, 1783	Calabria, Italy	50,000	
June 20, 1990	Gilan, Iran	50,000	7.7

Table modified from the U.S. Geological Survey.

+ Uncertain date of main event

APPENDIX E

The Deadliest Volcanoes

Deaths	Volcano	Year
92,000	Tambora, Indonesia	1815
36,417	Krakatoa, Indonesia	1883
29,025	Mount Pelée, Martinique	1902
25,000	Ruiz, Colombia	1985
14,300	Unzen, Japan	1792
9,350	Laki, Iceland	1783
5,110	Kelut, Indonesia	1919
4,011	Galunggung, Indonesia	1882
3,500	Vesuvius, Italy	1631
3,360	Vesuvius, Italy	79
2,957	Papandayan, Indonesia	1772
2,942	Lamington, Papua New Guinea	1951
2,000	El Chichón, Mexico	1982
1,680	Sofrière, St. Vincent	1902
1,475	Oshima, Japan	1741
1,377	Asama, Japan	1783
1,335	Taal, Philippines	1911
1,200	Mayon, Philippines	1814
1,184	Agung, Indonesia	1963
1,000	Cotopaxi, Ecuador	1877
800	Pinatubo, Philippines	1991
700	Komagatake, Japan	1640
700	Ruiz, Colombia	1845
500	Hibok-Hibok, Philippines	1951

Appendix F

The Highest Magnitude Earthquakes

	Location	Date	Magnitude
1	Chile	May 22, 1960	9.5
2	Prince William Sound, Alaska	March 28, 1964	9.2
3	Andreanof Islands, Alaska	March 9, 1957	9.1
4	Kamchatka	November 4, 1952	9
5	Sumatra	December 26, 2004	9
6	off the Coast of Ecuador	January 31, 1906	8.8
7	Rat Islands, Alaska	February 4, 1965	8.7
7	off the Coast of Sumatra	March 28, 2005	8.7
9	Assam-Tibet	August 15, 1950	8.6
10	Kamchatka	February 3, 1923	8.5
10	Banda Sea, Indonesia	February 1, 1938	8.5
10	Kuril Islands	October 13, 1963	8.5

Appendix G

The Frequency of Occurrence of Earthquakes

Descriptor	Magnitude	Average Annually
great	8 and higher	1
major	7–7.9	17
strong	6–6.9	134
moderate	5–5.9	1319
light	4–4.9	13,000 (estimated)
minor	3–3.9	130,000 (estimated)
very minor	2–2.9	1,300,000 (estimated)

APPENDIX H

Magnitude v. Ground Motion and Energy

Magnitude change	Ground motion change	Energy change
1	10.0 times	about 32 times
0.5	3.2 times	about 5.5 times
0.3	2 times	about 3 times
0.1	1.3 times	about 1.4 times

INDEX

Note: **Boldface** page numbers indicate main entries; *Italic* page numbers indicate illustrations; *m* indicates a map.